How It's Done

AN INVITATION TO SOCIAL RESEARCH

Emily Stier Adler
Rhode Island College

Roger Clark
Rhode Island College

Wadsworth Publishing Company
I(T)P® An International Thomson Publishing Company

Belmont, CA • Albany, NY • Boston • Cincinnati • Johannesburg • London • Madrid • Melbourne
Mexico City • New York • Pacific Grove, CA • Scottsdale, AZ • Singapore • Tokyo • Toronto

Sociology Editor: Halee Dinsey
Editorial Assistant: Jennifer Jones
Marketing Manager: Christine Henry
Project Editor: Jerilyn Emori
Print Buyer: Karen Hunt
Permissions Editor: Robert Kauser
Production: Robin Gold/Forbes Mill Press
Interior Designer: Lisa Mirski Devenish/Devenish Design
Photo Researcher: Terri Wright/Terri Wright Design
Copy Editor: Robin Gold
Illustrator: Patty Arnold
Cover Designer: Bill Stanton/Stanton Design
Cover Image: David Oliver/Tony Stone Images
Compositor: Wolf Creek Press/Forbes Mill Press
Printer: Webcom Ltd.

Printed in Canada
2 3 4 5 6 7 8 9 10

For more information, contact Wadsworth Publishing Company, 10 Davis Drive, Belmont, CA 94002, or electronically at http://www.wadsworth.com

International Thomson Publishing Europe
Berkshire House
168-173 High Holborn
London, WC1V 7AA, United Kingdom

International Thomson Editores
Seneca, 53
Colonia Polanco
11560 México D.F. México

Nelson ITP, Australia
102 Dodds Street
South Melbourne
Victoria 3205 Australia

International Thomson Publishing Asia
60 Albert Street
#15-01 Albert Complex
Singapore 189969

Nelson Canada
1120 Birchmount Road
Scarborough, Ontario
Canada M1K 5G4

International Thomson Publishing Japan
Hirakawa-cho Kyowa Building, 3F
2-2-1 Hirakawa-cho, Chiyoda-ku
Tokyo 102 Japan

International Thomson Publishing Southern Africa
Building 18, Constantia Square
138 Sixteenth Road, P.O. Box 2459
Halfway House, 1685 South Africa

Library of Congress Cataloging-in-Publication Data

ISBN 0-534-53325-6

This book is printed on acid-free recycled paper.

Contents

Preface

We'd like to invite you to participate in one of the most exciting, exhilarating and sometimes exasperating activities we know: social science research. We extend the invitation not only because we know, from personal experience, how rewarding and useful research can be, but also because we've seen what pleasure it can bring other students of the social world. Our invitation comes with some words of reassurance, especially for those of you who entertain a little self-doubt about your ability to do research. First, we think you'll be glad to discover, as you read How It's Done, how much you already know about how social research is done. If you're like most people, native curiosity has been pushing you to do social research for much of your life. This book is meant simply to assist you in this natural activity by showing you some tried-and-true ways to enlightening and plausible insights about the social world.

Special Features

Active Engagement in Research

Our second word of reassurance is that we've done everything we can to minimize your chances for exasperation and maximize your opportunities for excitement and exhilaration, in preparing this book. Our philosophy is simple. We believe that honing one's social research skills is analogous to honing one's skills in other enjoyable and rewarding human endeavors, like sport, art, or dance. The best way isn't simply to read about it: It's to do it and to watch experts do it. So, just as you'd hesitate to teach yourself tennis, ballet, or painting only by reading about them, we won't ask you to try learning the fine points of research methodology by reading our prose alone. We'll encourage you to get out and practice the techniques we describe. We've designed exercises at the end of each chapter to help you work on the "groundstrokes," "serve," "volleys" and "overheads" of social research. We don't think you'll need to do all of the exercises at home. Your instructor might ask you do some in class and might want you to ignore some altogether. In any case, we think that, by book's end, you should have enough control of the fundamentals to do the kind of on-the-job research that social science majors are increasingly asked to do, whether they find themselves in social service agencies, the justice system, in business and industry, government or graduate school.

The exercises reflect our conviction that we all learn best when we're actively engaged. Other features of the text also encourage such active engagement, including the Stop and Think questions that run through each chapter, encouraging you to actively respond to what you're reading.

Engaging Examples of Actual Research

Moreover, just as you might wish to gain inspiration and technical insight for ballet by studying the work of Anna Pavlova or Mikhail Baryshnikov, we'll encourage you to study the work of some accomplished researchers. Thus, we build most of our chapters around a research essay, what we call focal research, usually previously published, that is intended to make the research process transparent, rather than opaque. We have chosen these essays for their appeal and accessibility, as well as to tap what we hope are some of your varied interests: for instance, crime, gender, aging, attitudes towards rape, drug use and abuse, and others.

Behind-the-Scene Glimpses of the Research Process

These focal research pieces are themselves a defining feature of our book. In addition to such exemplary "performances," however, we've included behind-the-scenes glimpses of the research process. We're able to provide these glimpses because many researchers have given generously of their time to answer our questions about what they've done, the special problems they've encountered, and the ways they've dealt with these problems. The glimpses should give you an idea of the kinds of choices and situations the researchers faced, where often the "real" is far from the "ideal." You'll see how they handled the choices and situations and hear them present their current thinking about the compromises they made. In short, we think you'll discover that good research is an achievable goal, as well as a very human enterprise.

Clear and Inviting Writing

We've also tried to minimize your chances for exasperation by writing as clearly as we can. The goal of all social science, we believe, is to interpret social life, something you've all been doing for quite a while. We want to assist you in this endeavor and believe that an understanding of social science research methods can help. But unless we're clear in our presentation of those methods, your chances of gaining that understanding are not great. There are, of course, times when we'll introduce you to concepts that are commonly used in social science reseach and that may be new to you. When we do, however, we will try to provide definitions that make the concepts as clear as possible, definitions that are highlighted in the margin of the text and in the glossary at the end of the text.

Balance Between Quantitative and Qualitative Approaches

We think you'll also appreciate the balance between quantitative and qualitative research methods presented here. Quantitative methods focus on things that are measured numerically. ("He glanced at her 42 times during the performance.") Qualitative methods focus on descriptions of the essence of things. ("She appeared annoyed at his constant glances.") We believe both methodological approaches are too useful to ignore. Emblematic of this belief is the inclusion of a chapter (Chapter 16) on qualitative data analysis, following a chapter (Chapter 15) on quantitative data analysis. The presence of such chapters is another defining feature of the book.

Moreover, we will introduce you to some relatively new research strategies, such as using the Internet to refine ideas and collect data, as well as more conventional strategies.

Our aims, then, in writing this book have been (1) to give you first-hand experiences with the research process, (2) to provide you with engaging examples of social science research, (3) to offer behind-the-scene glimpses of how professional researchers have done their work, (4) to keep our own presentation of the "nuts-and-bolts" of social science research as clear and inviting as possible, (5) to give a balanced presentation of qualitative and quantitive research methods, as well as introduce recent technological innovations. Whether we succeed in these goals, and in the more important one of sharing our excitement about social research, remains to be seen. Be assured, however, of our conviction that there is excitement to be had.

Acknowledgements

We cannot possibly thank all those who have contributed to the completion of this book, but we can try to thank those whose help has been most indispensable and hope that others will forgive our neglect. We'd first like to thank all the students who have taken research methods courses with us at Rhode Island College for their general good-naturedness and patience as we've worked out ideas that are crystallized here, and then worked them out some more. We'd also like to thank our colleagues in the Sociology Department and the administration of the College, for many acts of encouragement and more tangible assistance, including released time.

We'd like to thank colleagues, near and far, who have permitted us to incorporate their writing as focal research and, in many cases, then read how we've incorporated it, and told us how we might do better. They are: Patricia A. Adler, Peter Adler, Richard Aniskiewicz, Benjamin Bates, Sandra Enos, Rachel Filinson, Paula Foster, Norma B. Gray, Mark Harmon, Michele Hoffnung, G. David Johnson, Rachel Lennon, Steven Messner, Leanna Morris, Donald C. Naylor, Gloria J. Palileo, Ian R. H. Rockett, Gordon S. Smith, Rose Weitz, David Wright, Earl Wysong. These researchers include those who've given us the behind-the-scenes glimpses of the research process that we think distinguishes our book. We'd like to extend special thanks to Sandra Enos, who composed the excellent instructor's manual accompanying the book.

We're grateful to some wonderful editors who have encouraged us. In particular, our thanks go to Joseph Terry, Sharon Adams Poore, Denise Simons, Eve Howard, Jerilyn Emori, Halee Dinsey, and Robin Gold. We've benefitted greatly from the comments of the social science colleagues who have reviewed our manuscript at various stages of its development. We are most grateful to Mary Archibald; Cynthia Beall, Case Western Reserve University; Daniel Berg, Humboldt State University; Susan Brinkley, University of Tampa; Jeffrey A. Burr, State University of New York, Buffalo; Richard

Butler, Benedict College; Daniel Cervi, University of New Hampshire; Charles Corley, Michigan State University; Norman Dolch, Louisiana State University, Shreveport; Craig Eckert, Eastern Illinois University; Pamela I. Jackson, Rhode Island College; Alice Kemp, University of New Orleans; Michael Kleinman, University of South Florida; James Marquart, Sam Houston State University; James Mathieu, Loyola Marymount University; Steven Meier, University of Idaho; Deborah Merrill, Clark University; Lawrence Rosen, Temple University; Josephine Ruggereo, Providence College; Kevin Thompson, University of North Dakota; Steven Vassar, Mankato State University; and Gennaro Vito, University of Louisville.

Finally, we'd like to thank our spouses, George L. Adler and Beverly Lyon Clark, for providing many of the resources we've thanked others for: general good-naturedness and patience, editorial assistance, encouragement and other tangible aids, including released time.

Emily Stier Adler

Roger Clark

We dedicate this book to our children:

Josh Adler, Rebecca Adler, Adam Clark, and Wendy Clark.

1 The Uses of Social Research

Gee. It all seems so much more organized than when I played," Janet thought to herself as she watched her younger brother, Sam, at his soccer game. "For me, it was always fun, always just a game. Sam seems to take it so seriously. I wonder what he gets out of it."

Janet was home for a day from college and had taken Sam to his game. As they drove home together, Sam, who'd been looking out the window at the sky, asked her, "How do we know the earth goes around the sun, and not the other way around?" Janet was momentarily stumped, but found herself answering, not entirely to her own satisfaction, "Because that's what astronomers tell us."

"But it IS how I know it," Janet said to herself, almost as if Sam had challenged her. Then, somewhat less defensively, she asked herself, "How do THEY know it?"

Janet's questions—"How do I know things?" and "How do scientists come to know things?" and even "What do kids get out of organized afterschool activities (like Sam's soccer)?"—are the kinds of things we'll be discussing in this chapter on the uses of social research.

Introduction

Studying social research methods is not so much about acquiring knowledge about the social world as it is about learning *how* to acquire knowledge about that world. You might ask yourself: Why should I subject myself to a course on such a curious topic? What good can come of my knowing how knowledge about the social world is arrived at? And aren't there faster ways of coming to such knowledge anyway—like taking a class in the area I'm interested in and learning what others have found out? These are all splendid questions. In fact, we think they're so good that we devote our first chapter to them.

We hope that quite a few of our readers go on and apply their knowledge of research methods as part of their professional lives. We know that many of our students will do so in a variety of ways: as graduate students in the social sciences, as social workers, as police or correctional officers, as analysts in state agencies, as advocates for specific groups or policies, as community organizers, as family counselors, to name but a few.

Even if you don't enter a profession in which you'll do research of the sort we discuss in this book, we still think learning something about research methods can be one of the most useful things you do in college. Why? Oddly, perhaps, our answer implicates another apparently esoteric subject: social theory.

social theory, a story about how and why people behave and interact in the ways that they do.

When we speak of **social theory,** we're not only referring to the kinds of things you study in specialized social theory courses, although we do include those things. In our view, social theories, like all other theories, are stories about how and why things are as they are. In the case of social theories, the stories are about why people "behave, interact, and organize themselves in certain ways" (Turner, 1991: 1). Such stories are useful, we feel, not only because they affect how we act as citizens—as when, for instance, we inform, or fail to inform, elected representatives of our feelings about matters like welfare, joblessness, crime, and domestic violence—but also because we believe that Charles Lemert (1993: 1) is right when he argues that "social theory is a basic survival skill."

Useful social theory, in our view, concerns itself with those things in our everyday lives that can and do affect us profoundly, even if we are not aware of them. We believe that once we can name and tell stories (or create theories) about these things, we have that much more control over them. At the very least, the inability to name and tell such stories leaves us powerless to do anything. These stories can be about why some people commit suicide and some don't, why some are homeless and some aren't, why some commit crimes and some don't, why some do housework and others don't, why some people live to be adults and others don't. They can come from people who are paid to produce them, like social scientists, or from people who are simply trying to make sense of their lives. Lemert (1993) reminds us that the title for Alex Kotlowitz's (1991) *There Are No Children Here* was first uttered by the mother of a 10-year-old boy, Lafeyette, who lived in one of Chicago's most dangerous public housing projects. This mother observed, "But you know, there are no children here. They've seen too much to be children" (Kotlowitz, 1991: 10). Hers is eloquent social theory, with serious survival implications for those living in a social world where nighttime gunfire is commonplace.

BOX 1.1

Research in the Real World

Although not all students who receive bachelor's degrees in sociology, justice studies, psychology, or related fields will be able to apply their research skills, many of them will. Here are just a few examples of what some of our graduates have done:

"When I worked for the Department of Children and Youth and their families, we conducted a survey of foster parents to see what they thought of foster care and the agency's services. The parents' responses provided us with very useful information about the needs of foster families, their intentions for the future, and the kinds of agency support that they felt would be appropriate."

Graduate employed by a state Department of Children, Youth and their Families

"As the Department of Corrections was under a court order because of crowding and prison conditions, it was important that we plan for the future. We needed to project inmate populations and did so using a variety of data sources and existing statistics. In fact, we were accurate in our projections."

Graduate employed by a state Department of Corrections

"I gave a questionnaire to a sample of teachers in local schools, asking them what they thought of their schools' 'zero tolerance' policies for drugs and weapons. I found out that, overall, the teachers were much more concerned with the lack of effective policies for dealing with disruptive students in the classroom than they were about either drugs or weapons in their schools."

Graduate employed as an elementary school teacher

"I'm working at a literacy program designed to help children in poverty by providing books for the preschoolers and information and support for their parents. I've realized that while the staff all think this is a great program, we've never really determined how effective it is. It would be wonderful if we could see how well the program is working and what we could do to make it even better. I plan on talking to the director about the possibility of doing evaluation research."

Graduate employed by a private pediatric early literacy program

We'll have more to say about social theory and its connection to research methods in the next chapter. But, for now, we'll simply say that we believe the most significant value of knowledge of research methods is that it permits a critical evaluation of what passes for knowledge when we and others develop social theory. This critical capacity should, among other things, enable us to interpret and use the research findings produced by others. Our simple answer, then, to the question about the value of a research methods course is not that it adds to your stock of knowledge about the world but that it adds to your knowledge of *how* you know the things you know, *how* others know what they know, and, ultimately, *how* this knowledge can be used to construct and evaluate the theories by which we live our lives.

Research Versus Other Ways of Knowing

Knowledge from Authorities

On March 8, 1996, we (Emily and Roger) learned from a TV news report that the unemployment rate had fallen the previous month from 5.8 to 5.5 percent and that the stock market had, as a result, plummeted 171 points (the third longest one-day fall in history). Fascinating news, but perhaps not quite as interesting as what we had learned many years ago, literally at our mothers' knees, that "In fourteen hundred and ninety-two, Columbus sailed the ocean blue," despite nearly everyone's belief that he was sailing in a direction that jeopardized his very existence.

Now, why do we think we "know" these things? Basically, it's because some authority told us so. In the first case, we heard it from a news anchor who, at least for the unemployment rate, was relying on a Labor Department report. In the second case, we relied on our moms, who were reporting a pretty commonly accepted version of America's "discovery." **Authorities,** like the television newscaster and our moms, and even the Labor Department official, are among the most common sources of knowledge for most of us. Authorities are socially defined sources of knowledge. Religion, government, education, family—all social institutions—have their authorities: people whose superior access to relevant knowledge is assumed by others. For most of us, most of the time, learning from authorities is good enough, and it certainly helps keep life simpler than it would be otherwise. Life would be ridiculously difficult and problematic if, for instance, we who live in the Western world had to reinvent a "proper" way of greeting people each time we met them. At some point, we're told about the customs of shaking or slapping hands, or saying "Hi," and move on from there.

authorities, socially defined sources of knowledge.

STOP AND THINK *What, do you think, are the major disadvantages of receiving our "knowledge" from authorities?*

Although life is made simpler by "knowledge" from authorities, sometimes such knowledge is inappropriate, misleading, or downright incorrect. Very few people today take seriously the "flat-earth" theories that have been attributed to Columbus' social world. More interesting, perhaps, is an increasingly accepted view that, our mothers' teachings to the contrary notwithstanding, very few people in Columbus' social world took the flat-earth view seriously either; that, in fact, this view of the world was wrongly attributed to them by late-nineteenth century historians to demonstrate how misguided people who accepted religious over scientific authority could be (for example., Gould, 1995: 38–50).

Knowledge from Personal Inquiry

personal inquiry, inquiry that employs the senses' evidence.

But if we can't always trust authorities, like our moms, or even experts, like those late-nineteenth century historians (and, by extension, even our teachers or the books they assign) for a completely truthful view of the world, who or what can we trust? For some of us the answer is that we can trust the evidence of our own senses. **Personal inquiry,** or inquiry that employs the senses' evidence for arriving at knowledge, is another common way of knowing.

STOP AND THINK *Can you think of any disadvantages in trusting personal inquiry alone as a source of "knowledge"?*

The problem with personal inquiry is that it, like the pronouncements of authorities, can lead to misleading, even false, conclusions. As geometricians like to say, "Seeing is deceiving." This caution is actually as appropriate for students of the social world as it is for students of regular polygons. This is because the evidence of our senses can be distorted. Most of us, for instance, developed our early ideas of what a "family" was by observing our own family closely. By the time we were six or seven, each of us (Emily and Roger), had observed that our own families consisted of two biological parents and their children. As a result, we concluded that all families constituted two biological parents and their children.[1]

There's obviously nothing wrong with personal inquiry ... except that it frequently leads to "knowledge" that's pretty half-baked. There are many reasons for this problem. One of them is obvious from our example: Humans tend to overgeneralize from a limited number of cases. Both of us had experienced one type of family and, in a very human way, assumed that what was true of our families was true of all human families. Another barrier to discovering the truth is the human tendency to selectively perceive what we've been conditioned to perceive. Thus, even as ten-year-olds, we might have walked into a commune, with lots of adults and many children, and not entertained the possibility that this group considered itself a family. We just hadn't had the kind of experience that made such an observation possible. And so on.

So neither relying on authorities nor relying on one's own personal inquiry is a foolproof way to the truth. In fact, there might not be such a way. But authors of research methods books (ourselves included) tend to value a way that's made some pretty astounding contributions to the human condition: the scientific method. In the next two subsections, we'll, first, discuss some relative strengths of the scientific method for knowledge acquisition and, second, give you some idea about what this knowledge is intended to do.

scientific method, a way of conducting empirical research following rules that specify objectivity, logic, and communication among a community of knowledge seekers, as well as the connection between research and theory.

Strengths of the Scientific Method

Specifying precise procedures that constitute the **scientific method** is a dicey business at best, but we'd like to suggest four steps that are often

[1]Of course, by the time that Roger, in his early forties, had adopted two children from another country, his ideas of "family" had changed many times.

involved, to a greater or lesser degree, and that suggest the relative emphasis placed on care and community that distinguishes science from other modes of knowing. An early step is to specify the goals or objectives that distinguish a particular inquiry (here care is paramount). A subsequent step involves reviewing a literature, or reading what's been published about a topic (here learning what a relevant community thinks is the goal). At some point it becomes important to specify what is actually observed (care again). A later step is to share one's findings with others in a relevant community so that they can scrutinize what's been done (community again). These steps or procedures will come up again throughout the course of the book, but we'd now like to stress some of the strengths that accrue to the scientific method as a result of their use.

The Promotion of Skepticism and Intersubjectivity

One great strength of the scientific method, over modes that rely on authorities and personal inquiry, is that, ideally, it promotes skepticism about its own knowledge claims. One way in which healthy skepticism is generated is through the communities of knowledge seekers. Each member of these communities has a legitimate claim to being a knowledge producer, as long as she or he conforms to other standards of the method (mentioned later). It is sometimes said that one major benefit of having groups of relatively equal knowledge producers working in the same general area is that their joint efforts contribute to the greater **objectivity** of the group. Objectivity refers to the ability to see the world as clearly as possible, free from personal feelings, opinions, or prejudices about what it is or what it should be. We're frankly not completely convinced by the argument that a community of relatively equal knowledge producers necessarily enhances objectivity,[2] but we do think it increases the chances of what is sometimes called **intersubjectivity.** Intersubjectivity refers to agreements about reality that come from the practice of comparing one's results with those of others and discovering that the results are consistent with one another. Equally important, when intersubjective agreement eludes a community of scientists because various members get substantially different results, it's an important clue that knowledge remains elusive and that knowledge claims should be viewed with skepticism.

objectivity, the ability to see the world as it really is.

intersubjectivity, agreements about reality that result from comparing the observations of more than one observer.

The Extensive Use of Communication

Another related ideal of the scientific method is *adequate communication* within the community of knowledge seekers, implicit in the scientific procedures of referring to previous published accounts in an area and of sharing findings with others. Unlike insights that come through personal inquiry, scientific insights are supposed to be subjected to the scrutiny of the larger community and therefore need to be as broadly publicized as possible. Communication of

[2]After all, just because most Western biologists were creationists (believing that each species was the independently created by a supreme being) before Darwin published *The Origin of the Species*, doesn't mean that creationism was objectively true.

scientific findings can be done through oral presentations (as at conferences) or written ones (especially through publication as articles and books). Increasingly, the computer revolution has provided new media (discussed in Chapter 12) for communication about research and the exchange of data. Once findings are communicated, they then become grist for a critical mill. Others are thereby invited to question (or openly admire) the particular approach that's been reported or to try to reproduce (or *replicate*) the findings using other approaches or other circumstances. Adequate communication thus facilitates the ideal of reaching intersubjective "truths."

Testing Ideas Factually

These communal aspects of the scientific method are complemented by at least three other goals: that "knowledge" be *factually testable*, be *logical*, and be *explicable through scientific theory*. Factual, or empirical, testability means scientific knowledge, like personal inquiry, must be supported by observation. But rather than simply using evidence to support a particular view, as we sometimes do in personal inquiry, scientific observation also includes trying to imagine the kinds of observations that would undermine the view and then pursuing those observations. Confronted with the idea that people of Columbus' day held a "flat-earth" view of the world, recent historians have consulted the writings of scientists of that day and earlier and found evidence of a pretty widespread belief that the earth was spherical—surprisingly similar to our beliefs today (Gould, 1995: 38–50). The pursuit of counter-examples, and the parallel belief that one can never fully prove that something is true (but that one can show that something is not true), is a key element of the scientific method.

The Use of Logic

The *Star Trek* character Spock embodied the desirability of logical reasoning in science. Logical reasoning involves making sure that conclusions follow from premises. One particular model of logical reasoning is called the *syllogism*, or reasoning from general propositions to more particular ones. Syllogisms start with one premise that is a statement about a general class of items (for example, "All social scientists are human"). Syllogisms proceed to a premise that is a statement about particular items (for example, "Roger and Emily are social scientists"). Syllogisms then move to conclusions about the particular items, based on the previous two premises (for example, "Roger and Emily are human"). Spock, like most scientists, approved of syllogistic and other forms of logical reasoning.

Also like many scientists, Spock was always quick to point out illogical reasoning. If young Roger or Emily had proposed the notion that all families consisted of two biological parents and their offspring and presented the reasoning—"We belong to such families. Therefore everyone must."—Spock might have responded, "That's illogical." One canon of the scientific method is that one must adhere, as closely as possible, to the rigors of logical thinking. Few practicing scientists invoke the scientific standard of logic as frequently as

Spock did, but fewer still would wish to appear as illogical as Roger and Emily were in their reasoning.

The relative strengths of the scientific method, then, derive from several attributes of its practice. Ideally, the method involves communities of relatively equal knowledge seekers (scientists) among whom findings are communicated freely for careful scrutiny. Knowledge claims are ideally subjected to factual tests as well as tests of logical reasoning. They're also supposed to be explicable in terms of scientific theory, something we'll have more to say about as we pursue some of the major purposes of scientific research, especially in the section on explanatory research.

The Purposes of Social Research

The major purposes of scientific research include *exploration, description, explanation, critique,* and *evaluation or action.* Although each research project can have several goals, let's look at the purposes individually.

Exploratory Research

exploratory research, ground-breaking research on a relatively unstudied topic or in a new area.

In research with an **exploratory** purpose, the investigator works on a relatively unstudied topic or in a new area. Because of the ground-breaking nature of the work, frequently part of the task is to develop some kind of plausible story, or theory, about the subject matter at hand.

An example of exploratory research is Terry Williams's (1989) work on the teenage cocaine scene. Williams began by asking the following questions about the kids who sold drugs: "How did they get into the cocaine business and how do they stay in it? How transient is their involvement—can they get out of the business? And where do they go if they do? What are the rewards for those who succeed?" (Williams, 1989: 1). For four years, two hours a day, three days a week, Williams spent time with eight young cocaine dealers in New York City. He discovered that, although dealers knew the work was hard and dangerous, they were nonetheless attracted to the underground economy by the chance for status and prestige, leaving the drug trade voluntarily only when they had a stake in something else—a spouse, a baby, an education, a job, and so on (Williams, 1989).

Like many other exploratory studies, Williams's didn't involve examining numerous cases: He studied only eight young cocaine dealers. But also like many other studies, his leads to a theory, or story: in this case, a theory of why youths enter and leave the cocaine business.

Descriptive Research

descriptive research, research designed to describe groups, activities, situations, or events.

In a **descriptive** study, a researcher describes groups, activities, situations, or events, with a focus on structure, attitudes, or behavior. Researchers doing descriptive studies typically know something about the topic under study before they collect their data, so the intended outcome is an accurate and precise picture. Examples of descriptive studies include the kinds of polls conducted by Gallup and Roper during political election campaigns, which are intended to describe how voters intend to vote, and the U.S. Census, which is designed to describe the U.S. population on a variety of characteristics.

A descriptive study is included as the focal research in Chapter 6 of this text. As you'll see, research by Bates and Harmon (1993) seeks to compare the ability of different kinds of samples to describe the attitudes of residents of one community, Lubbock, Texas, on a series of political issues. For example, using a scientific sampling procedure, the researchers found that 73 percent of the respondents from Lubbock and environs agreed when asked the question, "Should political candidates be allowed to spend as much of their own money as they want on their campaign?"

Explanatory Research

explanatory research, research designed to explain why subjects vary in one way or another.

The goal of **explanatory** research is to explain why subjects vary in one way or another. In a study that we present in Chapter 9, researchers Norma Gray, Gloria Palileo, and David Johnson (1993) ask themselves the explanatory question: Why are some people more likely than others to blame the victims of rape for the rape? At the individual level, the researchers knew, the tendency to blame rape victims might be explained by having personally known someone who had been raped, having personally experienced rape (or rape-like) assaults, having committed rape (or rape-like) assaults, and so on. But these researchers also believed that such intimate experience with rape situations couldn't possibly explain all the variation in people's willingness to attribute blame to rape victims. Even people who hadn't had such experiences, they knew from previous research, would vary in how much they attributed blame to victims.

Gray and her associates did what many social scientists do: They consulted a theory, or a story of how and why things are the way they are. The theory Gray and her associates consulted is called "attribution theory," one that tries to describe why people attribute any characteristics to other people.

In the next chapter, you will read about attribution theory in the researchers' own words, but for now we'll just mention a key element of the theory: It expects an observer, when deciding whether to attribute responsibility to the victim of any event, to compare his or her own social characteristics to those of the victim. The theory suggests that an observer is less likely to attribute responsibility to the victim if the observer's own social characteristics are similar to the victim's than if they are not. From attribution theory, Gray, Palileo, and Johnson derived an expectation about the way that gender and the willingness to blame rape victims go together. After surveying students in college classes, the researchers did indeed find that the women students in their study attributed responsibility to rape victims less often than the men did. In their work, Gray, Palileo, and Johnson describe something (rape blame attribution) that "goes with" something else (gender) and explain this connection through a theory (attribution theory). In doing so, they explain their knowledge claim through scientific theory, one of the ideals, you will recall, of the scientific method.

STOP AND THINK *Can you remember any of the other ideals of the scientific method?*

Critical Research

Before going much further, we want to stress that just because we've decided to distinguish various purposes of social research doesn't mean that any individual researcher will be glad to have her or his research tossed into one or another of our categorical bins. Indeed it might be more accurate to say that any individual piece of research can have many purposes than to imply that all research can be distinguished by a single purpose.

critical research, research with the goal of critically assessing of some aspect of the social world.

This caveat strikes us as particularly germane as we introduce **critical research,** the fourth purpose of research, for surely much good social science research entails (or could entail) critically assessing some aspect of the social world, along with whatever other purpose it might serve. Indeed, we suspect you might see that Williams's (1989) study of cocaine dealers, Bates and Harmon's (1993) work on political attitudes, and Gray and associates' (1993) effort to explain rape victim blame could all be used for the purposes of criticizing certain aspects of modern American society.

STOP AND THINK *Can you imagine an aspect of American society that might be criticized as a result of one of these studies?*

Of course, the critical implications of particular pieces of research become much more evident if the researchers themselves spell them out. Frequently, to do this, researchers use theory, or stories about the how and the why of things. In Chapter 11, we present the work of Patricia and Peter Adler (1994) on adult-organized afterschool activities for children. The Adlers spent a couple of years observing and interviewing the participants in such activities and devised a pretty fascinating theory about their effect on children. The Adlers' take on organized afterschool activities is actually fairly critical, both of the activities themselves and of the society in which they are embedded. Thus, the Adlers concluded that, because of the adult-imposed structure of these activities, organized afterschool activities deprive children of skills they might have developed if they had been left to organize play on their own: "goal-setting, negotiation, improvisation, and self-reliance" (Adler and Adler, 1994: 325). And even those qualities that the Adlers count as positive outcomes of participating in adult-organized activities are framed in ominous terms. Thus, they see such activities as instilling values that can be useful in the work worlds of adults—values like obedience, discipline, sacrifice, seriousness, and focused attention. They leave the reader wondering, however, whether such attributes are all that admirable when they say such things as, "Adult-organized activities may prepare children for passively accepting the adult world as given; the activities that children organize themselves may prepare them for creatively constructing alternative worlds." The implicit social criticism in their comments might be difficult for many Little League parents to hear.

Applied Research

basic research, research designed to add to our knowledge and understanding of the social world for the sake of that knowledge and understanding.

applied research, research intended to be useful in the immediate future and to suggest action or increase effectiveness in some area.

The examples we've used to illustrate exploratory, descriptive, explanatory, and critical research all illustrate a larger category of research that is defined by its purpose: **basic research.** Basic research is designed to add to our knowledge and understanding of the social world for the sake of that knowledge and understanding. In contrast, **applied research** aims to have practical results and produce work that is intended to be useful in the immediate future. Schools, legislatures, government and social service agencies, health care institutions, corporations, and the like all have specific purposes and ways of "doing business." Applied research, including evaluation research and action-oriented research, is designed to provide information that is immediately useful to those participating in institutions or programs. Such research can be done for or with organizations and communities and can include a focus on the action implications of the research. Evaluation research, for example, can be designed to assess the impact of a specific program, policy, or legal change and often focuses on whether a program or policy has succeeded in effecting intentional or planned changes. Three examples of applied research will be found in the text: In Chapter 7, we include Filinson's research which examines some consequences of a new federal policy that encourages age desegregation in public housing. Chapter 8 focuses on Adler and Foster's evaluation of the effectiveness of a school reading project on student caring. Finally, in Chapter 14, Wysong, Aniskiewicz, and Wright evaluate the long-term effects of a school-based drug education program on adolescent drug use.

Summary

We've argued here that, at its best, research exists to serve theory and to contribute to stories about the "how and why" of things. We distinguished the scientific approach of social research methods from two other approaches to knowledge about the social world: a reliance on the word of "authorities" and an exclusive dependence on "personal inquiry." We've suggested that the scientific method compensates for the shortcomings of these two other approaches in several ways. First, science emphasizes the value of communities of relatively equal knowledge seekers who are expected to be critical of one another's work. Next, science stresses the simultaneous importance of empirical testing, logical reasoning, and the development or testing of theories that make sense of the social world. We suggest that social research methods can help us explore, describe, explain, and critique aspects of the social world and help us engage in action oriented research.

SUGGESTED READING

Sagan, Carl. 1995. *The Demon-Haunted World: Science as a Candle in the Dark.* New York: Random House.

A splendid guide to the advantages of science over other ways of knowing.

EXERCISE 1.1

Ways of Knowing about the Weather

This exercise is designed to compare three ways of knowing about the weather.

Part 1: Knowledge from Authorities

The night before: See what the experts have to say about the weather for tomorrow by watching a TV report or listening to a radio newscast. Write what the experts said about tomorrow's weather (including the temperature, the chances of precipitation, and the amount of wind).

Knowledge from Casual Personal Inquiry

That day, before you go outside, look through only one window for a few seconds but don't look at a thermometer. After taking a quick glance, turn away, and then write down your perceptions of the weather outside (including the temperature, the amount and kind of precipitation, and the amount of wind).

Knowledge from Research

Although we're not asking you to approximate the entire scientific method (such as reviewing the literature and sharing your findings with others in a research community), you can use some aspects of the method: specifying the goals of your inquiry and making and recording careful observations.

Your research question is "What is the weather like outside?" To answer the question use a method of collecting data (detailed observation of the outside environment) and any tools at your disposal (thermometer, barometer, and so on). Go outside for at least five minutes and make observations. Then come inside and write down your observations of the weather outside (including the temperature, the amount and kind of precipitation, and the amount of wind).

Part 2: Comparing the Methods

Write a paragraph comparing the information you obtained using each of the ways of knowing. (For example, was there any difference in accuracy? In ease of collecting data? Which method do you have the most confidence in?)

EXERCISE 1.2

Ways of Knowing about Social Behavior

This exercise compares our recollections of our everyday world with what we find out when we start with a research question and collect data.

Part 1: Our Everyday Ways of Knowing

Pick any **two** of the following questions, and answer them based on your recollections of past behavior.

1. What does the "typical" student wear as footwear to class? (Will the majority wear shoes, boots, athletic shoes, sandals, and so on?)
2. While eating in a school cafeteria, do most people sit alone or in groups?
3. Of those sitting in groups in a cafeteria, are most of the groups composed of people of the same gender or are most mixed-gender groups?
4. Of your professors this semester who have regularly scheduled office hours, how many of them will be in their offices and available to meet with you during their next scheduled office hour?

Based on your recollection of prior personal inquiry, describe your expectations of social behavior.

Part 2: Collecting Data Based on a Research question

Use the two questions you picked for Part 1 as "research questions." With these questions in mind, collect data by carefully making observations that you think are appropriate. Then answer the same two questions, but this time base your answers on the observations you made.

Part 3: Comparing the Ways of Knowing

Write a paragraph comparing the two ways of knowing you used. (For example, was there any difference in accuracy? In ease of collecting data? Which method do you have more confidence in?)

EXERCISE 1.3

Social Science as a Community Endeavor

This exercise is meant to reinforce your appreciation of how important the notion of a community of relatively equal knowledge seekers is to social research. We'd like you to read any one of the "focal research" articles in subsequent chapters and simply analyze that article as a "conversation" between the author and one or two authors who have gone before. In particular, summarize the major point of the article, as you see it, and see how the author uses that point to criticize, support, or amend a point made by some other author in the past.

1. What is the name of the focal research article you choose to read?

2. Who is(are) the author(s) of this article?

3. What is the main point of this article?

4. Does the author seem to be criticizing, supporting, or amending the ideas of any previous authors to whom she or he refers?

5. If so, which author or authors is/are being criticized, supported, or amended in the focal research?

6. Write the title of one of the articles or books with which the author of the focal research seems to be engaged.

7. Describe how you figured out your answer to the previous question.

8. What, according to the author of the focal research, is the central idea of this book or article?

9. Does the author of the focal research finally take a critical or supportive position in relation to this idea?

10. Explain your previous answer.

Theory and Research

So," said Sam, "Which comes first? The chicken or the egg?"

Driving her younger brother Sam home from a Little League game, Janet realized she enjoyed his company a lot more than she had before she'd left for college the previous fall. She smiled at him while she paused at a red light.

Waiting for the light to change, Janet thought about Sam's riddle for a bit. It reminded her of something she'd learned in one of the courses she had just completed the previous semester, and she surprised him with her answer. "That's an interesting question. But, I have an even harder one for you. Which comes first in science—the theory or the research?"

Sam was stumped, which was only partially Janet's goal. Being a good older sister, she proceeded to tell him a little about the cyclical model of science that she'd learned about in her course. In her mind, it was just as interesting as chickens and eggs. Maybe by the end of this chapter, you'll think so also.

Introduction

Concepts, Variables, and Hypotheses

theory:
a story about how and why something is as it is.

concepts:
words or signs that refer to phenomena that share common characteristics.

In Chapter 1, we told you that, in our view, social "theory" are stories about the "hows" and "whys" of social life. Now we'll add that **theories** are almost all formulated (or told) through **concepts,** which are words or signs that refer to phenomena that share common characteristics. For example, in his work on teenagers in the cocaine business, Williams (1989) focused on such concepts as "class," "status," and "prestige" to understand the attractions of drug dealing.

STOP AND THINK

The concept of "class" can be sociologically defined as "a group of people with roughly equivalent status in an unequal society." How, if at all, does this definition differ from the way "class" might be used in everyday conversation?

Scientific exploration, description, and explanation often involve both simple and complex concepts, each of which requires measurement. Some concepts, like age, can be measured relatively easily through a question on a questionnaire: "How old are you?" Others require more work, as you'll see in Chapter 6 on measurement. Once scientists make measurements, they can describe their subjects (for example, respondents to a questionnaire survey) through variable characteristics, such as "opinion about political candidate spending" or "gender."

variable:
a characteristic that can vary from one subject to another or for one subject over time.

A **variable,** or variable characteristic, is a characteristic that can vary from one subject to another or for one subject over time. To be a variable, a concept must have at least two categories. For the variable "gender," for instance, we can use the categories "female" and "male." Scientists can use variables to describe the social world, just as you might use the variable "gender" to describe the composition of your research methods class.

Another frequent use of measures involves determining which variable characteristics of subjects "go together." The use of what is sometimes called "variable language," a language that focuses on variable characteristics of subjects, rather than the subjects themselves, becomes even more noticeable when scientists want to explain something. Thus, Gray, Palileo, and Johnson (1993), in the research mentioned in Chapter 1 and in the next section of this one, tell us that the female students in their sample were less likely to attribute responsibility to rape victims than were the male students. Using the categories of two variables, "gender" and "rape blaming behavior," they found that "females" were more likely to do "less blaming of rape victims," and "males" were more likely to do "more blaming of rape victims."

hypothesis:
a testable statement about how two or more variables are expected to be related to one another.

The finding supported an expectation that Gray and her associates had derived from attribution theory about the way the two variables "went together." They stated this expectation in the form of a **hypothesis.** A hypothesis is a statement about how two or more variables are expected to relate to one another. The hypothesis that Gray and her associates developed is this:

> Females will attribute less rape victim responsibility than will males (Gray, Palileo, and Johnson, 1993: 380).

Note that this hypothesis links two variable characteristics of people: gender and the attribution of responsibility to rape victims. You might also note

FIGURE 2.1

A Diagram of the Hypothesized Relationship Between Gender and Attribution of Responsibility.

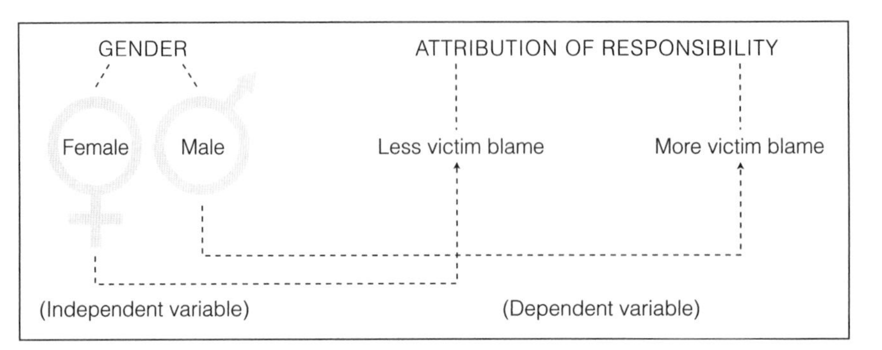

that, like many other social science hypotheses, this one doesn't use lead-pipe certainties (such as "If a person is a female, she will never attribute responsibility to rape victims, and if a person is male, he always will."). Rather, this hypothesis speaks in term of tendencies.

The business of research, you've probably surmised, can be complex and can involve some fairly sophisticated ideas (theory, hypotheses, concepts, variables—to name but four), all of which we'll discuss more later. Meanwhile, we'd like to introduce you to one more fairly sophisticated idea before moving on to a more general consideration of the relationship between research and theory. This is the idea that explanations involve, at minimum, the notion that change in one variable can affect or influence change in another. Not surprisingly, scientists have special names for the "affecting" and "affected" variables. Variables that are affected by change in other variables are called **dependent variables**—dependent because they depend on change in the first variable. **Independent variables,** on the other hand, are variables that affect change in dependent variables. Thus, when Gray and her associates hypothesize that females are less likely than males to attribute responsibility to rape victims, these researchers are implying that a change in an independent variable (gender) will cause change in a dependent variable (attribution of responsibility). Having an arrow stand for something like "goes with," hypotheses can be depicted using diagrams like the one in Figure 2.1.

dependent variable: a variable that a researcher sees as being affected or influenced by another variable (contrast with independent variable, below).
independent variable: a variable that a researcher sees as affecting or influencing another variable (contrast with dependent variable).

Explanatory hypotheses frequently imply that the researcher thinks that one (or more) variable(s) "cause(s)" variation in another. But causality is a hard thing to establish (see Box 2.1), and many social scientists steer clear of referring to it in their work.

STOP AND THINK *Suppose we find that a person's gender affects his or her attitude toward abortion. What are the two variables involved in this explanatory finding? Which variable is the independent variable? Which is the dependent variable?*

We want to assure you that at this point we don't necessarily expect you to feel altogether comfortable with all of the ideas we've introduced so far. They'll all come up again in the course of the book and, in the end, we expect you'll feel quite comfortable with them. For now we want to emphasize a major function of theory for research: to help explain research findings. Now let's turn to a more general consideration of the relationship between social research and social theory.

BOX 2.1

Social Science and Causality: A Word of Caution

The fact that two variables "go together" doesn't mean that change in one variable causes changes in another variable. In Chapter 12 we present a study by Steven Messner in which he shows that violent crime rates (a dependent variable) are lower in metropolitan areas where people tend to watch violent TV programs than in areas where they don't. Does this mean that watching violent TV programs "causes" less violent crime? Probably not.

In the social sciences we have a great deal of difficulty establishing causality for several reasons. Some reasons are purely technical, and we'll discuss them in greater detail in Chapter 8. One reason that establishing causality is difficult is that to show change in one variable causes change in another, we want to be sure that the "cause" comes before, or at least not after, the "effect." But many of our most cherished methods of collecting information in the social sciences just don't permit us to be sure which variable comes first. If, for instance, we do a questionnaire survey of students to see whether conflicts with friends "cause" a lack of self-esteem, we can (probably) find out which students have conflicts with friends and which have low self-esteem. We might even be able to show that these "go together," but we won't necessarily be sure whether conflict with friends or low self-esteem comes first because, given the nature of questionnaire surveys, we'd be collecting information about them at the same time.

Another technical problem with establishing causality is that even if we are able to hypothesize the correct causes of something, we usually can't demonstrate that something else isn't the reason why the "cause" and "effect" go together. We might be able to show, for instance, that members of street gangs are more likely to come from single-parent households than are nonmembers. However, because we're not likely to want or be able to create street gangs in a laboratory, we can't test whether other things (like family poverty) might create the conditions that make the "family structure" (an independent variable) and "street gang membership" (a dependent variable) go together. Doing experiments (discussed in Chapter 8) is most useful in demonstrating causality, but it is a strategy that frequently doesn't lend itself well (for practical and ethical reasons, discussed in next two chapters) to social science investigations.

Some reasons for the difficulty in establishing causality in the social sciences hark back to the question of theory, however. Many social scientists, for instance, raise the question of whether social theories are well enough developed to isolate proper causes. If they aren't, it's unlikely that we'll be able to find the proper causes of certain things, simply because our theories don't give us any idea of what they might be. A related problem is that there are often numerous possible causes of certain effects. We suspect, for instance, the list of "causes" you might give for being in college is extremely long. And the longer the list, the less likely it is that any theory will identify all the correct causes.

The Relationship Between Theory and Research

If social theories are stories about the way people act, interact, or organize themselves, then good social theories are those that enable people to articulate and understand something about everyday features of social life that had previously been hidden from their notice. We begin this section with two examples of social theories used by the social-scientist authors of focal research pieces that we use later in this text. Although the complete articles are included in later chapters, here we will excerpt the authors' descriptions of a theory they've employed or developed through their research. The first excerpt is from Norma Gray, Gloria Palileo, and G. David Johnson's essay (see Chapter 9) entitled "Explaining Rape Victim Blame: A Test of Attribution Theory" (1993).

STOP AND THINK *As you read each excerpt, see if you can guess whether it appears before or after the authors' presentation of their own research.*

Excerpt from "Explaining Rape Victim Blame"

by Norma B. Gray, Gloria J. Palileo, and G. David Johnson

We test hypotheses derived from attribution theory . . . in order to explain variations in rape myth acceptance. A basic assumption of attribution theory is that individuals will attribute causality for observed behavior, i.e., they will make decisions about who or what is responsible for behavior they observe. Individuals will attribute causality and responsibility either to the actor's "personal disposition" or to external (e.g., other persons, the environment, chance) causes . . . The attributions an individual makes about an actor's responsibility for his or her own behavior are influenced by (1) the social location of the observer and the observed (e.g., their class position, gender, and race); (2) the observer's previous experience (particularly in related activities); (3) the setting; and (4) the potential psychological benefit to be derived from alternative attributions (Shaver, 1975) . . .

[In particular] [w]e examine the association between the social characteristics of the observers and the attribution of responsibility from the . . . perspective . . . of the "defensive attribution" (Shaver, 1975) . . . model . . . The defensive attribution model states that the greater the similarity in social characteristics or experiences between observer and observed [victim] (e.g., if they are the same gender and/or the same race), the greater the likelihood that the observer will attribute responsibility to someone or something other than the observed victim. This attribution is motivated by the need for the observer to protect his or her self-esteem and to avoid self-blame if he or she should be similarly victimized in the future.

REFERENCE

Shaver, Kelly, 1975. *An Introduction to Attribution Processes.* Cambridge, MA: Winthrop.

Our second excerpt is from Patricia and Peter Adler's (1994) "Social Reproduction and the Corporate Other: The Institutionalization of Afterschool Activities" (see Chapter 11 for full article).

Excerpt from "Social Reproduction and the Corporate Other: The Institutionalization of Afterschool Activities"

By Patricia A. Adler and Peter Adler

In considering the benefits or harm generated by the infusion of adult control and the associated adult-oriented structures and values into the leisure of children and youth, we see elements of both. Our observations suggest that in progressing through the stages of afterschool activities, children learn several important norms and values about the nature of adult society. They discover the importance placed by adults on rules, regulations, and order. Creativity is encouraged, but acknowledged within the boundaries of certain well-defined parameters. Obedience, discipline, sacrifice, seriousness, and focused attention are valued; deviance, dabbling, and self-indulgence are not. Coordination with others, the organic model of working toward challenging and complex goals, stands as the ultimate model toward which young people are directed . . .

At the same time, the earlier imposition of adult norms and values onto childhood may rob children of developmentally valuable play, unchannelled and pursued for merely expressive rather than instrumental purposes. By participating in increasing amounts of adult-organized activities, children are steered away from goal-setting, negotiation, improvisation, and self-reliance toward acceptance of adult authority and adult pre-set goals. While these afterschool activities may prepare children for the formally rational, hierarchical, and, in Foucault's (1977) words, disciplined adult world, children's spontaneous play may teach different but important social lessons.

Adult-organized activities may prepare children for passively accepting the adult world as given; the activities that children organize themselves may prepare them for creatively constructing alternative worlds.

REFERENCE

Foucault, Michel. 1977. *Discipline and Punishment.* New York: Pantheon.

Deductive Reasoning

Gray, Palileo, and Johnson's description of "attribution theory" comes at the beginning of their article, before they present their own research; the Adlers' theory of what might be called the "functions and dysfunctions" of adult-organized afterschool activities for children comes at the end of their article. Note, however, that both theories attempt to shed light on what might have been a previously hidden aspect of social life. They do so with concepts (such as "social characteristics" and "self-blame" or "obedience" and "self-reliance") that are linked, explicitly or implicitly, with one another. These concepts are neither so personal as the one espoused by Lafeyette's mother at the beginning of *There Are No Children Here* nor as grandiose as some theories you might have come across in other social science courses. But they are theories, nonetheless.

What distinguishes these theories, and many others presented by scientists, from those presented by pseudoscientists (astrologers and the like) is that they are offered in a way that invites skepticism.[1] Neither Gray, Palileo, and Johnson nor the Adlers are 100 percent sure they've got it right. Nor would they want a theory that didn't "fit" with the observable world. For both sets of researchers, a major value of research methods is that they offer a way to test or build social theories that fit the observable social world. We'll spend the rest of this chapter discussing how research can act as a kind of "go-between" for theory and the observable world.

Many social scientists engage in research to see how well existing theory stands up against the real world. That is, they want to *test* at least parts of the theory. Gray, Palileo, and Johnson engaged in their research because of their interest in explaining why some people attribute blame to rape victims and others don't. They started, as we've suggested, with a general theory of why people attribute characteristics to others: attribution theory. From this theory, you'll recall, they deduced a testable hypothesis: that females would be less likely than males to attribute responsibility to rape victims.

Having made this deduction, Gray, Palileo, and Johnson then needed to decide what kind of observations would count as support for the hypothesis

[1]The implicit or explicit invitation to skepticism is what distinguishes scientific theories from pseudoscientific ones, according to Carl Sagan (1995) in *The Demon-Haunted World.*

FIGURE 2.2

An Illustration of the Logic of Theory Verification Based on Gray, Palileo, and Johnson (1993)

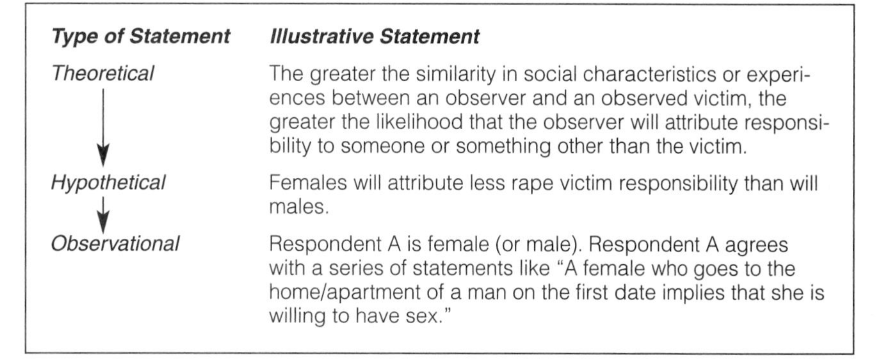

Type of Statement	*Illustrative Statement*
Theoretical	The greater the similarity in social characteristics or experiences between an observer and an observed victim, the greater the likelihood that the observer will attribute responsibility to someone or something other than the victim.
Hypothetical	Females will attribute less rape victim responsibility than will males.
Observational	Respondent A is female (or male). Respondent A agrees with a series of statements like "A female who goes to the home/apartment of a man on the first date implies that she is willing to have sex."

and what would count as nonsupport. In so doing they devised a way of measuring the variables of their hypothesis for the purpose of their study. **Measurement,** as we will suggest in Chapter 6, involves devising strategies for classifying subjects (in this case, students) by categories (say, female or male) to represent variable concepts (say, gender). To measure things such as the gender of their respondents, Gray, Palileo, and Johnson could have asked questions like "What is your sex?" Instead they used visual clues to give males and females different questionnaires. To measure whether the students might attribute blame to rape victims, they asked students to express their level of agreement with a series of prepared statements, such as "A female who goes to the home/apartment of a man on the first date implies that she is willing to have sex." Direct observation and responses to questions, then, were the ways in which Gray, Palileo, and Johnson observed the social world.[2]

measurement: the process of devising strategies for classifying subjects by categories to represent variable concepts.

Once they'd decided how they would measure important variables, Gray and her colleagues were prepared to make relevant observations to ascertain how particular students responded to questions. What we'd like to emphasize here is that in the process of testing attribution theory, Gray and her associates did what is typical of most such exercises: They engaged in a largely **deductive** process in which more general statements were used as a basis for deriving less general, more specific statements. Figure 2.2 summarizes this process.

deductive reasoning: reasoning that moves from more general to less general statements.

Theory testing rarely ends with the kinds of observational statements suggested in Figure 2.2. These statements usually are summarized with statistics, some of which we'll introduce in Chapter 15 on "Quantitative Data Analysis." The resulting summaries are often **empirical generalizations,** or statements that summarize a set of individual observations. Gray and her associates, for instance, report the empirical generalization that females were, indeed, less likely than males to attribute responsibility to rape victims.

empirical generalization: a statement that summarizes a set of individual observations.

[2]Answers to questions on a questionnaire might seem to be *indirect* ways of discerning things like attitudes. But much scientific observation is indirect. An x-ray of your tooth doesn't provide direct evidence of the presence or absence of a cavity, but, when read by a specialist, it can provide pretty conclusive *indirect* evidence.

As with Gray, Palileo, and Johnson's study, research is frequently used as a means of testing theory: to hold it up against the observable world and see if the theory remains plausible or becomes largely unbelievable. When researchers use their work to test theory in this way, they typically spend some of their effort deducing hypothetical statements from more theoretical ones, and then making observational statements that, when summarized, support or fail to support their hypotheses. In general, they engage in deductive processes, moving from more general to less general kinds of statements.

Inductive Reasoning

inductive reasoning: reasoning that moves from less general to more general statements.

In contrast with the more or less deductive variety of research we've been discussing is a kind that emphasizes moving from more specific kinds of statements to more general ones and is, therefore, a process called **inductive reasoning.** Many social scientists engage in research to develop or build theories of some aspect of social life that has previously been inadequately understood. Patricia and Peter Adler engaged in the research reported on in Chapter 11 because, after observing their own children's involvement in afterschool activities, they wanted a deeper understanding of those activities. Unlike Gray and her colleagues, who started with a particular theory, Adler and Adler focused initially on a series of individual observations (their own and those of other participants) about afterschool activities. They quote in the article from which chapter 11's focal research is derived, for instance, a twelve-year-old girl who claims to like her dance company because less capable dancers aren't admitted and therefore don't slow her progress:

> I really like being in the performing company. Having tryouts cut out all of the people who can't really dance, who aren't coordinated, and who just don't pay attention. For so many years my dance classes were filled with those kids and they dragged the class down . . . Now we learn much more because we can move faster, the classes are tough and the rehearsals are serious. (See Adler and Adler, 1994, page 321)

Having immersed themselves in many such observations, the Adlers then developed statements at a higher level of generality, statements that summarize a much larger set of observations and therefore can be seen as empirical generalizations. One generalization suggests that children who participate in adult-organized afterschool activities tend to experience predictable developmental changes that involve increasing involvement, commitment, and fervor with age:

> A broad range of adult-organized afterschool activities are available to youth. Three categories emerge, varying in their degree of organization, rationalization, competition, commitment, and professionalization: recreational, competitive, and elite . . . Not all young people follow the progression of the extracurricular career; some remain at more recreational levels or retreat from intense types of afterschool participation to more moderate and less demanding activities. However, these are

FIGURE 2.3

An Illustration of the Process of Discovery Based on Adler and Adler (1994)

Type of Statement	*Illustrative Statement*
Observational ↓	A twelve-year-old girl has a preference for a selective dance company.
Empirical Generalization ↓	Young people tend to follow extracurricular careers that move from purely recreational to more competitive.
Theoretical	Adult-organized afterschool activities prepare young people for the adult corporate world.

exceptions to the norm, and [this] depiction represents the path followed by most young people. (Adler and Adler, Chapter 11, page 275)

The Adlers might easily have left their study here, providing us with a kind of descriptive knowledge about children's participation in afterschool activities that we'd not had before. Instead they offered a theory of afterschool activities that is illuminating and, perhaps, disturbing. On the one hand, they point out how modern versions of afterschool activities foster values (obedience, discipline, sacrifice, and seriousness among them) that prepare boys *and* girls for futures in the adult corporate world. On the other, they suggest that these values are taught at the expense of others (such as goal-setting, negotiation, improvisation, and self-reliance). As a result, the Adlers made subtle and informed contributions to existing theories of socialization and stratification. The Adlers' movement from specific observations to more general theoretical ones suggests a different approach to relating theory and research than the one used by Gray and her associates. Figure 2.3 summarizes this alternate approach.

The Cyclical Model of Science

A division exists among social scientists who prefer to describe the appropriate relationship between theory and research in terms of theory-testing and those who prefer theory-building. Nonetheless, as Walter Wallace (1971) has suggested, it is probably most useful to think of the real interaction between theory and research involving a perpetual flow of theory-building into theory-testing, and back again. This circular or cyclical model of how science works was captured by Wallace in his now famous representation, shown in Figure 2.4.

The significance of Wallace's representation of science is manifold. First, it implies that if we look carefully at any individual piece of research, we might see more than the merely deductive or inductive approach we've outlined. We've already indicated, for instance, that Gray and her associates' research goes well beyond the purely deductive work implied by the "right half" of the Wallace model. Gray, Palileo, and Johnson do use statistical procedures that lead to empirical generalizations about their observations. (One such generalization: that females are less likely than males to attribute

FIGURE 2.4

Wallace's Cyclical Model of Science Theories

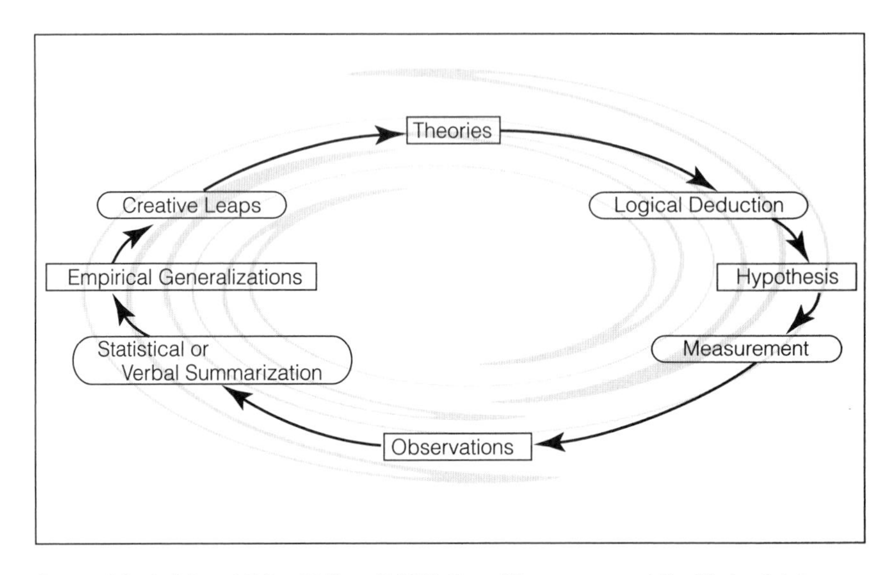

Source: Adapted from Walter Wallace (1971). Boxed items represent the kinds of statements generated by researchers during their research. Items in ovals represent the typical ways in which researchers move from one type of statement to the next. Reprinted with permission from: Wallace, Walter L. *The Logic of Science in Sociology*. (New York: Aldine de Gruyter) Copyright ©1971 by Walter L. Wallace.

responsibility to rape victims.) This led, in turn, to a renewed affirmation of their original theoretical perspective—what they call "attribution theory." Moreover, although much of the Adlers' work is explicitly inductive, involving verbal summaries of observations and creative leaps from empirical generalizations to theoretical statements shown on the "left half" of Wallace's model, the Adlers actually did quite a bit of hypothesis testing as well. Thus, in arriving at the inference that children who participate in afterschool activities tend to experience predictable developmental stages (from recreational to competitive to elite activities), the Adlers first entertained hypotheses, based on a few observations, that children typically move from, say, recreational to competitive activities, and then cross-checked these hypotheses against observations they made about other children.

A second implication of Wallace's model is that, although no individual piece of research need touch on all elements of the model, a scientific community as a whole might very well do so. Thus, for instance, some researchers might focus on the problem of measuring key concepts, whereas others use those measurement techniques in testing critical hypotheses. Gray, Palileo, and Johnson wanted to test the hypothesis that females are less likely than males to attribute responsibility to rape victims. In creating such a test, Gray, Palileo, and Johnson used measurement strategies developed by several other researchers, notably Martha Burt's (1980) technique for measuring "rape myth acceptance" and the

Koss, Gidycz, and Wisniewski (1987) measures of sexual aggression and victimization.

Most important, Wallace's model underscores the basic point that theory and research are integrally related to each other in the sciences. Whether the scientist uses research for the purpose of theory-testing, theory-building, or both, the practitioner learns to think of theory and research as necessarily intertwined. Perhaps even more than the "chicken and the egg" of the old riddle or the "love and marriage" of the equally old song, when it comes to theory and research, "you can't have one without the other."

Summary

In this chapter, we added to our discussion of theory by introducing the terminology of concept, variable, and hypothesis. Using these terms and two examples of actual research, we illustrated the relationships that most scientists, including social scientists, see existing between research and theory. Many of the nuances of the connections between theory and research have been captured by Walter Wallace's cyclical model of how science works. Wallace's model suggests, on the one hand, how research can be used to test theories or explanations of the way the world really is. On the other, the model suggests how research can be used as a tool to discover new theories. Moreover, Wallace's model suggests that an appropriate model of science not only encompasses both of the previous views of the relationship between theory and research but also the view that research and theory "speak" to each other in a never-ending dialogue.

SUGGESTED READING

Wallace, Walter L. 1971. *The Logic of Science in Sociology*. New York: Aldine de Gruyter.

Among other topics of interest, this classic includes the elegant discussion of the relationship between theory and research briefly discussed in this chapter.

EXERCISE 2.1

Hypotheses and Variables

1. Name a subfield in the social sciences that holds some interest for you (for example., marriage and family, crime, aging, stratification, gender studies, and so on).

2. Name a social problem you associate with this subfield (for example, divorce, domestic violence, violent crime, medical care, the cycle of poverty, sharing of domestic work).

3. Form a hypothesis about the social problem you've named that involves a guess about the relationship between an independent variable and a dependent variable (for example, and this is just an example—you have to come up with your own hypothesis—"children whose parents have higher status occupations will begin their work lives as adults in higher status occupations than will children whose parents have low status occupations").

4. What, in your opinion, is the independent variable in your hypothesis (for example, "parents' occupation")?

5. What is the dependent variable of your hypothesis (for example, "child's occupation")?

6. Rewrite your hypothesis, this time in the form of a diagram like the one in Figure 2.1, a diagram that connects the categories of your variables.

EXERCISE 2.2

The Relationship Between Theory and Research

Find a recent research article in one of the following journals: *Social Science Quarterly, American Sociological Review, American Journal of Sociology, Journal of Marriage and the Family, Criminology, The Gerontologist, Social Forces, Qualitative Sociology, The Journal of Contemporary Ethnography.*

Or find any other social science research journal that interests you.

Read the article and determine how the research that is reported is related to theory.

1. What is the article's title?

2. Who is (are) the author(s)?

3. What is the main finding of the research that is reported?

4. Describe how the author(s) relate(s) the research to theory.

Hint: do(es) the author(s) begin with a theory, draw hypotheses for testing, and test them? Do(es) the author(s) start with observations and build to a theory or a modification of some theory? Do(es) the author(s) work mostly inductively or deductively? Do(es) the author(s) use theory in any way?

5. Give support for your answer to question 4 with examples of the use of theory and its connection to observations from the research article.

3 Selecting Researchable Topics and Questions

Sandy[1] felt having a career was very important. Majoring in history, she had enjoyed her summer internships with the American Civil Liberties Union and in the office of the mayor of Los Angeles. Although she did not rule out an advanced degree in law or public policy, she took a job as the assistant to the director of public interest advertising at a law school after graduation. Despite her career goals, Sandy expected many of life's satisfactions to come from family. Even in her senior year, she felt pressure to marry sooner rather than later. As she said, "I would like to have kids before I'm older . . . and then there's the fact I have met somebody. It seems like so much of marriage comes down to a question of timing." She envisioned herself working continuously, with perhaps six months to a year off for childbearing, but was concerned about the unfairness of the "Mommy track." Sandy felt that only women get penalized for parenthood, and she resented it. As she said, "I don't want to be CEO of the top corporation. I don't ever want to be at the very, very top, so I don't think it will prevent me from anything. But I know it's always going to be there."

Sandy worked for a year before entering a two-year master's program in public policy. As planned, one month after leaving her job she married a man she had met in her senior year. As she said, "He'll be helping to put me through school. When I finish the program, we plan to move to my home state, where I hope to start a career in local or state government. My fiancé will be going to grad school after I start working, so I can return the favor of paying off tuition! In about five years we plan to have our first child." Sandy's progress toward combining career and family was considerably faster than most of the women who graduated with her, but her understanding of the difficulties women face have already led her to disclaim any desire to reach "the top." In their first year of marriage, neither she nor her husband had launched their careers, and Sandy believed that they both valued family above career and everything else (Hoffnung, 1996).

Researcher Michele Hoffnung has been studying women's lives for a number of years. She knows that college students like Sandy often have great expectations and she wonders what will happen to them after graduation. Will Sandy and the other young women establish long-term relationships, have children, share family responsibilities, and settle into careers as valued as those of their spouses? Will having children affect their careers? Will they be able to give their time equally to career and family as they expect? Will their careers become so engaging that they delay or forgo having children? To answer questions like these, Hoffnung designed an ambitious research project. We'll learn about her study in this chapter.

[1]Sandy is a pseudonym for one of the subjects in Hoffnung's study.

Introduction

In the first two chapters, among other things, we talked about concepts, variables, hypotheses, working inductively and deductively, the purposes of research, and the connections between research and theory. Reading those chapters, you might have asked yourself, "Theory and research *about what?* Where does the idea for a research project come from?" We'll admit that we skipped over the issues of selecting research topics and questions to present an introduction to the research process. Now that you're acquainted with some aspects of research, we'll begin at the beginning.

concepts,
words that refer to phenomena that share common characteristics.

In Chapter 2, we defined **concepts** as words that refer to phenomena that share common characteristics. Social scientists have studied many thousands of concepts. A brief list includes crime, social class, race, career expectations, family responsibilities, unemployment, parental divorce, religiosity, neighborliness, and community attitudes. In this chapter, we'll add that a **research topic** is a concept, subject, or issue that can be studied through research. Because all of the concepts listed can be studied through research, they are also research topics. Career expectations was the specific concept that intrigued psychologist Michele Hoffnung several years ago.

research topic,
a concept, subject, or issue that can be studied through research.

Beginning with a topic, a researcher will usually do some reading or consult with others in the field to figure out what it is that he or she wants to know more about. The next step is to develop a **research question,** which is a question about one or more concepts that can be answered through research. Hoffnung began with several research questions in planning the study described in this chapter. One of her questions was "Do women's career expectations change as they make decisions about marriage and motherhood?" Some other examples of research questions are the following:

research question,
a question about one or more concepts that can be answered through research.

- What is the psychological impact of unemployment?
- Does parental divorce have long-term effects on children?
- Which countries have the highest infant mortality rates?
- Which region of the United States has experienced the greatest population growth in the past decade?
- What has been the result of changing highway speed limits?
- Does the death penalty have an impact on the crime rate?
- Do neighborhoods that differ in social class also differ in neighborliness?
- Is the amount of violent crime in a community connected in some way to community attitudes?

Research questions can be about local or global phenomena, can be very specific or more general, can focus on the past, the present, or the future, and can ask basic questions about social reality or seek answers to social problems. Research questions are similar to hypotheses, except that a **hypothesis** presents an *expectation* about one or more social phenomena, but a research question does not. Like hypotheses, research questions can be arrived at inductively by generating a general question from specific observations, or by deducing a specific question from a more general understanding. Both

hypothesis,
a testable statement about how two or more variables are expected to be related to one another.

research questions and hypotheses can be "cutting edge" and explore new areas of study or they can seek to fill gaps in existing knowledge. Whereas research projects that have explanatory or evaluation purposes typically begin with one or more hypotheses, most exploratory and some descriptive projects start with research questions.

STOP AND THINK *We're willing to bet that by this point in your college career you've read dozens of articles (and most likely some books as well) that report the results of research. You can probably identify the research questions or hypotheses that the studies started with. But, think about some articles you've read in the recent past, and try to recall if there was any information about the actual genesis of each project. That is, do you know* why *those particular social or behavioral scientists did* those *particular research projects at that point in time?*

If the articles you've read are like most journal articles, they probably included a review of the literature on the research topic and made connections between the research question and at least some of the previous work in the field. In journal articles, research questions and hypotheses are presented as if they've emerged dispassionately (and, most often, deductively) simply as the result of a careful and thorough review of the prior work on a given topic.

As a result of the formal and impersonal writing style that is often used in academic writing, we often don't learn much about the beginnings and the middles of many projects. Although real research is "messy"—full of false starts, dead ends, and circuitous routes—we frequently get sanitized versions of the process. There are some wonderful exceptions,[2] but the "inside story" of the research process is rarely told publicly. More typically, written accounts of research projects are presented in formalized ways. Articles often include clearly articulated hypotheses, summaries of a great deal of work in the field that the study builds on, descriptions of carefully planned and flawlessly executed methods of collecting data, and analyses that support at least some of the hypotheses. Researchers will frequently comment on the limitations of their work, but only sometimes about problems encountered, mistakes made, and pits fallen into. Most articles published in scholarly journals present the research process as a seamless whole. There is little sense that, to paraphrase John Lennon, real research is what happens to you when you're busy making other plans. Nor is there a sense that researchers are multidimensional people with passions, concerns, and lives beyond that of "researcher."

Because our primary goal is to explain research techniques and principles, we've tried to do more than reproduce conventional social science writing in the excerpts of research that we've included. To give you a realistic sense of the research process, we've focused on small, manageable research projects of the kind being done by many scientists. In each case, we wanted to get "the story behind the story." We have been most fortunate to have found a wonderful group of researchers, each of whom was willing to give us a candid

[2] See, for example, some of the classics in social research including *Streetcorner Society* (Whyte, 1955), *Sociologists at Work* (Hammond, 1964), *Reflections on Community Studies* (Vidich, Bensman, and Stein, 1964) and *Tell Them Who I Am* (Liebow, 1993).

account. From them, we learned about the interests, insights, or experiences that were the seeds from which their projects grew, the kinds of practical and ethical concerns they had at the start, and the ways in which "real life" shaped their data collections and analyses. As a result, in each chapter, we are able to share at least a little of "the inside scoop."

It is likely that as you read this, researcher Michele Hoffnung is still working on her study of women's lives. As a result, it will be a while before we can learn what's happened to Sandy and the other women who participated in Hoffnung's study. But, although the "end of the story" has yet to be written, Hoffnung can tell us about the beginning.

FOCAL RESEARCH

Studying Women's Lives After College Graduation

by Michele Hoffnung[3]

The Research Questions

In 1992, I began a study of women's work and family choices that will take at least a decade to complete. Several factors led me to undertake this time-consuming research project. First and foremost, my personal experience as a woman committed both to career and motherhood gave me an understanding of what such a commitment entails. Second, my concern about the unequal status of women in the labor force led me to focus on the costs of the conflicting demands of work and family for women. Third, the scarcity of theories about women's adult development and longitudinal research using female samples presented a challenge to me as a feminist researcher.

Even as a girl, I always knew I wanted children. I loved babies. My mother was one of eleven children; my father one of eight. While I was one of three children, I wanted to have *at least* four. I also expected to go to college. During my college years, I became very interested in psychology. My success as a student led my professors to suggest I go to graduate school. I did and loved it. During my years at the University of Michigan, while I committed myself to an academic career, I also got married. My desire for motherhood remained unchanged; with my course work completed, I had my first child at the end of my third year of graduate school.

The first year of motherhood, I worked on my dissertation. Since my son was a good sleeper, since my husband was willing and able to share infant care except for nursing, and since we were able to hire a student to help us

[3]This article was written for this text and is published with permission.

for a couple of hours a day, I managed. But, the next year was far harder than I ever imagined. Starting as a new assistant professor was more taxing and less self-scheduled than researching and writing a dissertation. I found balancing a new career and caring for a baby extremely difficult. I spent three days at school and four days at home with my son, writing lectures during nap time and late into the night. By mid-year I was exhausted. Luckily I had an intersession in which to partially recover, so I survived.

What, I wondered, did women do who had less flexible jobs, husbands with less time or willingness to share family responsibilities, less money for some outside help, and no intersessions? The social science literature I read was proscriptive (saying what women should do or feel) rather than descriptive (describing what they *actually* did and felt). When my second child was out of infancy, I finally found the time to engage this question as a researcher. I conducted in-depth interviews with forty mothers of at least one preschool child to find out about their lives and discovered that mothers who were pursuing their careers were likely to carefully plan their families. This showed up particularly in their rigorous use of birth control, their systematic preparation for childbirth, and their careful selection of child care. They formed marital relationships with a less gender-stereotyped division of labor than the noncareer mothers; their husbands shared parenting. I also found that all of the women were grappling with the issue of employment; even at-home mothers were concerned with when and whether they would be able to work again (Hoffnung, 1992).

In my first study of women's lives, the women I interviewed were products of the fifties and sixties, when career aspirations were less fashionable for women than they are now. What, I wondered, were the realities for women growing up in more recent decades? In one of the few studies to assess women at more than one point in time, Elizabeth Almquist, Shirley Angrist, and Richard Mickelsen (1980) found that college seniors in 1968 underestimated the importance of work in their lives. As seniors, 40 percent of the women had wanted careers and family; the rest had wanted families. Seven years later, most of these women had exceeded their earlier work plans. Studies done during the eighties and nineties have indicated that a vast majority of college women plan to combine family and careers. When asked which roles they plan to have, the great majority of women say they want it all: marriage, motherhood, and career.

While we know that most college women say that they plan to have a career as well as a family, we have little data on the kinds of families they want or how they are going to balance career and family. When Kristine Baber and Patricia Monaghan (1988) questioned 250 college women, they found that these college women presented a contemporary vision for combining career and family. All planned careers, 99 percent planned to marry as well, and 98 percent also planned to be mothers. When asked how they would accomplish their goals, 71 percent said they planned to establish their careers before having their first child and the more career oriented the women, the later they projected their first birth. The most common plan was to return to work part time after the child was at least a year old. Such plans for "combining," if implemented, can have a negative impact on a woman's career development. We also know that, whether they originally planned to or not, most women are employed these days, even mothers of preschool children. Many women have jobs rather than careers; others are underemployed given their talents and training, often because of their family commitments.

My experiences, previous research, and current interests led me to plan a long-term study of a group of young women. I thought that starting at a time when they had plans to "do it all," and continuing for about ten years as they made life choices, would reveal patterns in their lives and the efficacy of various choices. For example, does the level of career commitment affect family choices? Does it affect work satisfaction?

My research goal is to see how women's ideas and expectations change as they become more educated, face the world of graduate and professional school and employment, and make the choices of whether and when to marry, whether and when to have children, and whether and how to combine career and motherhood. These are important questions for our time, as most mothers work outside the home, even though the family continues to have sole responsibility for infant and preschool care.

The Study Design

My earlier study was retrospective, with adult respondents recalling their childhood hopes and dreams; this one was to be prospective, asking young adults still in the relatively sheltered environment of college to look toward their futures. Since I am following the women over time, I can find out about the present rather than the past and watch them change and grow. My strategy is designed to avoid the problem of having respondents' memories altered by the choices they've actually made, but it requires a commitment of many years and creates a range of other methodological challenges.

In 1992–1993, I interviewed two hundred seniors at five New England colleges about their expectations and plans for the future regarding employment, marriage, and motherhood. Every spring since then I have followed their progress with a brief mailed questionnaire. This regular contact, I believe, has helped me maintain a high response rate over the years. Since the post office will forward mail only for one year, the annual contact helps to keep addresses current. I also have asked respondents to notify me if they move. Some have, especially those who use e-mail.

Collecting Data

My study requires that I recontact respondents over a period of many years with annual questionnaires and periodic phone interviews. After completing the initial in-depth interviews, I have had to interest the women enough in my project to commit to responding annually. Because American young adults move frequently, I have devised ways to contact them as they travel and relocate. In this over-time process, I see my relationships with the women as more ongoing, personal, and collaborative than is typical of more traditional and impersonal survey procedures. At the same time, I strive to maintain a value-free, open-ended environment in which they can honestly tell about their individual lives.

A major methodological concern was to select a sample that included white women and women of color from a wide variety of backgrounds. This was important because until recently most social science research, even research on women, has focused on members of the white middle-class. I wanted to recruit a sample that would be varied and accessible to me. I decided to select a stratified random sample of female college seniors from five different colleges and universities. While my strategy excludes the lower ranges of the socioeconomic spectrum, by including a small private college of mostly first generation college students, a large state university, two women's colleges, and an elite private university, I reached women with a wide range of family, educational, and economic backgrounds. Although each college required me to follow its own procedures to gain access to the senior class lists which would allow me to pick random samples of both white women and women of color, once obtained, the college permission gave me credibility when I called to ask the women to participate and meant that alumni associations would help me locate students in the years after graduation.

Since I could expect students to come to me only if I made it convenient, I needed to find an office in which to do my interviews in private and a phone for people to call if they couldn't come. One college was my home institution where I already had an office. At another college, I applied for and received a fellowship and office space at a research institute housed on campus. For the other three campuses, I used strategies based on the contacts I had. At one school, I called three feminist psychologists whom I had met professionally; one offered me her own (but phoneless) research space. At the second, I called the chairman of the psychology department and he provided me with unused space reserved for student projects. A campus phone in the hall enabled me to call interviewees, although I couldn't receive calls. In the third, I called the chairman of the history department who knew my husband and asked for help. His secretary provided me with the office and phone of a faculty member on sabbatical.

Even with the office problem solved, students needed to be called, recruited, scheduled, reminded, and if they did not show, called again and rescheduled. Each interview took about an hour, but many more hours went into phone calls and waiting for those who forgot or skipped their appointments. I quickly learned to make reminder phone calls, and I always called and rescheduled interviews for those who missed an appointment. As a result, most of the random sample I contacted completed the interview. I suspect a couple of women finally showed up to stop my phone calls!

In the initial interview, I used a set of questions that was consistent from respondent to respondent yet was flexible enough to allow respondents to explain themselves. I asked questions about their families of origin, their attitudes about work, marriage, motherhood, and their short- and long-term plans for the future. In the mailed surveys I've used in subsequent years (see Appendix C), I've asked questions about relationships, work, school, and a few open-ended questions about aspirations, feelings, and concerns. Notes and comments from the women have helped to shape the next year's questionnaire. I intend to use the questionnaire data to determine when to do the next interview. I want enough of the women to have started making decisions about family before I invest time and money in phone interviews.

Preliminary Findings

What have I found so far? As I anticipated, in their senior year, while only one woman was married and none was a mother, virtually all planned to have a career (96 percent) and to have children (99 percent). Only 87 percent expected to marry—less than in earlier studies of young women. Since in my sample, women of color were significantly less likely to expect to marry than white women, the smaller percentage may be due to my more culturally diverse sample of women (118 white, 25 African American, 20 Hispanic, and 36 Asian American).

One year after graduation, I asked "What is your dream for the next five years?" Of the 168 who responded, 42 percent had a work dream, such as the student who said "I hope to go to medical school . . . If I do not get into medical school, I hope to go to school in a health-related field and eventually serve under-served populations in the U.S." Another 55 percent had a split dream, such as the senior who said "My plan is to work for two to three more years, and then go back to school for a Master's degree, probably in urban planning. My 'dream' for myself would include a committed relationship with a yet unidentified wonderful man." Only one individual had a family dream, which was to marry and start having children, and four women (2 percent) had no specific goals, only the general sense that they wanted happiness, satisfaction, and good health.

Two years after graduation, 165 of the 200 sample responded to my survey. While 94 percent had firm or somewhat firm plans for work in the next five years, only 36 percent had plans for marriage (including the 13 percent who were engaged or married), and only 29 percent had plans to have a child in that time period (including one who had a child and one who was pregnant). Clearly in the first two years after college, these women, age twenty-five, on average, have more developed career plans than marriage and family plans. Only when more of them marry and have children will we begin to know how they will go about balancing these different areas of life.

REFERENCES

Almquist, E. M., Angrist, S. S., and Mickelsen, R. (1980). Women's career aspirations and achievements: College and seven years after. *Sociology of Work and Occupations, 7*, 367–384.

Baber, K. M., and Monaghan, P. (1988). College women's career and motherhood expectations: New options, old dilemmas. *Sex Roles, 19*, 189–203.

Hoffnung, M. (1992). *What's a mother to do? Conversations on work and family.* Pasadena, Calif.: Trilogy.

Sources of Research Questions

STOP AND THINK *Chances are today's newspaper includes several stories about social phenomena. Perhaps there's an article about the effects of divorce on children, a graph showing changes in the crime rate, an editorial on a proposed educational reform, or a story about how recently unemployed workers are coping with the loss of jobs. Do you have a personal interest in any of these topics? If you needed to design a research project for a class assignment, what topic might you select to study?*

In the focal research, Hoffnung told us the "story behind the story." Her description tells us that the selection of a research question is often the result of many factors, including personal interests, values, and passions; the desire to satisfy scientific curiosity; previous work on a topic (by the researcher in question or by others); the current political, economic, and social climates; being able to get access to data; and having a way to fund a study. Although each research project will have a unique history, some factors are common to most.

Values and Science

During the nineteenth century, the common thinking was that all science was "value free," but today it is more widely believed that values, both social and personal, are part of all human endeavors, including science. Values can subtly influence scientific investigations, and they are especially influential during the formulation or judgment of hypotheses (National Academy of Sciences, 1993: 342). Values come into play because, "At any given time, several competing hypotheses may explain the facts equally well, and each may suggest an alternate route for further research" (National Academy of Sciences, 1993: 342). However, social and personal values do not necessarily harm science.

The desire to do accurate work is a social value. So is the belief that knowledge will ultimately benefit rather than harm humankind. One must simply acknowledge that values contribute to the motivations and conceptual outlook of scientists. The danger comes when scientists allow values to introduce biases into their work that distort the results of scientific investigation (National Academy of Sciences, 1993: 343).

Personal Factors

Researchers often undertake studies to satisfy intellectual curiosity and to make a contribution to society and to social science. However, the selection of a specific research topic or question can be influenced by personal interests and experiences. It shouldn't surprise us that researchers often study topics with special significance for them. After all, research projects can take months or years to complete; having a strong personal interest can lead to the willingness to make the necessary investment of time and energy.

We've learned how Hoffnung's interest in career and motherhood influenced her work. Another example of the connection between experiences and research can be found in William Foote Whyte's recollection of the impetus for his classic study, *Street Corner Society*. Whyte recalls that one of his most vivid college memories was of a day spent visiting the slums of Philadelphia as part of a group of students. He remembered it "not only for the images of dilapidated buildings and crowded people but also for the sense of embarrassment I felt as a tourist in the district. I had the common young man's urge to do good to these people . . . I began to think sometimes about going back to such a district and really learning to know the people and the conditions of their lives" (Whyte, 1955: 281–282). Several years later, having been given a fellowship to pursue any line of research, Whyte's memory of the vague notion of studying a slum district became the genesis of a well-known research project.

Another important study came about through much more difficult personal circumstances. In 1984, anthropologist Elliot Liebow learned that he had cancer and a limited life expectancy. Not wanting to spend what he thought would be the last months of his life working at his government job, Liebow retired on disability, but found himself with a great deal of time on his hands (1993: vii). He became a volunteer at a soup kitchen and later at an emergency shelter for homeless women, and found he enjoyed meeting the residents. Liebow was struck by the enormous efforts that most of the women made to secure even elementary necessities, and he appreciated their humor and lack of self-pity (1993: viii). Liebow volunteered for two years, but then, out of either habit or training, during the last three years of his work at the shelter, he collected data as a participant observer—talking to, observing, and taking notes about the women and their lives. Before his death, Liebow completed a last contribution to helping others, the thoughtful and sympathetic analysis, *Tell Them Who I Am: The Lives of Homeless Women* (1993).

Other, much less dramatic examples of personal interests can be found in many research projects. Emily's first major study, for example, was a study of the division of labor and the division of power in marriage. She began the study in the first few years of her own marriage, as half of a two-career couple. And, perhaps because both of us, Emily and Roger, are parents who have read hundreds of books to children over the years, we've produced, separately and together, ten studies on children's literature (two of which we've included as focal research sections in this text).

Research and the Social, Political, and Economic Climates

Personal interests alone rarely account for the selection of a research topic. Knowledge is socially constructed, and social relations produce individual research projects and the enterprise of research itself. The development of a research question is affected by the specific time, place, and society in which the researcher lives (Acker, Barry, and Esseveld, 1991: 146). As social, political, and economic climates change, the kinds of research questions that social scientists pose and the kinds of questions that are acceptable and of interest to the research community and the larger society change with them.

STOP AND THINK *Some topics have been the focus of considerable research in the social sciences for a century or more, but others have received attention only within the past decade or two. Can you think of examples of topics that have been studied only recently?*

Some topics have been largely invisible to the general public and generally ignored by social scientists until recently. For example, before the 1970s, few studies focused on women and, when considered at all, data about women were analyzed using male frames of reference[4] (Westkott, 1979 ; Glazer, 1977). Hoffnung, for example, knew about quite a few long-term studies in her field of study, adult development. Although some of the research had been ongoing for as long as six decades and much was recent, very few of the major studies included women as subjects. Hoffnung found herself perplexed that the theories that she studied as a student focused on only men and their lives. In her professional work, she sought to redress the imbalance and to use the new theories being developed by feminist thinkers like Nancy Chodorow, Carol Gilligan, and Jean Baker Miller (Hoffnung, personal communication, 1997).

The same pattern is found in the sociology of the family, where until the 1970s, there were few studies of marital violence or child abuse. In 1974, for example, only nine articles on family violence were published in scholarly journals (Straus, 1992: 213). In the 1980s and 1990s, there was exponential growth of research on the topic, with 222 articles in 1988 alone (Straus, 1992: 213). Studies of the homeless have followed a similar pattern, with only the more recent studies paying attention to women's experiences (Liebow, 1993; Lindsey, 1997). These changes can be attributed to changes both in society and sociology: the growth of social activism in the 1960s championing oppressed groups, including children and women, the increases in violent crime, which sensitized people to violence, and the women's movement, which, in turn, created a new public consciousness about battering and rape (Straus, 1992: 215).

When researchers break new ground, they can find a lack of support for their projects. Helena Lopata, a well-established sociologist, encountered negative reactions from many in her social and professional circles when she decided to study the social role of widows. She pursued her interest despite the view that the subject was judged "depressing and as unworthy of serious sociological research while 'more important' scientific problems remained unstudied" (Lopata, 1980: 69).

Given the changes in the social climate in the past few decades, it's not surprising that the kinds of research questions being asked have more gender balance. However, some subjects continue to be neglected. One example is sexual orientation, which remains infrequently studied, especially in research on families (Allen and Demo, 1995). Nine journals that published

[4]The natural, no less than the social, sciences offer many interesting examples of such male bias. In one federally funded study examining the effects of diet on breast cancer, only men were used as sample subjects (Tavris, 1996).

more than 8,000 articles on family research between 1980 and 1993 included only 27 articles focused explicitly on gay men, lesbian women, or their families (Allen and Demo, 1995). The data led Allen and Demo to conclude that

> Sexist and heterosexist assumptions continue to underlie most of the research on families by focusing analysis on heterosexual partnerships and parenthood. Lesbians and gay men are thought of as individuals, but not as family members. This reflects the society-wide belief that "gayness" and family are mutually exclusive concepts. (1995: 112)

Topics and issues become the focus of studies not only when they become visible, but also when the phenomena become more common or a source of concern to the public and the scholarly community. We've seen, for example, more studies on the elderly as the proportion of the American population older than 70 has increased. Similarly, the impact of family structure on children and workplaces has received additional scholarly attention as the number of children living in single-parent households has increased. Perhaps the most obvious example is found in criminology. It isn't surprising that our society supports increasing numbers of studies on crime and criminal behavior. The United States now spends $60 billion a year on federal, state, and local criminal justice expenditures, $100 billion in costs to victims, and an inestimable amount in wasted lives and lives with the continued erosion of confidence in our social justice system (Earls and Reiss, Jr., 1994: 1). The current trends reflect a continuing pattern: Social scientists have always been concerned about social problems and have engaged in research to develop understandings of the problems and solutions.

Research Funding

Even a small study like Hoffnung's can be expensive, and the cost of doing research can run into hundreds, thousands, or millions of dollars. Many research projects are funded by private foundations, government agencies, local and state institutions, or corporate sponsors. The United States has the largest number of research foundations among the Western industrialized nations. Many of these foundations offer support for the social and behavioral sciences. In 1981, for example, the four largest contributors to social science research—the Carnegie, Ford, Rockefeller, and Sloan foundations—contributed 17 percent of the total of $160 million granted to all social science research that year (Hewa, 1993: 71).

Federal funding for behavioral and social science research grew in the middle part of the century. Although it has recently decreased—with the share of total federal research funding for these sciences falling from 8 to 4.5 percent (Smith and Torrey, 1996: 611)—federal agencies remain an important resource for social research. Funding research expresses a value choice—the judgment not just that knowledge is preferable to ignorance, but that "it is worth allocating resources to research against the other directions in which modern states expend, or could expend, resources" (Hammersley, 1995: 110).

Some researchers obtain funding in the form of grants, whereas others do work under contract for an agency or sponsor. Grants typically allow the researcher more control of the topic, methods of data collection, and analysis techniques than contracts do. A contract typically commits the researcher to conduct a study on a specific topic and often specifies timetables and predetermined methods. Although some sources are more restrictive than others, the availability of funding and economic support can influence a study, specifically the questions asked, the amount and kinds of data collected, and the availability of the research report. "[M]oney is not a free good, available for any scholarly purpose, and those with funds to dispense do so for purposes and conditions of their own choosing . . . Too often, the politics of research patronage—i.e., the choice of particular scholars to conduct particular studies likely to advance particular social objectives—is confused with its morality" (Orlans, 1967: 4).

A fascinating example of the effect of funding on research is the story behind the study, *The Social Organization of Sexuality: Sexual Practices in the United States* (Laumann, Gagnon, Michael, and Michaels, 1994). The project was influenced by a combination of factors including a major health crisis, public values, politics, and funding. The initial motivation for the project was a growing AIDS epidemic and a public health community with little data on sexual behavior because of the controversial nature of doing such research in our society (Laumann, et al., 1994: xxvii).

By the mid-1980s, coalitions of federal agencies had expressed support for a national survey of sexual practices. In 1987, the National Institute of Child Health and Human Development released two requests for proposals to design studies—one on adult sexual behavior and the other on adolescent sexuality (Laumann, et al., 1994: 39). Researchers at the National Opinion Research Center in Chicago won the contract and designed the adult study, but Reagan and Bush political appointees at the highest administrative levels of the Department of Health and Human Services refused to allow approval of even a narrowly focused survey of sexual practices. When it became clear that public funding would not be forthcoming, the researchers found financial support from a consortium of private foundations. With the funding in place, the researchers were able to complete the first study of human sexuality in decades that used a large, nationally representative sample. It is ironic that a *refusal* to sponsor the research ultimately led to a much more comprehensive study than had been proposed originally (Laumann, et al., 1994: 41).

Developing a Researchable Question

researchable question, a question that can be answered with research that is feasible.

Regardless of the sources of the idea for a project—scientific curiosity, personal experiences, social or political climate, the priorities of funding agencies, or even chance factors—once a researcher generates an idea for a topic, there's still work ahead. Several steps are needed to turn a research question into a **researchable question,** a question that is feasible to answer with research.

Reviewing the Literature

The accumulation of scientific wisdom is a slow, gradual process. Each researcher builds on the work that others have done and then offers his or her findings as starting points for new research. A **literature review** helps determine what has already been studied and what remains to be done. Reading, summarizing, and synthesizing at least a significant portion of the work that's preceded ours enables us to learn from it. We can look for warnings of possible pitfalls and try to reap the benefits of others' insights. A thorough review of the literature can help us generate a more cumulative social science, with each research project becoming a building block in the construction of an overall understanding of social reality.

literature review, reading, summarizing, and synthesizing existing work on a topic.

STOP AND THINK *Let's assume we've selected two-career families as our research topic, and we want to begin a review of the literature. Where should we start?*

Library Resources

Most reviews of the literature begin with a search of articles recently published in scholarly journals. More than 200 print journals publish sociological research (Hargens, 1991) and dozens of others include work in related fields. The citations and abstracts of articles from a great many appropriate journals are found in databases and indexes (such as Social Sciences Index, Sociological Abstracts, Social Science Citation Index, and Psychological Abstracts). These are widely available in library reference sections, both in print versions and computerized bibliographic databases, such as FirstSearch, InfoTrac, and Sociofile. Book and article citations can be found by searching using one or more **keywords,** which are the terms and concepts of interest. You'll generate numerous citations with a given keyword if it happens to be one that is commonly included in the titles or abstracts of articles. As we note in Box 3.1, the keyword "career" yielded more than 6,000 citations from one database. When using common words, multiple keywords should be used to narrow the focus. On the other hand, for very specific concepts, it might be necessary to try several different alternative keywords to find a sufficient number of citations.

keywords, the terms for the concepts you want to find sources for in a literature review.

You can locate additional citations in several ways. One index, the *Social Sciences Citation Index*, an international multidisciplinary index, also indexes the bibliographies and lists of references cited in articles. If you find the name of one or more established authorities on the topic you want to study, you can identify other work in the area by seeing who has cited the expert's work using this index.

A second method is to use the keyword(s) to locate a few good examples and then switch to the subject headings provided by some databases. For example, in doing the search described Box 3.1, using InfoTrac, Hoffnung could have first identified the article by Steil and Hay (1997) by using three keywords: family, career, and women. When she noted that "dual-career families" was one of the subject headings that the database listed for this citation, she could then have narrowed her search by using this subject heading when looking for additional citations.

A time honored method for locating additional citations is to check out the footnotes and references in articles that you have found useful. Working backwards in this way can help identify studies that other scholars considered important enough to include in their discussion of the literature.

Once a list of citations is generated, titles, descriptions, and abstracts can be used to help you decide if the article seems relevant to your own work. Many university libraries have electronic subscriptions to a few hundred journals and thousands of subscriptions to the print versions of academic journals (Knowlton, 1997: 18). If a library doesn't subscribe to a specific journal, you can usually request a copy of an article through inter-library loan with a complete bibliographic citation (which includes the name of the author, title of the article, name of the journal, volume, date of publication, and pages).

Books and government documents that report the results of research can also be located. Card catalogs and appropriate electronic databases can identify relevant volumes that are available through your local library system or through inter-library loan.

Although we can't show you all the work that results from a serious examination of the literature, we can share a small part of Hoffnung's review. Hoffnung began by looking for books and articles on women's life cycle issues research and other relevant topics. Box 3.1 describes some of the results of her efforts.

The Internet

Another source of material on a topic is the Internet and the World Wide Web. The Internet is used by researchers and others to share information through mailing groups (or listserv groups), newsgroups, and Web sites. Libraries, government agencies, businesses, and universities usually have Web sites on which they post a great deal of information. Although a huge number of sites exist on the Web, many with accurate and up-to-date information, there is no guarantee that the information on a given site will be accurate or useful. Gerald Santoro, the lead research programmer for Academic Computing at Pennsylvania State University, argues that rather than using an analogy of the Web as a library, we should see the Internet as a bookstore, with sites existing not because any independent judge has determined them worthy of inclusion, but because someone wants them there (Knowlton, 1997: 18). Web site information might be biased, obsolete, or simply not useful to scholars (Knowlton, 1997: 18). In addition, unlike the Library of Congress Subject Headings that libraries use, the Internet has no common set of headings or categories, it can be more difficult to locate appropriate material (Rodrigues, 1997: 36).

If students are careful to assess the quality of the information obtained on the Internet, they can find it a useful source of material. Two ways to do an Internet search are to use a search engine (a computer program that indexes information) or a subject directory (a categorized list of information) search (Rodrigues, 1997: 36). If you need help doing an Internet search, there are many books on the topic (for example, Campbell, 1995, Harmon, 1996, Kurland and John, 1997, Rodrigues, 1997, and Thompson, 1995). You can probably also get help from your college library or computer center.

BOX 3.1

The Literature Search Using Library Resources

Hoffnung began her literature search in the library. She used the card catalog and indices of journals. She used some keywords alone and others in combination to limit the citations generated. She was studying the lives of women college graduates using a life cyle approach, so it made sense for her to use words like "family" and "career" in her search for books and articles to review. Using InfoTrac, she found that the word "family" generated almost 20,000 recent references, with "women" and "career" identifying more than 30,000 and 6,000 respectively. However, when all three keywords, "family and career and women," were put together, that database listed 116 recent citations. Looking at titles, descriptions, and abstracts, Hoffnung was able to identify sources likely to give her theoretical frameworks, help her generate hypotheses, and suggest possible variables and methods of data collection and analysis techniques. The following are just a few of the useful references that Hoffnung identified.

Aryee, Samuel, and Vivienne Luk. 1996. Balancing two major parts of adult life experience, work and family identity among dual-earner couples. *Human Relations* 49 (April): 465–487.

Betz, Nancy. 1993. Women's career development. In *Women's lives through time: Educated American women of the twentieth century*, edited by K. T. Hubert and D. T. Schuster. San Francisco: Jossey-Bass.

Boath, E. H., A. J. Pryce, and J. L. Cox. 1995. Social adjustment in child-bearing women: The modified work, leisure and family life questionnaire. *Journal of Reproductive and Infant Psychology* 13 (3/4): 211.

Cassidy, Margaret L. and Bruce Warren. 1996. Family employment status and gender role attitudes: A comparison of women and men college graduates. *Gender & Society* (June): 312–329.

Koeningsberg, Judy, Michael S. Garet, and James E. Rosenbaum. 1994. The effect of family on the job exits of young adults: A competing risk model. *Work and Occupations* 21 (February): 33–63.

Schuster, Diane Tickton. 1993. Studying women's lives through time. In *Women's lives through time: Educated American women of the twentieth century*, edited by K. T. Hubert and D. T. Schuster. San Francisco: Jossey-Bass.

Steil, Janice M., and Jennifer L. Hay. 1997. Social comparison in the workplace: A study of 60 dual-career couples. *Personality & Social Psychology Bulletin* 23 (April): 427–439.

Yoon, Young-Hee, and Linda Waite. 1994. Converging employment patterns of black, white and Hispanic women: Return to work after first birth. *Journal of Marriage and the Family* 56 (February): 209–217.

BOX 3.2

Results of a Web Search

Using "family," "career," "women," and "research" as keywords and Yahoo, Alta Vista, and Infoseek as search engines, we generated many matches. Although it took some time to sort the wheat from the chaff, the following is a partial listing of the materials that might be useful to Hoffnung in her work:

Moms in Science Bibliography: a listing of books that focus on balancing careers in science with family responsibilities.

Success Patterns in Men's and Women's Career Transitions: A research report by Herminia Ibarra of the Harvard Business School.

Life Patterns, Personality and Self-Esteem in Gifted Family-Oriented and Career-Committed Women, 1969–1970: A description of data set currently available at Murray Center at Radcliffe from a research project completed by J. Birnbaum.

Women's Career Paths: An article by Diane Licht on the Advancing Women home page.

We did an Internet search on Hoffnung's topic using three of her keywords ("women," "family," and "career") and located more than 6,000 "matches." When we added the additional keyword "research," we located a much smaller number of sites. Box 3.2 shows a list of some of the most useful materials we found on her topic with an "online search."

A thorough literature search is a process of discovery. In addition to providing useful information, each article, book, or other material can direct you to new ideas and sources. Just as if you are following a path through the woods with only a general idea of the direction that you want to travel, you'll need a good compass and some common sense to find your way.

Using the Literature Review

When you read the focal research sections in the chapters to come, you'll find that most of the researchers have used the existing literature to develop their own research questions or hypotheses. Some researchers, however, postpone a complete literature review until after data collection so they can develop new concepts and theories without being influenced by prior work in the field. In both approaches, however, prior work in the field is used at *some point* in the research process.

Existing research helps put new work into context. What comes before determines if a study is breaking new ground, using an existing concept in a new way, expanding a theory, or replicating a study in a new setting or with a new group of people. A useful literature review provides a context for the concepts of interest, highlighting them in the same way that a frame displays a painting.

A review of the literature can be used to discuss the significance of a research question, share an overview of the thinking about a research question as it has developed over time, and present one or more theoretical perspectives on the topic. In addition, a literature review can include definitions of useful concepts and variables, identify testable hypotheses deduced from prior research, and suggest some useful strategies and methodologies. A logical endpoint of a literature review is one or more questions that can be answered with data from one or more observations or by other, more indirect kinds of information.

STOP AND THINK *Let's assume we've picked a topic and developed a research question. Let's also assume that as a result of the literature review, we have some ideas for a research strategy. If, say, we have one semester in which to do it, how can we determine if our study is "do-able"?*

Practical Matters

Even with a general idea of a research strategy, it's important to make sure the strategy is feasible before beginning the actual collection of data. No matter how important or interesting a question is, we don't have a researchable question unless it is feasible or practical to answer the question with research. **Feasibility** means considering three separate but related concerns: access, time, and money.

feasibility,
whether it is practical to complete a study given the access, time, and money available.

Access

access,
the ability to obtain the information needed to answer a research question.

Access is the ability to obtain the information needed to answer a research question. Although it's very easy to get some data, other kinds are much more difficult to secure. In her study of college seniors, Hoffnung had relatively easy access. Once the ethical issues of her proposed research had been reviewed and approved by committees explicitly designed to consider such concerns, Hoffnung was permitted by all five colleges to obtain a random sample of names and phone numbers of seniors. Hoffnung was able to cite the college's "stamp of approval," when she called to introduce herself, describe the project, and request participation. Relatively easy access and Hoffnung's personal and professional skills each contributed to the success of the project: Most of the women agreed to participate in the initial and later phases of the study. (Hoffnung's study design and data collection methods will be discussed in later chapters.)

STOP AND THINK *College students are the subjects of many studies in the social and behavioral sciences, partly because of convenient access. Are there other groups or situations that you think are also easily accessible? Any that you think are particularly difficult to study?*

People and information that are connected in some way to organizations and institutions are fairly accessible. Examples of groups that are fairly accessible include employees of a specific company, members of a union or

BOX 3.3

Groups or Records with Difficult Access

Group or Record	*Examples*
People without an institutional connection	Children 2 years of age Homeless people Atheists Americans who don't vote
People who are not easily identifiable	People with HIV Italian Americans Undocumented immigrants
People who don't want attention from outsiders	Anonymous philanthropists Wife batterers The Amish Members of gangs
People that are very busy or guard their privacy	Chief Executive Officers U.S. Senators Celebrities
Records that are private or closed	Private letters Diaries Tax records Census data for individual households

professional organization, patients in a hospital, residents of a nursing home, and children attending a summer camp. In each case, there is a place that allows us to easily meet or contact people to ask for their participation and an organizational structure that could facilitate this contact.

Most public information is easy to obtain and therefore practical to analyze. For example, statistics produced by the U.S. Department of the Census, and media, such as newspapers, children's books, and television commercials, are quite easy to access. On the other hand, some kinds of people or records pose more difficulty. (See Box 3.3 for some examples of difficult access.)

In some situations, gaining access is an ongoing process as people "check out" the researcher, find out if he or she is to be trusted, and then gradually let him or her into various parts of their lives and organizations. Susan Ostrander (1995), for example, describes this process in her study of upper-class women. She was able to find women who were willing to be interviewed and who then would refer her to others like themselves, often calling them on her behalf. Using this process, she was able to complete interviews with an adequate sample.

Other researchers face settings that are more problematic. In a study of prison work programs in India, N. Prabha Unnithan (1986) found access to the site, inmates, staff personnel, and records almost impossible to obtain. Although requests were made through the appropriate channels, all opportunities for direct examination of the daily management of prison work programs were denied. Although permission to visit the inmates' actual work and dining areas was initially granted, it was later withdrawn. As a result, the study had to depend solely on statistics gathered from administrative reports (Unnithan, 1986).

You'll notice the focal research selections in this text use many different methods of data collection with great variation in the accessibility of data. In Chapter 7, for example, we'll describe Rachel Filinson's analysis of attitudes of elderly residents of public and publicly subsidized housing. Because she is the Director of the Gerontology program at a local college, she found managers of the housing sites quite willing to generate lists of elderly tenants for her to contact. In Chapter 8, we present a study that one of us did with data from seventh grade students. Emily obtained access to the school by teaming up with someone with who had an existing connection to the school system. By working with social worker Paula Foster, whose responsibilities at a community mental health center included school liaison work, Emily was able to locate interested principals and teachers.

In Chapter 14, the focal research is an evaluation of the Drug Abuse Resistance Education (DARE) program in Kokomo, Indiana. Richard Aniskiewicz, one of the researchers, was recommended to the police department and schools as an evaluator by the first local DARE police officer who was a former student of his. The evaluation process of the DARE project in Kokomo continued for more than seven years after the initial contact.

In some chapters we present research projects with very easy access to data. In Chapter 13, we describe Roger's work with Rachel Lennon and Leanna Morris, an analysis of award winning children's books that the researchers obtained by borrowing them from local libraries. In Chapters 6 and 12, respectively, we highlight Ian Rockett and Gordon Smith's and Steven Messner's studies. In both cases, the researchers did not generate any new data, but instead used existing data, including official government statistics.

Time and Money

Research can be expensive and time consuming. In a study where subjects will be interviewed, salaries for interviewers and transcribers need to be budgeted. If mailed questionnaires are planned, costs include paper, printing, and postage. If historical documents are a source of data, there often are copying costs.

Some projects can be completed quickly and very inexpensively. In studies that use existing documents or books borrowed from libraries, the data collection costs will be negligible. In other studies, the use of questionnaires, especially those that can be handed out in groups, can be relatively inexpensive.

Projects that take many years to complete (like Hoffnung's) are much more costly than those with only one data collection. Similarly, projects that study large numbers of people or groups are more expensive than those with fewer people. Researchers can sometimes rely on donated materials and volunteer workers to make a study feasible. Other times, unanticipated costs and events can lead to a scaled down project.

research costs, all monetary expenditures needed for planning, executing, and reporting research.

time expenditures, the time it takes to complete all activities of a research project from the planning stage to the final report.

Although each method of data collection has unique costs and time expenditures, we can identify some general categories of expense. **Research costs** include the salaries needed for planning the study; any costs involved in pre-testing the data collection instruments; the actual costs of data collection and data analysis such as salaries of staff, payments to subjects for their time, rent for office space or other facilities; the cost of equipment (such as tape recorders and computers), and supplies (such as printing and paper costs for questionnaires); and other operating expenses. **Time expenditures** include the time to plan the study, the actual data collection process, and the data organization and analysis.

The focal research in this chapter is an example of an ambitious and a "do-able" study. For the past few years, Hoffnung has worked on this project at least several hours each week and expects to do so for years to come. By doing the interviewing herself, she was able to control the quality of the interviews and avoid the expense of hiring interviewers. In exchange, however, she invested a great deal of her own time in the project. Hoffnung obtained small research grants to cover expenses including postage, a tape recorder, tapes, and interview transcription. She kept costs to a minimum by employing work-study students as assistants for the follow-up questionnaires, by using her existing and donated office space, and by working on a computer she already owned.

Summary

Hoffnung's interest in studying young women's career and family expectations and the questions she asked about how college graduates change over time is one example of the wide variety of topics and questions that can be studied. Research questions can focus on one or more concepts, can be narrowly or broadly defined, and can be developed inductively or deductively. Questions can be generated from a number of sources, including personal interests and experiences and the social and political climate. The availability of funding can determine not only the size and scope of a project, but also the specific question, the timetable, and the methodology.

A review of the literature is essential at some point in the research process. Like other researchers, Hoffnung was able to locate useful books, articles, and reports using keywords and databases in her literature search. Using words like "career aspirations," "two-career family," and "role conflicts," she obtained many citations on her topics. Careful reading of previous work gave Hoffnung a foundation for generating a strategy for data collection and analyses.

Planning a study means considering the practical matters of time, money, and access to data. Hoffnung's research was feasible. She has been able to keep in contact with the women she interviewed and has had sufficient opportunity and resources to continue to collect and analyze data. In this work, as in all research, it's clear that no matter how important or interesting a research question is, unless the research plan is practical, the study will not be completed.

SUGGESTED READINGS

Bart, Pauline, and Linda Frankel. 1986. *The student sociologist's handbook*, 4th edition. New York: Random House.

This volume includes a useful section on library research that can help in conducting a review of the literature.

Hunt, Morton. 1985. *Profiles of social research: The scientific study of human interactions*. New York: Russell Sage Foundation.

Hunt provides a fascinating collection of noteworthy case histories of social research, including studies of the impact of educational segregation, evaluations of income maintenance projects, and experiments on the psychology of teamwork. This behind-the-scenes view shows the research process, "warts and all."

Rodrigues, Dawn. 1997. *The research paper and the World Wide Web*. Upper Saddle River, NJ: Prentice Hall.

This useful text discusses research strategies using the Internet. Web tools and access to library resources through the Internet are covered.

Straus, Murray. 1992. Sociological research and social policy: The case of family violence. *Sociological Forum* 7 (June): 211–237.

In this interesting analysis of the exponential increase in research on family violence since 1970, Straus argues that the cause of this increase in research on violence is the result of changes in American society and sociology as a discipline. In addition, he examines the patterns of use and neglect of research findings, and the dilemmas of mass media involvement in publicizing research findings.

EXERCISE 3.1

Selecting a Topic

1. Get a copy of a local or regional newspaper. List the headlines of at least five stories that deal with social behavior or characteristics.

2. Select one of the stories and hand it in with your work on this exercise.

3. Generate one research question on the topic of the article that could be answered by social research and state it.

4. Assuming you could get enough funding and had the time to do research to answer the question you stated, would you do the research?

Describe why you would or would not do the research, focusing on your personal interests, today's social and political climate, and the contribution you think the research would make to social science or society.

EXERCISE 3.2

Starting a Literature Review

Select one of the following topics from Hoffnung's study: career aspirations, self-concept, two-career family, family planning attitudes, role conflicts. Using a print index or an electronic database, find ten citations of original social research on this topic that are no more than five years old. (Be careful not to get an article that summarizes other people's work or is just a discussion of theory.) List the topic and then the five complete citations.

1. Write down the topic you used for your search.
2. List five citations.
3. Select two of the articles you listed that are available in your library and read them. Put an asterisk (*) next to the two citations you've read.
4. Compare and contrast the reviews of the literature in the two articles. For example, is the literature review in one article as extensive as the other? Do the articles mainly cite the same literature or do they review predominantly different work? To what extent does each review focus on theories and theoretical perspectives?
5. Do these studies start with research questions or hypotheses? List one research question or hypothesis from each article.
6. Are their research questions or hypotheses different? If they are different, why do you think this happened?

EXERCISE 3.3

Using the Internet and the World Wide Web

1. Select one of the following topics: two-career family, death penalty legislation, or immigrant assimilation. Write down the name of your topic.
2. Using a search engine on the Web, such as Yahoo, Alta Vista, or Infoseek, search for the topic. Write down the number of matches you had for the concept.

3. If you ended up with more than a few hundred matches, narrow down your topic in some way. Describe what you did.
4. Describe the kinds of information you found on your topic (or narrower topic).
5. Read three of the documents you obtained from the Internet or World Wide Web that you think would be useful if you were actually going to do research on the topic you selected. Write a paragraph about each of the three documents.

EXERCISE 3.4

Summing Up Practical Matters

Think about the following research questions and the suggested ways that a researcher might collect data. For each, give a cost and time estimate and your evaluation of whether access will be easy or hard.

Research Question	*Will access be easy or hard?*	*Time consuming or not?*	*What are some expenses?*
1. Do teachers' characteristics and teaching styles have an impact on amount of student participation in classes? (Suggested method: Observe 50 classes for several hours each.)			
2. What kinds of countries have the highest infant mortality rates? (Suggested method: Use government documents for a large number of countries.)			
3. Is the psychological impact of unemployment greater for people in their 20s or 50s? (Suggested method: Do a survey using a questionnaire mailed to 300 people who have contacted the unemployment office in one state.)			

(continued)

Research Question	*Will access be easy or hard?*	*Time consuming or not?*	*What are some expenses?*
4. Which region of the United States has experienced the greatest population growth over the past 20 years? (Suggested method: Use data from the U.S Bureau of the Census.)			
5. Are people who are more religiously observant more satisfied with their lives than those who are less observant? (Suggested method: Interview 100 people for approximately 1 hour each.)			

4 Ethics and Social Research

In 1984, medical sociologist Rose Weitz attended a chapter meeting of the American Civil Liberties Union where attorney Jane Aiken discussed discrimination against persons with AIDS and other stages of HIV disease. Although Weitz was already aware of the medical implications of HIV, the presentation gave her a better understanding of the political and sociological ramifications of the disease. When Aiken and Weitz met shortly after the presentation, Aiken lamented the paucity of social science data on HIV and urged Weitz to consider research on the topic. Aiken's encouragement became the impetus for Weitz's project, *Life with AIDS*.

Between 1986 and 1989, Professor Weitz conducted in-depth interviews with 37 men and women with HIV disease, re-interviewed 13 of them four to six months later, and conducted shorter interviews with 26 doctors who had treated persons with HIV. Weitz came to define the study as an opportunity to understand how people with HIV see their lives. Aiming for a holistic picture, she explored many issues, including how diagnoses are obtained, the way initial ideas about the consequences of HIV are developed, and the process by which lives, self-concepts, and relationships change in response to the illness.

Rose Weitz had completed her graduate education and been working as sociologist for eight years before beginning the research on HIV. She felt that her training had prepared her for a variety of research situations, but soon realized that she had not foreseen much of what awaited her. As Weitz reports in the focal research section of this chapter, interviewing persons with HIV forced her to confront a series of unanticipated personal, legal, and ethical dilemmas.

Introduction

Conducting Ethical Research

"This may be an exaggeration, but it seems like virtually every research project I have ever engaged in has had ethical dilemmas for which I was less than fully prepared" (Weitz, personal communication, 1996). Most social scientists begin their studies feeling prepared to behave ethically. At the end of their projects, at least some researchers have found themselves in agreement with Weitz's point of view.

Ethics refers to the abstract set of standards and principles that are used to determine appropriate and acceptable social conduct. In addition to judging research on methodological and practical grounds, it's important to evaluate the ethics of any research project. However, a researcher's initial confidence about the ease of doing ethical research is sometimes unwarranted. As we'll see in this chapter, unforeseen circumstances and unintended consequences at every stage of research—from planning to publication—can create ethical problems. Determining appropriate research behavior can be a complex topic and a source of heated debate. One issue is whether there is a set of universally applicable **ethical principles** or if it's better to take a case by case approach and consider circumstances of every research project when deciding what is, and what is not, ethical. Another challenge is the extent to which the ethical *costs* of the study should be measured against the study's *benefits* for the participants or others when making research decisions.

ethical principles in research, the set of values, standards, and principles used to determine appropriate and acceptable conduct at all stages of the research process.

STOP AND THINK *Do you know about any research project in the social sciences, medicine, or biology where the researcher did something that was ethically questionable? What was it about the study that raised questions for you?*

An Historical Perspective on Research Ethics

Some individual scholars have probably always been interested in defining appropriate research conduct, but not until the middle of the twentieth century did ethics become a concern of social science organizations as a whole. The first ad hoc committee of the American Psychological Association (APA) established in the late 1930s to discuss the creation of a code of ethics did not support its development. Not until the 1950s did the APA first develop and adopt a code of ethics (Hamnett et al, 1984: 18).

Similar trends can be found in other fields and in international social science organizations as well. The British Sociological Association approved a code of ethics in 1973, the American Anthropological Association first drafted its statement in 1967, and the members of the American Sociological Association (ASA) were so divided about creating a statement on ethics that it took more than eight years from the code being proposed to its adoption in 1971 (Hamnett et al, 1984). The most recent revision of the ASA Code of Ethics was passed in 1997. It covers sociologists' responsibilities and conduct in all areas and the rules and procedures of the ASA committee designed to investigate and resolve complaints of unethical behavior.

(Appendix D of this text contains the parts of the code most clearly related to research.)

The relatively recent impetus for disciplines to consider ethical issues stems from several factors. One factor was the reaction to the total disregard for human dignity perpetrated during World War II by researchers in concentration camps controlled by Nazi Germany. As a result of the Nuremberg trials, in 1946, a set of principles about research on human beings was adopted by the United Nations General Assembly (Seidman, 1991: 47). The first principle of the Nuremberg Code is that it is absolutely essential to have the voluntary consent of human subjects (Glantz, 1996).

There has also been increasing awareness of and distress among scientists, policymakers, and the general public concerning problematic biomedical research involving human subjects. For example, government sponsored research conducted during the 1950s studied the effects of radiation on human subjects. Funded by the Atomic Energy Commission, this research included legitimate medical research on using radioisotopes to diagnose or cure disease, as well as some experiments that were gruesome: Injecting people with "relatively short life expectancies" with plutonium and determining how much uranium was needed to produce kidney damage by using a sample of semicomatose patients with brain cancer (Budiansky, 1994). In a study of cancer done in 1963, three physicians at the Brooklyn Jewish Chronic Disease Hospital injected live cancer cells into 22 elderly patients without informing them that live cells were being used or that the procedure was unrelated to therapy (Katz, 1972).

One long-lasting research scandal ended as the result of public outrage. The Tuskegee study (labeled as such because the project was conducted in Tuskegee, Alabama,) was conducted by the United States Public Health Service to study the effect of untreated syphilis to determine the natural history of the disease (Jones, 1981). The 40-year research project was started in 1932, with a sample of 399 poor, black men with late stage syphilis. It ultimately included 600 men, all of whom were offered free medical care in exchange for their medical data, and none of whom were told that they had syphilis. More than 400 of the men were not offered the standard (and often fatal) treatment for syphilis in the 1930s. In later years, they were not provided with penicillin even though it was available by the late 1940s (Jones, 1981). Many of the men died of the disease, and some of them unknowingly transmitted it to wives and children (*The Economist* 1997: 27). As late as 1965, an outsider's questioning of the morality of the researchers was dismissed as eccentric (Kirp, 1997: 50). The project ended in 1972, but not until 1997 was an official governmental apology offered by President Clinton to the surviving victims and their families (Mitchell, 1997: 10).

The growing interest in ethics can also be connected to the social context of the second half of the twentieth century. Students in the late 1960s and early 1970s, some of whom became social scientists, questioned many previously accepted ideologies, including the value and morality of science. After becoming professionals in the field, these individuals continued to question assumptions about the responsibilities of social scientists both to the larger society and to the people they study (Hamnett et al, 1984: 26).

By the 1970s, social scientists could look to their professional associations for guidance in conducting ethical research. From 1971 on, the federal government had guidelines for projects receiving funds that typically included a mandated review by a committee explicitly designed to consider ethical issues (Warren and Staples, 1989: 270). These committees, typically called institutional review boards (IRB) or human subjects committees, now exist at almost every college, university, and research center. Before any federal money can be spent on research with human subjects, the research plan must be approved by an institutionally based committee of five or more members, at least one of whom is not affiliated with the institution (Edgar and Rothman, 1995). The central charge of an IRB is to verify that the benefits of proposed research outweigh the risks and to determine if subjects have been informed of the risks that accompany participation (Edgar and Rothman, 1995). The need to seek the approval of such a committee varies depending on the topic, the source of funding, and the specific institutions involved. Typically, research projects are subject to review when they deal with sensitive issues, can cause damage to participants, or when subjects can be identified. On the other hand, the Department of Health and Human Services makes secondary analysis, educational testing, routine surveys, and routine field observations exempt from prior review (Lee, 1993: 33). Researchers need to be familiar with the ethical codes of the organizations and should submit their proposals to appropriate review committees. Ultimately, the responsibility for ethical research behavior lies with the individual researcher.

Researchers like Rose Weitz who began their work in the 1980s and 1990s probably studied the topic of ethics at least briefly as part of their undergraduate and graduate education. At the start of a project, typically, a researcher will consider the general principles of ethical behavior and might foresee at least some of the ethical concerns to be faced. But as the following excerpt from Rose Weitz's book makes clear, the day-to-day process of doing research can bring unanticipated difficulties and the need to make hard choices.

FOCAL RESEARCH

Personal Reflections on Researching HIV Disease[1]

by Rose Weitz

When I began my research, I expected it to be ethically challenging and psychologically draining. In retrospect, however, I was terribly naive about the difficulties and stresses I would encounter.

[1]This is an abbreviated version of the chapter of the same name reprinted with permission from *Life with AIDS* by Rose Weitz, published by Rutgers University Press (New Brunswick, NJ), 1991.

Personal Dilemmas

One of my most serious problems has been the unusually burdensome sense of responsibility this research created. Probably all researchers struggle with doubts about their abilities to do justice to their topics, but the nature of my work has heightened this sense of responsibility enormously. If my research becomes widely read, it may affect health care policy, and hence significantly affect many individuals' lives.

For example, I have testified to a Governor's Task Force on AIDS and have learned that my testimony was influential in convincing the task force to oppose mandatory reporting of persons who test positive for HIV. More crucially, in terms of its emotional impact, many people I spoke with explicitly referred to their interviews as legacies. They participated in the project despite the emotional and sometimes physical pain it caused them because they believed I would use their stories to help others. Thus they gave me the responsibility of giving meaning to their lives and to their deaths.

I have also encountered personal difficulties because the research made the inevitability of death and the possibility of disability far more salient for me. Early on, I realized that at each interview I automatically compared my age to that of the person I was interviewing. I was frequently disturbed to find that they were much younger than they looked and were often younger than I. (I was 34 when I began the work.) Having to witness the pain of their lives was far more difficult than coping with the knowledge that they were dying, especially because the dying occurred out of my sight.

In addition to teaching me some truths about death and disability, the interviews forced me to recognize the irrational nature of my own and others' response to potential contagion. To control my fears during interviews, I often found myself repeating silently a calming litany about how HIV is and isn't transmitted. Meanwhile, I had to calm the fears of friends and acquaintances who wondered whether I had become a source of contagion. All I could do was provide information, and, fortunately, no one proved irrationally fearful.

Legal Dilemmas

I was aware of potential legal problems from the start. Arizona law requires all health care workers to report to the state any person who has HIV disease, and according to the Department of Health Services, I could legally be required to report anyone I interviewed. Recognizing that reporting someone could jeopardize their civil rights, I was unusually concerned about protecting the confidentiality of my records.

The need for confidentiality produced some unexpected difficulties. I wanted to make it easy for persons with HIV who were considering participating in this research to reach me. However, I could not give out my home phone number because my housemate might listen to our answering machine and unintentionally hear a message meant only for me. Similarly, I could not write names, addresses, or phone numbers in my pocket calendar because I could not risk having it subpoenaed and losing all the other daily records it contained.

Ironically, my chief worry was protecting the signed "informed consent forms" that my university's research ethics committee required me to obtain from each person I interviewed. Before I did the initial interviews, we were unable to reach an agreement on an alternative that would not require individuals to sign their names. (Before I did later interviews, however, we agreed that instead of a consent form, I would start each taped interview by reading a statement about informed consent and the person's response.)

Ethical Dilemmas

Early in my training as a sociologist, I learned the value of informed consent. Translating this principle into action, however, requires more than simply presenting potential interviewees with a description of one's purposes and research methods. Most of the people with HIV disease that I interviewed learned of the study through letters from nonprofit groups offering emotional and financial support to such persons. Others learned of it from doctors, friends, groups, or a newspaper. I soon discovered that many of the persons with HIV disease I was to interview assumed that I was a counselor, was working for one of the community organizations, or was a lesbian. I then had to decide how much responsibility I had to correct their assumptions, particularly when they were not explicit. Should I have assumed that they knew what a sociologist was, or should I have carefully differentiated sociology from social work? If an individual commented that I did not wear a wedding ring, should I have assumed that they were searching for a polite way of asking if I was a lesbian? If the person I was interviewing did not understand the first time I explained, how many times should I have stopped a given interview to explain that I did not work for a community organization?

Truly informed consent was even more rare among those who felt they could not afford to refuse to participate. Despite my disclaimers, many individuals obviously believed that I worked either for the state or a community organization for persons with HIV disease. As a result, they may have feared that they would jeopardize their access to services if they did not participate. Even if they believed they would not be punished for lack of cooperation, they may still have believed that they would be rewarded for helping. Several, for example, mentioned that they wanted their physicians to consider them exemplary patients so that the physicians would remember them when choosing patients for experimental treatment programs.

Truly free consent was even less likely when interviewing individuals in jail or prison. Because of the connection between HIV and drug use, many persons with HIV disease are imprisoned. As a result, I initially pursued the possibility of contacting persons with HIV disease through the health department of Arizona's prison system, and the first contacts with officials seemed promising. Even more than other persons with HIV disease, however, prisoners might have felt that they would be punished if they refused to participate. Moreover, I would have had no way to keep the administration from knowing who had refused because I would have had to arrange the interviews through prison officials. Consequently, I was relieved when negotiations fell through.

I also faced a broad range of ethical dilemmas when the persons I interviewed viewed me as an information or counseling source. Often persons I interviewed asked me questions about the disease, drugs, or doctors. A similar difficulty arose when I encountered persons with HIV disease whose knowledge of their illness was clearly deficient. I then had to decide whether to answer their questions even if doing so might jeopardize my relationship with influential physicians (who, for example, had not provided information about the side effects of certain drugs). I also had to decide whether to correct individuals' information if not asked. These questions posed minimal problems compared with those generated by implicit or explicit pleas for counseling. I had no answer for the individual who asked me how to convince her uninfected husband to use condoms. Nor was it clear to me how to respond when individuals told me that their disease was punishment for their sins, or that they felt suicidal or that their greatest grief came from losing their children but they would not contact them for fear of infecting them. Saying nothing did not feel ethical to me. Yet saying something required me to break professional norms for sociological research.

Searching for Solutions

The legal dilemmas posed by this research presented the fewest difficulties because the solutions were purely practical. To avoid writing confidential information in my usual daily calendar, I began carrying a small notepad with me. As is typical in this sort of research, all of my records and tapes were number-coded, and all identifying information was removed from the transcripts and resulting publications. I recorded data on each individuals' age, marital status, occupation and the like separately from the taped interviews, so that persons who transcribed the tapes would not have that information. I hoped this would make it more difficult for the transcriptionist to recognize the person I was interviewing, should the transcriptionist happen to know him or her. All papers with names or addresses were kept in one less-than-obvious location and have now been destroyed. In addition, I decided that if I were subpoenaed, I would refuse to turn over information. (Of course, such a decision is easy to make in the abstract, and I cannot predict what I might have done if actually faced with a jail sentence for contempt of court.) Finally, when some individuals, far from desiring confidentiality, told me that they would like me to use their real names in this "legacy" we were creating, I decided not to do so because the potential "stigma fallout" seemed an unfair burden to thrust on unsuspecting friends and relatives.

The remaining problems had no easy answers. I tried to answer all questions, whether direct or indirect, asked by those I interviewed. I also tried to clarify their understanding of such things as sociology versus social work and working *with* versus *for* a community organization. If someone still seemed not to understand after two or three attempts at explaining, I did not pursue the matter.

When it came to a choice between being ethical and being "sociological," I went with ethical every time. I felt a strong moral obligation to answer any question I was asked and to provide needed information even when not asked directly. I did try to wait until after the individual had finished stating his views before interjecting any information. And I occasionally requested that individuals not tell their doctors that I was their source.

Similarly, I believe it is wrong to listen to a person's feelings of guilt and self-deprecation without attempting to alleviate those feelings. Thus if a person told me he felt horribly stupid for not using "safe sex" until a couple of years ago, I heard him out and then gently suggested that until recently no one knew how HIV was transmitted. I also volunteered suggestions about sources of information and social support when it seemed warranted. I realize that these tactics may have diminished my credibility as a researcher, but I thought that was less important than diminishing my credibility as a person.

Finally, I did whatever I could to maintain my own mental health because I realized that I would be of little use to anyone if I allowed myself to burn out. So I tried to restrict my contact with those who have HIV disease to interviews. This tactic proved insufficient on two notable occasions—once when the family of someone I interviewed, recognizing that the interview had been an important experience for their relative, asked me if I would visit him on his deathbed, and the second time when students asked me to speak at a memorial vigil for an alumnus who had recently died of HIV disease.

While I now have one good friend with HIV disease, I am grateful for my sanity's sake, that, at the time I was conducting the interviews, I had no friends in the high-risk groups (so far as I knew). It also helped my state of mind that I did not know what any of the people I interviewed were like before their illness and that I did not have to see them as their health deteriorated. For this reason I was very reluctant to do follow-up interviews, even though I knew they would provide invaluable data; it took many months of soul-searching before I felt able to do so. Similarly, for a long time I avoided taking steps to commit myself to doing the research for more than a year at a time. Knowing that my commitment was finite made it more manageable psychologically.

At the time I began interviewing, few women in Arizona had been diagnosed with HIV. By 1989, however, I was able to locate women with various stages of HIV disease and interviewed twelve of them. I learned the hard way that it was much more difficult emotionally for me to interview women than men because I identified so much more strongly with the women. The summer I began interviewing women was the summer I began having nightmares about contracting HIV disease.

To protect my mental and physical health, I tried to schedule interviews to have time for exercise afterward and to satisfy my worries with popcorn rather than chocolate. I calmed other new anxieties by updating my will, obtaining disability insurance, and giving a friend a medical power of attorney. I also lost most of my inhibitions about discussing my research and my attendant emotions with friends and acquaintances, for I found it helped to share the burden.

Undoubtedly many other researchers have coped with similar problems over the years. Unfortunately, the norms of professional conduct do not allow easy and open discussion with colleagues about either psychological or ethical problems and the rarity of lawsuits leaves few of us sensitized to legal issues. As a result, many universities provide excellent technical training in research skills, but few prepare students for dealing with these nontechnical but equally crucial difficulties. I am sure that I reinvented the wheel in my attempt to deal with these dilemmas, but I found few good places to read about—let alone discuss—my concerns. Perhaps the stresses of this kind of research will pressure scholars to begin tackling these issues openly.

Principles for Doing Ethical Research

As Weitz tells us, she used a set of moral and social standards that includes both *prohibitions against* and *prescriptions for* specific kinds of behavior to guide her behavior in dealing with unanticipated situations and difficult moments. That is, she applied ethical principles to the decision-making process to create a moral compass. For most researchers, the essential core of acting ethically is their desire to safeguard the welfare of those who participate in their studies, protect the rights of colleagues, and serve the larger society. As we consider several specific ethical principles we'll examine the kinds of issues Professor Weitz and other researchers grapple with.

STOP AND THINK *What ethical standards do you think are most important when doing research? For example, if you were asked to observe friends at a party to study social interaction among college students for a class assignment, what would you do? Would you do the observation without first asking your friends for their permission?*

Principles Regarding Participants in Research

Protect Subjects From Harm

protecting study participants from harm, the principle that subjects are not harmed, physically, psychologically, emotionally, legally, socially, or financially as a result of their participation in a study.

First and foremost among ethical standards is the obligation to **protect study participants from harm.** This means that the physical, emotional, and psychological well-being of those who are research subjects must be maintained, both during the time that the data are collected and after the conclusion of the study. The researcher's actions should not put participants at risk of social ostracism or of negative repercussions from their family or community (Ringheim, 1995).

In a common social science method, data are collected by asking study participants questions. For example, Hoffnung, in the research described in Chapter 3, asks the same group of women a series of questions annually, including whether they are currently working for pay, if they are in school, and what their marital status is. Although answering questions like these might generate feelings of sadness or disappointment for some, they represent a minimal risk to the participants and little intrusion on their privacy, especially because the answers are held in confidence. When Hoffnung's research was reviewed by the institutional review boards at the colleges attended by the students she interviewed, it received approval. Researchers doing the kind of work Hoffnung did and those studying noncontroversial topics often have little difficulty with the ethical principle of "no harm to subjects."

A study that has been criticized as violating the "no harm" principle and other ethical standards is the Milgram experiment on obedience to authority figures. In this famous experiment, psychologist Stanley Milgram deceived his subjects by telling them that they were participating in a study on the effects of punishment on learning. The subjects were told they that would be "teachers," who, after watching a "learner" being strapped into place, were taken into the main experimental room and seated before an impressive shock generator with thirty switches ranging from 15 volts to 450 volts. The subjects were ordered by the experimenter to administer increasingly severe shocks to the "learner" whenever he gave an incorrect answer. They were pressed to do so even when the "learner" complained verbally, demanded to be released, or screamed in agony. The "learner," or victim was in reality an actor who actually received no shock at all; the purpose of the experiment was to see at what point, and under what conditions, ordinary people would refuse to obey an experimenter ordering them to "torture" others (Milgram, 1974).

Before the actual experiment, Milgram asked a number of psychiatrists, students, and middle-class adults to predict the results. They predicted that most people would refuse to give high-voltage shocks and refuse to give shocks if the "learner" demanded to be freed (Tavris and Wade, 1995: 342). Much to his surprise, Milgram found that although many of the hundreds of participants exhibited doubt and anxiety about administering the shocks, with "numbing regularity good people were seen to knuckle under to the demands of authority and perform actions that were callous and severe" (Milgram, 1974: 123).

Milgram's work was consistent with the ethical principles for psychological research at the time. The American Psychological Association's 1953 guidelines, for example, allowed for deception by withholding information from or giving misinformation to subjects when such deception was, in the researcher's judgment, required by the research problem (Herrera, 1997: 29).[2] Milgram did hold post-experimental debriefings where the participants were told that the "learners" had not received any shocks. When the results of the experiment were published, however, Milgram found himself in the middle of a controversy over ethics. Critics blamed him for deceiving his subjects about the purpose of the study and for causing his subjects a great deal of emotional stress (Baumrind, 1964).

Hearing about Milgram's work, you might feel that the study was justified because it made an important contribution to our understanding of the dangers of obedience. Or, you might agree with those who have vilified

[2]Herrera (1997: 29) says "the ethical legitimacy that the APA bestowed on deception would remain through the later versions of the ethics code, to the most recent, published in 1992."

Milgram's work because it violated the ethical principle of protecting subjects from harm.

Whatever your perspective on Milgram's study, his research points out that in many projects, the line between ethical and unethical methods is not always obvious. Often it is difficult to design a study in which there is absolutely no risk of psychological or emotional distress to participants. For example, for at least a few people, being interviewed about their marriages or their relationships with siblings can dredge up unhappy memories. Similarly, for some, talking about experiences with child abuse, sexual assault, divorce, or the death of a loved one can produce adverse effects or create difficulties in the groups to which they belong. It can be impossible to anticipate each and every consequence of participation for each and every subject in a study.

Possible negative reactions by subjects does *not* mean a study should be abandoned. Instead, a realistic and achievable ethical goal is for studies to create, *on balance,* more positive than negative results and to give participants the right to make informed decisions about that balance. Remember, for example, the participants in Weitz's study who, because of their desire to help others, agreed to be interviewed *despite* the emotional and sometimes physical pain it caused them. Even with some negative consequences, they wanted to be part of the study because they believed the outcome was important.

Some studies go beyond the protection from harm dictum and try to *help* respondents. Offering help can attract participants, keep their cooperation, and reciprocate for their contributions. For example, in a study on sexual assault, Bart and O'Brien (1985) offered a psychotherapy session at the research project's expense after each interview. In addition, each woman was given information about self-defense courses and legal assistance (Bart and O'Brien, 1985: 134).

Voluntary Participation and Informed Consent

voluntary participation, the principle that study participants choose to participate of their own free will.

informed consent, the principle that potential participants are given adequate and accurate information about a study before they are asked to agree to participate.

One of the ethical strengths of Weitz's study is that all of the interviewees were volunteers. Also important is that *before giving permission,* each participant was informed of the study's purpose, the kinds of topics the interview would cover, Weitz's identity and affiliation, and the names of the community groups with whom she was working. The ethical principles she was following were **voluntary participation** and **informed consent.** The goal of these principles is to ensure that all potential participants in a research project are accurately informed about the study, the method of data collection, and the availability of results *before* being asked to participate. Presenting inadequate information and misinformation and using pressure or threats to obtain consent violate ethical standards.

STOP AND THINK

Have you ever been asked by a faculty member whose class you were in to participate in a research project—to complete a questionnaire or participate in an interview? Did you feel any pressure or did you feel you could decline easily? Do you think that students in this kind of situation who agree to participate are true volunteers?

Although the ethical principles of voluntary participation and informed consent might be clear, their application isn't always obvious. As Weitz discovered, potential respondents sometimes made inaccurate assumptions about who she was and for whom she worked. Weitz did try to correct incorrect impressions, but she sometimes wondered when to stop explaining.

As a way of ascertaining if participants are really *volunteers*, researchers should ask the following questions: Do potential participants have enough information to make a judgment about participating? Do they feel pressured by payments, gifts, or authority figures? Are they competent to give consent? Are they confused or mentally ill? Do they believe that they have the right to stop participating at any stage of the process?

In the Weitz study, the sample was composed of self-selected volunteers because she did not contact any of them directly, but rather encouraged interested individuals to contact her. Not all studies use this kind of sample because using a self-selected sample prohibits the researcher from drawing conclusions about individuals, groups, or organizations beyond the sample. One useful way to get more representative samples is to approach individuals or groups rather than wait for volunteers. To be ethical, researchers must solicit for participation in noncoercive ways.

One way to provide potential respondents with information about the study is through a written informed consent form. Typically, surveys are exempted from the federal requirement for informed consent because of the minimal risk they are seen as creating for participants (Presser, 1994: 450), and some researchers rely on verbal rather than written permission. However, many researchers routinely use written statements of consent. Such a statement can provide potential participants with information about the purpose of the study, its benefits to participants and to others, the identity and affiliation of the researcher, the nature and likelihood of risks for participants, a description of how study results will be disseminated, an account of who will have access to the study's records and for what purposes, and a statement about the sponsor of the study. With such information, the benefits, risks, possible uses of the data, and results can be evaluated by individuals deciding about participation. Although Weitz preferred using a taped verbal permission statement before the interview to protect participants from being identified to state officials as having AIDS, early in the study she did use the informed consent form presented in Box 4.1. (Another example of an informed consent form is presented in Chapter 10.)

STOP AND THINK *Who should give informed consent in a study involving high school students under 18? The students? Their parents? Both?*

There is some debate about the issue of who can appropriately give informed consent. Of particular concern are children, adolescents, and individuals who are thought to be unable to act of their own free will in an informed manner. Most agree that, for minors, parental permission is important. However, there is debate about the age at which someone can decide for himself or herself to grant or withhold consent, and there is debate about who should be authorized to give consent for those judged unable to decide for themselves (Ringheim, 1995).

BOX 4.1

Informed Consent Form Used by Weitz

I, ____________________, in return for the opportunity of participating as a subject in a scientific research investigation, hereby agree to participate in an interview on the subject of having AIDS. I have been informed that the purpose of this research is to understand the experience of having AIDS and how it affects individuals' lives.

The research will consist of an in-depth interview, lasting approximately two hours. The potential benefits of this research are to give the public and the medical profession a better understanding of how AIDS affects individuals and to help government and medical officials better to design research, treatment and prevention programs. The benefits to me as an individual are the chance to express my feelings to a sympathetic and nonjudgmental listener. I understand, though, that some questions may be stressful. However, I understand that I have the right to withdraw from the interview at any time or to decline to answer any individual question or questions.

I have been informed that I can contact Rose Weitz, Department of Sociology, Arizona State University, with any questions about this research.

I have been informed that I will receive ten dollars ($10.00) for my participation in this study.

I knowingly assume the risks involved, and I am aware that I may withdraw my consent and discontinue participation at any time without penalty and loss of benefit to myself. I have received a copy of this completed consent form.

In signing this consent form, I am not waiving any legal claims, rights or remedies nor releasing the research investigator, the sponsor, the institution or its agent from liability or negligence.

Date: ___________ Signature: ______________________

Other studies raise different questions about consent. For a study on police interrogations, for example, researcher Richard Leo was required by his university's institutional review board to secure informed consent from the police officers, but not from the criminal suspects (Galliher and Galliher, 1995: 6). In another research project, Julia O'Connell Davidson obtained informed consent from a prostitute and her receptionist, but did not ask for permission from the prostitute's clients. Davidson says that the clients did not know a sociologist was listening to their conversations in the hall and watching them leave the house, nor did they consent to having the prostitute share information about their sexual preferences or physical and psychological defects (Davidson and Layder, 1994: 214). Davidson was not troubled by the uninvited intrusion into the clients' world, arguing that the prostitute had willingly provided knowledge that belonged to her.

BOX 4.2

Another Point of View

The following is from "The Ethics of Deception in Social Research: A Case Study" by Erich Goode:

" I intend to argue that it *is* ethical to engage in *certain kinds* of deception . . . the ethics of disguised observation [should be] evaluated on a situational, case-by-case basis . . . In *specific* social settings, some kinds of deception should be seen as entirely consistent with good ethics . . .

I strongly believe every society *deserves* a skeptical, inquiring, tough-minded, challenging sociological community. It is one of the *obligations* of such a society to nurture such a community and to tolerate occasional intrusions by its members into the private lives of citizens . . . I have no problem with the *ethics* of the occasional deceptive prying and intruding that some sociologists do. I approve of it as long as no one's safety is threatened" (Goode, 1996: 14 and 32).

Anthropologist Murray Wax (1995) notes that informed consent is quite difficult in field research, where the researcher observes daily life in groups, communities, or organizations. The researcher can encounter situations where his or her presence "is welcomed by one faction and regarded negatively by another; and where [the] ability to function may hinge on which faction wins the next tribal election, because the negative group may revoke the permission to study" (1995: 330). Wax argues that, in this type of situation, working in collaboration with participants can be more important than informed consent.

STOP AND THINK *How about observing people in public, such as watching crowd behavior at a sporting event? What research procedures would be ethical? Would you or could you ask for permission? What would you do if you saw illegal behavior? How about if you were asked by law enforcement personnel for the notes you took while watching the crowd?*

Anonymity and Confidentiality

Most researchers would argue it's unnecessary, impractical, or impossible to get informed consent when observing in public places. In addition, sometimes there is no simple distinction between "public" and "private" settings. For example, what about attending a meeting of Alcoholics Anonymous? Maurice Punch (1994: 92), a qualitative researcher, poses the question "Can we assume that alcoholics are too distressed to worry about someone observing their predicament (or that their appearance at A.A. meetings signal their willingness to be open about their problem in the company of others)?" He argues that observation in many public and semi-public places is acceptable even if subjects are unaware of being observed. The ASA Code of Ethics presented in appendix D supports such a position *if* the researcher has consulted with an authoritative body such as an IRB about the issue *and* decided after careful consideration that there is no need for consent.

anonymity, when no one, including the researcher, knows the identities of research subjects.

Even if informed consent is not obtained, the ethical principles of anonymity and confidentiality do apply. **Anonymity** is when it is not possible for anyone, including the researcher, to connect specific data to any particular member of the sample. Anonymity results when data are collected with no names, no personal identification numbers, and no information that could identify any subject. For example, even without names, a questionnaire returned by someone with the information that she is female, 35 years old, and the chief executive officer of a major software development firm located in Oregon might not be anonymous because her identity could easily be determined.

confidentiality, when no third party knows the identities of the research subjects.

Confidentiality is the issue of keeping the information disclosed by study participants, including their identities, from all other parties and hinges on the ability of the researcher to protect the data. This includes not allowing others, such as spouses, parents, teachers, school administrators, and others to have access to a participants' answers even if those persons have granted permission for the respondents to participate in the research (Ringheim, 1995: 1693). Respondents might be willing to report their views and experiences fully and honestly *because* of the assurance that the information will be treated as confidential (Laumann, Gagnon, Michael, and Michaels, 1994: 71). To keep material confidential, the data collected from all respondents must be kept secure. Identifying respondents by code number, for example, is a useful way to keep identities and data separate. When results are made public, confidentiality can be achieved by not identifying individuals—for example, by using pseudonyms or by grouping all the data together and reporting summary statistics for the whole sample. In addition, researchers can change identifiers like specific occupations or industries and names of cities, counties, states, and organizations when presenting study results.

The issues of anonymity and confidentiality are at the very heart of ethical research. One way to ensure the security of participant data is to destroy all identifying information as soon as possible. Such a step, however, eliminates the possibility of doing future research on the same individuals because a researcher who wants to collect data from the same participants more than once must be able to identify them. In the focal research in the previous chapter, for example, Hoffnung described the records of names, addresses, and phone numbers she has maintained to contact the members of her sample annually. Keeping identifying information means that confidentiality could be at risk because it is possible through subpoena to link a given data record to a specific person (Laumann et al., 1994: 72).

One very controversial study, *The Tearoom Trade,* by Laud Humphreys (1975), illustrates several ethical principles in doing research with human subjects—perhaps more by violation of than by compliance with these principles. This well-known study gives an account of brief, impersonal, homosexual acts in public restrooms (known as "tearooms"). Humphreys was convinced that the only way to understand what he called "highly discreditable behavior" without distortion was to observe while pretending to "be in the same boat as those engaging in it" (Humphreys, 1975: 25). For that reason, he took the role of a lookout (called the "watchqueen"), a man who watches the interaction and is situated so as to warn participants of the approach of anyone who was not a "regular" (p. 27).

Humphreys observed in restrooms for more than 120 hours. Once, when picked up by the police at a restroom, he allowed himself to be jailed rather than alert anyone to the nature of his research. He also interviewed 12 men to whom he disclosed the nature of his research and then decided to obtain information about a larger, more representative sample. For this purpose, he followed men from the restrooms and recorded the license plate numbers of the cars of more than one hundred men who'd been involved in homosexual encounters. Humphreys obtained the names and addresses of the men from state officials who were willing to give him access to license registers when he told them he was doing "market research." Waiting a year, and using a social health survey interview that he had developed for another study, Humphreys and a graduate student approached and interviewed fifty men in their homes as if they were part of a regular survey. Using this deception, Humphreys collected information about the men and their backgrounds.

Humphreys viewed his primary ethical concern as the safeguarding of respondents from public exposure. For this reason, he presented his published data in two ways: as aggregated, statistical analyses and as vignettes serving as case studies. The publication of Humphreys's study received front-page publicity, however, and because homosexual acts were against state law at the time, he worried that a grand jury investigation would result. Humphreys spent weeks burning tapes, deleting passages from transcripts and feeding material into a shredder (1975: 229), and resolved to go to prison rather than betray his research subjects. Much to his relief, no criminal investigation took place.

STOP AND THINK *Review the principles of ethical research that have been discussed in this chapter. Before we tell you what we think, decide which of the principles you believe Humphreys violated. Make some guesses as to* why *he behaved the way he did.*

Two principles that Humphreys violated are voluntary participation and informed consent. By watching the men covertly, tracking them through their license plates, asking them to participate in a survey without telling them its true purpose, and lying to the men about how they were selected for the study, Humphreys violated these principle several times over. After the study's publication, critics took him to task for these actions.

Humphreys's work provided insights about sexual behavior that had received little scholarly attention and about groups that were often the targets of hostile acts. Because of the usefulness of the work and his observational setting, Humphreys maintained that he suffered minimal doubt or hesitation. He argued that these were *public* restrooms and that his role, a natural one in the setting, actually provided extra protection for the participants (Humphreys, 1975: 227). But, reevaluating his work in the years after the study's publication, Humphreys came to agree with critics about the use of license plate numbers to locate respondents and approach them for interviews at their homes. He came to feel that he had put the men in danger of public exposure and legal repercussions. We will never know how many sleepless nights and worry-filled days the uninformed and unwilling research subjects had after the publication of and publicity surrounding Humphreys's book. But, at least no subpoena was ever issued; the data remained confidential, and no legal action was taken against the subjects.

Decades later, the Humphreys study continues to be a source of debate about where to draw the line between ethical and unethical behavior. Although many of its substantive findings have been forgotten, Humphreys's work has become a classic because of the intense debate about ethics that it generated. The ongoing debate and discussion about ethics that researchers like Humphreys and Weitz have engaged in helps others to understand more clearly the legal and ethical issues implicit in doing social research.

STOP AND THINK *Assume you have designed a study and feel confident that the participants will face minimal risks. Can you think of any other ethical responsibilities that should concern you?*

Ethical Issues Concerning Colleagues and the General Public

In addition to treating research participants ethically, there are responsibilities to colleagues and the public at large. The actions of an unethical researcher can affect the reputations of other social scientists and the willingness of potential research participants in the future, so the appropriate treatment of research participants is part of the responsibility to colleagues. Another central ethical principle is to produce accurate data and to report it honestly. The presentation of data without fraud, distortion, or omission; disclosure of the research's limitations; and emphasis of hypotheses *not* supported by the data are important responsibilities to colleagues and the general public.

If a study is funded by a sponsor, the researcher must be aware of the possibility of losing control over the study. Funding agencies and sponsors can influence research agendas and methodology by making awards on the basis of certain criteria. In addition, they can limit the freedom of researchers by imposing controls on the release of research reports. A researcher hired and paid by an organization will be expected by that organization to work in its best interest. That organization might attempt to keep data that are gathered from public disclosure (Homan, 1991 cited by Davidson and Layder, 1994: 56).

Sometimes a conflict of interest exists between the aims of the sponsor of the study and those of the larger community, especially if the sponsor is interested in influencing or manipulating people, such as consumers, workers, taxpayers, or clients (Gorden, 1975: 142). For example, in one city, a team of researchers representing a city planning commission, a community welfare council, a college, and several cooperating organizations, including local Boy Scout officials, studied the metropolitan area's social problem rate. At the conclusion of the study, the planning commission, the welfare council, and a local council of churches attempted to suppress those research findings that did not fit their groups' expectations or images. In each case, "it was one of the front-line field workers, viewed as 'hired hands' by the sponsoring agencies, who had to press for full dissemination of the findings to the public at large" (Gorden, 1975: 143–144).

Ethical Conflicts and Dilemmas

Most examples of unethical behavior in the social and behavioral sciences research literature are rooted in ethical dilemmas rather than in the machinations of unethical people. Many of the dilemmas involve *conflicts*—conflicts between ethical principles, between ethical principles and legal pressures, or between ethical principles and research interests.

Conflict Between Ethical Principles

In some research situations, researchers encounter behavior or attitudes that are problematic or morally repugnant. Although ethical responsibilities to subjects might prohibit us from reporting or negatively evaluating such behavior, other ethical principles hold us responsible for the welfare of others and the larger society. An example of this ethical dilemma was encountered by James Ptacek. Ptacek recruited a sample of wife batterers from a counseling organization devoted to helping such men, paid each participant a nominal fee, and told the men that participation would help the organization and that the interview was not a formal counseling session (1988: 135). In almost every interview he used a typical sociological interviewing style—staying dispassionately composed and nonjudgmental even when the men described bloody assaults on women. Even though Ptacek's goal was to "facilitate a narrative rather than continually challenge the men," he did worry about the moral dimension of his impartial approach (1988: 137). In the three interviews where men reported ongoing if sporadic violence, Ptacek switched from an interviewer role to a confrontational counselor role after the formal questions had been completed. He noted that "At the very least, such confrontation ensures that among these batterers' contacts with various professionals, there is at least one place where the violence, in and of itself, is made a serious matter" (Ptacek, 1988: 138). Facing a conflict between ethical principles, Ptacek gave a higher priority to the elimination of abuse than to the psychological comfort of study participants.

When the method of presenting findings can result in participants being identified, the researcher faces another dilemma. Conflict between reporting results and maintaining confidentiality confronted the authors of *Small Town in Mass Society: Class, Power and Religion in a Rural Community* (Vidich and Bensman, 1964). Vidich and Bensman have described their community study of a town in upstate New York as unintentional and unplanned, a byproduct of a separate organized and formal research project (1964: 314). One of the major problems they faced occurred because, during the data collection stage of the research, the townspeople of "Springdale" had been promised confidentiality and purely statistical analyses of the data. As the study progressed, the researchers came to focus on the town's political power structure and decided that the case study approach would be the most appropriate method of analysis. Identity concealment of townspeople described in the case study proved impossible. The use of pseudonyms and changing some personal and social characteristics to describe the members of the power hierarchy was not sufficient to keep their identities hidden, and, with the book's publication, the identities of those in the town's "invisible government" were recognizable to residents. The community leaders were distressed to be identified as part of such a government. Townspeople felt humiliated and were upset that their trust in the researchers had been betrayed (Vidich and Bensman, 1964: 338).

Vidich and Bensman (1964: 347) recognized that by publishing their analysis, they gave priority to the "ethic of scientific inquiry—the pursuit of knowledge for the sake of knowledge regardless of its consequences," rather than to the ethic of confidentiality. They came down on the side of honest reporting to colleagues and the public, but other researchers might have made different choices.

Conflicts Between Ethical Concerns and Legal Matters

STOP AND THINK *What would you do if someone you were interviewing told you that they were planning or had engaged in destructive or illegal behavior? Would you keep the information confidential or report it to someone?*

Social scientists collect data on a wide variety of topics. They can hear or observe things that can put them in situations that are difficult ethically or legally. For example, in 22 states, studying a topic such as child abuse or neglect means a researcher will face the legal requirement that anyone having reason to suspect maltreatment to report it to the authorities (Socolar et al., 1995: 580). This has significant implications for the informed consent process as there are consequences of participating, including legal risks for subjects and legal liabilities for investigators (Socolar et al., 1995: 580).

Although the researcher's duty on disclosure of child abuse, suicidality, or assaultive intent has not been tested in court, case law suggests researchers can be held responsible for safe-guarding participants in threatening circumstances against self-harm and harm by others (Bussell, 1994: 366–367). Some studies have tried to avoid a legal obligation to report suspected child maltreatment by collecting data anonymously or by having the subjects enter data directly into a computer, but many would argue that ethical obligations remain. A strong argument can be made on ethical grounds that the researcher should attempt to intervene when someone is in grave danger (Socolar et al., 1995: 581).

In other situations, researchers can face legal pressures to violate confidentiality. Under current rules of evidence, a social scientist has no legal right to refuse to turn over materials by claiming a privileged or confidential relationship with the participants (Leo, 1995: 124). In the focal research study, Weitz recognized that because health care workers in Arizona were required to report any person with HIV, she might have been asked to provide the Department of Health Services with the identities of the study's participants. Weitz took appropriate precautions, but remained nervous about the signed consent forms that identified interviewees as having AIDS. She was relieved that she did not have to chose between behaving ethically or facing legal consequences during the study.

Other researchers have not been as fortunate. For example, following the 1989 environmental disaster created when an oil tanker, the Exxon Valdez, ran aground off the coast of Alaska, research projects were conducted to assess the impact of the spill, including the social and psychological consequences. After the data were collected, some of the researchers and their associates were subpoenaed and deposed by Exxon, and one researcher, J. Steven Picou, was ordered to turn over all of the information he had gathered (McNabb, 1995:331). At least one researcher, Steven McNabb, had recognized the possibility of subpoena, and before its occurrence had destroyed all documents disclosing participant identities, addresses, and phone numbers and eliminated electronic paper trails and residence information from electronic data files. Although the case was settled before going to trial, the researchers learned that almost anything (files, contracts, agreements, schedules, diaries, notebooks, drafts, data sheets, journals, personal correspondence, scribbling, summaries, tapes of conversations, and the like) can be subpoenaed and that criminal prosecution can result if the material is not released (McNabb, 1995: 332).

Researcher Richard Leo learned the same lesson on a much smaller scale. Leo had spent more than 500 hours "hanging out" inside the criminal investigation division of a large urban police department in California—mostly observing police interrogation practices with the permission of the department and the individual detectives. Leo found himself subpoenaed by a suspect's public defender as a witness in a felony trial. Facing the threat of incarceration and feeling that his research notes would do no harm to the interests of the detectives, Leo complied with the judge's order and testified at the preliminary hearing (Leo, 1995: 128). He came to regret his decision because as a result of his testimony, the suspect's confession was excluded from the jury. He felt that he had betrayed his promise of confidentiality and spoiled the field for future police researchers (Leo, 1995: 130).

In a case with a different outcome, sociologist Rik Scarce went to jail to protect the confidentiality of participants in his research. After completing a study of radical environmental activists, Scarce remained in contact with some of them. At one point, one activist, Rod Coronado, stayed at Scarce's house for several weeks while Scarce was away. The time in question included the evening that the Animal Liberation Front took an action that resulted in the release of 23 animals and $150,000 in damages at a local research facility. Scarce was subpoenaed to testify about Coronado before a grand jury. Even after being granted personal immunity, Scarce refused to answer some of the questions because they called for information he had by virtue of confidential disclosure given in the course of his research (Scarce, 1995: 95). In a precedent-setting appeal, the courts decided that if the government has a legitimate reason for demanding cooperation with a grand jury, a social scientist does not have the right to refuse to testify. Scarce's refusal to comply resulted in his being jailed for 159 days in contempt of court.

Conflict Between Ethical Principles and Research Interests

There can be a conflict between following ethical principles and collecting useful data. Sometimes the choice is between using ethically compromised methods and not doing the research. We must ask at what point should the researcher refrain from doing a study because of the ethical problems. Did Milgram's work contribute so much to our understanding of human nature that the violations of ethics are worthwhile? Did the research by Humphreys make significant contributions? At what point should the Tuskegee study have been stopped?

At other times, the choice will be between methodologically valid data and ethically appropriate methods. We will explore some of these issues in future chapters as we discuss various study designs and methods of data collection. At this point, however, we'll present a few examples of conflict.

For example, if you were interested in studying a group or organization by making observations of behavior, you are likely to know that to be ethical, you'd need to receive permission from those you'd be observing. If, however, you also believe that those being observed are likely to change their behavior if they know they are being observed, then you will need to make a choice. Consider, for example, studies that have used "pseudo-patients," research confederates that go into mental hospitals feigning symptoms. Some argue that such research creates high-quality data and yields new knowledge by observing true-to-life situations that would be distorted if informed consent was requested (Bulmer, 1982: 629–630). Another example is Judith Rollins's (1985) study of domestic work. She collected data while doing domestic work for ten households, telling no one that she was doing research or even that she was a graduate student (Rollins, 1985). In describing her work, Rollins acknowledged feeling deceitful and guilty, especially when her employers were nice, but argued that her behavior was acceptable because there was no risk of harm or trauma to the subjects (1985: 11–12). At the end of her project, she concluded that it is appropriate for social scientists to evaluate each research situation separately, deciding what can be gained in exchange for deception (Rollins, 1985: 15).

Dilemmas can occur even when recruiting a sample of participants for a survey because telling every potential interviewee of every possible consequence guarantees a greater number of refusals to participate than does a more cursory description of the study. Ruth Frankenberg (1993), for example, originally told people, accurately, that she was "doing research on white women and race," and was greeted with suspicion and hostility. When she reformulated her approach and told white women that she was interested in their interactions with people of different racial and cultural groups, she was more successful in obtaining interviews. As we saw from Weitz's experience, even deciding when to correct the implicit assumptions that study participants make can become an ethical dilemma. Each researcher seeking access to data must decide how much detail to share with participants and how much effort should be expended in correcting assumptions about matters, including the study's purpose, sponsor, content of the data collection instrument, and the identity of the researcher.

Making Decisions: Maximize Benefit, Minimize Risk

> Although most social scientific research does not place subjects in situations that directly and overtly jeopardize their health and well-being, social research does involve some risk. Discomfort, anxiety, reduced self-esteem and revelation of intimate secrets are all possible costs to subjects who become involved in a research project. No investigator should think that it is possible to design a risk-free study, nor is this expected of researchers. Rather, the ethics of human subject research require that investigators calculate the risk-benefit equation, or balance the risks of a subject's involvement in the research against possible benefits of the project (both to the subject and to society) (LaRossa, Bennett, Gelles, 1994: 110).

Many colleges and universities have research review boards, and most professional organizations have codes of ethics. Examining these documents can make us aware of the ethical issues involved in conducting a study and provide guidelines for appropriate behavior. Ultimately, however, the responsibility for ethical research lies with each individual researcher.

To make research decisions, it's wise to use separate but interconnected criteria focusing on the ethics, practicality, and methodological appropriateness of the proposed research. In selecting a research strategy, we should ask these questions: Is this strategy practical? Is it methodologically appropriate? Is it ethically acceptable? When the answer to all three questions is yes, there is no dilemma. But, if we feel that it will be necessary to sacrifice one goal to meet another, we must first determine our priorities.

Weitz felt that when it came to a choice between being ethical and being "sociological," the ethical concerns should come first. She decided that, when faced with a choice between "purity of data" and helping (or at least not harming) her respondents, the latter was more important. So, when she thought it would help to give an honest response, she would give such a response instead of simply recording what an interviewee said, even if such behavior had the potential of biasing later responses.

Each researcher must think about the consequences of doing a given study as opposed to *not* doing the study and must consider *all* options and methods to find a research strategy that is ethical, practical, and likely to provide good quality data. In doing research, each of us must recognize that we are balancing our desires to obtain data we have confidence in and analyses we can share with the public against the rights of study participants.

Summary

Even though researchers often face unanticipated situations, most aspire to conduct their studies ethically and to apply the values and standards of ethical behavior to all stages of the research process. Recognizing the need to use feasible and methodologically acceptable strategies, we should strive for appropriate and acceptable conduct during the planning, data collection, data analysis, and publication stages of research.

Ethical principles involving study participants direct that data be collected anonymously or kept confidential, specify that individuals be adequately informed about the study before being asked to agree to participate, and, most important, mandate minimal negative consequences for the participants as a result of the research. When researchers face unusual ethical and legal dilemmas from the start, as Weitz did, they need to take special care when planning research. When researchers act ethically, participants will be able to make informed decisions about their involvement in research. Like those in the Weitz study, the participants in an ethical study will be able to feel confident that, on balance, their participation will result in more positive than negative consequences.

In addition to the treatment of participants, ethical principles require researchers to conduct themselves in ways that demonstrate that they are responsible to other scholars and to society as a whole. Scholars contribute an important service to their fields and the general public by providing and disseminating accurate information and careful analyses.

Many ethical dilemmas can be anticipated and considered when planning a research strategy. Institutional review boards and the codes of ethics of professional associations can serve as guidelines for individual researchers. Although Weitz did not find all the suggestions of her local IRB to be useful, this kind of review can identify many of the ethical issues involved at each stage of research.

Researchers will confront difficult situations if ethical principles conflict with each other, when legal responsibilities and ethical standards clash, or when collecting high-quality data means violating ethical principles. These situations—as when Weitz felt compelled to go beyond the traditional data collection role to break the norms of traditional sociological research by providing participants with information and suggestions—must involve careful decision making.

In facing the series of decisions that a research project will entail, a researcher must evaluate the rights and welfare of the participants of their studies against the rights and welfare of others and of the larger society. The best way to prepare for ethical dilemmas is to anticipate possible ethical difficulties early in the planning process and to engage in open debate of the issues.

SUGGESTED READINGS

Kimmel, A. J.1988. *Ethics and values in applied social research.* Newbury Park, CA: Sage.

Kimmel provides an overview of the subtle and complex ethical issues in social science, discussing moral dilemmas that can appear unresolvable. The goals of applied research seen within an ethical context and case studies of ethically controversial investigations are presented.

Lederer, S. 1995. *Subjected to science.* Baltimore: Johns Hopkins University Press.

This is a fascinating social history of human experimentation before the Second World War. Focusing on the professional and public debates on "human vivisection," that is, experiments on human beings undertaken for research rather than to benefit the individual subject, Lederer does a good job of documenting the complexity of the issue and the politics of medical research.

Socolar, R., D. K. Runyan, and L. Amaya-Jackson. 1995. Methodological and ethical issues related to studying child mistreatment. *Journal of Family Issues,* 16 (September): 565–586.

This article is especially useful for researchers considering doing work on topics that are as methodologically and ethically complex as the area that is the focus of this article, child abuse and neglect. The ethical and legal issues connected to subject recruitment, informed consent, confidentiality, and reporting are discussed.

EXERCISE 4.1

Knowing Your Responsibilities

All institutions that receive federal funds, as well as many others, have a committee or official group that considers all proposed research. Called an institutional review board, a human subjects review committee, or something similar, the group is charged with reviewing the ethics of research involving human subjects. It is likely that your college or university has such a committee.

Find out if there is a committee on your campus that evaluates and approves proposed research. If there is no such a group, check with the chairperson of the department that is offering this course, and find out if there are any plans to develop such a committee and describe those plans. If there is an institutional review board, contact the person chairing this committee, and get a copy of the policies for research with human subjects. Answer the following questions based on your school's policies. Attach a copy of the policies and guidelines to your exercise.

1. What is the name of the review committee? Who is the committee's chairperson?

2. Based on either written material or information obtained from a member of the committee, specify the rules that apply to research conducted by *students* as part of course requirements. For example, does each project need to be reviewed in advance by the committee or is it sufficient for student research for courses to be reviewed by the faculty member teaching the course? If the committee needs to review every student project for every course, how much time does the committee need to conduct such a review?

3. What are some of the advantages and disadvantages of your institution's procedures for reviewing the ethics of student research?

4. Based on the material you've received from the committee, describe at least three of the ethical principles that are included in the standards at your school.

EXERCISE 4.2

Ethical Concerns: What Would You Do?

In the following examples, the researcher has made or will need to make a choice about ethical principles. For each situation, identify the ethical principle(s), speculate about why the researchers made the decisions they did, then comment whether or not you agree with their decisions.

A. Professor Ludwig wants to see the connection between experiencing family violence and academic achievement among students. He contacts the principal of a local high school who agrees to the school's participation. At the next school assembly, the principal hands out Professor Ludwig's questionnaire, which includes questions about family behavior including violence between family members. The principal requests that the students fill out the survey and put their names on it. The following day, Professor Ludwig uses school records to obtain information about the each student's level of academic achievement.

   1. What ethical principle(s) do you think was(were) violated in this research?

2. Describe probable reason(s) Professor Ludwig violated the principle(s) you noted in your answer to question 1.

3. Describe your reactions to the way this study was done.

B. Professor Hawkins receives funding from both a state agency and a national feminist organization to study the connection between marital power, spouses' financial resources, and marital violence. Deciding to interview both spouses from at least 100 married couples, she contacts potential participants by mail. Writing on university letterhead stationary, she tells each couple she would like to talk to them about their marriage but does not specify the topics that the interview will cover.

1. What ethical principle(s) do you think was(were) violated in this research?

2. Describe the probable reasons Professor Hawkins violated the principle(s) you noted in your answer in question 1.

3. Describe your reactions to the way this study was done.

C. To study a radical "right-to-life" group, Professor Jenkins decides to join the group without telling the members that he is secretly studying it. After each meeting, he returns home and takes field notes about the members and their discussion. After several months of attending meetings, he is invited to become a member of the group's board of directors and accepts. One evening at the group's general membership meeting, one member of the board becomes enraged during the discussion of a local women's health clinic and shouts that the clinic's staff deserve to die. One evening that week, someone fires a gun through several windows of the clinic. Although no one is killed, several clinic staff members and one patient are injured. The police later find out that Dr. Jenkins has been studying the local group and subpoena his field notes. Professor Jenkins complies with the subpoena, and the board member is arrested.

1. What ethical principle(s) do you think was(were) violated in this research?

2. Describe the probable reasons Professor Jenkins violated the principles you noted in your answer to question 1.

3. Describe your reactions to the way this study was done and the researcher's response to the subpoena.

EXERCISE 4.3

Research Ethics

A number of studies in this text describe actual research projects. Read one of these focal research sections—Filinson (Chapter 7), Adler and Foster (Chapter 8), Gray, Palileo, and Johnson (Chapter 9), Enos (Chapter 10) or Adler and Adler (Chapter 11)—and answer the following questions.

1. Identify one or more ethical issues (such as harm to subjects, voluntary participation, informed consent, anonymity, or confidentiality) that the researcher(s) confronted.

2. Do you feel the researchers behaved ethically in doing this study? Why or why not?

3. Thinking about the balance between wanting to collect data and behaving ethically, do you agree with their choices? Why or why not?

Sampling

In July 1936, George Gallup, the force behind the now famous Gallup poll, made a prediction. Gallop predicted, in public and in print, that the then famous *Literary Digest* poll, to be taken later that year, would project Alf Landon, Republican nominee for the presidency, as the landslide winner of the November election. Moreover, he predicted that the projection would prove grossly incorrect—that, in fact, incumbent President Franklin Roosevelt would win the election . . . perhaps with as much as 54 percent of the vote.

The prediction solidified Gallup's reputation in opinion polling and helped bring the *Literary Digest* to ruin. Gallup was basically correct. The *Digest,* whose editors could merely splutter at Gallup's cheek in July, did project Landon's victory (by a margin of 57 to 43 percent) and did get it exactly wrong. This, after the *Digest* had, in its own words, predicted with "uncanny accuracy" the actual winners of the 1924, 1928, and 1932 Presidential elections—and several other national referenda to boot.

The fact that Gallup's own projection—that Roosevelt would take 54 percent of the voters' votes and 477 electoral votes—was off by a fair margin (Roosevelt actually got 61 percent and 523 electoral votes) was beside the point. He had played David to the *Digest's* Goliath—his own predictions had been based on a sample of less than 4000, while the Digest's sample was over 2 million—and he'd won!

How had Gallup been able to predict the *Literary Digest* findings months before they were made? Why were those findings so inaccurate? Gallup told all in the same article in which he made the July prediction (Converse, 1987: 120). He said that in surveying its readers (about 10 million strong in 1936), the *Literary Digest* would reach only middle- and upper-income respondents—those much more likely to vote Republican. And in relying on mail-in questionnaires (and hence only 2 of the possible 10 million voices), the *Digest* was hearing from a group that was even more likely to vote Republican.

Introduction

quota sampling, a nonprobability sampling procedure that involves describing the target population in terms of what are thought to be relevant criteria and then selecting sample elements to represent the relevant subgroups in proportion to their presence in the target population.

George Gallup's explanation of the *Literary Digest* poll's problems has hardly been improved on in 60 years. But neither the *Digest*'s demise nor Gallup's explanation of its polling failures has kept folks from doing the kind of sampling it employed. And even Gallup didn't advocate the kind of sampling that his organization, and many others, employ today when they study public opinion. Today's pollsters achieve more accurate predictions, using a technique known as **probability sampling** that we'll describe later. This technique permits the prediction of election results with extraordinary accuracy—usually within 3 percent of the actual result—with samples of only about 1500 individuals. Gallup used a sampling approach, called **quota sampling,** that yielded better results than the *Digest*'s mail-in sampling approach. We'll be talking more about quota sampling later, but suffice it to say that although it permitted Gallup to pick the winner of the Presidency in the 1936 election, it failed him, and other pollsters, when he used it to predict Dewey's victory over Truman in 1948 (see Figure 5.1).

Why Sample?

Sampling is a means to an end: to learn something about a large group without having to study every member of that group. Both the *Literary Digest* and the Gallup organization wanted to know how Americans were going to vote for president. Adler and Adler, whose work we'll feature in Chapter 11, wanted to know more about what happens in adult-organized afterschool activities for children. Why didn't the *Literary Digest* and Gallup simply survey the whole of the American voting-age population? Why didn't Adler and Adler study all afterschool activities? Put this way, one answer to the question "Why sample?" is obvious: We frequently sample because studying every single instance of a thing is impractical or too expensive. The *Literary Digest* and Gallup couldn't afford to contact every American of voting age. Adler and Adler couldn't afford to study all afterschool activities nor even figure out a way of identifying all such activities.

We sample because studying every single instance is beyond our means. But we also sample because sampling can improve data quality. Sandra Enos, whose study of how imprisoned women handle mothering we present in Chapter 10, wanted to understand individual women's approaches to mothering in an in-depth fashion, with a real appreciation of each individual woman's social context. Enos intensively interviewed a few incarcerated women, not only because studying the whole population of such women would have been expensive, but also because she wanted in-depth information about each of her subjects, rather than more superficial data on all.

In short, we sample because we want to minimize the number (or quantity) of things we examine or maximize the quality of our examination of those things we do examine.

STOP AND THINK *Using the logic of this section, when do you think sampling is unnecessary?*

FIGURE 5.1

Quota Sampling Led Gallup, and Others, Astray in 1948

Reprinted with permission of Bettman News Photos

We don't need to sample, and probably shouldn't, when the number of things we want to examine is small, when data are easily accessible, and when data quality is unaffected by the number of things we look at. For instance, suppose you were interested in the relationship between team batting average and winning percentage of major league baseball teams last year. Because there are only 28 major league teams, and because data on team batting averages and winning percentages are readily available (see any of this year's almanacs), you'd only be jeopardizing the credibility of your study by using, say, 10 of the teams. So why not examine all 28 of them?

element or sampling unit, the kind of thing a researcher wants to sample.

units of analysis, the kind of thing a researcher wants to analyze. The term is used at the analysis stage, whereas the terms element or sampling unit are used at the sampling stage.

We need to define a few terms before proceeding much further. First, we need a term to describe the kind of thing about which we want information—for example, a major league baseball team, an adult American of voting age, an afterschool activity. Social scientists define such things as **elements** or **sampling units** when they're engaged in sampling, and as **units of analysis** when they're engaged in data analysis. Sampling units or units of analysis can be people, organizations, institutions, collectivities, and so on. In a public opinion survey, the element or unit of analysis would be individual people. In the aforementioned study of major league baseball teams, it would be individual teams. In a study concerned with how the population size of a city is related to its expenditures on education, the element or unit of analysis would be the individual city. A crucial question to ask of any study that you read about or propose to do is this: What is the unit of analysis of this study or what kind of sampling unit do I want to get information about?

STOP AND THINK

Suppose you wanted to examine the hypothesis that the wealth of a nation is related to its kind of economy (for example, agricultural, manufacturing, or service-oriented). What kind of elements would you sample (or, alternatively, what kinds of units of analysis would you analyze)?

population,
the group of elements from which a researcher samples and to which she or he wants to generalize.
sample,
a number of individual cases that are drawn from a larger population.

Next we need a term for the group of elements that interests us theoretically—for example, all major league baseball teams last year, American adults of voting age, or afterschool activities. Social scientists define such a group as a **population.** Finally, we can formally define a **sample** as a subset of a population—used typically to gain information about or insight into the entire population.

STOP AND THINK

Suppose you're interested in describing the nationality of Nobel prize-winning scientists. What would an element in your study be? What would the population be?

Probability versus Nonprobability Samples

probability sample,
a sample that gives every member of the population a known (nonzero) chance of inclusion.
nonprobability sample,
a sample that has been drawn in a way that doesn't give every member of the population a known chance of being selected.

There are two basic kinds of samples: *probability* and *nonprobability* samples. **Probability** samples are selected in a way that gives every element of a population a known (and nonzero) chance, or probability, of being included. Most of the samples used by news agencies to predict the most recent presidential races were probability samples. They all aspired to give every adult citizen of the United States a (known) chance of being part of their sample. **Nonprobability** samples are those in which members of the population have an unknown (or perhaps no) chance of being included. The sample drawn by the *Literary Digest* was a nonprobability sample. The poll takers were interested in the population of adult Americans of voting age, but only surveyed readers of its magazine because the *Literary Digest* editors were willing to believe their readers were typical of Americans as a whole. Their technique, however, gave a huge number of Americans no chance of being included in the sample.

biased,
in sampling, the quality of being systematically unrepresentative.
generalizability,
the ability to apply the results of a study to groups or situations beyond those actually studied.

The advantage of probability over nonprobability samples is not, as you might hope, that they ensure that a sample is representative or typical of the population from which it is drawn. The sad news is that no sampling technique guarantees that. (We'll demonstrate this later.) Still, probability samples are *usually* more representative than nonprobability samples of the populations from which they are drawn. To use the jargon of statisticians, probability samples are less **biased** than nonprobability samples, where "biased" refers to being systematically atypical or unrepresentative of the population as a whole. And because probability samples increase the chances that samples are representative of the populations from which they are drawn, they tend to maximize the **generalizability** of results. Generalizability here refers to the ability to apply the results of a study to groups or situations beyond those actually studied.

Nonprobability samples do have some advantages, however. Although they are less likely than probability samples to be representative of any population, they are useful when one has limited resources, an inability to identify all members of a population (say, underworld or upperworld criminals or homeless people), or when one is exploring whether a problem exists or determining the nature of a problem, assuming it does exist (such as the problem of mothering from prison). We'll have more to say about these advantages when we describe some ways to select nonprobability samples.

This chapter's focal research piece, by Benjamin Bates and Mark Harmon, revisits the issue Gallup raised about the *Literary Digest* polls from a slightly different angle. Bates and Harmon examine whether "instant polls"—for example, using nonprobability mail-in responses to questionnaires appearing in a periodical or phone-in responses to questions asked on television news programs—can be counted on for an accurate picture of public opinion on particular issues. In this case, Bates and Harmon compare the results of nonprobability phone-in surveys to those obtained in a probability sample. Their interest in the topic was sparked by a lunch-time conversation. Harmon was working part-time as a news producer at a TV station that was conducting call-in polls, as well as teaching at Texas Tech, where Bates worked full-time. During their conversation, Bates expressed a highly critical view of call-in polls, so the two devised a plan for testing the adequacy of such polls. Bates and Harmon drew their probability sample with a technique called **random-digit dialing**. In so doing, they used random telephone numbers to select the people they talked to. We'll have more to say about this technique later in the chapter, but for now, let's see if the TV station's instant polls "hit the spot."

random digit dialing, a probability sampling procedure which puts interviewers in contact with interviewees by a computer that randomly selects area code, three digit exchange, and the next four digits, as needed.

FOCAL RESEARCH

Do "Instant Polls" Hit the Spot? Phone-In versus Random Sampling of Public Opinion[1]

by Benjamin Bates and Mark Harmon

Introduction

One of the more problematic developments in public opinion research in the last decade has been the development and the rapid proliferation of the phone-in poll. Since the inception of such polls, serious researchers have raised the specter of their bias and nonrepresentativeness.

The question of just how representative phone-in polls are is one that has just begun to be addressed by public opinion researchers. Previous examples of self-selection polls, such as the infamous *Literary Digest* poll, have

[1]This article is adapted by permission of the publisher from Bates and Harmon's (1993) "Do 'Instant Polls' Hit the Spot? Phone-In versus Random Sampling of Public Opinion," *Journalism Quarterly*, 70, 369–380.

given researchers strong grounds to doubt the validity and representativeness of self-response surveys on several grounds. Recently, researchers have argued that a major source of problems lies in the area of response bias (Cahalan, 1989; Squire, 1988).

Singer (1988) noted three potential sources of response bias in phone-in polls: (1) differential exposure to the channel or program through which the poll is announced, resulting in the likely exclusion of nonviewers; (2) the costs of responding, as when responding requires a long distance call or valuable time; (3) the opportunity for each participant to respond to a phone-in poll more than once. A highly motivated or highly organized viewer, such as one with strong views on a divisive emotional issue like abortion, thus could "sandbag" a poll by repeated dialing of the desired response, perhaps aided by an automatic redial button on a phone.

This study explores various alternative explanations of potential response bias in the phone-in news poll by comparing phone-in results to results of a concurrent random sample telephone poll. We will try to identify and measure the effects of particular sources of response bias by considering responses from selected portions of the random sample. We will also consider whether any differences exist between those who have had, and those who have not had, any previous experience with phone-in polls.

Methods

The data for this study were collected from two separate sources. Information on the "Phone-In Poll" results was obtained from a local television station in Lubbock, Texas, which ran the poll in conjunction with its evening local news program. The response to the phone polls was high: It may have been the equivalent of one call from every three to five TV households in the market exposed to the poll. Viewers were given only "yes" and "no" phone numbers to respond to given phone poll questions. These calls were answered and tallied by automatic answering machines.

This research was begun in February 1990 in conjunction with the first of a series of polls that were to be conducted throughout 1990 dealing with the upcoming elections and a few other local issues. The survey was designed to ask questions identical to those to be used by the station in their phone-in polls, as well as to acquire standard demographic information. Several of the February-March results, however, were lost before being reported to the researchers, and the station inverted the wording of one question, changing a "forbid" to an "allow." Due to these problems, additional data were collected the following October.

The researchers had to accommodate to the shifting sands of news judgment and newsroom policy. For instance, the station management cut back on its use of phone-in calls in October, halfway through the second data collection period. Thus, the researchers had available a total of five phone-in poll results for comparison, fewer than originally planned. To save space, only four of these comparisons are made in this report.

Our own data were taken from two local surveys of public opinion in the county of Lubbock, Texas. Lubbock County is by far the largest in the Area of Dominant Influence (ADI) for the station. It comprises 57.3 percent of all ADI television households. The remaining viewing is scattered among 17 sparsely populated counties.

The surveys used random digit dialing and a respondent randomization technique (asking to speak to the adult who had the most recent birthday) to gather a random sample of adults. Respondents were called on weekday evenings and weekend afternoons by volunteers from undergraduate classes.

A first survey was conducted in late February/early March, 1990. Undergraduate volunteers did the calling after a short training session on basic procedures. Attempts were made to contact 900 randomly generated telephone numbers. Of these, 259 were nonworking numbers, or numbers which did not connect to private residences. Another 128 numbers were not reached after four separate attempts. Of the remaining 513 numbers, 284 completed some or all of the survey, for a completion rate of 55.4 percent. Responses for some questions, however, may be lower. The contact and cooperation rates are virtually identical to the 1990 figures for the Southwest region as reported by Survey Sampling Inc. (1990).

A second survey was conducted during the first three weeks of October 1990. Attempts were made to contact 1,236 telephone numbers. Of these, 308 were nonworking numbers, or numbers which did not connect with a private residence. Another 185 were not reached after four attempts. Of the remaining 743, 426 completed some or all of the survey, for a completion rate of 57.3 percent. As a result of a clerical error, 36 surveys were completed using the wrong survey instrument. All were removed from the final survey, leaving a maximum sample of 390. Actual numbers for some questions may be lower.

The surveys collected information on general and specific public opinions about issues (four of which are identified in Table 5.1), media use information, election-related issues, and general demographics. Opinions on specific issues were measured by asking respondents to indicate strength of opinion about issues, using a scale from 1 to 5, where 1 meant that they strongly agreed with the statement and 5 meant that they strongly disagreed with it. Use of this scale, rather than the simple "Yes/No" response choice offered in the phone-in poll, permits the consideration of the relative strength of opinions. The difference between the answer options was deliberate: The authors intended to check whether the Yes/No format or strength of opinion may be one factor decreasing the reliability of TV call-in phone polls.

The likelihood of respondents being exposed to the phone-in questions was measured through the media use question. Those indicating that they tended to watch the appropriate local evening news program were labeled as those likely to be exposed to the phone-in poll. A predisposition to participate in phone-in polls was measured by asking respondents whether they had ever "voted" in a phone-in poll. The three suggested factors of strength of opinion, differential exposure, and predisposition were considered by breaking out the appropriate subsamples and considering responses within those subsamples.

The inclusion of the predisposition measure allowed some consideration of the distinctiveness of phone-in poll callers. Thus demographic information was also collected and used to identify differences between phone-in poll participators and nonparticipators.

Results

The basic comparisons of phone-in poll and random sample responses are given in Table 5.1. The table reports the percentage of positive responses to four questions for both the phone-in poll and the random poll. To try to identify potential sources of response error, the random poll results were broken down in several ways. First, to indicate whether intensity of opinion was related to response bias, extreme responses were separated and considered separately for all subsets of the data. The effects of predisposition toward participation were considered by breaking out those respondents who had previously participated in phone-in polls ("Previously PIPed"). The effects of differential exposure were investigated by breaking out those who reported regular viewing of the newscasts conducting the phone-in polls ("Viewers of Program"). Examination of the table indicates that the random sample results were quite different from the phone-in poll results. On the other hand, none of the various break-outs were very different from the random sample.

The two February/March phone-in poll items (Questions 1 and 2) dealt with issues of campaign finance reform. Random sample respondents tended to be more against such reform efforts than were phone-in poll respondents, with differences ranging from 15 to 23 percentage points. None of the possible mitigating factors (for example, predisposition toward participation phone-in polls) seemed to make much difference. Differences in the polls suggested that phone-in poll participants were more liberal, advocating campaign spending limits and public financing of elections.

The first of the October questions (Question 3) dealt with respondent perception of the campaign for governor then being waged in Texas. Respondents, like much of the national media, considered the race to be nasty and dirty. Random sample respondents, however, indicated a much higher level of disgust with the campaign, particularly among those with strong views. Once again, application of controls did not appear to have much effect on poll results.

The last question (Question 4) examined a broader political issue. The results indicated that phone-in callers were more conservative, coming out against a tax increase. Furthermore, respondents in our poll who had previously participated in phone-in polls and those who had viewed the newscast conducting the phone-in polls gave responses that were closer to the phone-in poll results, suggesting that these factors might explain some of the differences between our results and those of their phone-in counterparts.

While the nature of the procedure prohibited the identification of specific participants in the particular phone-in polls used in the study, we were able to identify those respondents in the telephone random samples who stated that they had, in the past, participated in a phone-in poll. Cross-tabulations of previous participation with a series of demographic and political variables

TABLE 5.1

Comparison of Poll and Survey Results (Ns in parentheses)

Data Source	*Question 1*	*Question 2*	*Question 3*	*Question 4*
Phone-In Poll	56.9	32.0	63.2	27.7
	(598)	(1164)	(269)	(1076)
Random Survey				
Full Sample	73.1*	15.0*	89.9*	52.3*
	(245)	(254)	(317)	(302)
Full Sample	74.6**	10.2**	92.0**	46.5**
	(189)	(206)	(250)	(230)
Previously PIPed	71.6*	16.1*	90.0*	43.4*
	(88)	(93)	(101)	(99)
Previously PIPed	71.4**	8.7**	94.9**	39.7**
	(70)	(69)	(79)	(78)
Viewers of Program	71.0*	12.9*	88.5*	51.9*
	(62)	(70)	(87)	(81)
Viewers of Program	76.1**	11.3**	94.1**	47.5**
	(46)	(53)	(68)	(59)

Notes:
Question 1 (February/March 1990): Political candidates should be allowed to spend as much of their own money as they want on their campaign.
Question 2 (February/March 1990): Tax money should be used to finance election campaigns.
Question 3 (October 1990): This year's governor's race has been very nasty and dirty.
Question 4 (October 1990): A tax increase will be necessary to reduce the Federal budget deficit.
Responses recorded on 5 point scale, where 1 = Strongly Agree and 5 = Strongly Disagree.
*Calculated as those giving 1,2 as response out of those giving 1,2,4 or 5.
**Calculated as those giving 1 as response out of those giving 1 or 5.

revealed a number of differences (see Table 5.2). The survey findings confirm many of the past suspicions about the weaknesses of instant, self-selected television polls in accurately reflecting public opinion.

The demographic characteristics of those who report they had "voted" in a telephone poll skew toward an upscale, white, older, and politically active group. Income, age, and contributing to a campaign or political action committee were all associated with "voting" in phone-in polls.

The profile of "upper income, white, older, and politically active" corresponds to one of two groups that have been identified (Robinson, 1971; Harmon, 1988) as audiences for local TV news. The other is a lower income, politically inactive, ethnically diverse group that watches a great deal of television—and will stumble across news but not seek it out.

TABLE 5.2

Characteristics of Respondents by Phone-In Poll Participation

Have you ever "voted" in a telephone poll before?		
	Yes	*No*
(February Poll)		
Percent Republican	41.4	42.2
Percent Democrat	24.2	20.2
Percent Independent	32.3	36.0
Percent Contribute to Campaign	47.1	32.0
Percent Contribute to PAC	26.3	11.9
Percent Plan to Vote	85.6	86.0
(October Poll)		
Percent Conservative	57.7	45.3
Percent Moderate	27.9	42.4
Percent Liberal	14.4	12.3
Percent Republican	50.4	49.4
Percent Democrat	20.0	23.6
Percent Independent	29.6	27.0
Percent Contribute to Campaign	52.5	26.4
Percent Voted in Primary	68.9	49.4
Percent Voted in Runoffs	56.2	36.4
Percent White	90.2	78.3
Percent Black	2.5	9.9
Percent Hispanic	6.6	10.7
Percent Income < $10,000	11.6	23.1
Percent $10,000 < Income < $20,000	13.4	21.9
Percent $20,000 < Income < $35,000	32.1	24.8
Percent $35,000 < Income < $50,000	17.0	13.6
Percent Income > $50,000	25.6	16.1

Political party preference and self-reported likelihood to vote in a general election are not strongly associated with previous "voting" in phone-in polls. However, indications of greater political activity—self-reported voting in a primary or run-off and making contributions to campaigns and other political groups—were associated with a greater likelihood of "voting" in a phone poll.

Political philosophy, as opposed to party identification, was related to phone poll participation. Moderates were less likely to "vote" than conservatives and liberals as moderate positions rarely are as motivating as views at the extremes. Rarely does one hear of a "raging moderate." Most phone polls, of course, permit no opportunity for a moderate or middle ground point of view as only a yes/no response is presented.

Conclusion

This study confirms the notion that results from phone-in polls clearly differ substantially from simultaneous random phone surveys. The study also sought to consider factors contributing to those differences. The results indicated that neither exposure nor previous participation in phone-in polls appeared to have consistent effects on the differences. Differences between the two poll types also did not seem consistently to reflect differences in political attitudes or identification.

It would appear, however, that the instant phone polls do attract those with "activist," pro-change points of view.

One factor that this study was not able to measure directly, but which could account for some of these differences, was multiple "voting" on phone-in polls. Those holding strong, activist positions are arguably more likely to engage in multiple "voting" behavior.

Simply put, so many things can go wrong with an instant phone-in poll that any time it reflects reality is purely by chance. And when the topic features an "activist" option, the poll is more likely to vent a heated dispute than shed light on public opinion.

REFERENCES

Cahalan, D. 1989. The *Digest* poll rides again. *Public Opinion Quarterly,* 53: 129–133.

Harmon, Mark. 1988. *Local television news gatekeeping.* Ph.D. dissertation, Ohio University.

Robinson, John. 1971. The audience for national TV news programs. *Public Opinion Quarterly,* 35: 403–405.

Singer, Eleanor. 1988. Surveys in the mass media. In Hubert O'Gorman's *Surveying Social Life.* Middleton, CT.: Wesleyan University Press.

Squire, P. 1988. Why the 1936 *Literary Digest* poll failed. *Public Opinion Quarterly,* 52: 125–133.

Survey Sampling Inc. 1990. *A survey researcher's view of the U.S.* Fairfield, CT.: Survey Sampling Inc.

Sources of Bias and Error in Sampling

STOP AND THINK *Bates and Harmon identify several ways in which TV phone-in polls are likely to be biased or nonrepresentative. But what, do you think, is the most fundamental reason that sampling, generally, involves bias and error?*

The overarching reason why sampling frequently involves bias and error is that the populations we're interested in aren't uniform or *homogeneous* in their characteristics. If, for instance, all human beings were exactly the same in every way, any sample of them, however small (say, even one person) would give us an adequate picture of the whole. The problem is that real populations tend to be heterogeneous, or varied in their composition—so any given sample might not be representative of the whole. Let's say, for instance, that, in all the universe, there are only four unicorns, aged 1 year, 2 years, 3 years, and 4 years, respectively. The average age for the total population of unicorns would be (1+2+3+4/4=) 2.5 years, but any sample of, say, 1 unicorn would misrepresent this average age (because no individual unicorn has an age of 2.5 years).

The primary reason for the typically large bias (or nonrepresentativeness) of nonprobability samples is that their composition relies so much on the judgment of a researcher to achieve the particular objectives of the research at hand. Imagine, for instance, that our researcher had only enough time to interview two unicorns. If, then, the speed of unicorns increased with age, and our researcher was pretty slow, he might only be able to catch up to (and therefore interview) the two smallest, and perhaps youngest, unicorns. This procedure would leave him with the inaccurate impression that unicorns average 1.5 years (1+2/2) of age, when we know they average 2.5 years. In a similar vein, one assumes, for instance, that the Lubbock TV station asked its viewers to call in "votes" largely because its management was interested in drawing attention to its news programs and partly because this sampling technique wasn't too costly. If the samples turned out to be representative, so much the better—but that wasn't the reason the TV station took the poll. Bates and Harmon list three potential reasons why the samples might have been unrepresentative (or biased): Some members of the relevant population aren't viewers or at least aren't viewers of the relevant news programs; some viewers of those news programs might find it too costly, in some sense, to respond (for example, perhaps they just don't feel strongly enough about the particular issues to pick up the phone); and some viewers might feel so highly motivated to participate that they phone-in more than one vote.

Unfortunately, bias can and frequently does occur in probability sampling as well. This can be seen by considering Bates and Harmon's probability sample. A distinction is useful here: between a population and a **study population.** A population, you will recall, is the group of elements that is of theoretical interest—the group to which we want to generalize our findings. In Bates and Harmon's study, the population was residents of the city of Lubbock and its immediate environs—the same one that the TV station used. A study population (sometimes called a **sampling frame**), on the other hand, is the group of sampling units from which the sample is actually selected.

study population or sampling frame, the group of sampling units or elements from which a sample is actually selected.

STOP AND THINK *Can you think of any ways in which Bates and Harmon's study population differed from its desired population?*

One of the more striking features of Bates and Harmon's survey was that it selected respondents through a process called random-digit dialing. Through random-digit dialing, interviewers are commonly put in telephone contact with interviewees by a computer that, first, randomly selects three-digit local exchanges and, then, randomly selects the next four digits as well. We'll be saying more about random selection a little later, but notice that one of the great advantages of this system is that random-digit dialing avoids the problem of unlisted telephone numbers (even if it can also create the problem of respondents warily and perhaps angrily wondering how you got their numbers), because all combinations of numbers in the relevant exchange areas can be selected. Indeed, such a technique does give everyone with a telephone in an area a chance of being contacted—the defining characteristic of probability samples. Unfortunately, however, not everyone has a telephone and because of this Bates and Harmon's study population (people in households with telephones) differed from their desired population (all residents of Lubbock and its environs). This is not much of a limitation in this day and age—in 1991 the Census Bureau estimated that 93.6 percent of American households had a phone (Census Bureau, 1991)—but it does create a known bias: Random-digit dialing surveys leave out those people who don't have telephones in their homes.

Depending on how you define Bates and Harmon's desired population, their sample might contain at least one other source of bias as well. Remember that they are attempting to assess the accuracy of a phone-in poll, the potential population of which would include all households reached by the relevant TV station. One could define the population of interest, then, as people residing in the station's broadcast region. Bates and Harmon confined their study population, however, to the county of Lubbock, which, to their knowledge leaves out "17 sparsely populated counties" that are reached by the TV station. They excluded the rural counties for two reasons: (1) it is a "toll call" from any of those counties to the station, so residents were less likely to respond to a phone-in poll than were residents of Lubbock County, for whom the station was a "local" call, and (2) the researchers' own budget would not permit "toll" calls to the rural counties (Bates, personal communication, 1996). Another known bias of their sample, therefore, especially if they were interested in generalizing their results to all those who could conceivably have watched the relevant TV news programs, is that people in those rural counties, however improbable their participation in the original polls, couldn't be included.

Notice that the biases described so far in both the TV station sample and the one collected by Bates and Harmon reflect researcher judgment. On the one hand, the station managers wanted a sample that was easily obtained and still likely to promote the viewing of its news programs. On the other, Bates and Harmon judged that only including households that had telephones and that were in Lubbock county would make their sample representative enough. Notice too, however, that there's a major difference between the biases introduced by these judgments. In the case of the station's poll, we can't say for sure what groups go unrepresented. Maybe everyone who could have watched the news program on a particular night did; maybe no one did. We just can't say. In the case of Bates and Harmon's sample, however, we can actually study some of the sources of bias: by determining, for instance, the percentage of households in Lubbock county that don't have telephones and by determining, as Bates and Harmon have, the percentage (57.3) of those television households in the station's Area of Dominant Influence that are found in Lubbock County. The capacity to study sources of bias is an important difference between probability and nonprobability sampling.

nonresponse bias, the bias that results from differences between those who agree to participate in a survey and those who don't.

Another source of bias in both nonprobability and probability samples is **nonresponse bias.** Nonresponse refers to the observations that cannot be made because some potential respondents refuse to answer, weren't "there" when contacted, and so forth. Nonresponse bias is the bias that results from differences between those who participate in a survey and those who don't. Nonresponse can involve large amounts of bias, as it was suspected to do in the case of the *Literary Digest*'s 1936 presidential poll, when about 8 million of the *Digest*'s 10 million readers simply didn't return their forms. Travis Hirschi (1969) has argued that nonresponse bias is also enormous in questionnaire studies of delinquency among school-aged adolescents given in school because delinquents are much more likely than their nondelinquent counterparts to have dropped out of school or to skip sessions when questionnaires are administered.

Nonresponse can plague even carefully collected, random samples, as it did with Bates and Harmon's. They, after all, report relatively large numbers of telephone numbers that were "not reached after four separate attempts," and report completion rates, on their two rounds of surveys of 55.4 percent and 57.3 percent, respectively—meaning, of course, that about 44.6 percent and 42.7 percent of those they did contact didn't respond for one reason or another. Although there is no real consensus about when nonresponse jeopardizes the value of probability sampling, high nonresponse rates (of, say, more than 50 percent) virtually ensure survey results that will differ significantly from what they would be if everyone responded.

STOP AND THINK *What kinds of people might not be home to pick up the phone in the early evening (when, say, Bates and Harmon's students made their calls)? What kinds of people might refuse to respond to telephone polls, even if they were contacted?*

If sampling is done from lists, as is frequently done in probability sampling, it is sometimes possible to study the nature of nonresponse bias, especially if the lists include some information about potential sampling units. Thus, in a study of married people, Karney and his colleagues (1995) were able to determine that people who did not respond to their questionnaire survey generally had less education and were employed in lower-status jobs than those who did respond. Karney and his colleagues used information from the marriage certificates that constituted their sampling frame.

Sampling Variability and Sampling Distributions: An Illustration

parameter, a summary of a variable characteristic in a population.

statistic, a summary of a variable in a sample.

One way to think about the effects of bias in samples is to think about two new terms: **parameter** and **statistic.** A parameter is a summary of a variable *in a population*. Because the 1936 election took place, we know that around 61 percent of the voting public favored Franklin Roosevelt over Alf

Landon for President. This percentage is a parameter of the population that voted for President in 1936. A statistic, on the other hand, is a summary of a variable *in a sample*. In the *Literary Digest*'s sample before the election that year, 57 percent of respondents expressed a preference for Landon. This percentage is the statistic that was used to make an estimate of and prediction about how the voting public would vote in the presidential election.

sampling error, the error that occurs when a sample statistic is not the same as the population parameter it's meant to estimate.

sampling variability, the variability in sample statistics that occurs when different samples are drawn from the same population.

It is clear, then, that not all samples provide statistics that are the same as the actual parameters. A sample statistic can misrepresent a population parameter (such a misrepresentation, by the way, is called **sampling error**) for any of the reasons we've previously mentioned: for bias introduced through human judgment in nonprobability sampling or through a mismatch between the theoretical and study populations in probability sampling or through systematic nonresponse. We'd like to focus here on another likely source of such misrepresentation: **sampling variability,** or the variability in sample statistics that can occur when different samples are drawn from the same population. The sources of this error are evident in nonprobability samples; whatever else they might be caused by they're likely to reflect the biases in the sample selection technique. Thus, the *Literary Digest* sample was likely to be much more Republican than the larger population at least insofar as *Literary Digest* readers (subscribers) were wealthy enough (in the middle of the Great Depression) to afford a subscription to a magazine (the *Digest*) and insofar as the wealthy were likely to be Republican.

We've already suggested that one advantage of probability samples is that they are more likely than nonprobability samples to be representative of relevant populations. That's the good news. The bad news is that, even probability samples entail sampling variability and, therefore, frequently involve another source of sampling error.[2] To demonstrate, we'd like you to imagine drawing random samples of two elements from that population of unicorns we mentioned earlier: the one with four members of one, two, three, and four years of age, with the names, let us say, of Leopold, Frances, Germaine, and Quiggers, respectively.

STOP AND THINK *How could you draw a sample of two unicorns, giving each unicorn an equal chance of appearing in the sample?*

simple random sample, a probability sample into which every member of a study population has had an equal chance of selection.

A sample in which every member of the population has an equal chance of being selected is called a **simple random sample.** Perhaps we should emphasize that the word "random" in the phrase "simple random sample" has a special meaning. It does not mean, as it can in other contexts, haphazard, erratic, or arbitrary; here it connotes an unbiased choice, almost lottery-like, of equally likely elements from a larger population. One way of drawing

[2]There are actually several other potential sources of error in probability sampling, including measurement error and sampling bias, or error that results from using a sampling approach that overrepresents a portion of the study population. We won't spend much time on any of these sources in this chapter but recommend the first part of Chapter 3 in Henry (1990: 34-46) for extremely lucid discussions of all possible sources of error in sampling.

TABLE 5.3 **Possible Samples of Two Unicorns with Each Sample's Average Age**

Possible Samples	*Average Age of Sample*
Leopold and Frances	1.5
Leopold and Germaine	2.0
Leopold and Quiggers	2.5
Frances and Germaine	2.5
Frances and Quiggers	3.0
Germaine and Quiggers	3.5

(selecting) a simple random sample of two of our unicorns would be to write the four unicorns' names on a piece of paper, cut the paper into equal-sized slips, each with one unicorn's name on it, drop the slips into a hat, shake well, and, being careful not to peek, pick two slips out of the hat. Let's say we have just done so and drawn the names of Frances and Quiggers (ages 2 and 4). We thereby achieved our goal of a sample of unicorns whose average age ([2 years + 4 years]/2 = 3 years) was *not* the same as the that of the population (2.5), hence proving that it is possible to draw a probability sample that misrepresents the larger population.

Of course, if we had selected a sample of Frances and Germaine, the average age of sample members ([2 years + 3 years]/2 = 2.5 years) would have been the same as the average age for the whole population of unicorns. (But we would have also proven that it is not only possible, as we did in our first attempt, to draw a sample that misrepresents the larger population, but also to draw probability samples from the same population that have different statistics (and, so, show sampling variability).

STOP AND THINK *How many possible unique samples of two unicorns could be drawn from our population?*

There are six possible unique samples of unicorns you could randomly select from our population. We list all of these, with their average ages, in Table 5.3.

sampling distribution, the distribution of a sample statistic (like the average) computed from many samples.

We could make another kind of visual display of these samples, called a **sampling distribution,** by placing a dot, representing each of the sample averages, on a graph big enough to accommodate all possible sample averages. A sampling distribution is a distribution of a sample statistic (like the average) computed from more than one sample. The sampling distribution of any statistic, according to probability theory, tends to hover around the population parameter, especially as the number of samples gets very large. Figure 5.2, a sampling distribution for the possible unique samples of two unicorns, demonstrates this property by showing that the sampling distribution of the average age for each possible pair of unicorns is centered around the average age for unicorns in the population (2.5 years).

FIGURE 5.2

The Sampling Distribution of Unique Samples of Two Unicorns

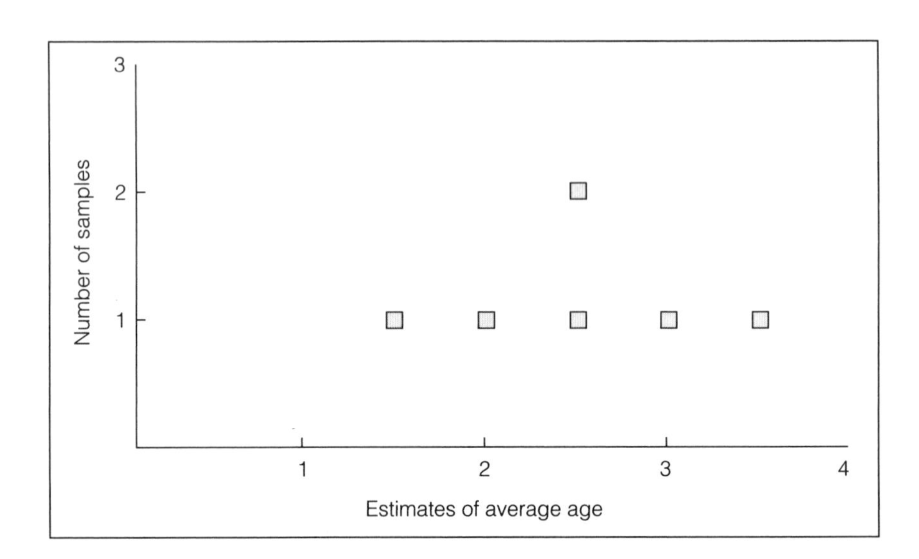

STOP AND THINK *You're obviously pretty imaginative if you've followed us up to here. But can you imagine the sampling distributions of all possible unique samples of one unicorn? Of four unicorns? What, if anything, do you notice about the possible error in your samples as the number of cases in the samples grow?*

The sampling distribution of the average ages of samples of one unicorn would have four distinct points, representing 1, 2, 3, and 4 years, all entailing sampling error. The sampling distribution of the average age for a sample of four unicorns would have one point, representing 2.5 years, involving no sampling error at all. Figure 5.2 suggests that the sampling distribution of samples with two cases generally has less average sampling error than that for samples with one case, but generally more average sampling error than the sampling distribution with four cases. This illustrates a very important point about probability sampling: Increases in sample size, referred to as n, improve the likelihood that sampling statistics will accurately estimate population parameters.

Another, more realistic, example of a sampling distribution would be the samples you might draw of your town's voters to see what percentage of them plan to vote for Candidate A. Suppose you wanted to estimate how well this candidate will do in the upcoming election for mayor. You might use your town's list of registered voters (if it were available) as a study population, select a random sample of 100 registered voters, and ask members of the sample whether they plan to vote for Candidate A or not. The possible results of your survey would be, of course, that anywhere from 0 to 100 percent of your sample would say they plan to vote for Candidate A. Suppose your first sample shows that 52 percent of your respondents say they will vote for Candidate A. Suppose your second sample shows that 57 percent say they will vote for Candidate A. Suppose you keep drawing additional samples of 100 respondents and decide to plot the new sample statistics on a summary graph, like the one shown in Figure 5.3. This would be

FIGURE 5.3

The Sampling Distribution of Samples of 100 Registered Voters

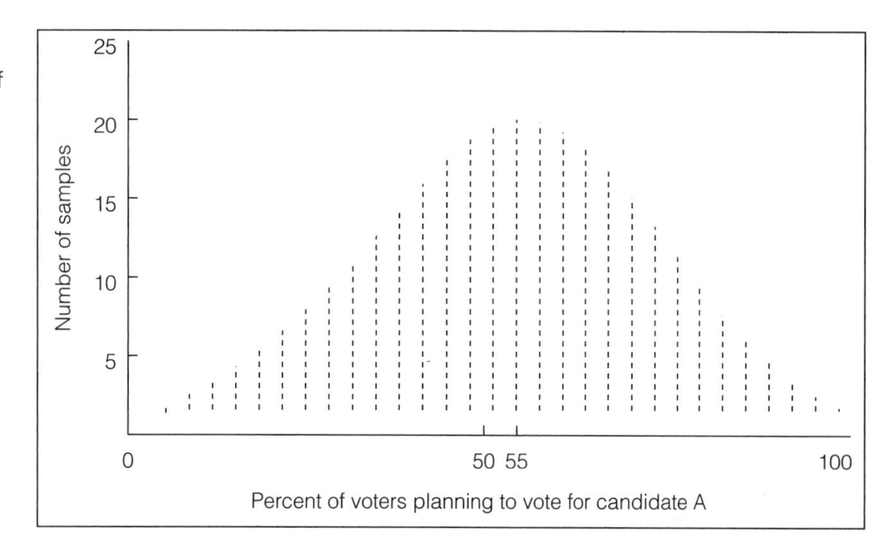

the sampling distribution of the variable "percent of registered voters planning to vote for Candidate A" for a large number of samples. Notice that, in this case, the sampling distribution seems to be centered around samples suggesting that 55 percent of the registered voters plan to vote for Candidate A.

A major point of this section has been to demonstrate that even when one employs a type of probability sampling (in this case, simple random sampling), sampling error can and frequently does occur. This means, unfortunately, that no sample statistic can be relied on to give a perfect estimate of a population parameter. But another point of the section has been to suggest something about the nature of sampling distributions (for example, that they tend to be centered around the population parameter when the number of samples is large), a topic to which mathematicians have given extensive consideration. The practical offshoot of this consideration is a branch of mathematics known as probability theory.

A tremendous advantage of probability samples is that they enable their users to use probability theory to estimate population parameters with a certain degree of confidence. Showing you enough about probability theory to convince you of this fact is beyond the scope of this book. We highly recommend a short book entitled *Practical Sampling* by Gary Henry (1990) for those of you who would like to pursue this or related topics. We would stress a point that is related to our previous discussion of sampling unicorns: The accuracy with which sample statistics estimate population parameters increases with the size of the sample. Suppose, for instance, you drew a sample of 100 residents that showed that 55 percent planned to vote for Candidate A. Based on our previous discussion, you'd know that there was a good chance that the population parameter isn't necessarily 55 percent as well. But you could strive to say, thanks to probability theory, something like this: There is a 95 percent chance that between 45 to 65 percent of all voters plan

confidence interval, a range of values within which the population parameter is expected to lie.

confidence level, the estimated probability that a population parameter will fall within a given confidence interval.

to vote for Candidate A. Then you'd be announcing what is known as a **confidence interval,** or range of values in which the population parameter is expected to fall, of 20 percent (between 45 and 65 percent). You'd also be announcing a **confidence level,** or an estimate of the probability that the population parameter falls in this range (in this case, 95 percent). The point is that probability theory tells us if we want to achieve a confidence interval of 20 percent with a confidence level of 95 percent, we need a sample of about 100. If you'd like to reduce the confidence interval from 20 to 10 percent (and keep the same 95 percent confidence level), you'd need a sample of 400. If you'd like that interval to drop to 6 percent (your sample percentage plus or minus 3 percent), you'd need a sample of about 1500. Choosing the correct sample size for a given task is a science, and one, again, for which we recommend Henry's (1990, particularly Chapter 7) book as an introduction.

Types of Probability Sampling

In the previous section we introduced you to simple random sampling, a type of probability sampling that is appropriate for many statistical procedures used by social scientists. Here we'd like to introduce you to another way of doing simple random sampling, as well as to other kinds of probability sampling and their relative advantages.

Simple Random Sampling

Simple random samples are selected so that each member of a study population has an equal probability of selection. To select a simple random sample, you need to have a list of all members of your population. Then you randomly select elements until you have the number you want—perhaps using the blind drawing technique we mentioned earlier.

To show you another way of drawing a simple random sample, we'd like you to consider the list (or sampling frame) of Pulitzer Prize Winners in Journalism for Commentary. The list of such award winners, only given since 1970, is relatively short at this point, and so is useful for illustrative purposes (though if, in fact, your study population included so few elements, you would probably want to use the entire population). Until recently, there were only 24 members of this population:

01. Marquis Childs	09. William Safire	17. Jimmy Breslin
02. William Caldwell	10. Russell Baker	18. Charles Krauthammer
03. Mike Royko	11. **Ellen Goodman**	19. Dave Barry
04. David Broder	12. Dave Anderson	20. Clarence Page
05. Edwin Roberts	13. Art Buchwald	21. Jim Murray
06. **Mary McGrory**	14. Claude Sitton	22. Jim Hoagland
07. Red Smith	15. Vermont Royster	23. **Anna Quindlen**
08. George Will	16. Murray Kempton	24. **Liz Baalmaseda**

FIGURE 5.4

Random Number Table Excerpt

5334	5795	2896	3019	7747	0140	7607	8145	7090	0454	4140
8626	7905	3735	9620	8714	0562	9496	3640	5249	7671	0535
5925	4687	2982	6227	6478	2638	2793	8298	8246	5892	9861
9110	2269	3789	2897	9194	6317	6276	4285	0980	5610	6945
9137	8348	0226	5434	9162	4303	6779	5025	5137	4630	3535
4048	2697	0556	2438	9791	0609	3903	3650	4899	1557	4745
2573	6288	5421	1563	9385	6545	5061	3905	1074	7840	4596
7537	5961	8327	0188	2104	0740	1055	3317	1282	0002	5368
6571	5440	8274	0819	1919	6789	4542	3570	1500	7044	9288
5302	0896	7577	4018	4619	4922	3297	0954	5898	1699	9276

STOP AND THINK *We have shown the female members of the population in boldface print. Because we've presented the population in this way, it is possible to calculate one of its parameters: the fraction of award winners who have been female. What is that fraction?*

Let us draw a simple random sample of 6 members from the list of award winners (that is, we want an *n* of 6). If we don't want to use the "blind" drawing technique, we could use a random number table. To do so, we would, first, need to locate a random number table, such as the one we've provided in Figure 5.4, a larger version of which exists in Appendix A of this book. (Such tables, incidentally, are generally produced by computers in a way that gives every number an equal chance of appearing in a given spot in the table.) Second, we'd need to assign an identification number to each member of our population, as we have already done by assigning the numbers 1 through 24 to the list of award winners (above). Now we need to randomly select a starting number whose number of digits is equal to the digits in your highest identification number. The highest identification number in our list was 24 (with two digits), so we need to select (randomly) a number with two digits in it. Let's say you did this by closing your eyes, pointing toward the page, and (after opening your eyes again) underlining the closest two-digit number—perhaps, the number 96 (see the "96") underlined in Figure 5.4.

Then, proceeding in any direction in the table, you could select numbers, discarding any random number that does not have a corresponding number in the population, until you'd chosen the number of random numbers you desire. Thus, for instance, you could start with 96 and continue across its row, picking off two digit numbers, then down to the next row, and so forth, until you'd located six of them (remember our desired *n* numbers).

STOP AND THINK *What numbers would you select if you followed the procedure described in the previous sentence?*

The numbers you'd select, in this instance, would be the following:

96, 30, **19**, 77, 47, **01**, 40, 76, **07,** 81, 45, 70, 90, **04**, 54, 41, 40, 86, 26, 79, **05,** 37, 35, 96, **20**

Of course, you'd only keep six of these: the ones in bold print in the previous array. These would be 19, 01, 07, 04, 05, 20. Incidentally, you'd be working with a definition of simple random selection that assumes that once an element (corresponding to, say, the number 19) is selected, it is removed from the pool eligible for future selection. So if, say, during your selection number "07" had appeared more than once, you would have discarded the second "07." We'll be using this definition later on in our discussions as well.

Your resulting sample of six journalists would be: Marquis Childs (case 01), David Broder (04), Edwin Roberts (05), Red Smith (07), Dave Barry (19), and Clarence Page (20).

STOP AND THINK *What fraction of this sample is female? Does this sample contain sampling error relative to its gender make-up?*

STOP AND THINK AGAIN *Pick a different simple random sample of six? (Hint: Select another starting number in the random number table and go from there.) Who appears in your new sample? Does this sample contain sampling error relative to gender make-up?*

Systematic Sampling

systematic sampling, a probability sampling procedure that involves selecting every kth element from a list of population elements, after the first element has been randomly selected.

Even with a list of population elements, simple random sampling can be tedious. A much less tedious probability sampling technique is **systematic sampling.** In systematic sampling, every kth element on a list is (systematically) selected for inclusion. If, as in our previous example, the population list has 24 elements in it, and we wanted 6 elements in our sample, we'd select every fourth element (because 24/6 = 4) on the list. To make sure that human bias wasn't involved in the selection and that every element actually has a chance of appearing, we select the first element randomly from elements 1 through k. Thus, in the example of the Pulitzer Award winners, you'd randomly select a number between one and four, then pick the case associated with that number and with every fourth case after that one on the list.

STOP AND THINK *Try to draw a sample of six award winners using systematic sampling. What members appear in your sample?*

In general, we call the distance between elements selected in a sample the selection interval (which is 4 in the award-winner example but is generally referred to by the letter "k"), which is calculated using the formula:

$$\text{selection interval (k)} = \frac{\text{population size}}{\text{sample size}}$$

We've mentioned the relative ease of selecting a systematic sample as one of the procedure's advantages. There is at least one other advantage worth mentioning: If your population is physically present (say, file folders in an organization's filing cabinets), you don't need to make a list. You need only determine the population size (that is, the number of folders) and the desired sample size. Then, you'd just need to look at every kth element (or folder) after you'd randomly selected the first folder. Ian Rockett, one of the authors of our Chapter 6 focal research, tells us that he has used a modified form of systematic sampling in a hospital emergency use study. A potential interviewee was the first eligible patient entering the emergency room after the hour. Eligibility criteria included being 18 years of age or older and able to give informed consent to participate in the study. In each hospital, interviewing lasted three weeks and covered each of three daily, eight-hour, shifts twice during the three-week period.

The only real disadvantage of systematic sampling over simple random sampling, in fact, is that systematic sampling can introduce a bias if the sampling frame is cyclical in nature. If, for instance, you want to sample months of several years (say, 1900 through 1999) and your months were listed, as they frequently are, January, February, March, and so on for each year, you'd get a very biased sample if your selection interval turned out to be 12 and all the months in your sample turned out to be Januarys. Imagine the error in your estimates of average annual snowfalls if you just collected data about Januarys and multiplied by 12! A word to the wise: When you're tempted to use systematic sampling, check your population lists for cyclicality.

Stratified Sampling

stratified random sampling, a probability sampling procedure that involves dividing the population in groups or strata defined by the presence of certain characteristics and then random sampling from each of the strata.

One disadvantage of simple random sampling is that it can yield samples that are biased in ways you want to avoid and can do something about. Take the earlier example of the Pulitzer Prize winners. Suppose you knew that approximately one-sixth of that population was female and you wanted a sample that was one-sixth female (unlike the one that we drew in the example). We could use a procedure called **stratified random sampling.** To draw a stratified random sample, we group the study population into strata (groups that share a given characteristic) and randomly sample within each stratum. Thus to draw such a sample of the prize winners, stratified by gender, we'd need to separate out the 4 female winners and the 20 male winners, enumerate each group separately, and random sample, say, a quarter of the members from each group.

STOP AND THINK *See if you can draw a stratified random sample of the prize winners, such that one female and five males appear in the sample.*

More complexity can be added to the basic stratifying technique. If, for instance, you were drawing a sample of students at your college, using a roster provided to you by the registrar, and the roster provided information not only about each student's gender but also about his or her year in school you could stratify by both gender and year in school to ensure that your sample wasn't biased by gender or year in school.

Stratified sampling can be done proportionately and disproportionately. The Pulitzer Prize sample you drew for the *Stop and Think* exercise was a proportionately stratified random sample because its proportion of females and males is the same as the proportion of females and males in the population. However, if you wanted to have more female prize winners to analyze (maybe as many as all of them), you could divide your population by gender and make sure your sample includes a number of females that is disproportionate to their number in the population.

STOP AND THINK *How would you draw a disproportionate stratified random sample of eight Prize winners, with four females and four males in it?*

Disproportionate stratified sampling makes sense when given strata, with relatively small representation in the population, are likely to display characteristics that are crucial to a study and when proportionate sampling might yield uncharacteristic elements from those strata because too few elements were selected. Suppose, for instance, you had a population of 100, 10 of whom were women. You probably wouldn't want to take a 10 percent proportionate sample from men and women, since you'd then be expecting one woman to "represent" all ten in the sample.

Pillemer and Finkelhor (1988), for example, wanted to estimate the prevalence of elder abuse in the Boston metropolitan area and drew a disproportionate random sample of elderly persons living in private residences. They knew that elder abuse was more common in households where the elderly live together with their children, even though the elderly in such households constitute no more than 10 percent of all elderly. Therefore, they oversampled elderly living with persons of a younger generation.

The only disadvantage of disproportionate stratified random sampling is computational. To correct, for instance, for the fact that females are overrepresented, and males underrepresented, in the sample of prize winners we asked you to draw for the *Stop and Think* question, you'd have to weight each female and male appropriately to calculate sample statistics to represent the population. Each female in the population would be in the sample, so you wouldn't need to "weight" the females. But because each male in the sample would "represent" five males, you could multiply their values by 5 when calculating certain statistics.

STOP AND THINK *As a practical, inane example of the value of weighting, let's suppose each of the females in your sample weighed 150 pounds, and each of the males weighed 200 pounds. Then the average weight of the 8 members in your sample would be 175 pounds. But this average wouldn't be the best possible estimate of the average weight of persons in your population, since females are overrepresented in the sample. To get a "weighted" average (and therefore a better estimate of the population average), you might multiply each of the female weights (150 pounds) by 1 and each of the male weights (200 pounds) by 5, then add all the weights up and divide by 24 in the following way:*

4 × 150 lbs. × 1 =	*600*	*lbs., or the total "weighted" weight for females.*
+ 4 × 200 lbs. × 5 =	*4000*	*lbs., or the total "weighted" weight for males.*
	4600	*lbs., the total "weighted" weight for everyone in the sample*
divided by 24 =	*191.67*	*lbs., the "weighted" average weight for sample members*

Don't be daunted by weighting problems. Most computer programs used to calculate sample statistics are equipped to handle them.

Cluster Sampling

cluster sampling, a probability sampling procedure that involves randomly selecting clusters of elements from a population and subsequently selecting every element in each cluster for inclusion in the sample.

Simple random, systematic, and stratified sampling all assume you've got a list of all population elements easily at hand and that the potential distances between potential respondents aren't likely to result in unreasonable expense. If, however, data collection involves visits to sites that are at some distance from one another, you might consider cluster sampling. **Cluster sampling** involves the random selection of groupings, known as clusters, from which all members are chosen for study. If, for instance, your study population is high school students in a state, you might start with a list of high schools, random sample schools from the list, ask for student rosters from each of the selected schools, and then contact each of those students.

STOP AND THINK *How might you cluster sample students who are involved in student organizations that are supported by student activity fees at your college?*

Multistage Sampling

multistage sampling, a probability sampling procedure that involves several stages, such as randomly selecting clusters from a population, then randomly selecting elements from each of the clusters.

Cluster sampling is actually much less frequently used than **multistage sampling,** which frequently involves a cluster sampling stage. Two-stage sampling, a simple version of the multistage design involves the random sample of clusters and then the random sampling of members of the selected clusters to produce a final sample. In reality this is what Ian Rockett and his co-author, Sandra Putnam, did in the survey of emergency room users mentioned earlier. The goal was to sample such users in the whole state of Tennessee in a fixed time period. The first step of the sampling process involved random sampling of one hospital in each of the seven designated health care areas of Tennessee; the second step was identifying individual cases, using the process described in the systematic sampling section.

Bates and Harmon actually used a similar form of two-stage sampling in their random sample survey of Lubbockians (described in the focal research piece). Bates and Harmon had a target population of adult residents in Lubbock County, but had no list of such residents. Their random-digit dialing put them in touch with residences that had phones (their clusters), but then they needed an unbiased way of contacting adults living at such residences. You might be tempted to ask your questions, for instance, of the person who answered the phone, especially if that person were an adult. But such a procedure could engender a bias in the sample.

STOP AND THINK *What biases might result from questioning the first adult to answer the phone?*

Some people who are more likely than others to answer phones are the able-bodied, those who have friends they like to chat with on the phone, those who aren't afraid that creditors might be calling, and so forth. Any number of biases might be engendered if you queried only those who answered the phone. But if you, as Bates and Harmon did, asked to speak to the adult who had the most recent birthday, you'd be much more likely to get an unbiased sample of the adults living in the residences you contacted. Lavrakas (1993, Chapter 4) provides an excellent discussion of this and other methods for selecting random respondents in random digit-dialing surveys. Thus, Bates and Harmon also used a relatively simple multistage sampling technique, one that involved random selection of residences with phones (through random-digit dialing) and random selection of adults within those residences (through the request for the adult with the most recent birthday).

The complexity of sampling designs tends to increase as the target population becomes more geographically dispersed and as lists become harder to find. Irene Hess (1985) has provided a detailed overview of procedures used by the Social Research Center for these and similar national household polls, involving six stages of sampling. The Center first divides the country into 2,700 primary units (clusters of counties or metropolitan areas), allocates these units into 74 strata, and takes a disproportionate stratified sample of units from each of these strata (the first stage). Then the Center divides each of the primary units into sub-units (cities, towns, and rural areas), allocates these sub-units into strata, and takes disproportionate stratified samples from each of the strata (the second stage). Third, the Center divides each sub-unit into sub-sub-units (usually blocks or "chunks") and systematically samples these sub-sub-units (the third stage). Then representatives of the Center "scout" these sub-sub-units to find all households in them, divide the sub-sub-units into sub-sub-sub-units and random sample these clusters (called sampling segments). This is the fourth stage. The fifth stage involves random sampling households (usually about 4) in the sampling segments, and the sixth stage involves deciding which member of each particular household should be interviewed so as to avoid bias.

You're probably as impressed as we are with the complexity of a sampling procedure like the one described in the previous paragraph. But we'd also like you to consider what it does for researchers. One big advantage is that it finally permits the researchers to focus their investigation on relatively well-defined areas. If, for instance, your goal were to reach 2400 households, this procedure would enable you to focus your investigation on (2400 households total/ 4 households per sampling segment) 600 sampling segments in the whole country. Moreover, you wouldn't need to start with a list of the whole population. In fact, in the example given above, you only need to start drawing up your list in the fourth stage (or once you've reached the block or "chunk" level).

The Social Research Center multistage design, like all probability sampling designs, is much better at ensuring representativeness than any nonprobability sampling design of a comparable population is. There are times, however, when nonprobability sampling is necessary or desirable, so we now turn our attention to some nonprobability sampling designs.

Types of Nonprobability Sampling Designs

We've already suggested that, although nonprobability samples are less likely than probability samples to represent the populations they're meant to represent, they are useful when the researcher has limited resources, an inability to identify members of the population (say, underworld or upper-world criminals), or when one is doing exploratory research (say, about whether a problem exists or about the nature of a problem, assuming it exists). Let us elaborate on each of these advantages as we describe some ways of doing nonprobability sampling.

Purposive Samples

purposive sampling, a nonprobability sampling procedure that involves selecting elements based on the researcher's judgment about which elements will facilitate his or her investigation.

For many exploratory studies and much qualitative research, **purposive sampling,** in any of a number of forms, is desirable. In purposive sampling, the researcher selects sampling units based on his or her judgment of what units will facilitate an investigation. One version of purposive sampling, sometimes referred to as *typical case sampling,* involves selecting cases that are judged, beforehand, to be normal or average. Especially early on in their study of afterschool activities, Adler and Adler (1994, see Chapter 11) sought activities that they viewed as typical so that they could gain a basic understanding of them. Another version, called *stratified purposeful sampling,* involves sampling from subgroups to facilitate comparisons between or among them. After she'd done her participant observation of a Saturday morning parenting program for women inmates at an Adult Correctional Institution (a convenience sample), Enos (see Chapter 10) deliberately decided to do intensive interviews with a small number of white, black, and Hispanic women. Both Adler and Adler, on the one hand, and Enos, on the other, illustrate what is true of many qualitative research efforts: They involve the sequential usage of different kinds of nonprobability sampling strategies. For a more comprehensive overview of such strategies, we recommend Kuzel (1992) and Patton (1990).

Quota Sampling

Quota sampling was the type of sampling used by Gallup, and others, so well in presidential preference polls until 1948. (You'll remember how much better Gallup's projection was about the 1936 election than the *Literary Digest*'s, which depended on a convenience sample of readers.) Quota sampling begins with a description of the target population: its proportion of males and females, of people in different age groups, of people in different income levels, of people in different racial groups, and so on.

STOP AND THINK *Where do you think Gallup and other pollsters might have gotten such descriptions of the American public?*

Until 1948, Gallup relied on the decennial censuses of the American population for his descriptions of the American voting public. Based on these descriptions, a researcher using quota sampling drew a sample that "represented" the target population in all the relevant descriptors (for example, gender, age, income levels, race, and so on).

Quota sampling at least attempts to deal with the issue of representativeness (in a way that neither convenience nor snowball sampling, discussed later, do). And quota samples are frequently cheaper than probability samples. But quota sampling is subject to some major drawbacks. First is the difficulty of guaranteeing that the description of the target population is an accurate one. One explanation given for the inability of Gallup and other pollsters to accurately predict Truman's victory in the 1948 presidential election is that they were using a population description that was eight years old (the decennial census having been taken in 1940), and much had occurred during the forties to invalidate it as a description of the American population in 1948. Another drawback of quota sampling is how the final case selection is made. Thus, the researcher might know that she or he needs, say, 18 white males, aged 20 to 30 years, and select these final 18 by how pleasant they look (in the process, avoiding the less pleasant looking ones). Pleasantness of appearance is likely to be one source of bias in the final sample.

Snowball Sampling

snowball sampling, a nonprobability sampling procedure that involves using members of the group of interest to identify other members of the group.

When population listings are unavailable (as in the case of "community elites," gay couples, upperworld or underworld criminals, and so forth), **snowball sampling** can be very useful. Snowball sampling involves using some members of the group of interest to identify other members. One of the earliest and most influential field studies in American sociology, William Foote Whyte's (1943) *Street Corner Society: The Social Structure of an Italian Slum*, was based on a snowball sample, as Whyte talked with a social worker, who introduced him to Doc, his main informant, who then introduced him to other "street corner boys," and so on. With snowball sampling, the disadvantage of not being able to generalize can be more than compensated for by the capacity to provide an understanding of a subject that had hitherto been concealed by a veil of ignorance.

Convenience Sampling

convenience sampling, a nonprobability sampling procedure that involves selecting elements that are readily accessible to the researcher.

A **convenience sample** (sometimes called an available-subjects sample) is a group of elements (often people) that are readily accessible to, and therefore convenient for, the researcher. Both the *Literary Digest* and the television station studied by Bates and Harmon used a convenience sample in their surveys of public opinion. Each asked their clients (readers or viewers) to respond to questions they could pose through their medium (in the magazine or during the TV news). Psychology and sociology professors sometimes use their students as participants in their studies. William Levin (1988), for instance, used students from three colleges to conduct an experiment about age stereotyping. He asked a third of the students to evaluate the characteristics (for example, the intelligence) of a man based on a picture of him when he was 25 years old, another third to evaluate the same man based on his picture when he was 52, and the third to assess him based on a picture of him when he was 73. Levin found that students were much more severe in their evaluations of the person when he appeared older than when he appeared either middle-aged or younger.

Convenience samples are relatively inexpensive, and they can yield results that are wonderfully provocative and plausible. Levin's results are important on both counts. But the temptation to generalize beyond the samples involved is dangerous and should be tempered with considerable caution.

Choosing a Sampling Technique

Choosing an appropriate sampling technique, like making any number of other decisions in the research process, involves a balancing act, often among methodological, practical, theoretical, and ethical considerations. If it is methodologically important to be able to generalize to some larger population, as in political preference polls, then choosing some sort of probability sample is of overwhelming importance. The methodological importance of being able to generalize from such a poll can clash with the practical difficulty of finding no available list of the total population and, so, dictate the use of some sort of multistage cluster sampling, rather than some other form of probability sampling (for example, simple random, systematic, or stratified) that depends on the presence of such a list. Or, however desirable it may be to find a probability sample of corporate embezzlers (a theoretical concern), your lack of access to the information you'd need to perform probability sampling (a practical concern) may force you to use a nonprobability sampling technique—and perhaps finally settle for those you can identify as corporate embezzlers and who volunteer to come clean (ethical).

You might find that your study of international espionage would be enhanced by a probability sample of the list of CIA agents your roommate has downloaded after illegally tapping into a government computer (methodological and practical considerations), but that your concern for going to prison (a practical consideration) and for jeopardizing the cover of these agents (practical and ethical) leads you to sample passages in the autobiographies of ex-spies (see Chapter 13 on content analysis) rather than survey the spies themselves. In another study your exploratory (theoretical) interest on the psychological consequences of adoption for non-adopted adult siblings may dovetail perfectly with your recent introduction to three such siblings (practical) and their obvious desire to talk to you about their experiences (practical and ethical).

Method, theory, practicality, and ethics—the touchstones of any research project—can become important, then, as you begin to think about an appropriate sampling strategy.

Summary

Sampling is a means to an end. We sample because studying every element in our population is frequently beyond our means or would jeopardize the quality of our data—by, for instance, preventing our close scrutiny of individual cases. On the other hand, we don't need to sample when studying every member of our population is feasible.

Probability samples give every member of a population a known chance of being selected; nonprobability samples do not. Probability sampling offers a better chance of drawing an unbiased or representative sample than does nonprobability sampling. Bates and Harmon's focal research demonstrates, for instance, that nonprobability samples (especially in the form of phone-in polls) can give very different pictures of public opinion than can probability samples and suggests that nonprobability sampling can result in serious biases for this purpose. Nonetheless, nonprobability sampling can be very useful when a researcher has limited resources, can't identify a total population to sample from, or is engaged in exploratory research.

Bias frequently occurs in nonprobability sampling because the researcher intentionally or unintentionally avoids or misses certain categories of the potential population. Bias can occur in probability sampling when the study population (the group from which sampling units are actually drawn) differs from the population of theoretical interest. It can also occur for several other reasons, especially if sampling units about which one can't get information (for example, because of nonresponse) differ in a consistent way from those about which one can.

Probability samples can also misrepresent the populations from which they are drawn because of sampling variability. Sampling variability, and the attendant possible error, can be reduced by taking larger samples.

We looked at five strategies for selecting probability samples: simple random sampling, systematic random sampling, stratified random sampling, cluster sampling, and multistage sampling. We then looked at four common strategies for selecting nonprobability samples: purposive sampling, quota sampling, snowball sampling, and convenience sampling. We emphasized that the appropriate choice of a sampling strategy always involves balancing theoretical, methodological, practical, and ethical concerns.

SUGGESTED READINGS

Converse, Jean M. 1987. Survey research in the United States: Roots and emergence 1890–1960. Berkeley: University of California Press.

A well-written, thoroughly researched history of survey research that involves a lovely history of sampling practice as well.

Henry, Gary T. 1990. *Practical sampling*. Newbury Park: Sage.

This compact book provides a reliable guide to sampling questions, from the mundane to the esoteric, with helpful illustrations.

EXERCISE 5.1

Practice Sampling Problems

This exercise gives you an opportunity to practice some of the sampling techniques described in the chapter.

The following is a list of the states in the United States (abbreviated) and their populations in 1990 to the nearest tenth of a million.

State	*Population*	*State*	*Population*	*State*	*Population*
AL	4.0	LA	4.2	OH	10.8
AK	0.5	ME	1.2	OK	3.1
AZ	3.7	MD	4.8	OR	2.8
AR	2.3	MA	5.9	PA	11.9
CA	29.8	MI	6.0	RI	1.0
CO	3.3	MN	9.3	SC	3.5
CT	3.3	MS	4.4	SD	0.7
DE	0.6	MO	2.6	TN	4.9
FL	12.9	MT	0.8	TX	17.0
GA	6.5	NE	1.6	UT	1.7
HI	1.1	NV	1.2	VT	0.6
ID	1.0	NH	1.1	VA	6.1
IL	11.4	NJ	7.7	WA	4.9
IN	5.5	NM	1.5	WV	1.8
IA	2.8	NY	18.0	WI	4.9
KS	2.5	NC	6.6	WY	0.5
KY	3.7	ND	0.6		

1. Number the states from 01 to 50, entering the numbers next to the abbreviated name on the list.

2. Use the random number table in Appendix A and select enough two-digit numbers to provide a sample of 12 states. Write all the numbers and cross out the ones you can't use.

3. List the 12 states that make it into your simple random sample.

4. This time take a stratified random sample of 10 states, one of which has a population of 10 million or more and 9 of which have populations of less than 10 million. List the states you choose.

5. How might you draw a quota sample of 10 states, one of which has a population of 10 million or more and 9 of which have populations of less than 10 million?

   a) Describe one way of doing this.

   b) Describe, in your own words, the most important differences between the sampling procedures used in question 4 and question 5a.

6. Generate a sample of 15 telephone numbers using the random number table in Appendix A. Use the following set of prefixes in equal proportion: 456-, 831-, 751-.

EXERCISE 5.2

Revisiting the Issue Raised by Bates and Harmon

Suppose you wanted to describe the differences between responses about whether students at your college or university regularly read the student newspaper in a write-in poll and a probability survey. (You might, incidentally, consider using a question like "Would you describe yourself as a regular reader of [name of the student newspaper]?" to measure whether a student does read the paper regularly.) Although you'll be planning a sampling strategy, we don't actually want you to go out and do the research you'll be considering here.

1. How might you use the college's newspaper itself to conduct a write-in poll?

2. How might you select a probability sample of, say, 100 students at your school to ask the same question?

3. Describe the differences you'd expect to find in the results of your two surveys (hint: pay special attention to the question of how biased or unrepresentative you'd expect each result to be).

4. Do you think the differences between the results of your two surveys would be greater or smaller than the differences that Bates and Harmon uncovered in the two surveys they compared? Discuss how you reached your conclusion.

EXERCISE 5.3

Examining Sampling in a Research Article

Pick a research article from any professional journal that presents the results of a research project involving some sort of sampling. Answer the following questions:

1. Name of author(s):

   Title of article:

   Journal name:

   Date of publication:

   Pages of article:

2. What kind of sampling unit or element was used?

3. Was the population from which these elements were selected described? If so, describe the population.

4. How was the sample drawn?

5. What do you think of this sampling strategy as a way of creating a representative sample?

6. Do(es) the author(s) attempt to generalize from the sample to a population?

7. Did the author(s) have any other purposes for sampling besides trying to enable generalization to a population? If so, describe this (these) purpose(s).

8. What other comments do you have about the use of sampling in this article?

Measurement

"But how did you know that Smythe had committed suicide, Holmes?" queried the bemused Watson.

"Elementary, my dear Watson," replied Holmes. "As I've told you before, once you've eliminated the impossible, whatever remains, no matter how improbable, is always the truth. In this case, the only other alternatives were death by natural causes, homicide, and accident. The case for natural causes could be dismissed at once, because we found Smythe standing, face down in the bird bath."

"But couldn't he have suffered a heart attack near the bath and fallen in?" asked Watson.

"Not without bruises or cuts to his head," observed Holmes. "And, as you very well know, since you performed the initial examination, there were none."

"But what about the case against murder, Holmes?"

"Again the absence of any sign of physical violence is absolutely compelling, since Smythe was a muscular 20 stones[1] and would surely have put up quite a struggle before submitting to such a murderous dousing," Holmes responded.

"Still, how can you be certain that it wasn't an accident?"

"I, of course, considered the possibility that Smythe might have paused to sip from the bird bath and, in his hurry to quench his thirst, sipped too much. In that case, however, he would surely have removed his face and pulled away from the bird bath, even as he spluttered to his death. But as you know, we found him standing, as I said before, face down in the bath.

"The clincher, Watson, was the note, in Smythe's handwriting, that we found attached to the bird bath. I would have thought even you, my trusted chronicler, would have guessed at suicide when you read,

"Through life I trod the straight and narrow.
"I never harmed an English sparrow.
"But now I cannot face the morrow
"So in this bath I'll drown my sorrow."

Determining that the cause of death is suicide is often not as cut and, er, dry as our version of Holmes makes it out to be. The focal research by Rockett and Smith in this chapter makes this clear. Indeed, sometimes whole nations approach the task of cause-of-death determinations in such a way, Rockett and Smith argue, as to cast doubt on the comparability of their suicide rates with other nations.

[1]A stone is roughly 14 pounds, making Smythe a hefty 280 pounds by conventional American calculations.

Introduction

measurement, classifying units of analysis by categories to represent variable concepts.

Measurement generally means classifying units of analysis (say, deaths) by categories (say, suicide or nonsuicide) to represent variable concepts (say, cause of death). Measurement, like sampling (see the previous chapter), is a process that frequently requires considerable care during the preparation of research as well as some attention when we read a research report. The building blocks of research are concepts, or ideas (like "suicide," "social class," "gender," and "self-esteem") about which we've developed words or symbols. Such concepts are almost always abstractions that we make from experience. Concepts are often not things that we can observe directly. What we, as researchers, do during the measurement process is to define the concept, for the purposes of the research, in progressively precise ways so that we can take indirect measurements of it. We might, for instance, begin with a dictionary definition (or some other form of theoretical definition [see later]) of suicide as "the act of killing oneself intentionally." Then we might decide to take any clear remnant of a deceased person's intention to take his life (for example, a note or a tape recording) as an indicator of suicide. **Indicators** are observations that we think reflect the presence or absence of the phenomenon to which a concept refers. Indicators can be comments or answers to questions, things we observe, or material from documents and other existing sources. "The game is afoot" for many scientists once they begin to look for indicators of important concepts, just as it was for Sherlock Holmes when he began to hunt for clues to help him solve a mystery.

indicators, observations that we think reflect the presence or absence of the phenomenon to which a concept refers.

Even concepts with widely accepted definitions and well-defined indicators, however, can elude consistent categorization or measurement from one place to another. Suicide is, in fact, one such concept. But why the variation? This is the kind of question that moved two epidemiologists (students of our collective health), Ian Rockett and Gordon Smith, to produce this chapter's focal research.

Before you read Rockett and Smith's article, however, there are two qualities of measures that you should know something about because of their importance to the authors: the qualities of **reliability** and **validity.** A measurement strategy is said to be reliable if its application yields consistent results time after time. When Roger measures his daughter with a yardstick, the yardstick provides reasonably reliable measurements of her height because it tells him she is about the same height today as she was yesterday. A measurement strategy is valid, if it measures what you think it is measuring. However reliable (and valid) Roger's yardstick is when measuring his daughter's height, it cannot measure validly her weight, or her intelligence, or her ambition—it's just not supposed to.

reliability, the degree to which a measurement strategy yields consistent results time after time.

validity, the degree to which a measure taps what we think it's measuring.

STOP AND THINK *Suppose you attempted to measure how prejudiced people are with the question "What is your gender?" Which quality of measurement, reliability or validity, is most apt to be offended by this measurement strategy? Why?*

We'll have more to say about reliability and validity later in the chapter, but now let's see how suicide can be misclassified from one country to the next.

FOCAL RESEARCH

Suicide Misclassification in an International Context[2]

by Ian R. H. Rockett and Gordon S. Smith

Introduction

Data misclassification has been a persistent and contentious topic in the international suicide literature (Zilboorg, 1936; Dublin, 1963). This paper asks whether official national suicide data are sufficiently reliable and valid to justify their use in international comparative studies. Are real differences and similarities in cross-national suicide rates obscured by artifactual differences (that is, differences in the way suicide is measured)? The paper moves from a consideration of potential sources of suicide misclassification to the presentation of techniques and data deemed useful in assessing the severity of the problem.

Manner of Death and Medicolegal Decision-Making

When an individual dies, the primary classification decision is whether death is attributable to *natural causes, accident, homicide, or suicide.* Most deaths are attributed to natural causes, but attributions of death to natural causes vary between about 85 and 97 percent of all deaths in industrialized nations. Whatever causes variation in the attribution of deaths to natural causes will have obvious effects on the attribution of deaths to other causes, including suicides.

A World Health Organization survey (1971) showed that decision-making about manner of death varies internationally and is itself a potential source of international variation in the classification of deaths as suicides. Private physicians, police officers, coroners, medical examiners, and the judiciary may all be involved in varying degrees. Most countries have either a coroner or medical examiner system, but while medical examiners are always medically qualified, coroners may have law degrees, medical degrees, or both.

Perhaps a key difference between the medical examiner and coroner systems is the standard by which decision-making is guided. Medical examiners are more likely than coroners to balance scientific probabilities. Coroners are more apt to employ a burden-of-proof standard and are therefore less

[2]This article is an abbreviated version of Rockett and Smith's publication of the same title in the *Proceedings of the International Collaborative Effort on Injury Statistics,* Volume 1. DHHS Publication No. (PHS) 95–1252. Hyattsville, MD: U.S. Department of Health and Human Services, 1995, Pp. 26-1—26-18. It is published with authors' permission.

apt to classify deaths as suicides, especially when doing so has implications for criminal liability, insurance payments, and the emotional well-being of survivors. Suicide rates in the United States are higher than they are in Great Britain and part of this difference may be due to the use of the medical examiner system in many U.S. states and the exclusive use of a formal coroners' court in Great Britain.

It is plausible then that diversity in medicolegal procedures and decision-making generates some cross-national variation in suicide reports.

Complications of Method

The capacity of medicolegal authorities to detect suicide also varies with the method of suicide used, and these methods may vary considerably across cultures (Monk, 1986, Kolmos and Bach, 1987, Rockett and Smith, 1993). Detection is easier when violent methods like hanging, shooting, and stabbing are used, but more difficult when nonviolent methods like drowning, poisoning, and gassing are employed. In fact, the last three methods, along with jumping from a height, lone driver vehicular crashes, and Russian roulette are called equivocal methods.

Suicide by drowning is particularly difficult to classify, especially when there have been no witnesses. But so is suicidal poisoning, especially when the substances involved are associated with abuse. When drugs of abuse result in death, it is especially difficult to determine intent: Some people may be truly ignorant about the line between a safe dose and an overdose.

Sociodemographic Characteristics

Variation in the age and gender make-up of populations may interact with the aforementioned decision-making and method-of-death differences to produce artifactual differences in suicide rates as well. Deaths of the elderly, for instance, are usually less thoroughly investigated than deaths of younger people, despite the fact that the elderly have higher suicide rates. Part of the under-investigation is probably due to the greater chance that the elderly will choose nonviolent, slower, and generally more subtle methods of suicide than their younger counterparts (McIntosh and Hubbard, 1988, Miller, 1979).

Recent data confirm the frequently reported finding that males commit more suicides than females, but also suggest that the difference varies enormously from one nation to another. The ratio of male to female suicide rates varied from a high of 7-to-1 in Iceland to a low of 1.3-to-1 in Singapore in our analyses of 30 nations for which data were available around 1990. Some of this variation, however, may be due to the greater likelihood that females choose less violent and less easily detected methods than males (Monk, 1986, Kolmos and Bach, 1987) and to the aforementioned international variation in decision-making about cause of death, which is likely to be associated with uneven scrutiny of equivocal cases.

Sociocultural Milieu

Like sociodemographic heterogeneity, sociocultural heterogeneity can be a source of artifactual differences in international suicide rates. Religion, for instance, has received serious attention from suicidologists at least since the work of the French sociologist, Émile Durkheim. A famous Durkheimian hypothesis is that adherents of religions that foster high degrees of social integration are less prone to suicide than counterparts whose religious affiliation encourages individualism or the pursuit of free inquiry. The social integration argument was used by Durkheim to explain lower reported suicide rates in Roman Catholic countries than in Protestant countries.

A plausible alternative to Durkheim's account, however, is that differences between Catholic and Protestant nations reflect variation in the social condemnation of suicide and the reluctance of physicians to certify a death as a suicide (Gibbs, 1961). Proponents of this view argue that suicide rates are socially constructed, and that the greater the condemnation of suicide, the more deficient the reporting. Whether social condemnation is related to religion and/or other factors, it may induce suicide victims to disguise the intent of their acts and encourage family, friends, and sometimes even medicolegal authorities to withhold or suppress crucial evidence.

Assessing Reliability

Let us mention two methods that have been used to assess the reliability of international suicide statistics. The first, an experimental approach, has been aimed at determining whether medicolegal officials from different nations made similar manner-of-death assignments to common cases. In one study, Danish and Scottish officials made such assignments for a sample of each other's cases, and no substantial differences were found (Ross and Kreitman, 1975).

A second approach to the reliability question has involved the comparison of suicide rates among immigrants in a particular country with rates in their countries of origin. In one study (Whitlock, 1971), for instance, a high degree of consistency was found between suicide rates for immigrants to Australia and their non-immigrant counterparts for 17 countries of origin.

On balance, the findings of studies using these and other methods give reason to believe that international suicide data are adequate from the standpoint of reliability. The validity of such data, however, is much more difficult to dismiss as a scientific concern.

Assessing Validity

The validity of suicide data can be examined from the complementary perspectives of sensitivity and specificity. Sensitivity refers to the degree to which suicides are correctly certified as suicides. Specificity measures the degree to which nonsuicides are certified as nonsuicides. Since suicides tend to

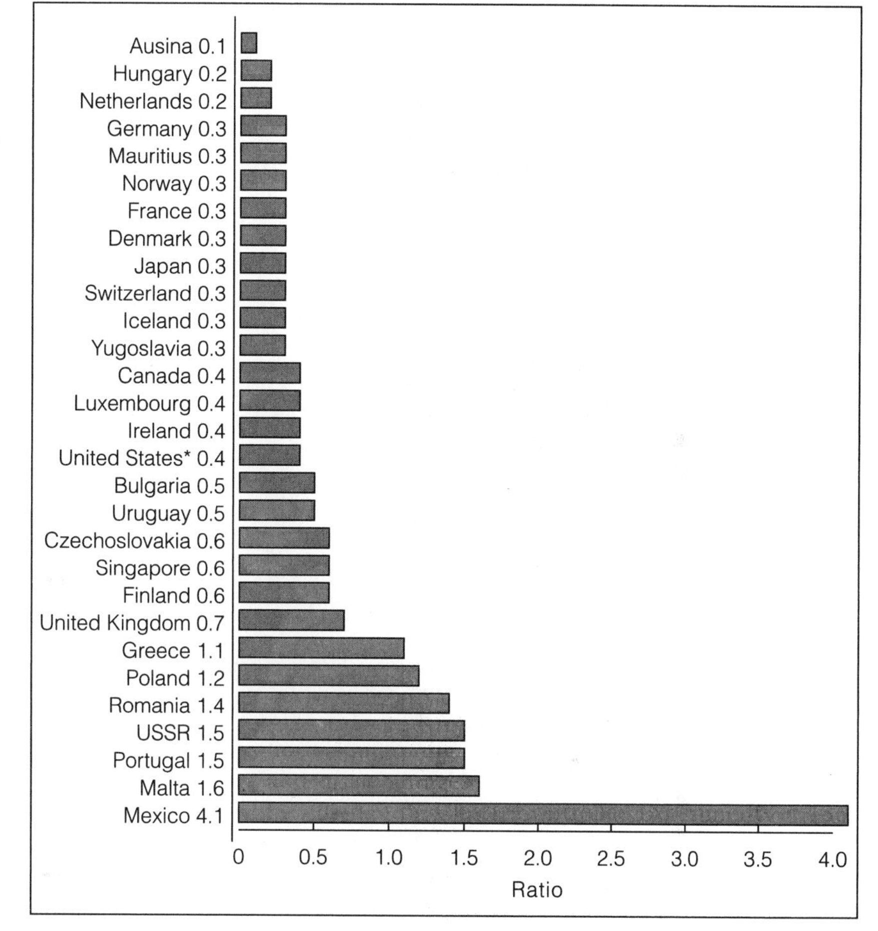

FIGURE 6.1

Ratio of Combined Deaths from Accidental Drowning, Accidental Poisoning, and Other Violence to Suicide by Country, 1990.

be undercounted, specificity is not likely to be a big problem—most nonsuicides get counted as nonsuicides. But sensitivity is problematic internationally. This is due to the interplay of forces already identified, such as sociodemographic characteristics of suicide victims, prevailing sociocultural milieu, and nature and training of medicolegal decision-makers and auxiliary staff. O'Carroll (1989) reports a range for sensitivity estimates—percentages of actual suicides that are certified as suicides—of 26 percent and 83 percent, with estimates concentrating between 56 and 71 percent.

Three categories of death are prime contenders for containing misclassified suicides: accidental poisoning, accidental drowning, and injury of undetermined intent. The mortality ratio of the combined death rate of these combined categories to the suicide rate is a guide to estimating an upper limit for various national suicide rates. We calculated such ratios for 29 countries and present them in Figure 6.1. The degree of potential suicide underenumeration varies directly with the magnitude of the ratio.

TABLE 6.1 **Annualized Accidental Drowning Rates by Age, Sex, and Country, 1979–1981***

	Age (years)													
	15–24		25–34		35–44		45–54		55–64		65–74		75+	
	M	F	M	F	M	F	M	F	M	F	M	F	M	F
Japan	2.9	0.2	2.5	0.3	2.5	0.3	3.5	0.6	4.5	1.1	7.9	3.8	18.1	13.5
Australia	3.3	0.4	3.1	0.3	3.2	0.4	3.9	0.8	3.6	0.6	3.2	1.0	3.9	1.5
France	3.5	0.4	3.2	0.4	3.2	0.6	3.3	0.6	3.7	0.9	4.2	1.6	5.4	2.0
New Zealand	4.8	0.6	3.4	0.3	1.8	1.1	2.8	1.3	2.8	1.2	1.9	1.9	--	1.0
Norway	3.5	0.3	4.2	0.3	4.3	0.1	5.3	0.6	4.5	0.5	4.2	1.7	4.7	0.9
Sweden	1.5	0.2	2.6	0.3	2.3	0.7	2.4	0.6	3.1	0.7	4.3	1.0	3.7	1.2
United Kingdom	1.5	0.2	1.2	0.2	1.0	0.2	0.9	0.4	1.1	0.4	1.1	0.6	1.9	1.2
United States	7.2	0.8	4.2	0.6	2.9	0.5	2.4	0.5	2.3	0.5	2.5	0.7	3.6	1.2

*Rates expressed per 100,000 population

**Accidental drowning deaths for Northern Ireland in 1981 were not reported by WHO. For these calculations, they are estimated as the annual average for 1979 and 1980.

Source: Rockett IRH. Smith GS. Covert suicide among elderly Japanese females: questioning unintentional drownings. *Social Science and Medicine* 36(11): 1993: 1467-1472.

The ratio of the rates for the selected combined injury categories to suicide reveal a range from 0.1 for Austria to 4.1 for Mexico. In the (unlikely) event that all combined injury deaths are misclassified suicides, reclassification would increase the Austrian suicide rate by only 10 percent, but would increase the Mexican rate by more than 400 percent. Thus suicide under-enumeration is much less likely in Austria than in Mexico.

Drowning and Elderly Japanese Females[3]

Typical suicide methods vary by nation, and this too has important implications for differential misclassification. The case of elderly Japanese females is suggestive. We argue that many more of these females than is currently acknowledged commit suicide by drowning, a method that like *harikari* historically had symbolic legitimacy in Japan (Iga et al., 1978).

Elderly Japanese of both sexes register comparatively high suicide rates, with male rates exceeding those of females (Rockett and Smith, 1989). Accidental drowning rates among the elderly in Japan are very high in comparison

[3]The material presented in this section is from a previously published source: Rockett, I.R.H., and Smith, G.S. 1993. Covert suicide among elderly Japanese females: questioning unintentional drownings. *Social Science and Medicine,* 36(11): 1467–1472.

TABLE 6.2

Percentage Drowning as Method of Suicide by Age and Sex, Japan: 1979–1981

Age (years)	*Male*	*Female*	*Both Sexes*
15–24	3.3	5.8	4.1
25–34	4.4	10.1	6.2
35–44	3.9	11.1	6.1
45–54	3.4	12.0	6.1
55–64	4.3	13.1	8.0
65–74	4.8	14.5	9.8
75+	5.7	17.4	12.2
Total	4.1	12.5	7.3

Source: Rockett IRH, Smith GS. Covert suicide among elderly Japanese females: questioning unintentional drownings. *Social Science and Medicine* 36(11): 1993:1467–1472.

with those in seven other countries known for ease of water access (see Table 6.1). Moreover, the proportion of suicides attributed to drowning among elderly females in Japan is about two-and-a half times what it is among elderly males (see Table 6.2).

We hypothesize that suicide of elderly females is relatively underenumerated because of misclassification of suicidal drowning as accidental drowning. Support for this hypothesis is implicit in our finding that, between ages 25–34 and 75 and older, the ratio of drowning suicides to accidental drownings declined by 81 percent for females as compared with only 49 percent for males. The differential decline might simply result from the greater retention of swimming abilities among elderly males, but this explanation seems implausible and has not been demonstrated.

We offer two arguments, in addition to the empirical evidence of the ratio shift, to support the drowning suicide misclassification hypothesis. First, Japanese females have a six–year advantage over males in life expectancy at birth and are at much greater risk of being widowed, and living alone. These differences reduce the chances that older female victims of suicide will have survivors to assist medicolegal authorities in their deliberations. Second, it seems plausible that elderly females may be inclined to disguise their suicides to protect their families against the social stigmatization that is increasingly associated with suicide in Japanese society (for example, Iga et al., 1978).

The drowning suicide misclassification hypothesis needs additional testing. If substantiated, however, it would underscore the caution that researchers should exercise if tempted to accept uncritically as valid the magnitude, and even the existence and direction, of observed age- and sex-specific differentials in cross-national suicide rates.

Conclusion

This paper has presented evidence that suicide data used in analyses involving more developed countries are reasonably reliable. The same cannot be said of their validity, or more precisely their sensitivity.

Suicide is widely acknowledged as a public-health problem, although an underenumerated one. Identifying high-risk groups, understanding etiology (causes), and designing and implementing effective prevention programs ultimately depend upon obtaining an accurate and detailed description of magnitude.

Finally, greater international use should be made of the psychological autopsy, which involves followback interviews with family, friends, and acquaintances to specifically identify antecedents of possible suicides. If psychological autopsies were implemented in all or a random sample of equivocal fatal injury cases, they would assist in the refinement of estimates of true suicide rates.

REFERENCES

Barraclough, B. M. 1973. Differences between national suicide rates. *British Journal of Psychiatry* 122: 95–96.

Dublin, L. I. 1963. *Suicide: A sociological and statistical study.* New York: Ronald.

Gibbs, J. 1961. Suicide. In Merton, R., and R. Nisbet (eds.). *Contemporary social problems.* New York: Harcourt, Brace and World, pp. 281–321.

Iga, M., Yamamoto, J., Noguchi, T., and Koshinga, J. 1978. Suicide in Japan. *Social Science and Medicine* 12A: 507–516.

Kolmos, L., and Bach, E. Sources of error in registering suicide. *Acta Psychiatrica Scandinavica.* Supplement 336 76: 22–41.

McIntosh, J. L., and Hubbard, R. W. 1988. Indirect self-destructive behavior among the elderly: a review with case examples. *Journal of Gerontological Social Work* 13: 37–48.

Miller, M. 1979 *Suicide after sixty: The final alternative.* New York: Springer.

Monk, M. 1986. Suicide. In Last, J. M. (ed.). *Public health and preventive medicine.* Norwalk, CT: Appleton-Century Crofts, pp. 1385–1397.

O'Carroll, P. W. A. 1989. A consideration of the validity and reliability of suicide mortality data. *Suicide and life-threatening behavior* 19: 1–16.

Rockett, I. R. H., and Smith, G. S., 1989. Homicide, suicide, motor vehicle crash and fall mortality: United States' experience in comparative perspective. *American Journal of Public Health* 79: 1396–1400.

Ross, O., and Kreitman, N. A. 1975. A further investigation of differences in the suicide rates of England and Wales and Scotland. *British Journal of Psychiatry* 127: 575–582.

Whitlock, F. F. 1971. Migration and suicide. *Medical Journal of Australia* 2: 840–848.

World Health Organization. 1974. *Suicide and attempted suicide.* Brooke, E. (ed.) Public Health Papers 58. Geneva: World Health Organization.

Zilboorg, G. 1936. Suicide among civilized and primitives races. *American Journal of Psychiatry* 92: 1347–1369.

Conceptualization and Operationalization

Ian Rockett claims he was drawn to the question of the validity of suicide data by what he calls an "aha" experience. He was peering at two graphs: one showing suicide rates for Japanese women; the other, accidental drowning rates for the same group. The lines on the graphs seemed to follow the same path, and Rockett said to himself, "I'll bet some of those drownings are suicides." This led Rockett and his co-author to pursue the Japanese case more intensively and, subsequently, to the central question of the current essay: Could suicide misclassification lead to problems in the cross-national comparability of suicide rates?

Concepts, words or signs that refer to phenomena that share common characteristics.

In juggling ideas like "suicide" and "accidental drowning" in this fashion, Rockett was playing with those building blocks of all scientific, indeed of all, thinking: **concepts**—words or signs that refer to phenomena that share common characteristics. Whenever we refer to ideas like "suicide," "accidents," "power," "self-esteem," or "gender," we are using such concepts. But let's notice something important: We each develop our basic understandings of concepts through our own idiosyncratic experiences. You might have initially learned what an "accident" is when, as a child, you spilled some milk, and a parent assured you that the spilling was an accident. I might have heard of a friend's bicycle "accident." Someone else might have had a baby brother who had an "accident" in his diaper. The point is we might all have reasonably similar ideas of what an accident is, but we come to these ideas in ways that permit a fair amount of variation in just what we mean by the term.

conceptualization, the process of clarifying just what we mean by a concept.

operationalization, the process of defining specific ways to infer the absence, presence, or degree of presence of a phenomenon.

So beware: Concepts are *not* phenomena; they are only symbols we use to refer to (sometimes quite different) phenomena. They are abstractions that permit us to, among other things, communicate with one another, but they do not, in and of themselves, guarantee precise and lucid communication—a goal of science. In fact, in nonscientific settings, concepts are frequently used to evoke multiple meanings and allusions, as when poets write or politicians speak. If, however, concepts are to be useful for social science purposes, social scientists should use them with clarity and precision. The two related processes that social scientists use to create such clarity and precision are called **conceptualization** and **operationalization.**

conceptual definition, defining a concept through other concepts. Also a theoretical definition.

Conceptualization is the process of clarifying just what we mean by a concept. This process usually involves providing a theoretical or **conceptual definition** of the concept. Such a definition describes the concept through other concepts. Again, we might use Webster's dictionary definition (or conceptualization) of "suicide" as "the act of killing oneself intentionally," a definition that itself employs the concepts of "killing" and "intention." Or we might define "power," using Max Weber's classic formulation, as the capacity to make others do what we want them to do despite their opposition—a definition that refers, among other things, to the concepts of "making others do" and "opposition." Or we might define "self-esteem," using Hudson's (1982) definition, as "the capacity to view oneself as worthwhile and loveable."

STOP AND THINK *What concepts does Hudson use in his conceptual definition of self-esteem?*

In all cases, we would be defining concepts through other concepts: ideas (like "intention," "opposition," and "loveableness") that themselves might deserve further clarification, but which nonetheless substantially narrow the possible range of meanings a concept might entail. Thus, for instance, when Hudson speaks of self-esteem as the capacity to see oneself as worthwhile or loveable, he's implicitly suggesting that self-esteem doesn't necessarily involve the "conceit" or "egotism" that others might associate with it. In so doing, he narrows and clarifies the meaning of his concept—the goal of conceptualization.

As we've said before, however, we often can't perceive the real-world phenomena to which our concepts refer, even after we've provided relatively unambiguous conceptual definitions of what we mean by the concepts. "Power" and "self-esteem" are very hard to observe directly, and even such a "concrete" concept as "suicide" can elude our direct perception. Someone found dead in a body of water might have been drowned (died by homicide), or accidentally drowned (died by accident), or died as result of his or her own effort (died by suicide)—or even died of natural causes. When the phenomena to which concepts refer can't be observed directly, we might make inferences about their occurrence (or lack or degree thereof). The process of defining specific ways to infer the occurrence of specific phenomena is called operationalization. Operationalization usually involves providing **operational definitions,** or declarations of the specific ways in which the existence or the degree of existence of phenomena described by concepts are to be determined in a specific instance. Thus we can conceptualize suicide to be something like "the act of killing oneself intentionally," but various students of suicide, indeed different nations as a whole, might employ different personnel and procedures for classifying deaths quite differently. And Rockett and Smith's review of the literature suggests that nations actually do employ very different personnel and procedures. They find, for instance, that coroner systems, employing a burden-of-proof standard, are less likely to classify deaths as suicides than are medical examiner systems, which use a balance-of-probabilities standard.

operational definition, declarations of the specific ways in which the absence, presence, or the degree of presence of a phenomenon will be determined in a specific instance.

STOP AND THINK *Why would a burden of proof standard which relies on demonstrating guilt beyond a reasonable doubt lead to a smaller number of deaths classified as suicides than the balance-of-probabilities standard would?*

Whether it is a national agency deciding how deaths will be classified, or a researcher deciding how she will recognize whether one nation has a higher suicide rate than another, it is important to be explicit about one's operational definition.

An Example of Conceptualization and Operationalization

How you conceptualize and operationalize a variable can have enormous consequences for your study. Say, for instance, you'd like to update Harris and Associates' (1975) finding that, when income is controlled, life satisfaction of older people is equal to or greater than that of younger people. To do so you'd need to conceptualize and operationalize at least three variable characteristics of your subjects: their income, their life satisfaction, and their age. Let's say you want to measure the key variables by responses that individuals give to questions on a questionnaire. But what questions?

For the sake of simplicity, let's focus on two of these variables, age and life satisfaction. Gerontologists (see Atchley, 1994, pp. 6–10) will tell you there are several ways to conceptualize age: as chronological age (time since birth), functional age (the way people look or the things they can do), and life stages (adolescence, young adulthood, adulthood, and so on). Clearly, how you conceptualize age will make a huge difference for the questions you ask (chronological age is obviously not the same as functional age), but even once you conceptualize age in a particular way (say, as chronological age), your problems aren't over: You still have to operationalize age. You could, for example, operationalize chronological age by answers to any of the following questions:

1. Are you old?
2. How old are you? Circle the letter of your age category:

   a. 25 or younger b. 26 to 64 c. 65 or older
3. How old are you?

And many others. But some of these might be much more useful than others. For instance, if Harris and Associates had found that people 65 years and older had greater life satisfaction than others (when income is controlled), question 1 (Are you old?) couldn't possibly yield data that would enable a proper test.

STOP AND THINK *Why not?*

So the way you operationalize a variable needs to be relevant to your research concern. Questions 2 and 3 would permit you to operationalize age in relevant ways, but they are clearly not equivalent. One way in which they differ is in number of categories they permit. Question 2 creates three categories; Question 3 allows as many categories as the respondents provide.

Let's examine another, slightly more complicated, example of operationalization. Based on the discussion so far, you can probably think of ways to operationalize, through questions in a questionnaire, both age and income. But what about life satisfaction? Operationalizing life satisfaction is more complicated than operationalizing age, in part because conceptualizing it is difficult. Does it refer to one's satisfaction with one's health, one's family, one's friends, or what? One broad conceptualization refers to one's "satisfaction with life as a whole" (Horley, 1984), but perhaps the safest thing to say about attempts to conceptualize life satisfaction is that there's no general agreement about how it's best done.

Let's say you're satisfied, though, with "satisfaction with life as a whole" as a conceptualization of life satisfaction. How do you operationalize such a definition? Again, the social science literature provides quite a few options, but few are as simple as asking the respondent "Are you satisfied with life as a whole?" Such a simple question would miss the mark on at least two major counts: (1) Its very simplicity makes it too vague to justify its use, and (2) it fails to tap many of the ways in which one might be satisfied or dissatisfied with life. Thus, for instance, when they measured life satisfaction, Harris and Associates (1975) used a set of questionnaire items that they, in turn, had borrowed from earlier authors. The Harris instrument asks respondents to agree or disagree with the following 18 statements:

Positive Statements

I expect some interesting and pleasant things to happen to me in the future.

As I look back on my life, I am fairly well satisfied.

Compared to other people my age, I make a good appearance.

The things I do are as interesting to me as they ever were.

I've gotten pretty much what I expected out of life.

I have made plans for things I'll be doing a month or a year from now.

I am just as happy as when I was younger.

As I grow older, things seem better than I thought they would be.

I have gotten more of the breaks in life than most of the people I know.

I would not change my past life even if I could.

These are the best years of my life.

Negative Statements

My life could be happier than it is now.

In spite of what some people say, the lot of the average man is getting worse, not better.

When I look back over my life, I didn't get most of the important things I wanted.

I feel old and somewhat tired.

Compared to other people, I get down in the dumps too often.

This is the dreariest time of my life.

Most of the things I do are boring or monotonous.

index,
a composite measure that is constructed by adding scores from several indicators.

Harris and Associates made an **index**—or a composite measure that is constructed by adding scores given to individual items—of these statements using the following scoring procedure:

- 2 points were given for each agreement with a positive statement or disagreement with a negative statement.
- 1 point was given for each "not sure" or no answer.
- 0 points were given for each disagreement with a positive statement or agreement with a negative statement.

Source: Harris and Associates (1975)

STOP AND THINK *Just for fun, why don't you answer the 18 statements and then calculate a "life satisfaction" score for yourself, using Harris and Associates' index? Average scores for Harris and Associates' respondents were around 26.4. How do you compare?*

Composite Measures

composite measure, a measure with more than one indicator.

Indexes, and other **composite measures,** like Harris and Associate's life satisfaction measure, abound in the social sciences. All composite measures, or measures composed of more than one indicator, are designed to solve the problem of ambiguity that is sometimes associated with single indicators by including several indicators of a variable in one single measure.

The logic of composite measures was at work in this chapter's opening passage, in which Holmes explained to Watson how he deduced Smythe's cause-of-death. Holmes did not simply point to a single indicator (say, the absence of signs of violence) to justify his deduction. He claimed to have used several indicators (the absence of signs of violence, Smythe's size, the fact that his face remained in the bird bath until he was found, and, of course, the presence of the suicide note) in deducing Smythe's suicide. The effort to confirm the presence (or absence or the degree of presence) of a phenomenon through the use of multiple indicators occurs in both qualitative and quantitative research and is typified by Harris and Associate's life satisfaction measure.

In qualitative research (see especially Chapters 10 and 11 on Qualitative Interviewing and Observation Techniques), a researcher might or might not need to prepare measurement strategies in advance of, say, an interview or observation (see, for example, Miles and Huberman, 1994: 34–39). But the careful qualitative researcher is no less likely than Holmes was to use multiple indicators of key concepts. Thus a qualitative interviewer might judge the "emotional tone" of a respondent not only by how she responds to a question like "How are you today?," but also by how she speaks and by her appearance.

In quantitative research, especially that involving questionnaires (see Chapter 9), however, there has been much sophisticated thinking about the preparation of composite measures in advance of data collection. A full discussion of this thinking is beyond the scope of this chapter, but we would like to introduce you to it briefly. Like Harris and Associate's life satisfaction measure, all strategies for making composite measures of concepts on questionnaires involve "pooling" the information gleaned from more than one item or question. Two such strategies, the index and the scale, can be distinguished by the rigor of their construction. An index is constructed by adding scores given to individual items, as occurs in the computation of the Harris and Associate's index of life satisfaction. It is not necessary that the items "go together" in any systematic way because indexes (sometimes called indices) are usually constructed with the explicit understanding that the given phenomenon, say "life satisfaction," can have different ways of showing itself in different elements of the population. A concept that refers to a phenomenon that is expected to have different ways of showing up or manifesting itself is said to be **multidimensional.** The Consumer Price Index, for instance,

multidimensionality, the degree to which a concept has more than one discernible aspect.

summarizes the price of a variety of goods and services (for example, food, housing, clothing, transportation, recreation, and so on) in the United States as a whole in different times and for different cities at a given time. It is an index (of a multidimensional concept) because one wouldn't necessarily expect the price of individual items (say, the price of food and the price of recreation) necessarily to go up and down together. The Consumer Price Index is, however, a good way of gauging changes in the cost of living over time or space.

STOP AND THINK *Look back at the list of items in the Harris and Associates index for life satisfaction. Would you, for instance, necessarily expect respondents who agreed with the statement "These are the best years of my life" to agree with the statement "I would not change my past life even if I could"? If not, then these items are probably intended to measure distinct aspects, or dimensions, of life satisfaction.*

scales, complex measures whose construction often involves efforts to focus on unidimensional concepts, sometimes in the effort to ensure reliability and validity.

Scales, on the other hand, are usually constructed to measure a single dimension of a phenomena by combining several items. In Chapter 9, on questionnaires, for instance, we feature an article by Gray, Palileo, and Johnson that attempts to explain why some people are more likely to blame rape victims than others. Gray and her colleagues needed a measure of the degree to which respondents were likely to ascribe blame to rape victims and used an eight-item scale involving commonly held rape myths. Respondents were asked if they strongly agree (5), agree (4), are undecided (3), disagree (2), or strongly disagree (1) with statements that included these three:

1. A female who goes to the home/apartment of a man on the first date implies she is willing to have sex.
2. Many females who get raped have done something to provoke it.
3. Any healthy female can successfully resist a rapist if she really wants to.

Gray and colleagues borrowed their items from a longer list created by Burt (1980), so they had reason to believe in advance that their items "went together," and after they collected data from their respondents they were able to confirm this belief. Respondents who agreed with any one of the statements were, in fact, much more likely than those who did not agree with other statements.

Indexes and scales are not hard to make up, though assuring yourself of their validity and reliability (see later sections) can be a problem. You're frequently well advised to find an index whose validity and reliability has already been tested if one exists for the variable you're trying to measure. (See Miller, 1991, or Fischer and Corcoran, 1994, for a list of indexes that are often used in the social sciences.) Keep your eyes peeled during your literature review and you might see ways (or references to ways) that others have operationalized the concepts you're interested in. If you think the same approach might work for you, try it out (and remember to cite your source when you write up your results). No sense reinventing the wheel. If you'd like advice about creating indexes or scales of your own, we recommend either Babbie (1995) or Frankfort-Nachmias and Nachmias (1992) as a source.

Level of Measurement

Measuring a concept (for example, mode of death, or age) means providing categories for classifying subjects. But what of the nature of these categories? In this section we'd like to emphasize two rules and four possibilities for category construction. The rules should stand as self-checks any time you engage in measuring a variable concept; the possibilities have important implications.

exhaustiveness, the capacity of a variable's categories to permit the classification of every unit of analysis.

The first rule for constructing variable categories is that they should be **exhaustive.** That is, the categories should enable you to classify every subject in your study. Rockett and Smith's claim that the primary manner-of-death decision for those who would classify deaths is whether they are attributable to natural causes, accident, homicide, suicide, or, implicitly, "undetermined." In so saying, they are suggesting that all deaths can be classified into these five categories.

STOP AND THINK *Can you think of a manner of death that couldn't be "captured" by the net offered by Rockett and Smith? If you can't, their categories are exhaustive.*

Of course, the aim of creating exhaustive categories of manner-of-death doesn't require that one devise five categories. Indeed, one could obviously cover it with just two: say, suicide and nonsuicide. But you really wouldn't want to measure manner-of-death by using only one of the categories suggested by Rockett and Smith (say, "suicide"). (Generally speaking, in fact, you want your measure to involve at least two categories.) Nor would you want to use only accident and homicide as your categories for manner-of-death. Your measure wouldn't permit all deaths to be classified.

mutual exclusiveness, the capacity of a variable's categories to permit the classification of each unit of analysis into one and only one category.

The second rule for variable construction is that the categories should be **mutually exclusive.** This means you should be able to classify every subject into one and only one category. A death can be classified as a suicide, but if it is, it can't be the result of a homicide, an accident, a natural cause, or undetermined. A person can fall into the category "65 or older," but if she does, she can't also fall into the "26 to 64" category.

STOP AND THINK *Suppose you asked the question "How old are you?" and then told your respondent to circle the letter associated with his or her age category, where the categories were "a. 25 or less, b. 25 to 65, c. 65 and older." Which of the rules for variable construction would you be offending? Why?*

levels of measurement, types of categorization used in variable creation (namely, nominal, ordinal, interval, and ratio levels).

Thus, variable categories should be exhaustive and mutually exclusive, but they can be related to one another in other ways, through their **level of measurement** or their type of categorization. Let's look at four levels of measurement: nominal, ordinal, interval, and ratio.

Nominal Level Variables

nominal measure, a level of measurement that describes a variable whose categories have names.

Nominal (from the Latin, *nomen*, or name) level variables are variables whose categories have names or labels. And any variable worth the name is nominal. If your variable is "cause of death" and your categories are "suicide," "homicide," "natural causes," "accident," and "undetermined," your variable is a nominal level variable. If your variable is "gender" and your categories are "male" and "female," your variable is nominally scaled. Nominal level variables permit us to sort our data into categories that are mutually exclusive, as the variable "religion" (with categories of "Christian," "Jewish," "none," and "other") would. Other examples of nominal-level variables would be marital status, if its categories were "never married," "married," "divorced," "widowed," and "other," or national origin, if the categories included the nations in which people were born.

A nominal level of measurement is said to be the lowest level of measurement, not because nominal level variables are themselves low or base but because it's virtually impossible to create a variable that isn't nominally scaled.

Ordinal Level Variables

ordinal measure, a level of measurement that describes a variable whose categories have names and whose categories can be rank-ordered in some sensible way.

Ordinal (from the Latin, *ordinalis*, or order or place in a series) level variables are variables whose categories have names or labels and whose categories can also be rank-ordered in some sensible fashion. The different categories of an ordinal level variable represent more or less of a variable. If racism were measured using categories like "very racist," "moderately racist," and "not racist," the "very racist" category could be said to indicate more racism than either of the other two. Similarly, if one measured social class using categories like "upper class," "middle class," and "lower class," one category (say "upper class") could be seen to be "higher" than the others.

All ordinal level variables can be treated as nominal level variables, but not all nominal level variables are ordinal. If your variable is "gender" and your categories are "male" and "female," you'd have to be pretty sexist to say that one category could be ranked ahead of the other. Gender is a nominal level variable, but not an ordinal level variable. On the other hand, if your categories of age were "less than 25 years old," "between 25 and 64," and "65 or older," you could say that one category was higher or lower than another. Age, in this case, is ordinal, though it could also be treated as nominal. Ordinal level variables guarantee that categories are mutually exclusive and that the categories can be ranked, as the variable "team standing" (with the categories of "1st place," "2nd place," and so on) does in many sports.

STOP AND THINK *What, do you think, is the scale of measure of the life satisfaction index that Harris and Associates (1975) used? Why?*

interval measure, a level of measurement that describes a variable whose categories have names, whose categories can be rank-ordered in some sensible way, and whose adjacent categories are a standard distance from one another.

Interval Level Variables

Interval level variables are those whose categories have names, whose categories can be rank-ordered in some sensible way, and whose adjacent categories are a standard distance from one another. Because of the constant distance criterion, the categories of interval scale variables can be meaningfully added and subtracted. SAT scores constitute an interval scale variable, because the difference between 550 and 600 on the math aptitude test can be seen as the same as the difference between 500 and 550. As a result, it is meaningful to add a score of 600 to a score of 500 and say that the sum of the scores is 1100. Similarly, the Fahrenheit temperature scale is an interval level variable because the difference between 45 and 46 degrees can be seen to be the same as the difference between 46 and 47 degrees.

All interval level variables can be treated as ordinal and nominal level variables, even though not all ordinal and nominal level variables are interval level.

Ratio Level Variables

ratio measure, a level of measurement that describes a variable whose categories have names, whose categories can be rank-ordered in some sensible way, whose adjacent categories are a standard distance from one another, and one of whose categories is an absolute zero point—a point at which there is a complete absence of the phenomenon in question.

Ratio level variables are those whose categories have names, whose categories can be rank-ordered in some sensible way, whose adjacent categories are a standard distance from one another, and one of whose categories is an absolute zero point—a point at which there is a complete absence of the phenomenon in question. Age, in years since birth, is often measured as a ratio level variable (with categories like 1, 2, 3, 4 years, and so on) because one of its possible categories is zero. Other variables like income, weight, length, and area can also have an absolute zero point and are therefore ratio level variables. They are also interval, ordinal, and nominal level variables. For practical purposes, however, interval and ratio level variables are similar. But this raises the question of what practical purposes there are for learning about levels of measurement in the first place, and that's the topic of our next section.

STOP AND THINK *When a coroner has to classify a death as a suicide or not, what level of measurement is she dealing with? Why? When a cross-national researcher compares the suicide rate for males in the United States (19.9 per 100,000 in 1990) to that of males in the Soviet Union (37.4 per 100,000 in 1990), what level of measurement is he dealing with? Why?*

The Practical Significance of Level of Measurement

Perhaps the most practical reason for being concerned with levels of measurement is that frequently researchers find themselves in the position of wanting to summarize the information they've collected about their subjects and to do that they have to use statistics (see Chapter 15). An interesting thing about statistics, however, is that they've all been designed with a particular level of measurement in mind. Thus, for instance, statisticians have given us three ways of describing the average or central tendency of a variable: the mean, the median, and the mode. The mean is defined as the sum of a set of values divided by the number of values. The mean of the ages 3 years, 4 years, and 5 years is (3+4+5)/3=4 years. It's pretty clear from the definition of the mean that it's been designed to summarize interval-level variables. If age were measured only by categories "under 25 years old," "25 to 64," and "65 and older," you couldn't hope to calculate a mean age for your subjects. The median, however, has been designed to describe ordinal level variables. It's defined as the middle value, when all values are arranged in order. You could calculate a median age if your (five) respondents reported ages of, say, 3, 4, 5, 4, and 4 years. (It would be 4, because the middle value would be 4, when these values were arranged in order—3, 4, 4, 4, 5). You could calculate a median because all interval level variables (as age would be in this case) are ordinal level variables. But you could also calculate a median age even if your categories were "under 25 years old," "25 to 64," and "65 and older."

STOP AND THINK *Suppose you had five respondents and three reported their ages to be "under 25 years old," one reported it to be "25 to 64," and one reported it to be "65 and older." What would the median age for this group be?*

A third measure of central tendency or average, the mode, has been designed for nominal-level variables. The mode is simply the category that occurs most frequently. The modal age in the previous Stop and Think problem would be "under 25 years old" because more of the 5 respondents fell into this category (three) than into either of the other categories. The modal age for a group with ages 3, 4, 5, 4, 4, would be 4, for the same reason. But you can also calculate the modal gender if, of your five respondents, 3 reported they were female and 2 reported they were male. (It would be "female.") And you wouldn't have been able to calculate either a median or mean gender.

STOP AND THINK *Suppose a coroner classified 2 deaths as suicides and 157 deaths as nonsuicides. Which of the measures of central tendency could you use to describe the cause-of-death classification of this coroner? Why?*

The upshot is that the scale of measurement used to collect information can have important implications for what we can do with the data once they've been collected. In general, the higher the level of measurement, the greater the variety of statistical procedures that can be performed. As it turns out, all statistical procedures scientists use are applicable to interval (and, of course, ratio) data; a smaller subset of such procedures is applicable to ordinal data; a smaller subset still to nominal data.

What's the practical implication of all this? Perhaps the most important implication is that, given a choice and other things being equal, you'd probably want to measure a variable in a higher level of measurement than in a lower one. You can always, for instance, take information collected at the interval level (say, age in years) and reduce it to an ordinal level (say, age in three categories, "under 25 years," "25 to 64," and "65 and above") and use statistics that are appropriate for the ordinal level. The flexibility of future data analyses is maximized when variables are measured at higher levels.

Quality of Measurement

Our discussion of level of measurement suggests one standard for judging the quality of a measurement scheme: its precision or exactness. Generally speaking, a variable, like age, that is measured as an interval variable is more precisely measured than one that is measured as an ordinal variable. It is more precise to say, for instance, that a person is 83 years old than to say that he is "65 or more." Generally speaking, more precise measures are preferable to less precise ones. This is another reason for choosing higher levels of measurement when we can.

Reliability

Let's return to the two qualities of measurement that figured so prominently in Rockett and Smith's article on suicide misclassification: reliability and validity. You'll recall that the reliability of a measurement strategy is its capacity to produce consistent results when applied time after time. A yardstick consistently tells Roger that his four-year-old daughter is about 36 inches tall, so it's pretty reliable (even if it's possibly inaccurate—we'd have to check whether its first inch or so have been sawed off). There are lots of ways of checking the reliability of a measurement strategy. One is to do what Roger's done with his yardstick: to have it measure (essentially) the same object(s) two or more times in close succession. This **test-retest** method is not mentioned by Rockett and Smith, but one could easily imagine how it might be employed to check the reliability of one nation's suicide classification system. One could ask medicolegal officials in one country to reassess their own manner-of-death assignments for a sample they'd already examined. If essentially the same assignments were made both times, one could assume the method was reliable.

test-retest method, a method of checking the reliability of a test that involves comparing its results at one time with results, using the same subjects, at a later time.

The major problems with test-retest methods of reliability assessment are twofold. First, the phenomenon under investigation might actually change between the test and the retest (for example, Heise, 1969). Roger's daughter might actually experience a growth spurt between today's measurement and tomorrow's. Second, earlier test results can influence the results of the second (for example, Nunnally, 1964). If medicolegal officials were asked to reassess their own manner-of-death assignments for a sample they'd already examined, for instance, they might be tempted to make similar assignments the second time just to show how good their first ones were.

alternative-form method, a method of checking the reliability of a measure that involves comparing its results with those received using a different form of the measure.

The **alternative-form method** is one way to deal with the problems of the test-retest method. The alternative-form test differs from the test-retest method in that a different form of the first test is administered on the second testing. Thus Rockett and Smith speak of one study (Ross and Kreitman, 1975) that compared what Danish and Scottish officials made of one another's cases. This study implies that the Danish and Scottish officials use similar, though not the same, methods and that their tests are, in effect, alternative forms of one another. When Danish officials obtain essentially the same results as Scottish officials do, they can be seen as confirming the reliability of the Scottish approach, and vice versa.

split-half method, a method of checking the reliability of several measures by dividing the measures into two sets of measures and determining whether the two sets are associated with each other.

Closely related to the alternative-form method is the **split-half method,** which relies on making more than one measurement of a phenomenon at essentially the same time. This method is grounded in the same logic as the complex measures (indexes and scales) mentioned earlier: We frequently want to take more than one measure of a phenomenon. The scale Gray and her colleagues used to measure the degree to which respondents tend to blame rape victims, for instance, used eight questionnaire items, each meant to tap the same attitude. The split-half method of checking the reliability of this scale would entail randomly assigning those eight items to two sets of four and using statistics to see if the two sets corresponded in the way they classify respondents to the survey. Gray and colleagues report, for instance, that the statistic they used, Cronbach's alpha, is .84. Cronbach's alpha is actually a way of estimating the average correspondence of all split-half combinations of items. An alpha of .70 or greater suggests that items in a scale "go together" very well, so an alpha of .84 indicates high split-half reliability.

Another common way of checking the reliability of a measurement strategy is to compare results obtained by one observer with results obtained by another using exactly the same method. We'll see this **interobserver reliability** or **interrater reliability** method figure extensively in the focal research of later chapters. In Chapter 8, on experiments, Adler and Foster indicate that their independent codings of student writing by how much "caring" is expressed were reliable because they agreed on codes about 95 percent of the time. In Chapter 13, on content analysis, Clark, Lennon, and Morris report that, when trying to assess the traits of individual characters in children's books using a pre-defined measurement scheme, they agreed 76 percent of the time.

Interobserver (or interrator) reliability method, way of checking the reliability of a measurement strategy by comparing results obtained by one observer with results obtained by another using exactly the same method.

STOP AND THINK

Suppose you found a way of measuring homophobia with ten questionnaire items and that you wanted to use those questions in a telephone survey relating homophobia to knowledge about AIDS. Describe a way in which you might check the reliability of the homophobia measure you were using.

Validity

Rockett and Smith are more impressed with the reliability of strategies used to measure suicide than they are with their validity. Validity refers to how well a measurement strategy taps what it should be measuring. You might wonder, how can you possibly tell whether a strategy is valid? After all, most measurement in the social sciences is indirect. Asking someone whether she thinks that she makes a good appearance compared with other people her age, for instance, isn't exactly the most direct way of determining whether she has high life satisfaction, even though this question figures into some of the most frequently used measures of life satisfaction in the gerontology literature (Harris and Associates, 1975; Liang, 1984; Neugarten, Havighurst, and Tobin, 1961).

Perhaps the most satisfying (and maybe the best) check on validity is this question: Does this measurement strategy *feel* as if it's getting at what it's supposed to? If a measure seems as if it measures what you and I (and maybe a whole lot of others) believe it's supposed to be measuring, we say it

face validity,
the degree to which a measure seems to be measuring what it's supposed to be measuring.

has **face validity,** and we can feel pretty comfortable with it. The answer to the question "How old are you?" seems like a pretty good measure of age, and therefore has high face validity. On the other hand, the question "What, in your opinion, is the proper role of the military in foreign affairs" would yield a poor measure of age, and therefore has low face validity. And, generally, if most people who know something about a concept (say, life satisfaction) feel that a given set of questions (say, those used by Harris and Associates) "gets at" that concept in an important way, the measure is said to have face validity. (In fact, if they also agreed that the set of questions asked by, say, Harris and Associates covered the usual range of "content" implied by the concept of "life satisfaction" [for example, appropriate zest for life, appropriate mood tone, etc.], they will also have affirmed the set's **content validity.**)

content validity,
how well a measure covers the range of meanings associated with a concept.

The obvious problem with face validity is that it is so subjective. Its virtue is that it is simple to use and simple to understand and that, once you "look behind" more objective procedures, they can look a bit subjective too. Establishing **criterion** or **predictive validity** for instance, involves comparing the results of your measurement scheme with behaviors you'd expect to be associated with it (Carmines and Zeller, 1979; Nunnally, 1978). If, for instance, you compared the SAT scores of high school seniors with their performance in college (that is, something those scores should "predict to") and found that students with higher test scores did better in college, then you could say the SAT test had predictive validity.

criterion or predictive validity,
how well a measure is associated with behaviors you'd expect it to be associated with.

STOP AND THINK

Can you see why checking the results of an SAT test against a criterion such as performance in college still involves an element of subjectivity? (Hint: What does it mean to do "well" in college?; how does one measure such performance?; doesn't judging the validity of SAT scores by this criterion, in turn, depend on the criterion's face validity?)

A major difficulty in validating social science measures using behavioral criteria, however, is the problem of finding behaviors that they obviously "predict to" (Carmines and Zeller, 1979). Many of our indicators are meant to measure abstract concepts (for example, life satisfaction or self-esteem) that are valued primarily because of their expected association with other abstract concepts (for example, happiness or sense of control). As a result, social scientists tend to pay more attention to the **construct validity** of their measures than to their predictive validity. Construct validity refers to how much a measure of one concept is associated with a measure of another concept that some theory says it should be associated with. If, for instance, you believe, after reading Arlie Hochschild's (1989) *The Second Shift*, that the marital satisfaction of wives is augmented when their husbands share domestic chores (such as childrearing, cooking and house-cleaning), then you can reasonably check one's measure of "sharing of chores" by seeing whether it is related to some plausible measure of "wife's marital satisfaction." Rockett and Smith checked the construct validity of suicide measures used in different countries. Rockett has confided that he put two propositions from the epidemiological literature together and came up with a

construct validity,
how well a measure of a concept is associated with a measure of another concept that some theory says the first concept should be associated with.

"theory" that led him to expect that the ratio of combined deaths from accidental drownings, accidental poisonings, and injuries of undetermined intent to suicides would not vary much from nation to nation if suicides were being validly measured in all nations. The first of these widely accepted propositions is that most nonsuicides get counted as nonsuicides, but that many suicides don't get counted as suicides. The second proposition is that the suicides that don't get counted as suicides are most likely to get classified as accidental poisonings, accidental drownings, and injuries of undetermined intent. From these two propositions, Rockett deduced that if the ratio of accidental drownings, accidental poisonings, and injuries of undetermined intent to suicides varied greatly from one nation to another, it was an indication that suicides were being disproportionately misclassified in some nations and hence not validly classified across all nations. In fact, he found this ratio to vary enormously and so concluded that the validity of cross-national measurement is a problem.

Whether or not the logic of Rockett and Smith's argument makes sense to you, we think you'll agree with our general point that the logic of construct validation retains a degree of subjectivity. You first have to accept the theory that leads to the expectation of an association between one indicator and another, then you must believe that the second indicator has been well measured (just as you must believe in criterion validation that the predicted behavior has been well measured). Nonetheless tests like Rockett and Smith's do provide some evidence (if, perhaps, no certain proof) that a concept is being validly (or invalidly) measured, where no other evidence had existed before.

STOP AND THINK *What are two ways you might check the validity of Harris and Associates' index of life satisfaction outlined earlier?*

Summary

Measurement means classifying units of analysis by categories to represent a variable concept. Measuring abstract concepts involves conceptualizing, or clarifying, what we mean by them, and then operationalizing them, or defining the specific ways we will make observations about the concepts in the real world.

The first step of measurement is to construct a conceptual definition of a concept. The second step is to operationalize the concept by identifying indicators with which to classify units of analysis into two or more categories. The categories should be exhaustive, permitting you to classify every subject in the study, and mutually exclusive, permitting you to classify every subject into one and only one category. Depending on the complexity of the concept involved, you can use simple or complex measures of it.

There are three significant levels of measurement (nominal, ordinal, interval-ratio) in the social sciences. In general, the data analyses that can be used with a given set of variables depend on the level of measurement of the variables.

The best measurement strategies are reliable and valid. (Rockett and Smith, for instance, are less impressed with the comparative validity of strategies used to measure suicide than with their reliability.) Reliability refers to whether a measurement strategy yields consistent results. We can examine the reliability of a measurement strategy with a test-retest, alternative-form, split-half, or interrater reliability check. Validity refers to whether a measurement strategy measures what you think you are measuring. We can examine the validity of a measurement by considering its face validity—by seeing how we and others feel about its validity—or by examining its predictive or criterion-related validity—by seeing how well results we obtain with it correlate with some behavior we think it should predict to—or by investigating its construct validity—by seeing how well it correlates with some other measure that a theory leads us to believe it should correlate with.

SUGGESTED READINGS

Carmines, E. G., and R. A. Zeller. 1979. *Reliability and validity assessment.* Beverly Hills, CA: Sage.

This will be one of the thinnest books on your library's shelves, but it's full of information about checking the reliability and validity of measurements.

Fischer, J. and K. Corcoran. 1994. *Measures for clinical practice: A sourcebook.* New York: Free Press.

An excellent source of scales and indexes used in clinical practice.

Jary, D., and J. Jary. 1991. *The HarperCollins dictionary of sociology.* New York: HarperCollins.

A good source of agreed-upon meanings of sociological concepts.

Miller, D. C. 1991. *Handbook of research design and social measurement.* Newbury Park: Sage.

Part 6 presents an extraordinary list of scales and indexes used in the social sciences. Recommended if you want conventional operational measures of social status, group structure and dynamics, organizational structure, community, social participation, leadership in work organizations, morale and job satisfaction, powerlessness or achievement, marital adjustment, personality, and others.

EXERCISE 6.1

Seeing Measurement at Work

This exercise challenges you to examine critically measurement procedures as they have been executed in research by other social scientists. Read one of the "focal research" pieces in chapters, 7, 8, 9, 12, or 13 of this book. Select one of the concepts in this research and describe how that concept is conceptualized and operationalized. Decide whether the level of measurement was nominal, ordinal, or interval-ratio. Assess the reliability and validity of the measure.

1. Which piece of "focal research" did you choose?

2. What concept did you focus on?

3. Did the author(s) conceptualize this concept clearly?

   If so, what is the conceptual definition?

4. How did the author(s) operationalize the concept?

5. Was the resulting variable measured at the nominal, ordinal, or interval ratio level? Explain your answer.

6. Assess the reliability of the measurement used by the author(s).

7. Assess the validity of the measurement used by the author(s).

EXERCISE 6.2

Putting Measurement to Work

This exercise gives you the opportunity to develop a measurement strategy of your own.

1. Pick an area of the social sciences you know something about (for example, aging, marriage and the family, crime and criminal justice, social problems, social class, what have you).

2. What is a concept (for example, attitudes towards the elderly, marital satisfaction, delinquency, class) that receives attention in this area and that interests you?

3. Describe how you might conceptualize this concept.

4. Describe one or more measurement strategies you might use to operationalize this concept.

5. What is the level of measurement of this concept given the operationalization you suggested?

6. Discuss the reliability of your measurement strategy.

7. Discuss the validity of your measurement strategy.

EXERCISE 6.3

Possible Alternatives to Suicide Classification: A Small Cross-National Study

This exercise gives you a chance to follow up on the chapter's focal research in a way that Rockett and another co-author, Billy M. Thomas, have already done (unpublished manuscript). Rockett and Smith found that some countries might misclassify the suicides of older women as, say, accidental poisonings, accidental drownings, and "other violence" (or injuries of undetermined intent). Thomas and Rockett have examined the hypothesis that countries with lower suicide rates will have higher combined rates of accidental poisonings, accidental drownings, and "other violence." The following data (from the World Health Organization's [1992] *World Health Statistics Annual*) permit you to examine this hypothesis for four nations around 1990 with cause-specific mortality rates for women 75 and older.

Country	*Suicide Rate**
Italy	9.5
Norway	7.9
France	1.1
Mexico	0.8

**All rates are presented as the annual number of deaths due to a given cause per 100,000 women 75 years and older.*

1. Rank these four countries (1 through 4) from the country that has the highest suicide rate to the one that has the lowest. List them in order from highest to lowest.

Country	*Accidental Poisoning Rate*	*Accidental Drowning Rate*	*Other Violence Rate*	*Sum of these Rates*	*Rank of Country's Combined Rate*
Italy	1.8	1.1	1.5		
Norway	1.1	0.5	1.1		
France	4.3	1.3	8.4		
Mexico	4.1	1.3	4.6		

2. Add up the rates attributed to accidental poisoning, accidental drownings. Ranking the four countries from the one that has the highest combined rate to the country that has the lowest, list the names of the countries and their combined rates in order from the highest rate to the lowest rate.

3. a) Which two countries rank highest in suicide rate?

   b) Which two countries rank lowest in the combined "alternatives" rate?

c) Which two countries rank lowest in suicide rate?

d) Which two countries rank highest in the combined "alternatives" rate?

4. Based on the data you've been provided, how would you evaluate the hypothesis with which you began this exercise? Does it seem to be largely correct or largely incorrect? Why?

5. How do you think Rockett and Smith might tie your findings into their discussion of the validity of suicide classification? (Hint: How might they use the concept of sensitivity to explain these results?)

EXERCISE 6.4

Suicide as Cause of Death by Age: An Internet Exercise

Find the web page of the Centers for Disease Control and Prevention (www.cdc.gov). Answer the first four questions using information on this web page. (Hint: you might want to pursue, first, the "Data & Statistics" option, then the "Health Statistics" options, then the "Fastats" option, then the "deaths/mortality" option, then the "ten leading causes of death by age, race, and sex" option). Then speculate a bit and answer the last question.

1. What is the number one cause of death for all people in the U.S.?

2. What is suicide's rank as a cause of death for all people in the U.S?

3. What is the number one cause of death for 15-24 year olds in the U.S.?

4. What is suicide's rank as a cause of death for 15-24 year olds in the U.S.?

5. Working inductively, try to think of reasons for the discrepancy between suicide's rank as a cause of death for 15-24 year olds and its rank in the total population. Do you think that part of the discrepancy might be due to measurement difficulties? What else could account for the difference among age groups?

7 Cross-Sectional, Longitudinal, and Case Study Designs

Ralph Thomas,[1] aged 78, lives in a small apartment in the city-owned Irma C. Rudolph Housing for the Elderly. Almost all of their neighbors were senior citizens when Ralph and his wife moved to the public housing complex twelve years ago. Since his wife's death, there have been changes in the composition of the residents. In addition to the seniors, Ralph's neighbors now include a middle-aged woman with mild retardation, a legally blind 25-year-old man, and a man in his forties who's been diagnosed with chronic mental illness and alcoholism.

Ralph often complains to his daughter about the quality of life in his housing complex. Sometimes he hears loud arguments between residents, and he's seen one neighbor urinating in the bushes on occasion. He is disturbed when some residents congregate in the courtyards between buildings after dark and create noise and litter. Although he likes many of the services that the housing complex provides and he wants to continue living independently, Ralph and his daughter have discussed the possibility that he will move into a room in her house.

What has happened to the environment in the Irma C. Rudolph Housing for the Elderly in recent years? Are Ralph's feelings shared by other residents? If the situation has deteriorated, what has caused the change? These are the kinds of questions that gerontologist Rachel Filinson had when she began the study that is featured in the focal research section of this chapter.

Filinson wanted to describe the current environmental quality of public and publicly subsidized housing for the elderly, but she also wanted to explain the causes and effects of environmental quality. She selected a specific study design and method of data collection to answer her research questions. In this chapter, we'll see the important connection between research questions and the selection of a study design.

[1] Ralph Thomas is a composite, created from the backgrounds and experiences of several senior citizens.

Introduction

In the first six chapters of this book, we discussed the planning of a research strategy. We talked about the purposes of research, theories, hypotheses, concepts, variables, research questions, and ethical considerations. We also addressed selecting populations and samples, measuring concepts, and the practical concerns of time, money, and access to data. At this point, we're almost ready to focus on actually *doing* research. But, before considering ways to collect and analyze data, we'll describe the last part of the planning process—selecting a study design.

Study Design

study design, a research strategy specifying the number of cases to be studied, the number of times data will be collected, the number of samples that will be used, and whether or not the researcher will try to control or manipulate the independent variable in some way.

causal hypothesis, a statement that theorizes that the independent variable causes or affects the dependent variable.

A **study design** results from several interconnected decisions about research strategies. To do research, data must be collected at least once, but it can be collected two, three, or more times. Therefore, one important decision concerns the *number of times* to collect data. A second decision focuses on the *number of samples* needed. For studies seeking to test a hypothesis, the third decision is related to the independent variable. Recall that if a researcher wants to test a **causal hypothesis,** then he or she is speculating that one variable, the independent variable, can be identified as the *cause* of another variable, the dependent variable. If a researcher has constructed a causal hypothesis, then it's necessary to decide if it is appropriate to *try to control or manipulate the independent variable* in some way. In the study we will be describing in Chapter 8, for instance, Emily and a colleague decided it would be useful for one group of students to have the experience of reading and discussing stories involving "caring" behavior while a second group of students had a different experience. The independent variable was whether or not students were exposed to stories focusing on caring, which was a variable that the researchers thought it best to "control." The research design described in Chapter 8 is an experiment, a study design in which the "control" of an independent variable is integral.

Ethical and practical matters are important influences on study design choices. Ethical issues can be raised, for example, if a researcher wants to study a sample more than once, because keeping the names and addresses of study participants might put them at risk of being identified by others. Similarly, a researcher's ability to control or manipulate an independent variable is an important responsibility because it can affect people's lives.

Researchers need to consider if a study is practical—that is, "do-able" with the available resources. Because it is almost always easier, less time consuming, and less expensive to collect data once, from one sample, and to measure rather than control or manipulate independent variables, using costly research methods should be justified. There must be important methodological considerations when making study design choices that are less practical or raise more ethical concerns.

Connections Between Purposes of Research, Theory, and Study Design

Decisions about study design are based partly on the purposes of the research, the researcher's interest in testing causal hypotheses, and the more general use of theory. In Chapter 1, we described five purposes of research: exploration (breaking new ground), description (describing a specific group or groups), explanation (explaining why cases vary on one or more variables), critique (critically assessing an aspect of the social world), and practical applications (producing findings useful for the immediate future). In Chapter 2, we defined theory as a plausible story about the way people act, interact, or organize themselves. In contrasting the deductive and inductive approaches to theory, we presented the deductive approach as starting with a theory from which a testable hypothesis (a statement of how two or more variables are expected to relate to each other) is generated. We compared this with inductive research, which starts with the actual collection of data before moving to empirical generalization and, sometimes, to theory building. Our point in this chapter in recalling the various purposes of research, as well as the distinction between deductive and inductive approaches to research, is to note that these purposes and approaches have important implications for research design decisions.

Connections between theory and design can be found in all the focal research projects presented in this text. At this point, however, we'll limit ourselves to comments about the theory-design connections in two of them: Hoffnung's research on college graduates (Chapter 3) and Weitz's study of people living with HIV disease (Chapter 4). Michele Hoffnung wanted to describe and explain changes in women's lives as they moved from college campuses into the workplace and as they shifted from transitional to more permanent social relationships. She used a deductive approach by relying on well-established developmental theories to frame her research questions. As we'll see in this chapter, she selected a longitudinal or overtime design to be able to describe and explain the changes in the lives of one sample of college graduates.

Rose Weitz began her study at a time when there was little social science research on AIDS. Exploring a new area, Weitz didn't begin with a theory or a hypothesis but instead worked inductively. To meet the goal of developing an understanding of people's self-images, relationships and daily lives within the context of HIV disease, Weitz talked to people with HIV and looked for patterns in their experiences. Because of the ethical problems involved in keeping records of participants' names, the practical problems of the impending disability or death of many of the study's participants, and the fact that the interview process was emotionally and psychologically difficult for both the interviewer and the respondents, Weitz decided to interview the majority of her study's participants only once.

STOP AND THINK *Suppose you are a researcher interested in comparing how elderly tenants living in different kinds of housing feel about their housing environment. What do you think* *the purpose of your research would be? Do you think you'd construct a hypothesis? Do you think you'd need to collect data more than once?*

In the focal research for this chapter, gerontologist Rachel Filinson compared different housing environments. Filinson wanted to know if elderly residents living in housing limited to senior citizens perceived the environmental quality as better than did those seniors living in housing with age-mixed residents. Filinson was interested in the patterns and the causes of differences in environmental quality. Like many scholars, Filinson had an important topic, but a "shoe string budget." We'll see how her research strategy enabled her to answer her research questions with limited resources.

FOCAL RESEARCH

The Effect of Age Desegregation on Environmental Quality for Elderly Living in Public/Publicly Subsidized Housing[2]

by Rachel Filinson

Introduction

Public housing for the elderly has increasingly altered its demographic profile to include substantial proportions of nonelderly, often handicapped persons. The changes reflect the impact of a number of trends including a serious shortage in public housing stock available to younger populations (Nelson, 1991), the continuing deinstitutionalization of the mentally ill, broadened legal definitions of "handicapped" to include the mentally ill and substance abusers, and legislated mandates for programs to be fully accessible to the handicapped.

Public housing authorities have responded by making more of their stock of "elderly" housing available to the nonelderly. For example, according to the Providence Housing Authority's socioeconomic profile for January, 1990, the proportion of nonelderly in facilities for the elderly under their aegis ranged from 21 percent to 49 percent, with over one half of these persons being mentally or physically disabled (Providence Housing Authority, 1990). Although housing authorities did not undertake systematic aggregate analyses in earlier years, there is overwhelming consensus that "elderly" housing was predominantly occupied by the elderly in the past.

The situation is more ambiguous for housing subsidized through the U.S. Department of Housing and Urban Development (HUD) through direct

[2] This is an abbreviated version of the article of the same name published in *Journal of Aging & Social Policy*, 1993, Vol. 5, No. 3, 77–93 with permission.

loans for construction (Section 202) or subsidized to a fair market rent (through Section 236 or Section 8). These housing units are not required to serve nonelderly if their application to HUD specified the intent to serve only the aged population. Precedent setting lawsuits and the Cranston-Gonzales National Affordable Housing Act (1990) allows project owners to limit admission to persons of similar disabilities and assures the future separation of housing provided for the elderly and for the disabled.

Subsidized housing for the elderly has been able to retain its age homogeneous character, with only 5 percent nonelderly in Section 202 "elderly" housing nationwide (Select Committee on Aging, 1989). The wedge for nonelderly residents to gain entry into subsidized housing may arise from legal questions concerning tenant selection criteria, such as the policy of screening applicants on health-related issues or the ability to function independently.

Even without a legal mandate, intergenerational conflict over scarce resources may push subsidized housing to consider opening their doors to nonelderly in greater numbers. Those with vacant apartments have already done so voluntarily. The fallout of such a change in states like Rhode Island would be significant since the majority of elderly and handicapped housing units are privately owned with Section 8 rental assistance (Hedge, 1990). A liberalized admissions policy would likely further exacerbate the competition for limited places, especially in metropolitan areas in the Northeast, where elderly and handicapped housing are scarcest and waiting lists longest (Select Committee on Aging, 1989).

Socio-environmental theory posits that the individual well-being of older adults depends on their interaction with their environment, which is in turn influenced by age homogeneity (Gubrium, 1973). However, there is only unscientific evidence to suggest that the trend toward age-mixed populations is detrimental to the elderly. The popular media has highlighted an emerging crisis in public housing (Crombie, 1992). Service providers to the elderly in Rhode Island have gone so far as to suggest bringing a class action suit of elder abuse against the housing authorities (Human Services Conference, 1990), while managers and administrators have acknowledged that age desegregation policies may compromise the living conditions of the elderly if no services are added (O'Rourke, 1991).

What is not clear is how the elderly tenants themselves feel about the changes in their living environment. This research attempts to fill the gap by surveying a random sample of elderly tenants about their satisfaction with their housing environment. By comparing elderly living in public housing (where the population is already mixed) to those elderly persons living in Section 202 housing (where the population is still predominantly homogeneous), the impact of mixed housing on environmental quality can be assessed.

Method

The sample consisted of 25 elderly persons from three of the seven sites for elderly housing operated by the Providence (Rhode Island) Housing Authority and 25 elderly persons from three of the 38 subsidized (with rental assistance for low-income residents) housing sites in Providence. The three public housing sites were selected because they had a substantial mixed population and had a social worker on site, a "best-case" scenario for age-mixed housing. The number of elderly units in subsidized (Section 202) housing facilities in Providence ranges from four to 204; the three facilities chosen were those primarily associated with housing the elderly and were located in an urban area similar to that of the public housing sites.

Interviews with housing managers prior to sample selection confirmed that public housing sites could validly be used as a proxy for age desegregation and subsidized housing sites as a proxy for age segregation. In the public housing sites, between 44 percent and 70 percent of the residents were younger than 65 and between 20 percent and 48 percent were under 55. None of the subsidized housing sites had more than 30 percent of residents younger than 65, and two had negligible proportions of younger residents.

The managers generated lists of elderly tenants from which random samples were systematically selected and contacted by the interviewers. Tenants who refused to be interviewed were replaced with the next person on the list. The final selection of eight or nine tenants from each site represented approximately 13 percent of the total elderly population of each building. The overall response rate was 74 percent, but it varied from 64 percent to 89 percent for the six housing sites.

The interview schedule was adapted from the Multiphasic Environmental Assessment Procedure (Moos and Lemke, 1984) and designed to examine a number of dimensions of environmental quality. These included the following:

1. *Privacy*. To what extent do tenants feel they are able to control others' access to them, to be isolated and physically distant if they choose?
2. *Stimulation*. To what extent does the environment encourage expression and activity, stimulating tenants' understanding and identification of a place and ease their orientation to it?
3. *Legibility*. To what extent does the environment facilitate or impair tenants' understanding and identification of a place and ease their orientation to it?
4. *Accessibility*. To what extent have barriers impeding the tenants' ability to utilize their physical space been reduced?
5. *Territoriality*. To what extent do tenants feel they have control over physical space and conditions of life?
6. *Social networks*. To what extent does the environment promote relationships with others sharing it? To what extent is a feeling of community fostered by the environment?

The interview was structured so that subjective perceptions of alienation, isolation, voluntary reductions in activity, and so forth were first elicited, followed by probes into the respondents' accounting of detriments in the quality of their environment. Open-ended questions were used to see if respondents spontaneously attributed them to desegregation.

Information was also collected on basic demographic data, features of the housing unit, and measures of health and social support factors that could differentiate public housing and subsidized housing tenants and influence their environmental satisfaction independent from the population mix. In addition, the interview explored the provision or coordination of services by the housing facility, since the presence of services could foster positive attitudes and have a significant explanatory value.

Comparability of Public Housing and Subsidized Housing Tenants

Demographic and Housing Unit Variables

There were no statistically significant differences between the public housing and subsidized housing respondents on marital status (one half being widowed), gender (most being female), age, broad occupational category, or speaking a language other than English. There was more ethnic diversity among public housing tenants but the differences were not significant; in both subsamples, three quarters of the respondents were white. Statistically significant differences emerged on the variables of education, being a native-born American, being a recipient of Medicaid, and being a recipient of SSI. Fewer years of schooling, being foreign born, receiving Medicaid, and receiving SSI were more often correlated with being tenants of public housing. There were no significant differences between the two subsamples with respect to living alone, number of years living at the site, reasons for moving to the site, and size of the unit.

Health and Social Interaction Variables

Most respondents in both groups did not require assistance in any area, except for help with shopping which was needed by nearly one half of the respondents. The two groups were not distinguishable with regard to visual impairment, but subsidized housing tenants were more likely to report hearing problems.

A striking difference between the groups arose with respect to social interaction. Public housing tenants were less likely to have frequent interaction with children and friends, and they less frequently attended a religious service or party. Statistically significant differences were not found with respect to interaction with other relatives, eating out, going to a movie, volunteer activities, going on a trip, or attending a senior center, but there was a consistent pattern of greater activity enjoyed by the subsidized housing tenants.

Dimensions of Environmental Quality

The two kinds of housing tenants did not differ with respect to privacy, stimulation, and legibility, and they were fairly similar in their perceptions of accessibility. The majority of all respondents felt they had enough privacy, were stimulated by their environment, and had no difficulty in getting around their housing site. Most reported few difficulties in getting in or out of the building and apartment, although access problems were more common among subsidized housing units.

The two groups were also similar with respect to ease in making friends and the number of friends they had who lived at the housing site. However, subsidized housing tenants reported a larger number of confidantes and more involvement in assisting and being assisted by other residents than did public housing tenants.

Differences between the two groups, however, were evident in territoriality—the dimension which includes measures of disturbing behaviors by other residents and measures of the respondents' subjective evaluation of control over their environment. Being a public housing tenant was significantly correlated with being subjected to disturbing behavior by other residents, such as verbal abuse, physical attack, drunkenness, and behaviors associated with taking too much medication. Public housing respondents were less likely than subsidized housing respondents to feel that they had enough influence or that rules and regulations were effectively implemented so as to prevent or solve troublesome episodes. They were more likely to report arguments between tenants which either could not be resolved or led to warnings of eviction.

Can these negative findings for public housing tenants be attributed to problems created by age desegregation? Is it possible that younger residents were able to increase their numbers in public housing because it had become increasingly undesirable? In other words, could age desegregation be the *effect* and environmental decline be the *cause*? Could those who had either the social or economic resources to move elsewhere already have been skimmed from the elderly population residing in public housing? The qualitative data from interviews indicate that behavior and attitudes changed *in response* to a social climate regarded as intimidating or unpleasant because of the presence of younger "intruders." Following is a sampling of the comments offered during interviews with public housing tenants in which problems created by younger tenants are explicitly mentioned:

Respondent (R) 2: The housing was built for the elderly. Now people in their twenties are here. They throw every Tom, Dick, and Harry here.

Interviewer: Have you ever tried to do anything about [it], for instance, complaining to the manager?

R 2: No, what can they really do?

R 22: [I'm] afraid to go out at night because of strange people . . . Two or three a.m., fire alarm goes off—happened five times . . . should be all elderly. We would feel safer.

R 18: As quick as maintenance does work, it is destroyed . . . Just last week [a resident attacked another] and the man needed stitches . . . People are intoxicated, particularly on elevators. I think a lot of the younger ones are either on drugs or from [the mental health center]. Disrespectful to older people.

Interviewer: Has any resident intentionally taken too much medicine?

R 22: Yes, the drug addict in Room 409 [refers to younger tenant] sells.

Interviewer: Have you ever tried to do anything about [these behaviors], for instance, complaining to the manager?

R 22: I complained about noise and drugs, but don't know the result . . . I'm moving to [ ].

Interviewer: Do you feel you have enough say over what goes on in your building?

R 2: No, I'm fearful of the people living here ... Wild place.

Overall Assessment of Environmental Quality

Questions asking respondents to rate their environments were used to gauge the tenants' global perceptions of their environment. Significant differences were found in answers to three of the questions. A much larger proportion of public housing tenants specifically mentioned that the housing did not meet their expectations because of the presence of younger adults (60 percent as opposed to 8 percent); identified younger or young mentally ill adults as "not belonging in their environment" (48 percent versus 0 percent); and/or recommended categories of persons for whom the housing should be reserved, such as elderly or elderly and physically handicapped (22 percent as opposed to 16 percent).[3]

In sum, although both kinds of housing target low-income elderly, public housing tenants appear to be more disadvantaged economically and in terms of social support. They also differed (in an unfavorable direction) from their subsidized housing counterparts on two dimensions of environmental quality—territoriality and social networks. Without specific prompting, they identified the younger or handicapped (viz., young, mentally ill residents) at their sites as the source of some of their dissatisfaction.

Impact of Social and Economic Resources

Before it can be concluded that age desegregation explains the differences between public housing and subsidized housing tenants' satisfaction with the environment, it is essential to eliminate the confounding effects of social and economic resources. Can the poor outcomes for public housing tenants in regard to social networks and territoriality stem from their lesser social supports and financial resources? Furthermore, what role does service provision or coordination provided by the housing site have on these outcomes?

Correlations between environmental outcome variables regarding social networks and territoriality and, respectively, social support (frequency of interaction with friends), economic status (receipt of SSI and/or Medicaid) and receiving services (meals, homemaker, transportation, a visiting nurse or case management) failed to be statistically significant. In effect, neither the characteristics of the residents nor service provision can account for the differences in environmental satisfaction between public housing and subsidized housing tenants.

Discussion

The findings of this study are preliminary in light of the small sample. More large-scale studies will be needed to confirm the findings that the current

[3] Each of these relationships is significant at the .05 level.

trend of mixing young, handicapped persons with elderly tenants is associated with a decline in the quality of the housing environment for the elderly. Whether there are other, unexamined elements differentiating the two kinds of housing also requires further study.

A longitudinal study of the process of environmental decline in elderly housing would contribute to our understanding of the changes in environmental conditions. The lesser resources of public housing tenants and the greater vacancy rates, despite the high overall demand for such housing, suggests that there may be a vicious cycle that begins when public housing for the elderly becomes a dumping ground for the vulnerable aged and leads to its undesirability for the less-at-risk elderly. The negative perception of a particular site results in turnover, allowing the younger population to gain increased access. Their presence further erodes the image of the site.

The evidence from this study has definite policy implications for those housing units in transition, which may become desegregated. Public housing sites for the elderly in Providence, for example, were jewels in HUD's crown at one time; yet deterioration in their quality, once begun, proceeded swiftly. Stress on effective management alone could not prevent or alleviate conflicts stemming from the mix of generations. The Providence Housing Authority has taken a pro-active stance in attempting to anticipate the changing population. It has upgraded existing services and sought new ones, but finding funds to support a disenfranchised group with little political clout (poor, older people) has been difficult. Many older residents remain frustrated that the housing authority cannot do the one thing many of them want—turn the clock back to restore a restricted living environment of tenuous legality.

REFERENCES

Crombie, D., 1992. Youth influx troubles elderly in high rises. *Providence Journal,* January 24, p. C1–2.

Gubrium, J., 1973. *The myth of the gold years: A socio-environmental theory of aging.* Springfield, IL: Charles C. Thomas.

Hedge, S., 1990. Housing needs assessment of persons with disability in Rhode Island. Prepared for RI Housing & Mortgage Finance Corporation, unpublished manuscript.

Human Services Conference Needs for the '90s, 1990 Transcript of conference of July 2, 1990, sponsored by the City of Providence and Brown University.

Moos, R. H., and Lemke, S., 1984. *Multiphasic environmental assessment procedure manual.* Palo Alto: Social Ecology Laboratory, Department of Veteran's Affairs and Stanford University Medical Centers.

Nelson, K. P., 1991. Housing policy and housing problems over the life course. Paper presented at the annual meetings of the Eastern Sociological Society, Providence, RI.

O'Rourke, S. J., 1991. Fair housing and the dilemmas of the public housing administrator, a conference sponsored by the Providence Housing Authority and the Taubman Center for Public Policy, Providence, RI.

Providence Housing Authority, 1990. Annual Report, unpublished manuscript.

Select Committee on Aging, U.S. House of Representatives, 1989. *The 1988 National Survey of Section 202 Housing for the Elderly and Handicapped.* Washington, D.C.: U.S. Government Printing Office.

Study Design Choices

STOP AND THINK *Did Filinson start her data collection with a research question she wanted to answer? Do you think the purpose of her study was exploratory, descriptive, explanatory, critical, or applied?*

Rachel Filinson's focus was on the elderly. She tells us that in previous decades, access to public and publicly subsidized housing for the elderly was limited to those over a specific age. She notes that changes in the population needing public housing and in federal housing policies led to significant numbers of nonelderly in public housing previously reserved for the elderly.

Beginning with existing theories about environmental quality, Filinson worked deductively and hypothesized that elderly residents of predominantly age-homogeneous public housing would be more positive in their assessments of environmental quality than would residents of age-mixed public housing. Filinson's work is mostly descriptive and explanatory, but has an applied and critical focus as well. One of her goals was to describe and to compare residents of different kinds of housing. In addition, she evaluated and was implicitly critical of the unintended consequences of legal changes for senior citizens in public housing. Perhaps most important, she wanted to know if age homogeneity or heterogeneity was the cause of other differences between the groups.

Filinson selected her study design and method of data collection to allow her to answer questions in a practical and ethical manner. Deciding that an in-person interview would be best for this population, she obtained a small research grant of $1,000 and hired two interviewers already employed in the field of gerontology.

STOP AND THINK *How many times did Filinson collect data? How many different samples did she select? Did she try to control or manipulate the independent variable in any way?*

Filinson collected data once from one sample of fifty residents of public and publicly subsidized housing in Providence. She did not try to control or manipulate the independent variable in any way. The elderly tenants in her sample lived either in relatively age-segregated or mixed-age housing before Filinson began her study. Although other options were available, we'll see that Filinson chose a cross-sectional study design.

Cross-Sectional Study Design

cross-sectional study, a study design in which data are collected for all the variables of interest using one sample at one time.

The **cross-sectional study** is probably the most commonly used design in social science research. In the cross-sectional design, data are collected about one or more variables for one sample at one point in time, even if that "one time" lasts for hours, days, months, or years. Filinson's study is an example of the cross-sectional design. Although it took two interviewers several months of part-time work to complete the 50 interviews, the data about each population element (each elderly resident) in the one sample were collected only once.

Another example of the cross-sectional design is the almost annual survey conducted by the National Opinion Research Corporation (NORC). Using a probability sample of approximately 2,000 U.S. households, the General Social Survey (GSS) uses a personal interview that covers a broad range of topics. Typically, it takes interviewers two months to collect all of the survey data for this large sample (Davis and Smith, 1992: 5). Although the sample is much larger than Filinson's, the design is the same.

Cross-sectional designs are widely used because they are useful for describing samples or populations on a number of variables and because they are usually less expensive and simpler to implement than studies with more than one sample or data collection. Cross-sectional studies often lend themselves to statistical analyses of the patterns of relationships among variables. In this type of statistical analysis, the sample can be divided into two or more categories of an independent variable and then compared on a dependent variable.

Filinson used a cross-sectional analysis to identify a pattern or connection between an independent and a dependent variable at one point in time. She divided the sample into two groups of elderly—those living in age-segregated housing and those living in mixed-age housing—to identify differences and similarities between the groups in basic demographic data, health and social support factors, and environmental satisfaction. We will describe this kind of analysis in more detail in Chapter 15. At this point, however, note one of Filinson's findings: residents of mixed-age housing were more dissatisfied with territoriality (the dimension of environmental quality concerned with disturbing behaviors by other residents and control over physical space) than were those living in age-segregated housing.

Studies like Filinson's that have at least one independent and one dependent variable and that use a cross-sectional design can be diagrammed[4] as follows:

		Time 1
One Sample (divided up into categories of the independent variable during analysis)	category 1 of an independent variable	measure of a dependent variable
	category 2 of an independent variable	measure of a dependent variable

[4] This diagram and the others used in this chapter and Chapter 8 have been influenced by those presented by Samuel Stouffer (1950) in his classic article "Some Observations on Study Design."

Cross-Sectional Studies and Causal Relationships

Cross-sectional designs are sometimes used to examine causal hypotheses. A causal hypothesis is a testable expectation that an independent variable is a cause of a dependent variable. Let's take a simple example of a causal hypothesis and hypothesize that the sight of a bone makes your dog salivate.

causal relationship, a nonspurious relationship between an independent and dependent variable with the independent variable occurring before the dependent variable.

Typically, scientists look to see if three conditions are met before saying a **causal relationship** exists, and our example would need to satisfy these conditions. First, in a causal relationship, the independent and dependent variable vary with each other. In the case of your dog and the bone, we'd want to see the dog generally salivating in the presence of a bone and generally failing to salivate in its absence. Second, scientists want to make sure that the independent variable occurs before the dependent variable does. In the bone-salivation relationship, for instance, we'd need to make sure that the bone appeared before, rather than after, salivation. (This is to make sure that the bone causes salivation, rather than salivation "causing," in some sense, the bone.)

spurious relationship, a noncausal relationship between two variables.

Finally, scientists want to make sure that a relationship is not **spurious,** or caused by the action of some third variable that makes the first two vary together. We'll illustrate spuriousness with a different example. Consider, for instance, that in some regions there is a relationship between two autumnal events: leaves falling from the trees and the shortening of days. One might be tempted, after years of observing leaves falling while days shorten (or, put another way, days shortening while leaves fall) to hypothesize that one of these changes causes the other. But, the relationship between the two events is really spurious, or noncausal. Both are caused by another factor: the change of seasons from summer to winter. It is perfectly true, that in some regions, leaves fall while days shorten, but it is also true that both of these events are due to the changing of the seasons, and hence neither event causes the other. Therefore, the two events are spuriously related.

criteria for causal relationships, criteria needed to support a causal relationship between two variables. These include the occurrence of the independent variable before the dependent variable, a pattern or relationship between the independent and dependent variables, and support for the conclusion that the apparent relationship is not caused by the effect of a third variable.

The reason we mention the criteria for determining causation in the context of our discussion of research design is simple: Certain research designs are better for examining the **criteria for causality** than others are. Thus, for instance, although it is possible to test causal hypotheses with cross-sectional designs, these designs are not ideal for such tests. One data collection can tell us about the current state of two or more variables (such as marital status and self-esteem), but can make it difficult to disentangle their time-order. In one data collection, it is possible to ask questions about past behavior, events, or attitudes, but such retrospective data might involve distortions and inaccuracies of memory.

In the focal research example, Filinson wanted to examine a causal hypothesis relating kind of residence to level of environmental quality. With a causal hypothesis, she needed to see if the two variables were associated with each other, to ascertain if the independent variable occurred before the dependent variable, and to determine if the relationship was spurious, or caused by the action of some third variable that made the first two vary together.

Filinson found a relationship between the variables—public housing tenants were more likely than subsidized housing residents were to be subjected to disturbing behavior by other residents. To figure out the time order of these variables, she noted the comments of the elderly residents of public housing units who said that their attitudes changed in response to a social climate made less pleasant by the presence of younger "intruders." To consider the possibility that the variables were spuriously (noncausally) related, she analyzed the possible confounding effects of social and economic resources on the original hypothesis and found that, for instance, the wealth of her respondents couldn't be used to "explain" away the relationship between residence and environmental quality since wealth itself wasn't related to residence. (See Chapter 15 for details about this type of analysis.)

As Filinson points out, however, the time order would have been better established by using an alternative study design, one that enabled the collection of data more than once. If a group of elderly residents living in age-segregated housing had been studied both before and after their housing environment became desegregated, we would know whether or not the characteristics and perceptions of the residents changed after age-integration. Despite her argument that age desegregation occurred first, causing a decline in environmental quality, Filinson notes that the actual order might be reversed. It is possible that age desegregation is the result of greater vacancy rates occurring *after* a decline in environmental quality.

The cross-sectional design is an appropriate choice of study design for studying causal hypotheses when the time order between the variables is easy to determine (as it would be, say, in the case of a relationship between gender and age at marriage because it's clear that a person's gender comes before their age at marriage) and when large samples allow the use of sophisticated statistical analyses to control for possible spurious variables. In addition, the cross-sectional design is very useful for describing a sample on one or more variables.

Longitudinal Designs

STOP AND THINK *Many students will graduate from your college or university in the next few years. Think about the students you know. Can you make some predictions about what will happen to them over the next decade? Which students do you think will pursue graduate education? Get married? Have satisfying careers? Be actively involved in their communities? Think about what you know that might help you make predictions about their futures. If you wanted to find what actually happens to your classmates over the next decade, you could design a study. Which students would you want to select as a sample? Do you think you would need to collect data more than once?*

The study discussed in Chapter 3 by psychologist Michele Hoffnung began with questions similar to those ones we've just posed. Hoffnung wanted to find out what happened to college women in the first decade after graduation. Her study had both descriptive and explanatory purposes. She wanted to describe the women's lives over the course of a decade and to examine the connections between their college experiences (such as having attended an all women's school or a coeducational one) and their later lives.

internal validity, agreement between a study's conclusions and what is actually true.

Hoffnung could have located students who had graduated from a variety of schools a decade before and asked them to recall their undergraduate experiences and their expectations for life after college. However, doing her study using a cross-sectional design with retrospective questions would have been problematic for the **internal validity** or accuracy of the study's conclusions. Doubt about the validity of the study's conclusions would result, in part, because of reliance on people's answers about what they felt or did years before. Not only do memories fade, but current realities affect recollections and interpretations of past experiences and attitudes.

The Panel Study

panel study, a study design in which data are collected about one sample at least two times where the independent variable is not controlled by the researcher.

Hoffnung's actual study design was a **panel study**, a choice that requires following a sample over time. During the 1992–1993 academic year, Hoffnung selected a stratified random sample of 200 female college seniors from five different colleges and universities and interviewed them about their current lives and their expectations for the future. In each subsequent year, she has mailed a questionnaire to the same women asking about their current situation (including marriage, relationships, family, education, and careers) and their future plans. Hoffnung plans to re-interview the sample within the next few years.

The major advantages of the panel design are the ability to track sample members over time, to collect current rather than retrospective data, and to determine the order of events and experiences. Hoffnung will be able find out about the women's lives each year as they encounter workplaces, graduate schools, communities, marriage, and motherhood. She will be able to examine relationships between variables and meet the temporal requirement for establishing causality—that an independent variable actually occurs before a dependent variable.

Two examples of large panel studies are the Health Retirement Study and the Panel Study of Income Dynamics. In 1992, the health panel interviewed 12,600 individuals ages 51 to 61 about topics that included their work, retirement plans, health, and financial status. The Institute for Social Research (ISR) at the University of Michigan plans to re-interview these individuals every two years to provide insights into why people retire and how they cope with declining health in later life (American Sociological Association, 1997:3). The Panel Study of Income Dynamics has been studying people for more than 20 years, answering a variety of questions, among them the issue of the extent to which people escape from poverty (Jensen and McLaugline, 1997).

longitudinal research, a research design in which data are collected at least two different times, such as a panel, trend, or cohort study.

The **longitudinal,** or over-time, nature of the panel allows the documentation of patterns of change and the establishment of time order sequences. Although Hoffnung plans to collect data many times, a panel study requires only two data collections. And, although typically only one sample is used, it is possible for a panel study to use two or more groups. For example, we could study one sample of students as they move from their freshman to their senior years in college, or we could use samples of students from two or more colleges or universities. We could consider the students from the different schools to be one sample (as Hoffnung is doing), or keep them separate to compare students from different schools over time. A visual image of the minimum panel design, is diagrammed as follows:

	time 1	time 2
one sample	independent and dependent variables measured	independent and dependent variables measured

Data are collected at least twice in a panel study, so the information about the independent variable at one point in time can be related to changes in the dependent variable at a later time. This is useful for studying the impact of an independent variable (such as type of school, college graduation, marriage, birth of a child, illness, and so on) on other variables (such as professional success or salary). Hoffnung, for example, will be able to analyze the differences over time in jobs, salaries, and the like among graduates of the five schools to test the hypothesis that women from single-sex schools achieve greater occupational success than those graduating from coeducational institutions. Similarly, if she hypothesizes that women who marry or have children within the first few years after graduation are less likely to excel in the job market than women who do not marry or have children, she could, through appropriately timed data collections, determine the nature of these relationships.

STOP AND THINK *While contacting each person in the sample two or more times has advantages, it has disadvantages as well. What do you think are some of the difficulties facing a researcher who collects data more than once from the same sample?*

Although the ability to see change over time is very important, there are offsetting costs and difficulties in doing a panel study. The most obvious concerns are time and money. Assuming identical samples and no inflation, a study with three data collections takes much longer and costs three times as much as a one-time study. For three data collections, with the same resources, a researcher could select a sample only one-third as large as a sample in a cross-sectional study. Therefore, it's important to consider the need for a longitudinal approach to establish time-order or improve data validity before incurring extra expenses or decreasing sample size.

Subject identification and retention are two other issues for panel studies. Anonymity cannot be promised to participants in a panel study as it is essential to have identifying information, including names and addresses, to re-contact them.[5] In our highly mobile society, keeping track of respondents is difficult, even when the participants are highly motivated. **Panel attrition,** the loss of subjects from the study because of disinterest, death, illness, or moving without leaving a forwarding address, can become a significant issue. In Hoffnung's study, the first follow-up in 1994 had a panel attrition rate of 15 percent as 170 of the 200 women in the initial sample completed questionnaires that year. Each year, Hoffnung has found some "lost" members of the sample, but others have "disappeared." In 1995 and 1996, 84.5 percent of the original sample responded; in 1997 and 1998, it was 83 percent (Hoffnung, 1998). It is important

panel attrition, the loss of subjects from a study because of disinterest, death, illness, or inability to locate them.

[5] The ethical issues concerning anonymity and confidentiality were covered in Chapter 4. We will be discussing anonymity and confidentiality as they relate to the validity of data in our discussion of questionnaires and interviews in Chapters 9 and 10.

for Hoffnung and other researchers using a panel design to determine if those who continue to participate are different in systematic ways from those who do not (less geographically mobile, different values, and so on).

Hoffnung's work illustrates two other matters that are not uncommon in panel studies—changes in the methods of data collection and measurement techniques. Hoffnung cut costs by substituting a questionnaire survey for interviews for the second through sixth data collections. She also changed some questions to make the questionnaire salient for women in their mid-twenties. In a study with repeated measurements, changes in responses result from real changes in respondents' experiences, attitudes, and so on, or from changes in the questions asked, question order and format, specific interviewers, interview settings, and the like.

panel conditioning, the effect of repeatedly measuring variables on members of a panel study.

Another concern is **panel conditioning,** the effect of repeated measurement of variables. Participants in panel studies tend to become more conscious of their attitudes, emotions, and behavior with repeated data collections. This awareness can be at least partially responsible for differences in reports of attitudes, emotions, or behavior (Menard, 1991: 39).

Interpreting cause-and-effect relationships also means sorting out the effects of the time or era from other variables. Hoffnung, for example, is studying one group of women who, for the most part, were in their early twenties in the early 1990s. In analyzing the women's attitudes, expectations, and behavior, she will not be able to disentangle the effects of age from the effects of being raised in the 1970s and 1980s.

STOP AND THINK *Can you think of any situations where you'd want to see changes over time and would want to do a longitudinal study, but would prefer to select a new sample each time rather than re-contact the original sample?*

The Trend Study

There are situations when selecting a new sample is useful, as, for example, when a researcher is interested in identifying changes over time among a large population, such as registered voters in the United States. Although it would be useful, it is likely to be impossible or very expensive to re-locate the same individuals for each data collection to track changes in the attitudes and opinions of a very large sample over a number of months or years. It would be much more practical to select a new random sample each time. In addition, selecting new samples permits you to collect data anonymously, which can be a more ethical and valid way to ask questions about embarrassing, illegal, or deviant behavior.

trend study, a study design in which data are collected at least two times selecting a new sample each time.

The longitudinal design that calls for the selection of a new sample from a population for at least two data collections is called a **trend study.** A diagram of a two-time trend study looks like this:

	time 1	time 2
first sample	measure independent and dependent variables	
second sample		measure independent and dependent variables

STOP AND THINK *As a example, let's think about investigating changes in attitudes toward people with AIDS. Suppose we did a trend study by studying the attitudes of a specific population two times, selecting a new sample each time. What would be the advantages and disadvantages of this strategy?*

Trend studies avoid panel attrition and panel conditioning, save the expense of relocating sample members, and enable the researcher to collect data anonymously. Furthermore, trend studies are useful for describing changes in a population over time. For these reasons, studies of voter preferences and studies of public opinion often use the trend design. The General Social Survey (GSS) that we described as an annual cross-sectional study, can also be seen as a trend study if two or more of the individual surveys are taken together. The focus of the GSS is replication, the literal replication of some 600 questions over the years since 1972 (Davis and Smith, 1992:1), making it possible to monitor trends and constants in attitudes, behaviors, and attributes of Americans.

Trend studies do have specific limitations. Because there is no way to compare a specific person's response at two different times, the trend identifies *aggregated* changes, not changes in individuals. Because people move in and out of populations, changes in attitudes, opinions, or characteristics might reflect the changes in the composition of the population rather than actual changes in the variables.

As an example, think about doing a trend study on attitudes and goals of college students using a new sample of students every three years. Each time you collected data, there would be a largely new group of people on campus. Differences between the first and second data collections (three years apart) could mean that attitudes and expectations of students had actually *changed*, but could also be the result of having *different* students in the population or of having a *new* sample.

Research on attitudes toward AIDS by Richard Seltzer (1993) illustrates the uses and limitations of trend studies. From the sampling frame of all residences in the United States with a telephone, different random samples of more than 2,000 adults were selected in 1985 and 1987. Each sample was asked about their support for repressive measures against people with AIDS, knowledge of AIDS transmission, and attitudes towards homosexuality. The attitudes of the 1987 nationally representative sample were somewhat different from those of the 1985 sample. In 1985, for example, 42.3 percent of the sample favored a law to make it a criminal offense for a person with AIDS to have sex with another person and 29.8 percent agreed that "AIDS is a punishment God has given homosexuals for the way they live." Two years later, there was a 4 percent increase in the percent agreeing with the statement about the law, but almost no change in the statement about God (Seltzer, 1993: 87). We don't know if there would have been no change, the same change, or a more extreme change in the attitudes toward a repressive law if the *same* people interviewed in 1985 had been re-interviewed in 1987 (that is, if a panel study had been done). In addition, it is possible for "no change" to result for the aggregate even if individual members of the population change in opposite directions between the data collections. The trend study design allows for the description of overall social change, even though it can not identify how many individuals have changed or pinpoint the causes of change. True to it's name, the trend study is useful for describing trends in a population.

The Cohort Study

cohort study,
a specific kind of trend study that studies a cohort over time.
cohort,
people born within a given time frame or experiencing a life event at approximately the same time.

A **cohort study** is a special kind of trend study. The population selected in this design is a **cohort**, a group of people born within a given time frame or experiencing a given life event, such as graduating from college or getting a divorce, at approximately the same time. This means although people can "exit" from a cohort, no one can join after its formation.

In a regular trend study, if we study the same population over time (for example, all married men in Rhode Island, all university students in France, all registered voters in Canada), the population itself changes as people move in and out. Doing a trend study is like attempting to hit a target that is somewhat different each time we collect data. We can see samples in a trend study as analogous to scooping cups of water from a stream—the water is different each time one takes a sample.

In a cohort study, a new sample is picked from essentially the same population each time data is collected; one population is followed over time. The researcher is aiming at a moving, or rather aging, target. (If it helps, consider the analogy of taking water samples from a swimming pool—aside from some evaporation, it's the same body of water each time.) If in 1990, 1995, 2000, and 2005, we surveyed different random samples of the cohort of Americans born in 1970 about their attitudes toward marriage and divorce, we would be doing a cohort study that would allow us to describe changes in the 1970 birth cohort. As they move through their twenties and into their thirties, we might find this cohort becoming more liberal, more conservative, or remaining constant in their attitudes toward divorce and child custody arrangements. As in other trend studies, the differences between times could be caused by "real changes" or by the selection of different samples.

A cohort study can examine an entire generation, such as the "Baby Boomers" or "Generation X." One continuing cohort study is the National Longitudinal Study (NLS) Mature Women's Cohort, a project that has followed a large group of women since 1967, starting when they were between 30 and 44 years old. Periodic data collection has focused on the sample's labor force experiences and has been used to study a large number of research questions. Elizabeth Hill (1995), for example, looked at the data for the 3,422 women who answered the survey in 1984, when the sample was between 47 to 61 years old. Hill was able to compare the women's experiences over time by examining the data collections that occurred between 1967 and 1984. One of Hill's questions was whether there were any differences between the women who had reported that they had received at least one form of training—formal schooling, on-the-job training, or some other kind of training—after the usual schooling age and those women who had not received this form of training. She found that women who obtained training any time between 1967 and 1984 had received larger wage increases in those years than had women who had not received training. Furthermore, in 1984, most of the women who had obtained training were still working whereas the others were out of the labor force (Hill, 1995).

The Case Study

case study,
a research strategy that focuses on one case (an individual, a group, an organization, and so on) within its social context.

The last nonexperimental design that we'll describe is the **case study**, a design with a long and respected history in the social sciences. A case study takes one individual, group, organization, or the like and analyzes it within its social context. The case under study might be examined over a brief period of time or for months or years. Typically, the case study relies on several data sources, is conducted in great detail, and results in an in-depth, multifaceted investigation of a single social phenomenon (Orum, Feagin, and Sjoberg, 1991: 2).

Community studies, such as the classic analyses *The Urban Villagers* (Gans, 1982) and *Middletown* (Lynd and Lynd, 1956), and anthropological ethnographies (also called field studies), like *Occasions of Faith* (Taylor, 1995), a contemporary analysis of Donegal county, Ireland, are examples of case studies. The *Cocaine Kids*, Terry William's (1989) study of one group of teenage cocaine dealers is an especially absorbing illustration of the case study design. Williams "hung out" with the group for four years to understand their daily lives, perceptions of the future, and relationships with families and friends.

The case study approach also lends itself to analysis of a series of cases. Sociologist Judith Rollins (1985) used this design in her research on the relationship between domestic workers and their employers. What distinguishes the case study from a cross-sectional design with a small sample is that the case study approach typically does not analyze relationships between two or more variables.

Case study analysis is somewhat paradoxical. On the one hand, case studies "frankly imply particularity—cases are situationally grounded, limited views of social life. On the other hand they are something more ... they invest the study of a particular social setting with some sense of generality" (Walton, 1992: 121). The studies that have become social science classics tell us about specific people and places, but also provide a sense of understanding about general categories of the social world. Using the case study approach, "Goffman (1961) tells us what goes on in mental institutions, Sykes (1958) explains the operation of prisons, Whyte (1943) and Leibow (1967) reveal the attractions of street corner gangs, and Thompson (1971) makes food riots sensible" (Walton, 1992: 125).

A disadvantage of the case study approach is its limited generalizability. With samples of one or a few, we cannot know if what is observed is unique or typical. On the other hand, an advantage of the case study is its "close reading" of social life and its attention to the broader social context (Orum, Feagin, and Sjoberg, 1991: 274). The case study can be very important in generating new ideas and theories (Orum, Feagin, and Sjoberg, 1991: 13) and can be particularly appropriate for exploratory and descriptive purposes.

Summary

STOP AND THINK *Two researchers, Drs. Baker and Miller, are interested in studying changes in college students' attitudes toward graduate school over time. Dr. Baker decides to do a cross-sectional study with a sample of freshmen and seniors at one university. Based on his interviews, Baker notes that the seniors have more positive opinions toward attending graduate school than the freshmen do. In contrast, Dr. Miller selects a panel design. She interviews a sample of freshman at one university and re-interviews them three years later when most are seniors. Miller finds that students' attitudes as seniors are more positive toward graduate school than they were as freshman. Both Dr. Baker and Dr. Miller conclude that student attitudes toward graduate education change, becoming more positive as students progress through four years of college. In whose conclusions do you have more confidence, Dr. Baker's or Dr. Miller's? Why?*

Studies seek to answer different kinds of research questions with different purposes. Researchers might be interested in exploring new areas, generating theories, describing samples and populations, seeing patterns between variables, critically assessing some aspect of the social world, testing a causal hypothesis, or, as in the example of Dr. Baker and Dr. Miller, documenting changes over time.

The purposes or reasons behind the research directly influence the specific choice of study design. If, for example, we were interested in describing student attitudes toward graduate school, the choice of a cross-sectional design would be effective and sufficient. A cross-sectional design, like the study by Dr. Baker, allows us to describe the student attitudes toward graduate school and to *compare* the attitudes of freshman and seniors. On the other hand, if we want to study *changes* in attitudes, the panel study, although more expensive and time consuming, would be more useful. A panel design would allow us to ask about current attitudes at two times, rather than depend on retrospective data, to document the time order of variables (year in school and attitudes toward graduate school) and to analyze any changes in the attitudes of individuals, rather than changes in the aggregate.

STOP AND THINK *What about doing a cohort study of the attitudes toward graduate school of the class of 2002? What would it look like? What research questions could it answer?*

Judging which design is most appropriate for a given project involves a series of decisions. The answers to the following questions provide a basis for selecting one of the five study designs covered in this chapter or one of the experimental designs that will be discussed in Chapter 8:

1. Is it possible and useful to analyze only one case or a small number of cases or does the sample need to be larger?
2. Is it possible and useful to collect data more than one time from the same sample?
3. Is it possible and useful to select a new sample each time if data will be collected more than once?
4. If there is a causal hypothesis, is it possible, useful, and ethical to try to control or manipulate the independent variable in some way?

The focus of this chapter has been on the strengths and weaknesses of the cross-sectional, panel, trend, cohort, and case study designs as related to the issues of internal validity and generalizability. We have postponed our discussion of designs involving manipulation or control of an independent variable by the researcher, but will cover them in the next chapter.

When selecting a study design, it is important to balance methodological concerns with ethical considerations and practical needs. Although the following summary focuses on the methodological issues, we recognize that all three criteria are important considerations when creating an appropriate research design.

Summary of Case, Cross-Sectional, and Longitudinal Designs

Study Design	*Design Features*	*Design Uses*
CASE STUDY [one or more cases]	In-depth analysis of one or a few cases (including individuals, groups and organizations) within their social context	Useful for exploratory, descriptive and, with caution, explanatory purposes; useful for generating theory and developing concepts Very limited generalizability
CROSS-SECTIONAL STUDY	One sample of cases studied one time with data about one or more variables	Useful for describing samples and the relationship between variables; Useful for explanatory purposes especially if the time order of variables is known and sophisticated statistical analyses possible
PANEL STUDY	One sample of cases is studied at least two times	Useful for describing changes in individuals and groups over time (including developmental issues) Useful for explanatory purposes
TREND AND COHORT STUDIES	Different samples from the same population are selected	Useful for describing changes in populations.

Cross-sectional study diagram:

One Sample (divided up into categories of the independent variable during analysis)		time 1
	category 1 of an independent variable	measure of a dependent variable
	category 2 of an independent variable	measure of a dependent variable

Panel study diagram:

	time 1	time 2
one sample	variables measured	variables measured

Trend and cohort studies diagram:

	time 1	time 2
first sample	measure several variables	
second sample		measure the same variables

SUGGESTED READINGS

Feagin, J. R., A. M. Orum, and G. Sjoberg, 1991. *A case for the case study.* Chapel Hill, NC: University of North Carolina Press.

An impressive collection of essays describing the nature and applications of the case study. The authors make a convincing argument for increasing the use of the case study in studies of social life.

Miller, D. C., 1991. *Handbook of research design and social measurement* (5th edition). Newbury Park, CA: Sage Publications.

In addition to describing quite a few ways to measure a variety of concepts, Miller has a useful overview of study design and research strategies.

Stouffer, S., 1950. Some observations on study design. *American Journal of Sociology* 55 (January): 356–359.

An oldy but a goody. This is a classic discussion of study design choices and internal validity.

EXERCISE 7.1

Identifying Study Designs

Find an article in a social science journal that reports the results of actual research. (Be sure it is not an article that only reviews and summarizes other research). Answer the following questions about the article.

1. List the author(s); title of article; journal; month and year of publication; and pages.
2. Did the researcher(s) begin with at least one hypothesis? If so, write one hypothesis and identify the independent and dependent variables. If there was no hypothesis, but there was a research question, write it down.
3. Identify the main purpose(s) of this study as exploratory, descriptive, explanatory, critical, or applied.
4. What is the population and the sample in this study?
5. Was there only one or just a few cases, or was a larger sample used in this study?
6. How many times were data collected in this study?
7. If data were collected more than once, how many samples were selected?
8. Did the researchers try to control or manipulate the independent variable in some way?

9. Did the researchers use a case, cross-sectional, panel, trend, or cohort study design? If so, which one? Give support for your answer.

10. Was the design useful for the researcher's purposes and goals or would another design have been better? Give support for your answer.

EXERCISE 7.2

Evaluating Samples and Study Designs

Janice Jones, a graduate student, designs a research project to find out if women who are very career-oriented are less likely to have children than are women who are less career-oriented. Jones selects a random sample of women who are corporate employees and hold jobs at the vice presidential or presidential level. Interviewing the women, she finds that their mean age is 45 and that 35 percent are mothers, 55 percent are childless by choice, and 10 percent are childless involuntarily. She concludes that strong career orientations lead women to chose childlessness.

1. Discuss some of the challenges to the internal validity of Jones's conclusions.

2. Think of a different way this study could have been done. Specifically describe a sample and a study design that you think could be useful in doing research to answer Jones's question.

EXERCISE 7.3

Environmental Quality on Campus

In the focal research, Filinson examined the impact of mixed-age housing on perceptions of environmental quality. We can take this concept and apply it to other contexts. Assume that you are interested in studying environmental quality in college dormitories. You decide to follow Filinson's example by constructing a series of questions to use to assess the perceptions of environmental quality of the dormitories at Ivy Tower College.

The student body at Ivy Tower College is approximately 50 percent male. The campus has 5 single-gender and 10 mixed-gender dorms. Before selecting a specific research question and study design, consider your study design options.

Design a cross-sectional, a trend, and a panel study using dormitory students at Ivy Tower College as the population you'd study. Describe the specific sample(s) you would use, the number of times you would collect data, and specify a research question that you could answer with each design.

1. Cross-sectional study:
   a. Describe a cross-sectional study you could do.
   b. Specify a research question you could answer with your study.
2. Panel study:
   a. Describe a panel study you could do.
   b. Specify a research question you could answer with your study.
3. Trend study:
   a. Describe a trend study you could do.
   b. Specify a research question you could answer with your study.

8 Experimental Research

Amanda[1] listened carefully as her mother finished reading about the narrow escape of Laura Ingalls Wilder's grandfather from a panther. The chapter in *The Little House in the Big Woods* was exciting, and Amanda could visualize the Ingalls family as her mother read the concluding passage. "When Pa told this story, Laura and Mary shivered and snuggled closer to him. They were safe and snug on his knees, with his strong arms around them . . . they were not afraid. They were cozy and comfortable in their house made of logs, with the snow drifts around it and the wind crying because it could not get in by the fire" (Wilder, 1971: 44).

As her mother closed the book, Amanda thought about the Ingalls family and wondered what life would be like if she lived in earlier times. Amanda was surprised to hear her mother give voice to these feelings. "I'm glad that there aren't any panthers in our neighborhood. But, I'm sure you'd be as courageous as Laura and Mary. I'm glad you like *The Little House* books," Mother continued. "I loved them when I was a girl. They were a window on another world and showed me families pulling together through difficult times."

Many use literature as Amanda and her mother do—for entertainment and to find out about other times and places. But can books do more? Can children's books teach values? Change attitudes? Reinforce ways of behaving?

Some educators and therapists believe that reading can be more than an academic or recreational activity. They advocate the use of literature-based approaches to help children deal with a variety of developmental changes and social and psychological issues. But, do these approaches work? What kind of impact beyond the academic might a literature-based curriculum have? What could it accomplish? These were some of the questions that were the start of a research project conducted in a public school by one of us and a colleague. Because of our interest in a causal hypothesis, we considered another kind of study design, the controlled experiment.

[1] Amanda and her mother are composites, created from conversations with several children and their parents.

causal hypothesis, a testable expectation about an independent variable's effect on a dependent variable.

explanatory research, research that seeks to explain the cause of a phenomenon, and typically asks "what causes what?" or "why is it this way?"

Introduction

In the last chapter we considered some of the practical, ethical, and methodological bases for selecting specific research strategies. We'll continue our discussion in this chapter by considering study designs that are especially useful for research with an explanatory purpose, that is, research that seeks to explain *why* cases vary on one or more variables. **Explanatory research** begins with a **causal hypothesis** or a statement that one variable, the independent variable, *causes* another, the dependent variable.

STOP AND THINK

Explanatory research can be conducted on any area of social life. If we were interested in studying some aspect of children's literature, we could consider many possible questions or hypotheses. Think of one *causal hypothesis that concerns the impact of children's literature on something else.*

Causal Hypotheses and Experimental Designs

We can construct any number of causal hypotheses about children's literature. You might think of others, but here are three:

1. Reading books in which the main character faces and solves a problem helps children become better problem solvers.
2. Reading books with characters that have a variety of ethnic and racial backgrounds increases children's knowledge of the cultures of those racial and ethnic groups.
3. Reading books in which the main characters care for and about others increases a child's own caring behavior and attitudes.

criteria for causal relationships, criteria needed to support a causal relationship between two variables. These include the occurrence of the independent variable before the dependent variable, a pattern or relationship between the independent and dependent variables, and support for the conclusion that the apparent relationship is not due to the effect of a third variable.

Each of these hypotheses is a causal hypothesis. To see if there is support for any one of them, we'd need a research strategy designed to meet the **criteria for causality.** Causality can not be directly observed but, instead, must be inferred from other criteria. As we discussed in the Chapter 7, factors that support the existence of a causal relationship include these:

1. An independent and dependent variable that are associated with, or vary in some way, with each other.
2. An independent variable that occurs *before* the dependent variable.
3. A relationship between the independent and dependent variable that cannot be explained by the action of a third variable that makes the first two vary together.[2]

To research any hypothesis, it would be necessary to make the series of choices we've discussed in the previous chapters. We'd need to find a way to

[2] If there was an apparent relationship between two variables caused *only* by the relationships of each of the variables with a third variable, the original relationship would be spurious.

measure the variables, select a population and sampling strategy, determine the number of times to collect data, and establish the number of samples to use. We might decide that it is appropriate to use one of the study designs that we presented in Chapter 7 and design a panel, trend, cohort, case, or cross-sectional study. However, before making a final decision, we should consider the possibility of using an **experimental design.** Although most of the study designs we've already considered permit a researcher to determine if independent and dependent variables are associated or correlated with each other, experimental designs are especially helpful for determining time order and minimizing the effects of third variables.

experimental design, a study design that calls for the control or manipulation of the independent variable in some way.

The Classic Experiment: Data Collection Technique or Study Design?

A critical set of decisions in any study is about how to measure variables. That is, for each variable, the researcher has to figure out how to classify each unit of analysis in the sample into one of two or more categories. The uniqueness of the experimental design becomes apparent when we think about measurement because in a traditional experiment, the independent variable is not measured at all! Instead, in experiments, the independent variable is typically "introduced," "manipulated," or "controlled" by the researcher. That is, the independent variable does not simply occur naturally but is the result of an action taken by the researcher.

In each of the three hypotheses we listed earlier, the independent variable is "type of books read by children." If we designed a panel or cross-sectional study for any of these hypotheses, we would need to measure this variable in some way. We could measure the type of books read by school children by asking a sample of students to list every book they'd read in the past 6 months. Alternatively, we could ask libraries and book stores for lists of their most popular children's or young adult books.

On the other hand, if we decided to do an experiment, we'd need to do something different. Rather than measuring the kinds of books read by students, we'd need to find a way to have some types of books read by some students and other types of books read by others. In other words, we'd need to create a situation where we could *control* the placement of sample members into the categories of the independent variable.

Controlling the placement of sample members into categories of the independent variable is the unique feature of the classic experimental design. If we use this control, and also measure the dependent variable two times, the experimental strategy allows us to see the effect of placement in one category or another of the independent variable by comparing the two measurements of the dependent variable.

In the next few chapters, we will be describing methods of data collection, including questionnaires, interviews, observations, and the use of available data. Because of the unique way that independent variables are handled in an experiment, some textbooks define the experiment as a method of data collection or as a mode of scientific observation. Although we recognize the usefulness of this approach, we want to highlight the experiment's special treatment of the independent variable as a unique *design* element. For this reason, we include experiments in our discussion of designs and see this chapter as a natural transition between the consideration of study designs and the presentations on data collection techniques that follow.

The History of One Experiment

Several years ago, we asked the following research question: How is adolescence, or the transition from childhood to adulthood, portrayed in children's literature? We designed a research project and, based on our reading of a sample of books geared toward adolescents, completed an analysis of the images of adolescence in these books (Adler and Clark, 1991). In reviewing the literature for our study, we examined research on children's books in a wide variety of disciplines and were intrigued to discover an area of study previously unfamiliar to us—bibliotherapy, a multidisciplinary field at the intersection of education, social work, counseling, and social science.

Advocates of bibliotherapy encouraged the use of literature-based approaches as part of therapy and to help children and adults deal with normal developmental transitions. Many of its supporters claimed bibliotherapy to be a wonderfully useful technique, so you can imagine our surprise at finding little to support this assertion. We found almost no research that tested the effects of literature on behavior, attitudes, or social adjustment.

After our research was completed, Emily had a conversation about the use of bibliotherapy in school settings with Paula Foster, a social worker employed in a local community mental health agency. This conversation helped crystallize a new research interest. Further discussion led to the development of a specific hypothesis and plans for a research project to test it. This chapter's focal research is the result of that conversation.

FOCAL RESEARCH

A Literature-Based Approach to Teaching Values to Adolescents: Does it Work?[3]

by Emily Stier Adler and Paula J. Foster

Introduction

The intentional teaching of morals and values was common in American public education for almost three hundred years and had widespread

[3] This article was adapted from Adler and Foster's article of the same name published in *Adolescence* 32 (Summer, 1997):275– 286, with permission.

support until the 1930s. After decades of a "hands off" approach to the teaching of values, there have been calls for schools to return to explicit moral or character education in recent years (Kilpatrick, 1992; Wynne and Ryan, 1993). One proposed curriculum revision suggests the use of literature to "help youngsters grow in courage, charity, justice and other virtues" (Kilpatrick, 1992: 268).

At the same time, advocacy of literature-based approaches has increased in educational psychology and social work practice. The literature-based approach, sometimes called bibliotherapy, has been championed for numerous uses—a way to teach children to relate moral principles to real life (Dana and Lynch-Brown, 1991), an effective deterrent to substance abuse (Bump, 1990; Pardeck, 1991), a guide to self-understanding and improved self-concept (Calhoun, 1987; Hebert, 1991; Lenkowsky and Lenkowsky 1978; Miller, 1993), and as a way to help children deal with parental divorce (Early, 1993). The advocacy has occurred despite the caution that the approach has been awarded a "scientific respectability it does not have," and that the application of the technique "far outstrips the tight validating studies supporting its use" (Riordan, 1991: 306). In fact, our review of the literature found no studies of bibliotherapy or other literature-based approaches to teaching attitudes, values, or character traits that used an experimental design to evaluate the effectiveness of the method. It is this evaluation which our study seeks to provide.

The Hypothesis

We were interested in determining the impact of a literature-based curriculum on student values. Constructing a causal hypotheses, we designed a specific curriculum for use as our independent variable and identified a specific value, the value of caring for others, as the dependent variable. We hypothesized that the use of a literature-based curriculum would increase students' belief in the value of caring for others. We went into the field—the real world of students, classrooms, and teachers—in order to test our hypothesis.

The Sample

A middle school in an urban community with a fairly homogeneous population of approximately 40,000 (94 percent were non-Hispanic and white, 3 percent Hispanic, 2 percent Asian, and 1 percent African American) was selected because administrators were willing to allow a new curricular approach to be implemented on a small scale. The school principal and the teachers of one large seventh grade class agreed to work with us. We did not select the community, school, or students randomly and recognize that our results can not be generalized beyond the students in our sample.

The class was team-taught, with two teachers, 57 students, and a room which could be divided into two rooms with a movable partition. The class structure varied between the teachers teaching all 57 students in one space, the teachers working with many small groups, and two separate classrooms, each with a teacher. The students were an average of 13.1 years old when the study began. There were 31 girls and 26 boys; 50 students were white and 7 were Asian.

Using the characteristics of gender, race, age, and academic ability, the teachers matched the students into pairs and created two approximately equal groups. One of the groups was randomly selected to be the experimental group while the other became the control group.

The Independent Variable: The Reading Project

The independent variable was the ten-week reading project developed for this study. In consultation with the teachers, we selected three books with appropriate themes and designed class exercises and discussions to accompany them.[4] The theme of each book is the importance of caring for others. The main characters make choices and ultimately decide to care for others, even at the expense of achieving personal goals. In all three books, the main characters are adolescents, two are female and one is male, two are white and one is Asian.[5] Several times each week for ten weeks, the students in the experimental group participated in classroom discussions and exercises designed to reinforce the theme of caring for others. During the same class time, the students in the control group worked with the other teacher in the other half of the divided classroom space, reading and discussing the regular seventh grade literature (books with animal main characters, such as *Call of the Wild).*

[4]Activities for students in the experimental group were based on methods presented in *Webbing With Literature* (Bromley, 1991). In "webbing," a core concept is selected and emphasized by using "webs" or lines connecting the concept to information. In the experiment, the concept of caring was the one highlighted in the discussion of each book. Class exercises included writing favorite quotes and feelings in a journal, making a group collage that expressed the theme and the feelings in each novel, and webbing how the characters in the books were connected to each other (by marriage, birth, friendship, etc.).

[5] *Friends Are Like That* by Patricia Hermes (New York: Scholastic Inc., 1984) is the story of Tracy, an eighth grader who weighs popularity against supporting her best friend, Kelly, a nonconformist and a social pariah. Tempted to become part of the school's "popular" crowd, Tracy instead chooses friendship. *Red Cap* by G. Clifton Wisler (New York: Lodestar Books, 1991) is the true story of Ransom Powell, who in 1862, at age thirteen, joined the Union Army and was captured with his company and sent to Camp Sumter, the Confederate prison at Andersonville. Under terrible conditions, the prisoners in Ransom's company help and support each other in ways that a family might. Ransom survived imprisonment because of this care. *The Clay Marble* by Minfong Ho (New York: Farrar Straus Giroux, 1991) is set in Cambodia in the early 1980s. Twelve-year-old Dara and her family are among the thousand who are forced to flee from their villages to a refugee camp on the border of Cambodia and Thailand. Her family meets and becomes emotionally attached to another family. Even though some family members die, the two units are as one at the story's end.

The Dependent Variable: Caring for others

The dependent variable, support for the value of caring for others, is defined conceptually as the extent to which students "see and respond to need" and advocate "taking care of the world by sustaining the web of relationship so that no one is left alone" (Gilligan, 1982: 62). The extent to which the students supported this value was measured both before and after the reading project by a set of three essays on caring for others. Students in both groups completed the essays during the two weeks before and the two weeks after the ten week reading project. The teachers used class time and asked the students to write the essays during the language arts class. The students were told that there were no "right" answers and were asked to express their opinions or make up a story without worrying about what they were "supposed to say." Student essays were identified by number. Due to absenteeism, some students did not complete all six essays.

After considering several additional essay questions, we used three questions to measure caring attitudes:

1. John and Bill both have jobs in a movie theater. The manager is deciding on the work schedule for the weekend. Both John and Bill want Friday night off. John wants to spend the evening visiting his grandmother who is sick and in the hospital, and Bill has plans to go out with his friends. The manager tells them that he needs one of them to work on Friday night, but that they can either decide between themselves who will work or he'll flip a coin to decide. What do you think should happen and why? (GRANDMOTHER)
2. Susan reads in the local newspaper about a family that lost all their belongings when there was a fire in their home. She reads that they didn't have much insurance to replace their belongings. She decides to help out by sending them her whole allowance of $5.00 a week for the next three weeks. Describe how you feel about Susan's behavior. (ALLOWANCE)
3. Sometimes people have small families. Do you think that friends can be like families? In what ways? In what ways are they different from families? Write about your opinions. (FRIENDS)

Each essay was read for content by both researchers and coded into one of three categories. For GRANDMOTHER and ALLOWANCE, each was coded from most supportive for caring for others to least supportive of caring for others. For GRANDMOTHER, answers that defined John's and Bill's needs as equivalent were defined as least supportive of caring (for example, "Bill and John should split the hours in half that Friday," or "The manager should flip a coin since it's fair and quicker."). Answers that gave some priority to John's visiting his grandmother were coded as moderately supportive of caring (for example , "I think Bill should work because he should understand that John really needs to see his grandmother this week. Next week John will work and let Bill have the time off"). Essays that identified John's desire to visit his grandmother as having more priority than Bill's spending time with friends were classified as most supportive of the value of caring. For example, "If Bill has any decency, then he should work instead of John. John seeing his sick grandmother is more important than going out with friends. Friends last forever, but grandparents don't."

Inter-coder reliability was 95 percent for GRANDMOTHER. On this essay, 63 percent of the sample (57 percent of the experimental group and 68 percent of the control group) was classified as most supportive of caring for others at the pretest.

Essays in response to the ALLOWANCE question were judged to be least supportive of the value of caring if the student disagreed with Susan's behavior. Examples include the student who said, "I wouldn't do that. Susan's five dollars wouldn't help very much anyway," the one who said, "I don't care. I think she was nice, but stupid to do that," and the response, "Susan's behavior is not logical. If she read the newspaper I read, then she would be broke. I doubt $15.00 will help the poor family anyway." Essays that were partially enthusiastic about Susan's behavior or felt she would get some sort of reward for it were coded as being somewhat supportive of caring. The student who said "She is very kind, but I don't think she should give all her money away. She should get a group of people and start a collection for the family so Susan won't have to give all her money" was placed in this category. If the student felt that Susan had done the right thing and expected nothing in return, the answers were classified as most supportive of caring. Illustrative of this response was the student who said "I think Susan is doing a great thing because most people wouldn't care. I also think she has a big heart to give up her allowance for three weeks. That is a really great thing to do."

Inter-coder reliability was 95 percent for this essay. Seventy-five percent of the sample (78 percent of the experimental group and 71 percent of the control group) was classified as most supportive of caring before the introduction of the literature project.

For the essay FRIENDS, student answers were coded into three categories: friends and families seen as very different from each other, friends and families seen as somewhat similar, and friends seen as serving the same emotional and support functions as families do, even if the people live in different households. An example of the first category is the essay that said, "I don't think friends are like family. Friends are people to hang around with and have fun with. You can have fun with family but it's different. I wouldn't consider my friends my family." An example of the second category, seeing some similarities between the two units, is the student who wrote, "In some ways friends can be like family, sometimes you can trust them and they can care for you. In other ways, they aren't like family because they can't know you as well as family members do." Illustrative of the essays that were coded as seeing the units as serving the same emotional and support functions are these two: "I think friends can be like family. You can like them as well as family and they can like you in return. You can trust them even if you haven't grown up with them" and "I do think friends can be like families. They can do special things for you, or just show you they care. The only way they are different is that they are not blood related. There could be other differences, but it doesn't matter as long as they care for you."

The inter-coder reliability for this essay was 94 percent. At the pretest, 59 percent of the sample (50 percent of the experimental group and 71 percent of the control group) thought that friends can provide the same sorts of emotional and social supports as families.

Findings

To determine the effect of the reading project, the essays each student wrote before and after completing one of the ten-week curriculums were compared. Based on the two essays, each student was classified as becoming more supportive of caring for others, less supportive, or showing no change in values. Most students in the experimental and control groups were consistent in their values at both time 1 and time 2 and showed no change in support for the value of caring (see Table 8.1). For the students who did change, the results for two of the essays were in the hypothesized direction.

There was a statistically significant difference between the two groups for the FRIENDS essay. Thirty percent of the experimental group and none of the control group became more supportive of the belief that friends can be like family in providing emotional and social support to people. For GRANDMOTHER, the results approached statistical significance, with 20 percent of the control group and none of the experimental group becoming less supportive of caring for others after the reading project, and slightly more (3 percent) of the experimental group becoming more caring. For ALLOWANCE, the results, which were not statistically significant, were nevertheless not supportive of the hypotheses. Twenty-three percent of the control group and only 8 percent of the experimental group became more supportive of caring values after the completion of the reading project. Analyses controlling for gender are very similar to the results with no control variable and also demonstrate some support for the hypothesis that a literature-based curriculum can have an impact on caring values.

Discussion and Conclusion

The strongest support for the hypothesis that a literature based curriculum can influence values is the finding that the experimental group became significantly more committed to the view that friends can fulfill many of the same emotional and social support functions as families. The data for one of the other two essays also showed changes in the hypothesized direction. The experimental group remained more supportive of the caring value for the GRANDMOTHER essay, providing weak support for the hypothesis. However, for the ALLOWANCE essay, the control group became more supportive of caring than the experimental group, a finding that does not support the effectiveness of the reading project.

Several issues, including methodological ones, are important for interpreting the results of the research. The specific indicators used to measure the value of caring can be questioned. The large percentage of the student essays at time 1 which advocated caring values could be a source of concern. Honesty might be an issue as students could have written the essays

TABLE 8.1 **Comparison of Experimental and Control Groups in Changes in Support for the Value of Caring for Others**

		Experimental Group	*Control Group*
FRIENDS	Friends Seen Less Like Family for Emotional and Social Support	0 (0%)	2 (10.5%)
	No Change	16 (69.6%)	17 (89.5%)
	Friends Seen More Like Family for Emotional and Social Support	7 (30.4%)	0 (0%)
	(N = 42, χ^2 = 9.7, d.f. = 2, p = .013)		
GRANDMOTHER	Less Support for Caring	0 (0%)	4 (20%)
	No Change	18 (81.8%)	13 (65%)
	More Support for Caring	4 (18.2%)	3 (15%)
	(N = 42, χ^2 = 4.8, d.f. = 2, p = .09)		
ALLOWANCE	Less Support for Caring	1 (3.8%)	2 (9.1%)
	No Change	23 (88.5%)	15 (68.2%)
	More Support for Caring	2 (7.7%)	5 (22.7%)
	(N = 48, χ^2 = 2.99, d.f. = 2, p = .22)		

that they thought their teachers wanted, rather than sharing their own beliefs. Another concern is variability. With such a large proportion of students being categorized as "most caring" at time 1, only a relatively small increase in caring was possible. A final methodological concern is the amount of missing data. With a large number of students not completing the essays both times due to absences, as many as sixteen cases were omitted from each analysis.

Even with the methodological weaknesses, one interpretation of the findings is that there were changes in caring values as a result of reading and discussing selected literature. It is clear that any changes were small, perhaps because values are not quickly or easily affected by a modest change in the curriculum. We can assume that these students, like other American children and teens, watch an average of three hours of television per day (Time, 1995: 74), engage in many out-of-school activities, and have other curricula materials that present values explicitly and implicitly. Therefore, it should not be surprising that the ten-week reading project in this study had, at best, a modest effect. Teaching values and developing character is a complex and time-consuming task. If the value system presented in popular culture and the rest of the school curriculum does not include a great deal of support for positive values and pro-social behavior, substantial character or moral education with small, discrete projects may not be feasible. Schools interested in character education should consider a multiyear curriculum starting in the early grades with frequent reinforcement. The curriculum could include the reading and discussion of books with specific values as well as community service projects, the development of rules within the school, the inclusion of families in curriculum development, and other techniques. While small reading projects may be able to have some effect on some values, only when part of a larger framework will they be able to provide the kind of socialization that the proponents of moral and character education advocate.

REFERENCES

Bromley, K. 1991. *Webbing with literature: A practical guide.* Boston: Allyn & Bacon.

Bump, J. 1990. Innovative bibliotherapy approaches to substance abuse education. *Arts in Psychotherapy* 17: 355-362.

Calhoun, G. 1987. Enhancing self-perception through bibliotherapy. *Adolescence* 22 (Winter): 939–943.

Dana, N. Fichtman and C. Lynch-Brown. 1991. Moral development of the gifted: Making a case for children's literature. *Roeper Review* 14 (Sept.): 13–16.

Early, B. P. 1993. The healing magic of myth: Allegorical tales and the treatment of children of divorce. *Child and Adolescent Social Work Journal* 10 (2) (April):97–106.

Gilligan, C. 1982. *In a different voice.* Cambridge, MA: Harvard University Press.

Hebert, T. P. 1991. Meeting the affective needs of bright boys through bibliotherapy. *Roeper Review* 13 (June): 207–212.

Kilpatrick, W. 1992. *Why Johnny can't tell right from wrong.* New York: Simon & Schuster.

Lenkowsky, B. E. ,and R. S. Lenkowsky. 1978. Bibliotherapy for the LD adolescent. *Academic Therapy* 14 (1978): 179–185.

Miller, D. 1993. The Literature Project: Using literature to improve the self-concept of at-risk adolescent females. *Journal of Reading* 36 (March): 442–448.

Pardeck, J. 1991. Using books to prevent and treat adolescent chemical dependency. *Adolescence* 26 (Spring): 201–208.

Riordan, R. J. 1991. "Bibliotherapy Revisited." *Psychological Reports* 68: 306.

Time. 1995. Special Report: the State of the Union. Jan. 30: 74.

Wynne, E. A., and K. Ryan. 1993. *Reclaiming our schools* New York: Macmillan.

The Reading Project as an Experiment

The reading project asked whether an independent variable had *caused* a change in or *had an effect* on the dependent variable. As such, it was an explanatory study with a causal hypothesis. The decisions about research strategies, such as sample selection, measurement of variables, and collecting data more than once, were influenced by the explanatory purpose of the research. Analysis of the reading project points to some of the benefits and disadvantages inherent in the use of the controlled experiment, a study design that is very useful for testing causal hypotheses.

STOP AND THINK *Review the focal research article and think about the hypothesis and design strategy. The dependent variable, caring, was measured twice. Why do you think this was a useful choice? The independent variable was the reading and discussion of specially selected literature. Do you think this variable was "measured" or "introduced" in this study?*

To test the hypothesis that participating in the reading project would affect "caring," it was important to determine changes in the dependent variable. For this reason, student support for caring was measured two times—before and after the implementation of the reading project. In experiments, the terms for the before and after measurements of the dependent variable are the **pre-test** and the **post-test.**

pre-test,
the measurement of the dependent variable that occurs before the introduction of the stimulus or independent variable.

post-test,
the measurement of the dependent variable that occurs after the introduction of the stimulus or the independent variable.

stimulus,
the experimental condition of the independent variable that is controlled or introduced by the researcher in an experiment.

placebo,
a possible, simulated treatment of the control group, which is designed to appear authentic.

internal validity,
the extent to which the conclusions of a study are true. Specifically in an experiment, the extent to which the observed changes in a dependent variable are caused by the introduction of an independent variable.

In this study, the independent variable—whether a student participated in the literature-based curriculum focused on caring or the regular curriculum—was not "measured." Instead the literature-based curriculum was "introduced" to the experimental group and the control group received the regular seventh grade reading curriculum. The ability to control the timing of the independent variable and to determine who is exposed to each condition are central elements in the experimental design.

In experiments, the control group is exposed to *all* the influences that the experimental group is exposed to *except* for the **stimulus** (the experimental condition of the independent variable). The researcher tries to have the two groups treated exactly alike, except that instead of a stimulus, the control group receives no treatment, an alternative treatment, or a **placebo** (a simulated treatment that is supposed to appear authentic). For maximum internal validity, the experimental design tries to eliminate systematic differences between the experimental and control group beyond introducing the independent variable. Remember that **internal validity** is concerned with factors that affect the internal links between the independent and dependent variables, specifically factors that support alternative explanations for variations in the dependent variable. In experiments, internal validity focuses on whether the independent variable or other factors, such as prior differences between the experimental and control groups, are explanations for the observed differences in the dependent variable.

If, as in the reading project, the data analysis reveals at least some association between the independent and dependent variables, then the advantages of the experimental design become evident. An experiment offers evidence of time-order because the dependent variable is measured both *before* and *after* the independent variable is introduced, and *only one group* is exposed to the stimulus in the interval between the measurements. In addition, the method of selecting the experimental and control groups tends to minimize the chances of preexisting systematic differences between the two groups.

In the reading project study, two of the three indicators of the dependent variable showed small changes in the hypothesized direction. In addition, the time-order of the variables is known. However, demonstration of change in the dependent variable and a time-order in which the independent variable precedes changes in the dependent variable offer *necessary,* but not *sufficient* support for a cause and effect relationship. It is possible that another, unobserved factor brought about the change in the dependent variable. In the reading project, for example, perhaps one or more significant events occurred in the community, school, or classroom during the weeks that students were reading and discussing the books. It could be that new staff members or students arrived at the school in those weeks. Or maybe a new television show, game, or movie became very popular during the experimental time period. We might be concerned that these or other factors influenced student caring.

Concern about the effects of other factors is the rationale for one of the essential design elements in the controlled experiment—the use of two groups, an experimental and a control group. In the focal research, although the students in the class might have been exposed to new students, staff, games, movies, or television programs, we know that only the experimental group was directly exposed to the stimulus. With two groups and the ability to control who is exposed to a stimulus, researchers using an experimental design can be fairly confident of their ability to discern effects of the independent variable.

STOP AND THINK *Can you think of any challenges to the assumption that there were no systematic differences between the two groups except for their participation in the reading project?*

There are several possible challenges to the assumption that there were no systematic differences between the two groups. It is possible, for example, that the two teachers were different in a way that could have affected the results—perhaps one was a better teacher, more caring or more enthusiastic. Another possibility is that the kinds of exercises used by the reading project encouraged closer student and teacher interaction than did the traditional class work. It could also be that the two groups were unequal in some important way to begin with even though the teachers matched them on age, race, gender, and academic ability. For instance, we don't know if there were more students in the experimental group who were avid readers or if there were more in the control group who had recently moved to the community. Using matching to select experimental and control groups in this and other studies has clear limitations. In hindsight, both Emily and her coauthor believe that an alternative method of selecting the two groups might have been better for internal validity. In the next section, we'll explore several options for selecting groups in experiments and then consider design strategies that can enhance internal validity.

classic controlled experiment, an experimental design with two or more randomly selected groups (an experimental and control group) in which the researcher controls or "introduces" the independent variable and measures the dependent variable at least two times (pre-test and post-test measurement).

Experimental Designs

Classic Controlled Experiments

The focal research is an example of the kind of experiment called the **classic controlled experiment.** This kind of experiment is the mostly highly

FIGURE 8.1

Use of probability sampling to Select Groups

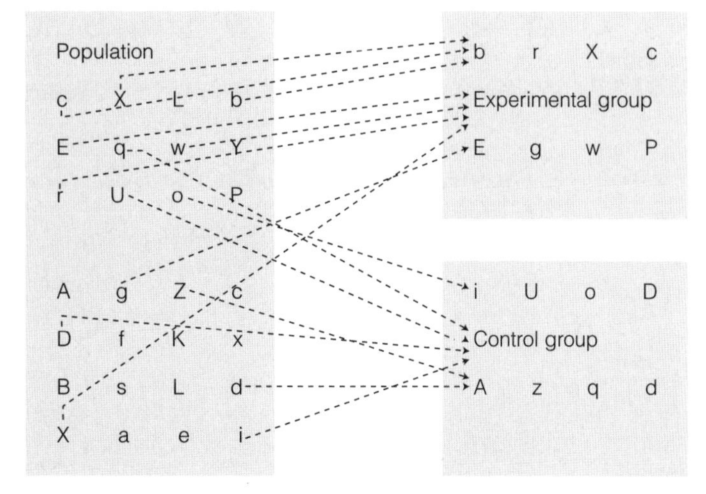

recommended of all experimental designs (Campbell and Stanley, 1963). It uses two samples or groups—the experimental and the control group—that can be selected in one of several ways. The preferred method of picking the two groups is to select an appropriate population and, using **probability sampling**, selecting two random samples. Figure 8.1 illustrates the use of probability sampling to pick experimental and control groups. This method gives the greatest generalizability of results because selecting the samples randomly from a larger population increases the chances that the samples represent the population.

probability sampling, a sampling technique that gives every member of the population a known chance of being selected.

However, participants in experiments are rarely selected from a larger population. Instead, participants are more typically a group of people who have volunteered or, as in this chapter's focal research, have been selected by someone in the setting. In such cases, the sample does not represent a larger population.

random assignment, a technique for selecting experimental and control groups from the sample to maximize the chance that the groups be similar at the beginning of the experiment.

matching, assigning members of the sample to groups by matching up members of the sample on one or more characteristics and separating the pairs into two groups with one group randomly selected to become the experimental group.

There are two ways to sort a sample into experimental and control groups to minimize preexisting systematic differences between the groups. The first is to assign members of the sample to the experimental and control groups using **random assignment**, the process of randomly selecting experimental and control groups from the study participants. This can be done by flipping a coin to determine which subject is assigned to which group, by assigning each subject a number and using a random number table to select members of each group (experimental or control), or by assigning such numbers and putting odd-numbered subjects into one group and even-numbered ones into another. Figure 8.2 illustrates the use of random assignment to create two groups.

Another way to assign members of the sample to experimental and control groups is through **matching**. Using this method, the sample's members are matched on as many characteristics as possible so that the whole group is "paired up." The pairs are then split, forming two matched groups. (See Figure 8.3.) Using this technique, one group is randomly selected to become the experimental group while the other group becomes the control group.

FIGURE 8.2

Use of Random Assignment to Select Groups

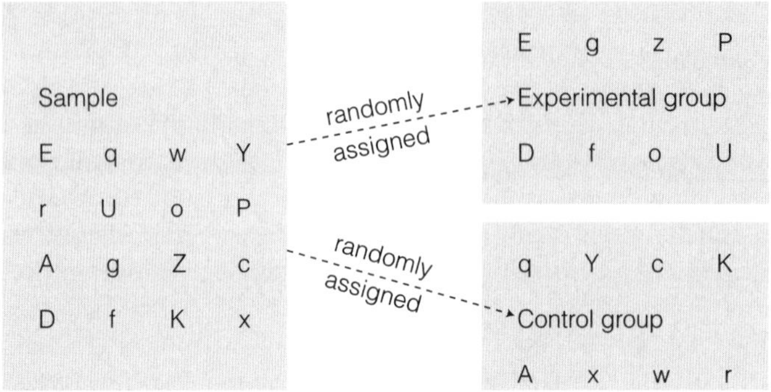

Sample

E q w Y

r U o P

A g Z c

D f K x

randomly assigned → Experimental group: E g z P / D f o U

randomly assigned → Control group: q Y c K / A x w r

FIGURE 8.3

Use of Matching to Select Groups

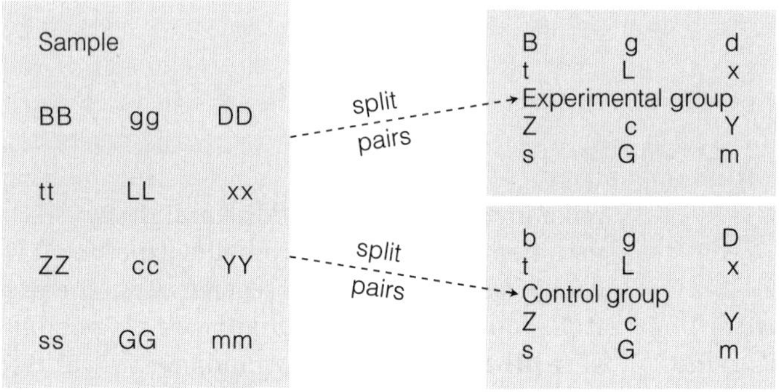

Sample

BB gg DD

tt LL xx

ZZ cc YY

ss GG mm

split pairs → Experimental group: B g d / t L x / Z c Y / s G m

split pairs → Control group: b g D / t L x / Z c Y / s G m

independent variable, a variable that a researcher hypothesizes to be affecting or influencing another variable.

dependent variable, a variable that a researcher hypothesizes is being affected or influenced by another variable.

The classic experiment has several elements that distinguish it from other study designs:

1. The study uses at least one experimental and one control group, selected using one of the strategies discussed earlier.
2. The dependent variable is measured at least two times in both the experimental and control groups. The first measurement is before and the second is after the independent variable is introduced. (These measurements are called the pre-test and the post-test.)
3. The **independent variable** is introduced, manipulated, or controlled by the researcher between the two measurements of the dependent variable. While the experimental group receives the stimulus, the control group gets nothing, an alternative treatment, or a placebo.
4. The difference in the **dependent variable** as measured at pre-test and post-test is calculated for the experimental group and for the control group. The differences in the dependent variable for the experimental and control groups are compared.

A diagram of the controlled experiment is in Figure 8.4.

FIGURE 8.4

Classic Controlled Experiment

Groups	Time		
	1 (pre-test)		2 (post-test)
Experimental	Measure dependent variable	Stimulus	Measure dependent variable again
Control	Measure dependent variable	*	Measure dependent variable again

*The control group receives no stimulus, receives an alternative treatment, or a placebo.

Field Experiments

field experiment, an experiment done in the "real world" of classrooms, offices, factories, homes, playgrounds, and the like.

An important design feature of experiments is their setting. The focal research was conducted in a regular classroom. As such, it was a **field experiment,** or an experiment that takes place where people congregate naturally—schools, supermarkets, hospitals, prisons, workplaces, summer camps, and so on. In a field experiment, the situation can be a natural, "real life" setting, but, as in the reading project, the researchers typically have the ability to select participants, have some control over the setting, and can decide which subjects constitute the experimental and control groups.

The field experiment is useful for several reasons: The setting is natural, the experimental event (such as a curriculum, treatment, marketing strategy, administrative decision, and so on) is more the way things really work, and the researchers can use samples that are closer to the actual students, patients, workers, customers, and so on that they want their results generalized to. The "realness" of field experiments is one reason that they typically have better **generalizability** (also called external validity) than other kinds of experiments.

generalizability, the ability to apply the results of a study to groups or situations beyond those actually studied. (Also referred to as external validity.)

Although working in real settings has advantages, it also presents difficulties. In the reading project, for example, the researchers could not match the two groups of students by social class even though they wanted to. The teachers could not identify parental occupations, incomes, or levels of education, and the researchers did not have access to students' files. In field studies, the researchers might be guests of the institutions or community settings where the studies are conducted; they might not be able to control as much of the experiment as would be ideal. Sometimes researchers are even asked to leave the setting before the completion of the study (Hessler, 1992: 176).

STOP AND THINK *In the Adler and Foster field experiment, the teachers did not tell the students that they were part of an experiment. Both teachers presented the materials similarly—such as the reading curriculum they would be covering for ten weeks. What were the advantages of not telling the students that they were part of an experiment? What were the disadvantages?*

Laboratory Experiments

laboratory experiment, an experiment done in a setting, such as a university or medical laboratory, that allows the researcher control over many of the conditions.

Another, perhaps more common, type of experiment is the **laboratory experiment.** The literature of psychology, social psychology, education, medical sociology, delinquency and corrections, political sociology, and industrial sociology is replete with reports of laboratory research, much of it conducted in universities with students as subjects.

One recent laboratory experiment that used undergraduates as subjects was designed to expand Milgram's experimental research on authority (see Chapter 4 for a discussion of Milgram's work). Brief and his associates (1995) conducted an experiment to test the hypothesis that when organizational members are instructed to do so, they will use race as a selection criterion in a hiring decision. Seventy-six volunteer participants from business courses took part in a simulated work situation by playing the role of chief financial officer for a hypothetical Mexican restaurant chain. Each was asked what they would do about a variety of managerial decisions, including evaluating candidates for jobs and making decisions about salary offers.

Brief and colleagues (1995) measured the dependent variable by asking the students to rate the quality of eight job candidates (three of whom were black) on a scale from 1 to 5 and to select three candidates for the final interview. The sample was randomly assigned to one of three groups, and two of the groups were given instructions from the president of the company to consider each applicant's race, education, and experience when making hiring recommendations. One-third of the students got the stimulus of a "pro-white" message, another third got the stimulus of a "pro-black" message, and the rest became the control group and were given instructions to consider education and experience only. Based on both the ratings of the black job candidates and the number of black candidates selected for a final interview by the three groups, the researchers concluded that there was support for Milgram's ideas about the influence of legitimate authority (Brief, Buttram, Elliot, Reizenstein, and McCline, 1995).

In the Brief example, we can see that, compared with field experiments, lab experiments typically allow the researcher more control over the setting and make it easier to use probability sampling and random assignment of sample members. On the other hand, although realistic situations can be created in labs and subjects can be kept in the dark about the real purpose of the study,[6] participants in laboratory experiments typically know they are involved in research. If a setting's artificiality and the staging of an experiment contribute to changes in participants' attitudes and behaviors, this will hamper a study's internal validity.

In addition, because researchers conduct studies to gain greater understanding of the world beyond the laboratory setting, it's important to evaluate a study's external validity or the ability to generalize the results from the lab to the "real world." The following issues must be considered when you evaluate the generalizability of the results of a lab experiment:

[6] See Chapter 4 for a discussion of the ethical issues involved in human subject research.

1. Was the situation very artificial or did it approximate "real life?"
2. How different were study participants from other populations?
3. To what extent did the participants believe that they were up-for-inspection, serving as guinea-pigs or play-acting or have other feelings that would affect responses to the stimulus (Campbell and Stanley, 1963: 21).
4. To what extent did the researcher communicate his or her expectations for results to the subjects with verbal or nonverbal cues?

experimenter expectations, when expected behaviors or outcomes are communicated to subjects by the researcher.

double-blind experiment, an experiment in which neither the subjects nor research staff who interact with them knows the memberships of the experimental or control groups.

The issue of **experimenter expectations** can be handled by using a design called the **double-blind experiment,** which involves keeping both the subjects and the research staff that directly interacts with them from knowing which members of the sample are in the experimental or control groups. In this way, the researcher's expectations about the experimental group are less likely to affect the outcome.

Internal Validity and Experiments

We've already noted that the classic experiment can handle many concerns about internal validity by regulating and identifying systematic differences between the experimental and control groups. The following are some other advantages of the controlled experiment that have been identified by Campbell and Stanley (1963):

maturation, the biological and psychological processes that cause people to change over time.

testing effect, the sensitizing effect on subjects of the pre-test.

history, the effects of general historical events on study participants.

selection bias, a bias in the way the experimental and control or comparison group were selected that is responsible for preexisting differences between the groups.

1. The experiment is useful for separating the effects of the independent variable from those of **maturation,** the biological and psychological processes that cause people to change over time, because, in experiments, both the experimental and control groups are maturing for the same amount of time.
2. The experiment can control for some aspects of the **testing effect,** the separate effects on subjects that result from taking the pre-test, because both groups take the pre-test.
3. The experiment can control for some aspects of **history,** the effects of general historical events not connected to the independent variable that influence the dependent variable, because both groups are influenced by the same historical events.
4. The methods of selecting the control and experimental groups are designed to minimize **selection bias,** or preexisting differences between the groups because of the way they have been selected.

Extended Experimental Design

We will consider one final internal validity issue—the interaction effect between the measurement of the dependent variable and the introduction of the stimulus. An illustration of this effect would be if, in the Adler and

FIGURE 8.5

Solomon Four-Group Experiment

Groups	Time		
	1 (pre-test)		2 (post-test)
Experimental	Measure dependent variable	Stimulus	Measure dependent variable again
Control	Measure dependent variable		Measure dependent variable again
Experimental with no pre-test		Stimulus	Measure dependent variable
Control with no pre-test			Measure dependent variable

Foster study, the writing of essays about caring alerted students to issues of caring for others so that the reading project curriculum was *more* effective in the experimental group than it would have been without the use of the pre-test. Because there were no students who read the books on caring without first completing the essays, the experimental design cannot distinguish the effects of the pre-test from the effects of reading special literature.

Solomon four-group design, a controlled experiment with an additional experimental and control group that each receive a post-test only.

An experimental design that was developed to address this internal validity concern is the **Solomon four-group design,** which is illustrated in the Figure 8.5. This controlled experiment adds an extra experimental and control groups that are treated just like the regular groups except that they do not have pre-test. The compensation for the extra effort in time and expense of having two experimental and two control groups is the ability to determine the main effects of testing, the interaction effects of testing, the needed measurement of the dependent variable, and the combined effect of maturation and history (Campbell and Stanley 1963: 25).

STOP AND THINK *What would Adler and Foster have had to do differently if they had decided to use the Solomon four-group experimental design instead of the classic controlled experiment?*

Quasi-Experiments

STOP AND THINK *The classic controlled experiment includes a control and an experimental group and pre-test and post-test measurements of both groups. Can you think of situations where it wouldn't be possible to include both aspects of the controlled experiment? Can you envision any alternative experimental designs?*

FIGURE 8.6

Comparison Group Quasi-Experiment

Groups	Time		
	1 (pre-test)		2 (post-test)
Experimental	Measure dependent variable	Stimulus	Measure dependent variable again
Comparison	Measure dependent variable	*	Measure dependent variable again

*The control group receives no stimulus, receives an alternative treatment, or a placebo.

Comparison-Group Quasi-Experiments

quasi-experiment, an experimental design that is missing one or more aspects of the classic controlled experiment.

In situations where it is not possible to use a controlled experiment, with two equivalent groups, two data collections, and the ability to manipulate the independent variable, an alternative is the **quasi-experiment.** The quasi-experiment is similar to the controlled experiment, except that it is missing one or more aspects of the classic experimental design.

In one common quasi-experimental design, there are two groups, both of which are given a pre-test and post-test, but the groups are *not* selected by random selection, randomization, or matching to have pre-experimental equivalence. There is an experimental group, but uses a comparison group in place of the control group (see Figure 8.6).

An example of this quasi-experimental design is found in a study on the effects of personal responsibility and choice on the well-being of elderly residents of a nursing home (Langer and Rodin, 1982). The researchers used the ambulatory and communicative patients of a facility with four floors as the sample. Residents on two of the floors were the experimental group, and those on the other two floors were the comparison group. The members of the experimental group received staff communications focusing on taking personal responsibility for caring for themselves, on making choices in arranging and decorating their rooms, and on deciding how to spend their time; the other group was told about the nursing home and the activities that were scheduled (Langer and Rodin, 1982: 73). Although group selection did not involve random selection or matching, all the other conditions of an experiment were met: control of the independent variable (the staff communication), the use of two groups, pre-testing and post-testing both groups on the dependent variable by observing the behavior of residents (by asking them questions and asking nurses to rate patients). The selection of the experimental and comparison groups raises the following challenge to the internal validity of this quasi-experiment: Was there some sort of pattern in the placement of residents on the four floors of rooms? Could preexisting differences in the two groups interact in some way with the independent variable in some way to create the outcomes on the dependent variable?

FIGURE 8.7

Post-Test Only Quasi-Experiment

Groups		Post-test
Experimental	Stimulus	Measure dependent variable
Control		Measure dependent variable

Another example of a quasi-experiment is Shingles' (1973) study of the effects of Project Head Start on the mothers of children in the program. This study's sample was composed of mothers who received Aid to Dependent Children and who had preschool children. The experimental group was selected from mothers who had recently enrolled one or more of their children in St. Paul's Head Start program. The comparison group was composed of mothers selected at random from a list of St. Paul ADC parents with one or more children the same age as the Head Start children. Both groups of women were interviewed twice, 18 months apart, to ascertain changes in feelings of personal and political alienation (Shingles, 1973: 243). The purpose of the study was to determine if changes in feelings of alienation were *caused by* enrollment in Head Start. But, because of the voluntary nature of participation in Head Start, the researcher could not select two groups using either randomization or matching and could not "control" the independent variable. Once again we see that the question for internal validity is this: Is the comparison group really equivalent to the experimental group?

"Post-Test Only" Quasi-Experiments

Another commonly used quasi-experimental design has no time 1 measurement (no pre-test), typically because it is either not possible to do a pre-test or because of a concern that using a pre-test would sensitize the experimental group to the stimulus (see Figure 8.7).

An example of a quasi-experiment with no pre-test was conducted by Bryan and Test (1982). To test the hypothesis that observing altruistic role models would increase altruistic behavior among observers, they conducted the quasi-experiment, "Coins in the Kettle," on a busy street one mid-December day. They stationed a female research associate dressed in Salvation Army cape and hat in front of a Salvation Army kettle on the sidewalk of a large department store. The researchers' stimulus for the independent variable of altruistic role models was a male research assistant dressed as a white-collar worker who contributed a coin to the kettle periodically. The pedestrians who passed the kettle within 20 seconds of the coin being deposited became the experimental group, and those who passed 40 to 60 seconds later (and who were assumed not to have seen the altruistic role model) became the control group (Bryan and Test, 1982). To measure the dependent variable of altruism, the researchers compared the percentages of individuals in the experimental and control groups who contributed coins. At the end of the study, they concluded that the presence of helping models significantly affected behavior. In this study it was not possible to measure the dependent variable twice as altruistic behavior or the lack of altruism could not be observed until the person passed the kettle and either contributed money or not.

The major challenge to the internal validity of the "post-test only" quasi-experimental design is the unverifiable assumption that a difference between the experimental and control groups on the dependent variable represents a *change* in the experimental group caused by the independent variable. Without a pre-test, there is no way of determining if the post-test measures represent a change in the experimental group or a continuation of a preexisting difference between the groups.

The use of a quasi-experiment means greater caution in reaching the conclusion that a causal hypothesis has been supported. However, when the complete controlled experiment is not an option, the quasi-experiment provides some of the advantages of the experimental design in the "real world." In settings where there is no possibility of selecting a true control group or no opportunity for a pre-test, the quasi-experiment can be a useful alternative. It can handle some of the challenges to internal validity, such as maturation, the testing effect, history, and some kinds of selection bias.

After we've had a chance to consider all the methods of data collection, such as questionnaires, interviews, observations, and the use of available data, we'll consider the use of experimental designs in one kind of social research-evaluation research, which is research designed to assess the influence of social programs, policies, and laws. In Chapter 14, we'll read about the use of experimental and quasi-experimental designs in evaluation research.

Experimental and Other Designs

Although the experimental study design is best suited for explanatory research and is a useful choice when you are considering causal hypotheses, many research questions cannot and should not be studied experimentally. For descriptive and exploratory research purposes, when large samples are needed, and in situations where it is not practical, possible, or ethical to control or manipulate the independent variable, other study designs might be more appropriate. As we discussed in Chapter 7, the panel, trend, cross-sectional, and case study designs are more useful for many research purposes.

STOP AND THINK *Can you think of a causal hypothesis that does not lend itself to being studied with an experimental design? What is your hypothesis? What other design could be used to test your hypothesis?*

Social scientists often construct hypotheses using background variables like gender, race, age, or level of education as independent variables. Let's say that we hypothesize that level of education affects the extent to which young adults are involved in community service and political activity; more specifically, that increases in education, the independent variable, cause increases in the dependent variables (amounts of community service and political activity). We can't control how much education an individual receives, so we'd need to use a nonexperimental design to examine these relationships. For example, we could do a panel study and observe changes in the variables over time by starting with a sample of 18-year-olds. Following the sample for a number of years, we could measure the independent variable (years of schooling completed) and the dependent variables (involvement in community service and political activities) at least twice. We'd be able to see the pattern or relationship between the variables and their time-order. We could use statistical analyses (discussed in Chapter 15) as a way to disentangle possible "third variables."

Panel studies have been widely used to study the cause and effect relationship between behaviors like drinking coffee, smoking tobacco, and eating a diet rich in red meat and the onset and progression of illnesses like heart disease and cancer. Health researchers can't and wouldn't randomly select groups of people and force them to drink coffee, eat red meat, or smoke cigarettes for a number of years while prohibiting the members of other randomly selected groups from doing so, but they can follow people who make these choices themselves. Panel studies like the Nurses Health Study have followed people for two decades. Every two years, more than 110,000 registered nurses have filled out a medical history and a detailed questionnaire about daily diet and major life events, which they send along with blood and toenail samples to the researchers at Boston's Brigham and Women's Hospital (Knox, 1997). The study is ethical as the researchers make no attempt to intervene in subjects' behavior. It is practical as well, having cost the federal government only $16 million to date (Knox, 1997: B5). It has contributed many important insights about the connections between behavior, life style, and health.

Let's turn to another nonexperimental design, the cross-sectional study, and consider the hypothesis that gender affects the degree of social isolation among senior citizens. Obviously we can't "control" or randomly assign gender. Using a sample of senior citizens, however, we could measure this independent variable and any number of dependent variables, including social isolation. Because the time-order of these independent and dependent variables is clear (the gender of a senior citizen obviously occurs before his or her social isolation), if we found an association between the variables, we could then look for other support for causality. We could, for example, use the statistical analyses techniques we will cover in Chapter 15 to control for other variables that might make the original relationship spurious.

Summary

This chapter adds to the researcher's toolbox by describing the study design options of the classic controlled experiment, an extended experimental design and two quasi-experimental designs. Because experiments allow the researcher to control the time-order of the independent and dependent variables and to identify changes in the dependent variable, they are very useful for testing causal hypotheses.

Experiments can be done as field experiments in the "real world," or they can be conducted in laboratories. Although the field experiment might be less artificial, the laboratory experiment can provide a greater ability to regulate the situation. Experiments typically use small samples and have limited generalizability, although some use large probability samples with greater generalizability.

Experimental designs can be extended by adding extra experimental and control groups or can be truncated into a quasi-experiment by eliminating one or more aspects of the classic design. Some design choices are more practical than others, whereas others have greater internal validity.

Ethical considerations and the practical concerns of time, cost, and the feasibility of controlling the independent variable often lead a researcher to select other designs. Designs like the panel or cross-sectional study often have greater generalizability and can be considerably more practical. However, the tradeoff for selecting a nonexperimental design might be concerns about a study's internal validity if the researcher cannot document time order or control for preexisting differences between groups.

Each study design has strengths and weaknesses. As we've previously noted, it's important to consider a project's goals given each design's costs and benefits and to balance methodological, ethical, and practical considerations when you are selecting a research strategy. The following summary of experimental designs should be considered along with the summary of the designs in Chapter 7 when you select and evaluate the appropriateness of specific research designs.

Summary of Experimental Designs

Study Design	***Design Features***	***Design Uses***
Classic controlled experiment	1. One experimental and one control group 2. The dependent variable is measured at least two times in both groups (before and after the independent variable is introduced) 3. The independent variable is manipulated, or controlled, by the researcher	For explanatory research testing causal hypotheses
Solomon four-group experiment	Same as the classic experiment with the addition of post-test only control and experimental groups	Same as the classic experiment but also can sort out the effects of pre-testing
Quasi-experiment	Like the classic experiment with one or more design elements omitted, such as no pre-test or the use of a comparison rather than a control group	Same as the classic experiment when the complete design is not practical or ethical with an increased concern about internal validity

SUGGESTED READING

Campbell, D. T. and J. C. Stanley. 1963. *Experimental and quasi-experimental designs for research.* Chicago: Rand McNally.

Campbell and Stanley do an excellent job of covering factors that can jeopardize internal validity and generalizability in their classic discussion of the controlled experiment and quasi-experiments.

EXERCISE 8.1

Experiments in Social and Psychological Research

Find an article in a psychology or social science journal that reports the results of an experiment. Answer the following questions about the article.

1. List the author(s), title of article, journal, month and year of publication, and pages.
2. What is one hypothesis that the research was designed to test? Identify the independent and dependent variables.
3. Describe the sample.
4. How many groups were used and how were they selected?
5. Was there a pre-test? If so, describe it.
6. How was the independent variable (or the stimulus) manipulated, controlled or introduced?
7. Was there a post-test? If so, describe it.
8. What kind of experimental design was used (a classic controlled experiment, the Solomon four-group, a quasi-experiment, or some other kind)? Give support for your answer.
9. Comment on the issues of internal validity and generalizability as they apply to this study.

EXERCISE 8.2

Conducting a Quasi-Experiment

What to do:

First check with your professor about the need to seek approval from the Human Subjects Committee, the Institutional Review Board, or the committee at your college or university concerned with evaluating ethical issues in research. If necessary, obtain approval before completing this exercise.

In this quasi-experiment, you will be testing the hypothesis that when there is a "conversational ice breaker," people are more likely to make conversation with strangers than when there is no "ice breaker."

Working in "the field," select a situation with which you are familiar from your daily life where strangers typically come into contact with each other—such as in an elevator, in a doctor's waiting room, waiting for a bus, standing in the check-out line in a supermarket, and so on.

Select one item to use as a "conversational ice breaker." (Make sure it is something that could be commented on but is not something that could be perceived as threatening, hostile, or dangerous.) Some examples of "ice breakers" that you can use are carrying a plant or a bouquet of flowers, holding a cake with birthday candles on it, pushing an infant in a stroller, carrying a fishbowl with water and tropical fish in it, or wearing a very unusual item of clothing or jewelry.

Over the course of several hours or days, put yourself in the same situation six times: three times with your "ice breaker" and 3 times without it. To the extent that you can control things, make sure that as much as possible, you keep everything else under your control the same each time (such as how you dress, the extent to which you make eye contact with strangers, whether you are alone or with friends, and so on).

Immediately after leaving each situation, record field notes, which are detailed descriptions of what happened. Each time include information about the number of strangers you interacted with, a description of the interaction between you and the stranger(s), and record any comments made by the stranger(s).

What to Write:

After completing your experiment, answer the following questions:

1. What situation did you use for your experiment?
2. What "ice breaker" did you use?
3. Describe all comments you received and any interactions you had in the three situations where you had an "ice breaker."
4. Describe all comments you received and any interactions you had in the three situations where you did not have an "ice breaker."
5. Does your data support the hypothesis that an "ice breaker" is more likely to foster conversation and interaction with strangers?
6. Comment on the internal validity and generalizability of your study.
7. Comment on the experience of doing this exercise.

EXERCISE 8.3

A Review of Other Study Designs

There will be situations in which it won't be practical, ethical, or feasible to do a controlled experiment. Think about what you would do if, like Adler and Foster, you had a hypothesis about the effect of reading some sort of literature on students' attitudes or values, but could not use an experimental design. Construct a specific hypothesis that has reading a specific kind of literature as the independent variable, and design a study using a nonexperimental design (see Chapter 7 for these designs).

1. State your hypothesis, identifying the specific independent and dependent variables you have selected.
2. Identify, describe, and diagram a study design you could use to examine your hypothesis.
3. Describe how you would measure your independent and dependent variables in this study.
4. Comment on the issues of internal validity and generalizability of the study you propose.
5. Compare the internal validity, generalizability, and practicality of your proposed study design to the kind of study conducted by Adler and Foster.

9 Questionnaires and Structured Interviews

It is increasingly obvious that college campuses are not immune to "real world" problems, including sexual assault. A social scientist interested in studying assault, attitudes toward rape, and prevention strategies doesn't need to look further than his or her own workplace for an appropriate research setting.

The dean of students at one university was interested in getting an estimate of the occurrence of sexual assault on campus and in evaluating the effectiveness of the school's rape prevention workshops. When the dean asked three sociologists to conduct a survey to answer these research questions, they took the opportunity to include measures that could be used to test hypotheses derived from attribution theory.

As we'll see in the focal research section of this chapter, Norma Gray, Gloria Palileo, and G. David Johnson thought it best to collect quite a bit of data about a fairly large sample of students. They decided to do a survey about student attitudes and behavior using a group-administered questionnaire. After Gray, Palileo, and Johnson collected and analyzed their data, they presented a report to the school's administration, which used it to develop and initiate a new peer education program focusing on rape prevention and the control of sexually transmitted diseases. In addition, Gray, Palileo, and Johnson wrote several articles for social science audiences, one of which we've included in this chapter.

Introduction

Have you ever received a phone call asking you to participate in a poll? Has someone in your household completed the questionnaire that the U.S. Bureau of the Census sends to every residence in all 50 states every 10 years? Most of us are familiar with polls and questionnaires. In fact, it's hard to avoid these methods of data collection because hundreds of thousands of surveys are conducted annually (Bradburn and Sudman, 1988: 3). In fact, for decades, **questionnaires** and their first cousins, **structured interviews,** have been the most widely used methods of data collection in the social sciences (Brown and Gilmartin, 1969; Bradburn and Sudman, 1988). Some examples of reports that used one of these methods are *The Current Population Survey* (U.S. Bureau of the Census, 1993), *British Social Attitudes: Special International Report* (Jowell, Berry, and Goldman, 1989) and *Sex in America* (Michael, Gagnon, Laumann, and Kolata, 1994).

questionnaire, a data collection method in which respondents read and answer questions in a written format.

structured interview, a data collection method in which an interviewer reads a standardized interview schedule to the respondent and records the respondent's answers.

survey, a study in which the same data are collected from all members of the sample and analyzed using statistics.

Research projects that use structured interviews and questionnaires are sometimes referred to as surveys because although these methods of data collection can be used in experiments or in case studies, they are most commonly used in surveys. In a **survey** the same data are collected (most often a specified set of questions, but possibly observations or available data) from one or more samples (usually individuals, but possibly other units of analysis), and the data are analyzed statistically. Typically, surveys use large probability samples and the cross-sectional study design (data collection about many variables from one sample at one time). However, surveys can also employ non-probability samples and trend and panel designs.

The questionnaire and the structured interview both use a standardized set of questions and rely on answers from the sample. Because the study participants, who are most often called **respondents,** are most typically asked questions about themselves, asking questions is also called the **self-report** method. The researcher can take the answers as facts or can interpret them in some way, but the information provided by the respondents is the basic data in this method.

respondent, the participant in a survey who completes a questionnaire or interview.

self-report method, another name for questionnaires and interviews as respondents are most often asked to report on their own characteristics, behavior, and attitudes.

interview schedule, the set of questions read to a respondent in an interview.

There are two major types of self-report method, distinguished by the procedure used to ask questions. In questionnaires, the respondents read the questions themselves and answer them in writing; an **interview schedule** is read aloud by an interviewer who records respondents' answers.

Questionnaires and interviews are similar in some ways to our everyday social interactions. When we meet new acquaintances, we often ask and answer questions about ourselves. In the same way, researchers use questions to elicit information about the backgrounds, behavior, and opinions of people in a study's sample. And, just as we usually do in our personal conversations, researchers try to figure out if the answers they've received are honest and accurate.

STOP AND THINK *Suggest a list of topics that you think you could ask questions about using a questionnaire. For which of these topics would you have the most concern about the accuracy of the answers?*

Gray, Palileo, and Johnson wanted to find out some things about their sample of college students. Specifically, they were interested in knowing to what extent their respondents accepted common myths about rape, if there were any differences of opinion between women and men, how many women had been sexually victimized, and how many men were sexual assailants. In addition, they wanted to explore some of the causes of some of the attitudes. As you will see, these researchers were interested in doing a study that had both descriptive and explanatory purposes.

In reviewing the literature about attributing blame for criminal acts, Gray, Palileo, and Johnson discovered two very different theories or models about the way that blame is assigned. To discover patterns in the acceptance of rape myths and to see which of the competing theories about the causes of these patterns received support, they collected data with the survey described in this focal research.

FOCAL RESEARCH

Explaining Rape Victim Blame: A Test of Attribution Theory[1]

by Norma B. Gray, Gloria J. Palileo, and G. David Johnson

Introduction

"Rape culture" refers to the set of beliefs that members of a social collectivity hold about rape, its causes, and the relative responsibility attributed to assailants and victims. It often has been argued that the prevailing rape culture in the United States contributes to its high incidence of rape. The victims of rape are frequently blamed for their own victimization, perhaps more frequently than the victims of any other crime. Such beliefs inhibit the reporting of rape, limit the provision of social support to victims, and thus facilitate the commission of the crime by assailants.

In this paper we examine victim-blaming attitudes with data from a probability sample of students at a southern university. Our objectives are to measure the pervasiveness of rape myths and examine sources of variation in their acceptance.

We test hypotheses derived from attribution theory (Heider, 1958; Kelley, 1971) in order to explain variations in rape myth acceptance. A basic assumption of attribution theory is that individuals will attribute causality for observed behavior, i.e., they will make decisions about who or what is responsible for the behavior they observe. Individuals will attribute causality

[1] Adapted from an article in *Sociological Spectrum*, 1993, Vol. 13 (4), pp. 377–392, by N.B. Gray, G. J. Palileo and G. D. Johnson, Taylor & Francis, Inc., Washington, D.C. Reproduced with permission.

and responsibility either to the actor's "personal disposition" or to external (e.g., other persons, the environment, chance) causes (Thorton, Robbins, and Johnson, 1981). The attributions an individual makes about an actor's responsibility for his or her own behavior are influenced by (1) the social location of the observer and the observed (e.g., their class position, gender, and race); (2) the observer's previous experience (particularly in related activities); (3) the setting; and (4) the potential psychological benefit to be derived from alternate attributions (Shaver, 1975).

Previous Research

Much of the research on rape victim blaming has focused on characteristics of victims. Victims dressed in sexy attire are more likely to be blamed than those more modestly dressed (Lewis and Johnson, 1989; Cahoon and Edmonds, 1989). Those with histories of sexual activity are more likely to be held responsible (Cann, Calhoun, and Selby, 1979). The setting of the rape is also associated with the attribution process. Terry and Doerge (1979), for example, found that victims were more likely to be blamed when the rape occurred outside the home.

We examine the association between the social characteristics of the observers and the attribution of responsibility from the opposing perspectives of the "defensive attribution" (Shaver, 1975) and "need for control" (Walster, 1966) models. The defensive attribution model states that the greater the similarity in social characteristics or experiences between observer and observed (e.g., if they are the same gender and/or the same race), the greater the likelihood that the observer will attribute responsibility to someone or something other than the observed victim. This attribution is motivated by the need for the observer to protect his or her self-esteem and to avoid self-blame if he or she should be similarly victimized in the future.

The need for control model, in contrast, assumes that, typically, individuals will seek to maintain a more predictable, and therefore more controllable, environment by discovering underlying dispositions in an actor's behavior (Heider, cited in Shaver, 1975). Consequently, observers attribute responsibility to the observed (including victims of crimes) in order to maintain a psychological sense that they, the observers, have control over their own lives. According to this model, if the observers attributed causality to external events, they would have to confront their lack of control over their own fates.

Previous sexual victimization has also been found to be modestly correlated with rape myth acceptance (Burt, 1980), suggesting that individuals who have experienced sexual violence are more likely to identify with other victims and, therefore, are more likely to avoid victim blaming.

There are a number of variables, not addressed in the current literature, that we also predict will be associated with victim blaming. We predict that victims of nonsexual violent crimes will also be more likely to identify with rape victims and, therefore, more likely to avoid victim blaming. Conversely, we predict that individuals who have committed acts of aggression, particularly sexual aggression, will be more likely to attribute blame to justify their own aggression.

Another characteristic we examine is risk-taking behavior. The defensive attribution model predicts risk takers are more likely to identify with rape victims. Individuals who take chances they regard as reasonable (such as walking across campus alone at night) will not hold others to a behavioral standard to which they, themselves, do not adhere. On the other hand, the control model makes the opposite prediction: Risk takers will be more likely to blame the victim. According to this theory, individuals need to maintain a sense of control over their lives and, unless they hold victims responsible, they will need to admit their vulnerability since they have taken similar chances.

Hypotheses

We test four hypotheses from the defensive attribution model:

1. Females will attribute less rape victim responsibility than will males.
2. Females who have experienced sexual victimization will attribute less victim responsibility than females who have not.
3. Males who have committed acts of sexual aggression will attribute greater victim responsibility than males who have not.
4. Males and females who have been victims of nonsexual crimes will attribute less responsibility to rape victims than those who have not.

We also construct a hypothesis from the perspective of each of the two opposing theories and use the data to choose between them.

5a. From the defensive attribution model: High risk takers will attribute less responsibility to rape victims than will more cautious individuals.

5b. From the need for control attribution model: High risk takers will attribute more responsibility to rape victims than will more cautious individuals.

Methodology

The data were collected in January 1991 from 1,177 students at a southern university with an enrollment of approximately 11,000. The sampling frame consisted of Monday 10:00 A.M. and 6:00 P.M. classes, which were the university's peak scheduling hours. Instructors from 58 out of the 63 classes selected by simple one-stage cluster probability sampling allowed our proctors to administer the survey questionnaires, yielding a total of 511 males and 666 females (more than 99 percent completion rate). (The women's and men's questionnaires are found in Appendix B of this text).

Rape victim blame was measured with an eight-item scale of commonly held rape myths, developed from the much larger Burt (1980) scale. A factor analysis revealed that the eight items had a single underlying dimension. Consequently, an eight-item additive index was created with an acceptable level of inter-item reliability.

Previous sexual victimization (asked only of female subjects) was measured by a series of "yes/no" questions on various forced sexual acts, modified with permission from Koss, Gidycz, and Wisniewski (1987). A respondent who said "yes" to any of the items was classified as having been sexually victimized.

Previous sexual aggression (asked only of male subjects) was measured by a similar series of "yes/no" questions on various forced sexual acts committed by the respondent within the past year. A man who said "yes" to any of the items was classified as having been a sexual victimizer.

Nonsexual victimization was measured by two "yes/no" questions concerning assault and mugging. These two crimes were selected because they most closely resembled the experience of rape victimization.

A risk-taking behavior scale included six questions that formed an additive index with acceptable inter-item reliability. In addition, two questions were asked about alcohol consumption. Frequency of drink was measured by an ordinal scale from 0 (none in the past year) to 4 (more than twice a week). Quantity of drinking was measured on a scale from 0 (none) to 4 (more than 6 cans of beer or 5 glasses of wine or spirits).

Additional variables included age, gender, race, marital status, class standing (freshman through graduate school), current residence (on-campus, off-campus), knowledge of rape prevention, and national residence (citizen or resident versus noncitizen and nonresident).

Results

The sample's mean age was 24.7, with the students fairly evenly split between all four undergraduate years and 13 percent in graduate school. The sample was 84 percent white, 8 percent black, 5 percent Asian, 1 percent Hispanic, and 2 percent American Indian or other. Ninety-four percent were U.S. citizens or residents, 76 percent lived off-campus, 76 percent were unmarried, and 57 percent were female.

The possible range of scores for the rape myths acceptance scale was from 8 (total rejection of myths) to 40 (total acceptance of myths). The mean score for the women in the sample was 13.0 (s.d. = .17) and for men it was 17.5 (s.d. = .27), indicating that overall the sample tended not to blame the victim.

We first analyzed the strength of relationships between two variables for the total sample and for the males and females separately. As predicted by Hypothesis 1, we found gender to be strongly associated with rape victim blaming, with males more likely than females to hold rape victims responsible ($r = .42$, $p < .05$). The findings also confirm the predicted relationship between previous sexual victimization and victim blaming. Women who had experienced sexual victimization are less likely to accept rape myths ($r = .09$, $p < .05$). Also as predicted, males who have committed sexual aggression are more likely to blame rape victims ($r = .21$, $p < .01$). Nonsexual victimization, however, is not related to rape myth acceptance for either gender.

There is significant interaction with gender in the results for risk taking. Among females, there is a negative and significant association between risk taking and victim blaming ($r = -.10$, $p < .05$), a finding more consistent with a defensive attribution model. For males, the relationship is positive ($r = .07$; increased risk taking is associated with increased myth acceptance), but is not statistically significant.

An interaction exists between gender and alcohol consumption as well. For males, increased alcohol consumption is associated with increased myth acceptance, but for females, increased alcohol consumption is associated with reduced victim blaming.

The other variables related to myth acceptance in bivariate analyses are national residence (i.e., noncitizens and nonresidents are more likely to accept myths), rape knowledge (i.e., the more knowledgeable are less likely to accept myths), and class standing (i.e., senior students are less accepting of myths).

However, after conducting a stepwise multiple regression analysis (a method of statistical analysis designed to control for the other variables under consideration when considering a relationship between two variables), only a few of the original relationships remain significant. As expected, gender remains the most important predictor of rape myth acceptance. Nationality, rape prevention knowledge and class standing are also significantly related to the dependent variable in the analysis of the whole sample.

For the female subsample, only two predictors of myth acceptance remain after multivariate analysis: frequency of drinking and degree of risk taking. Each relationship is negative, suggesting that the defensive attribution model best fits the data from the women.

For the males, several independent variables continue to be related to myth acceptance. As predicted, previous sexual aggression is associated with higher rape myth acceptance while previous nonsexual victimization ("ever assaulted") is associated negatively with victim blaming. Quantity of alcohol consumption is positively associated with the dependent variable. In addition, foreign students and nonwhites are more likely to accept rape myths. Class standing and degree of rape prevention knowledge are negatively related to myth acceptance among males.

Conclusions

Most of the hypotheses are at least partially confirmed. As anticipated, gender is the most powerful predictor of rape victim blaming, with females substantially less likely to accept rape myths. This finding is fully consistent with a defensive attribution model, supporting the interpretation that females are better able to identify with victims.

Many of the relationships that applied to the female sample in the initial analysis became statistically insignificant when controlling for the other variables. In the multivariate analysis, only risk-taking behavior (including frequency of alcohol consumption remains) was a significant predictor for females, with females who take risks more likely to empathize with victims, presumably because they are sensitive to their own vulnerability.

For males, a greater amount of the variance in victim blame attitudes is explained by the independent variables of past sexual aggression, previous nonsexual victimization, and alcohol consumption. Males, particularly those who have been sexual aggressors, are more likely to identify with the assailant, as indicated by their greater acceptance of rape myths. Males who take risks are less likely to empathize with rape victims, thus protecting their own subjective sense of invulnerability.

Rape prevention knowledge and class standing help explain variation in the male sample, suggesting that education influences attitudes about rape. Males who report knowledge of rape prevention are less likely to blame rape victims. In addition, even controlling for age, males who have been in college longer are less likely to embrace rape myths. Perhaps there is reason to hope that rape culture can be altered by education and knowledge about rape and its prevention.

REFERENCES

Burt, M. R. 1980. Culture myths and supports for rape. *Journal of Personality and Social Psychology* 38: 27–230.

Cahoon, D., and E. Edwards. 1989. Male and female estimates of opposite sex first impressions concerning females' clothing styles. *Bulletin of the Psychonomic Society* 27: 280–281.

Cann, A., L. Calhoun, and J. Selby. 1979. Attributing responsibility to the victim of rape: Influence of information regarding past sexual experience. *Human Relations* 32: 57–67.

Heider, F. 1958. *The psychology of interpersonal relations.* New York: Wiley.

Kelley, H. H. 1971. *Attribution in social interaction.* Morristown, NJ: General Learning.

Koss, M. P., C. Gidycz, and N. Wisniewski, 1987. The scope of rape: Incidence and prevalence of sexual aggression and victimization in a national sample of higher education students. *Journal of Consulting and Clinical Psychology* 55: 162–171.

Lewis, L., and K. Johnson. 1989. Effects of dress, cosmetics, sex of subject and causal influence on attribution of victim responsibility. *Clothing and Textile Research Journal* 8: 22–27.

Shaver, K. 1975. *An introduction to attribution processes.* Cambridge, MA: Winthrop.

Terry, R., and S. Doerge. 1979. Dress, posture and setting as additive factors in subjective probabilities of rape. *Perceptual and Motor Skills* 48: 903–906.

Thorton, B., M. Robbins, and J. Johnson. 1981. Social perceptions of the rape victim's culpability: The influence of respondent person-environmental causal attribution tendencies. *Human Relations* 39: 225–237.

Walster, E. 1966. Assignment of responsibility for an accident. *Journal of Personality and Social Psychology* 3: 73–79.

The Uses of Questionnaires and Structured Interview Schedules

With hypotheses to test, researchers Gray, Palileo, and Johnson decided to measure quite a few variables of interest. Among other things, they wanted to know about students' beliefs about rape, their background characteristics, their experiences as victims and, for men, the extent to which they had been sexually aggressive. After considering the possible options, the researchers selected a questionnaire as their data collection strategy.

Questionnaires and structured interviews are appropriate methods of data collection for a range of topics. As the examples in Box 9.1 demonstrate, questions can be asked about a wide variety of topics. It is very common, for example, to use a self-report method to collect information about *attitudes, beliefs, values, goals, and expectations,* although data on these topics can sometimes be inferred from observed behavior. Similarly, although data about *social characteristics* and *past experiences of individuals* and *information about organizations or institutions* might be accessible through records and documents, they are suitable topics for questions.

Questions can be used to measure a person's level of *knowledge* about a topic. They are also frequently employed to obtain information about *behavior*, even though observational methods will often yield more valid data. If, like Gray and her colleagues, you are interested in finding out about behavior that typically occurs without observers (such as sexual aggression or victimization) or if you want to find out about behavior that would be unethical, impractical, or illegal to observe (such as voting behavior or getting tested for AIDS), then a self-report method might be the only data collection option.

STOP AND THINK *In addition to the one presented in this chapter, several other studies in the focal research sections of the book have collected data by asking questions. Can you remember any of the studies that have used this method? Do you have any concerns about the validity of the self-report data collected in any of the focal research studies?*

The self-report method has a wide variety of applications, including public opinion polls, election forecasts, market research, and publicly funded surveys on health, crime, and education. Questionnaires and structured interview schedules can facilitate the collection of the same information from large samples, even if the samples have been selected from regional, national, or multinational populations. The resulting analyses are often used by local, state, federal, and international organizations and are frequently the focus of articles in the mass media. For example, each year the U.S. Bureau of the Census for the Department of Justice surveys a representative sample of approximately 50,000 Americans using the National Crime Victim Survey (Lynch, 1996), which provides useful estimates of crime victimization. Researchers and policymakers also benefit from the National Longitudinal Survey of Youth, sponsored by the U.S. Department of Labor, which provides information about children's cognitive abilities, school attitudes, and peer relationships. The General Social Survey and the International Social Survey Programs are two surveys which allow the documentation and

BOX 9.1

Topics and Examples of Questions

Questions can be asked about attitudes.
An example: Some businesses have policies about sexual harassment. How do you feel about your company having a policy concerning sexual harassment in the workplace? (check one)

☐ strongly oppose
☐ mildly oppose
☐ neither oppose nor favor
☐ mildly favor
☐ strongly favor

Questions can be asked about organizations.
An example: Does your current place of employment have a policy concerning sexual harassment?

☐ no
☐ yes →

If yes, do you know what the policy covers?
☐ no
☐ yes

Questions can be asked about a belief.
An example: Businesses have reported an increase in the number of charges of sexual harassment in recent years. In your opinion, what proportion of these charges are justified? (check one)

☐ all or almost all
☐ more than half but less than most
☐ about half
☐ less than half but more than a few
☐ very few or none

Questions can be asked about behavior.
An example: Have you ever filed a charge of sexual harassment against a co-worker at your current place of employment?

☐ yes
☐ no

Questions can be asked about a social characteristic or attribute.
An example: What is your gender?

☐ male
☐ female

comparison of attitudes and behavior over time and cross-nationally. The large number of surveys conducted in this country and others make possible information gathering from mass publics in some semblance of their natural settings. The data that surveys supply can help inform policy and provide a context for decision making. Moreover, the data can serve as a counterbalance to the political pressure groups or cultural elites that claim to speak for the public (Converse, 1987: 415).

Although surveys are very widely used, there are important concerns about the *validity* of the information that they generate. The use of surveys is based on the implicit *assumption* that people will answer based on their core values and beliefs. However, responses might represent casual thoughts rather than deep feelings and might reflect what the respondent heard on the news most recently rather than personally salient answers (Geer, 1991; Kane and Schuman, 1991: 82). In addition, people in more positive moods have been found to give more favorable answers to questions than do those who are feeling more negative (Martin, Abend, Sedikides, and Green, 1997).

Answering interview questions and completing a questionnaire are social acts in the same way that attending a class or browsing in a shopping mall are. It is essential to recognize that the social context of the interview and the task required of the respondent will affect the responses given (Marsh, 1982: 56), and that our views and the views of those we question are influenced by the surrounding culture.

It is fairly simple to design questions that ask respondents to describe their attitudes toward abortion, to share their views on laws that ban assault weapons, or to recall the number of sex partners they had in the past twelve months. Accurate answers require an understanding of what is being asked, having knowledge or an opinion about the topic, being able to recall events from months or years before precisely, and intending to answer honestly.

response errors, responses to a questionnaire or interview that include inaccurate information.

Many factors contribute to **response errors,** which are responses that include inaccurate information. Faulty memory, the desire to present a good image, and trying to answer in a way that fits the respondent's assessment of researcher's expectations can all contribute to error. For example, comparisons of actual vote counts to answers to the survey question, "Did you vote?" have found sizable percentages, often as high as 25 or 30 percent of the sample, who claim to have voted, but did not (Abelson, Loftus, & Greenwald, 1992: 138). The self-reports are gathered within weeks of the election, so we can assume that, in this case, a major cause of a false report is not poor memory, but rather the desire to be seen as a good citizen.

Recall of prior events by respondents has been found to be influenced by the length of time that has elapsed between the event and the data collection, the importance of the event, the amount of detail required to be reported, and the social psychological motivation of the respondents to answer the question (Groves, 1989: 422). In one study, patients of a health maintenance organization were asked if they had specific procedures, such as a flu shot, in the recent past. When the self-reports were compared to medical records, the median overreporting of procedures was 13 percent when asked about the past six months, and 4 percent for two months (Loftus, Klinger, Smith, and Fielder, 1990: 343). Things like gender role stereotypes also influence responses. When studying the play activities of fifth grade students, Lever (1981) compared answers to the question "Where do you usually play after school, indoors or outdoors?" with activity logs that the children kept and found that boys overestimated and girls underestimated their outdoor play. Although survey data can be valuable, it's important to examine the specific questionnaire or interview carefully before accepting the results of analyses.

Survey Participation

response rate, the percentage of the sample contacted that actually participates in a study.

nonresponse bias, the differences between those who agree to participate in the survey and those who do not.

One critical issue for surveys is the **response rate,** the percentage of the sample that can be contacted and, once contacted, agrees to participate. **Nonresponse bias** is the bias that results from the differences between those who participate in the survey and those who do not. If a substantial percentage does not participate, the study's *external validity* or *generalizability* is limited. The response rate is affected by the number of people who can not be reached, the number who refuse to participate, and the number who can not perform the tasks required of them (for example, those who are ill, do not speak the interviewer's language, or don't have the necessary skills to complete a questionnaire) (Fowler, 1988: 45). Of all survey methods, in-person surveys of the general public tend to have the highest response rate, with 80 percent typically participating except in large cities (Dillman, 1978). But the response rates of studies using self-report methods vary greatly. In his review of studies using questionnaires, Miller found return rates varying between 24 percent and 90 percent, with the kind of population, length of the survey, and use of follow-ups being important factors (1991: 145–146). Illustrating this variation are the Bureau of the Census National Health Interview Surveys, which typically get 95 percent of selected households to participate (Fowler, 1988: 48); the National Election Studies, which recently had less than 80 percent of the sample participating (Steeh, 1989: 36); and some published studies that report response rates of 10 to 20 percent (Bradburn and Sudman, 1988: 104).

Many longitudinal studies report that refusal rates have grown during the past 20 years (Steeh, 1989: 36). For example, the National Crime Survey reported about 3 percent of eligible persons declining to provide interviews in the mid-1970s, which increased to 6 percent by the mid-1980s (Groves, 1989: 149). The National Election Study went from a 7 percent refusal rate in 1952 to more than 20 percent in 1980 (Steeh, 1989:36).

One concern about refusal rates is bias and the nonrandom nature of the differences between those who do and do not participate. Studies have found higher refusal rates among older persons, nonblacks, and those with less than a high school education (Bradburn and Sudman, 1979; Groves, 1989; Kaldenberg, Koenig, and Becker, 1994; Krysan, Schuman, Scott, and Beatty 1994). If participants and nonparticipants differ in social characteristics, opinions, attitudes, or behavior, then, assuming a probability sample, generalization from the sample to the larger population becomes problematic.

STOP AND THINK *Have you ever been asked to complete a questionnaire or an interview? What factors did you consider when deciding whether or not to participate? Think about what you would do if you were on the other side. What would you do to try to convince someone to participate in a survey?*

Participation in surveys can best be understood within a social exchange context. That is, potential sample members decide about cooperation after thinking about the costs and benefits. Respondents are asked to give up their time, engage in interactions controlled by the interviewer, and take the risk of being asked to reveal embarrassing information. Even though they are almost always assured that the information will be kept anonymous or confidential, potential survey participants might worry about privacy and lack of control over the information. Or, researchers hope that those contacted might feel that participating will be interesting and that their information will be useful to science and society.

Response rates can be maximized by focusing on the interesting aspects of participating, providing small gifts or monetary inducements to the respondent, thanking respondents for their time, and making it clear that the research is legitimate and has a bona fide sponsor. The costs of participating (such as time and possible embarrassment) should be minimized.

Gray, Palileo, and Johnson had a very high response rate in the focal research—99 percent of those present and approximately 85 percent of those enrolled in the classes selected for the sample completed the questionnaire. After receiving permission from faculty members, Gray and associates hired graduate students and advanced sociology majors to make unannounced visits to each of 58 classes to request student participation. The students were told about the survey, that it was anonymous, sponsored by the dean of students, and would take approximately 20 to 30 minutes to complete. Students were asked to give the questionnaire serious consideration and, if willing, to complete it without discussion. The monitors answered questions about the survey, told participants how to obtain the results, and provided referral information for those who wanted help in dealing with any victimization experiences. The study's sponsorship by the university, the fact that the students were a "captive audience," the college's intention to use the results for campus crime prevention programs, the relative brevity of the questionnaire, and the anonymity of the responses were all likely to have contributed to the high response rate.

Choices of Method

Kinds of Questionnaires

Gray, Palileo, and Johnson distributed a questionnaire to all students attending one of the participating classes on a specific day. If you want to conduct a survey using a questionnaire and if, like Gray, Palileo, and Johnson, you can locate a group setting for an appropriate sample (such as church-goers, club members, students, and the like) and obtain permission to use the setting to recruit participants, then you might want to use their specific method of data collection, the **group-administered questionnaire**. Administering a questionnaire to a group of people who have congregated gives the researcher the opportunity to explain the study and answer respondents' questions, typically provides the researcher with some control over the setting in which the questionnaire is completed, allows the respondents to participate anonymously (which can aid in getting honest answers to "sensitive" questions), and usually results in a good response rate.

group-administered questionnaire, questionnaire administered to respondents in a group setting.

Group-administered questionnaires are typically inexpensive. With Gray, Palileo, and Johnson volunteering their time to develop the instrument and analyze the data, the study expenses, covered by university funds, were $200 to print the questionnaire and $1,000 to pay the proctors. The cost of about $1 per completed questionnaire demonstrates the bargain that this method of data collection can be.

On the other hand, there are drawbacks to group-administered questionnaires, beyond the fact that there might be no group setting for the population the researcher wants to study. The extra pressure to participate that people might feel when in a group setting raises an ethical concern about voluntary participation. In addition, there is a limit to the length of time that groups will allot to data collection.

If mailing addresses (including unit designations for multi-unit structures) are available, another way to administer questionnaires is to send them to respondents' homes or, less commonly, to their workplaces. The **mailed questionnaire** is the most commonly used survey method (Dillman, 1991), with many advantages to recommend it. Hoffnung, in the panel study of college graduates described in the focal research in Chapter 3, used a brief mailed questionnaire for some data collections. The 1997 version is included in Appendix C.

mailed questionnaire, a questionnaire mailed to the respondent's home or place of business.

A similar but less commonly used technique is the **individually administered questionnaire**, which is implemented by hand delivering a questionnaire and picking it up after completion. This method typically is more expensive and has a better response rate than the mailed questionnaire but is otherwise similar.

individually administered questionnaire, questionnaire that is hand delivered to a respondent and picked up after completion.

Mailed questionnaires are fairly inexpensive and reasonably effective. They cost about one-fifth as much as phone surveys and one-tenth as much as in-person interviews, have acceptable response rates (getting more than 50 percent of the sample to respond is usually possible) (Dillman, 1991), and permit respondents to check records and to re-read questions and answer categories at their leisure. In addition, such surveys are completed privately and put less social pressure on respondents to conform to the expectations of others or to try to please the interviewer. In one study, for example, Sudman found that 36 percent of personally interviewed respondents and only 23 percent of the questionnaire sample said they were very happy when asked "Taken altogether, how would you say things are today—would you say that you are very happy, pretty happy or not too happy?" (1967: 57). In another survey, Gano-Phillips and Fincham (1992) found married couples expressing greater satisfaction with marital quality over the phone than in their written responses. Krysan, Schuman, Scott, and Beatty. (1994) found mail respondents were more likely than phone respondents to indicate that they were having problems in their neighborhoods and less likely to express support for affirmative action (both for women and blacks) or for open housing laws. Aquilino's (1994) results were that questionnaires led to higher rates of reported cocaine use than did phone interviews, especially among minorities, although this was not true for reports of alcohol use.

Questionnaire Concerns

Questionnaires require good reading and writing skills and typically get lower response rates than interviews. In addition, there is some evidence whites are more likely to respond to them than blacks (Krysan et al. 1994).

cover letter, the letter accompanying a questionnaire that explains the research and invites participation.

Without an interviewer, questionnaires depend on the **cover letter** to persuade the respondent to complete and return it to the researchers. The cover letter provides a context for the research and information about the legitimacy of the project, the researcher, and the sponsoring organization. A good cover letter will include the following:

1. A statement of the purpose of the study
2. A section about how the respondent was selected
3. A discussion of benefits of participation
4. An appeal for the potential participant to respond
5. Instructions about how to return the questionnaire
6. If appropriate, a statement about the confidentiality or anonymity of the responses

The letter used by Adler, Bates, and Merdinger (1985) in their study of school policies and programs for pregnant teenagers and teenaged parents is included in Box 9.2. It includes each element of an appropriate cover letter except for the statement of confidentiality. The researchers did not promise confidentiality because the information about each school district was public and was to be included in a resource guide published by the state's Advisory Commission on Women.

The physical appearance of the questionnaire can affect the response rate. The format should be neat rather than careless and leave enough room to answer each question. Clear and detailed instructions should be provided for each section and must include statements about if one or more answers is appropriate, where to write the answers, and the desired method of answering (such as putting an x in a box, underlining or circling an answer, writing an answer in the space provided, and so on). The question order should be easy to follow and should begin with interesting and nonthreatening questions, leaving difficult or "sensitive" questions for the middle or end.

Response rates can be improved by including a return envelope with first class postage (Armstrong and Lusk, 1987) and by sending follow-ups (Miller, 1991). A reminder postcard can be sent a week or two later, then a second questionnaire can be sent three weeks after the original. If phone numbers are available, respondents should be called to be sure the questionnaire was

BOX 9.2

Example of a cover letter

Advisory Commission on Women in RI
220 Elmwood Avenue
Providence, RI 02907

Dear [superintendent of schools],

The Advisory Commission on Women in Rhode Island is very interested in assessing the educational needs of the women in the state. One specific group of women we need to know more about is adolescent parents. We have begun to collect information from state and local agencies concerning resources currently available to teenagers who are either pregnant or are parents. However, we feel that educators will be able to tell us most accurately about the needs of this population, as well as what services are available.

We are therefore writing to all school superintendents with the hope they can provide us with information about the policies in their school systems. In addition, we are contacting all secondary school principals in the state about the services available in their schools. In this way we will have information about both system-wide policies and school-based programs.

Would you take a few minutes to complete the brief questionnaire enclosed? Your input would be invaluable in providing us with an overall view of education policies from a superintendent's perspective. Please return it in the enclosed self-addressed envelope provided. We intend for the information sharing to be a two way process. We anticipate sending you a directory of services which we will compile.

With many thanks for your help.

Sincerely,

Jane Smith,
Executive Director

delivered and to encourage participation. A third follow-up can include a new cover letter and the questionnaire. Response rates tend to increase by 20 to 25 percent after the first follow-up with smaller returns after additional efforts (Dillman, Sinclair, and Clark, 1993). In Hoffnung's study, typically 50 percent of the mailed questionnaires were returned within a month and another 35 percent were returned after follow-up letters or phone calls.

Unless everyone in the sample will be getting a follow-up letter, however, the researcher needs to know who has not returned the questionnaire. This can be accomplished by using some sort of identification—a name or a number on the questionnaire or the return envelope. But, this means that replies will not be anonymous. Although respondents can be assured of confidentiality, some respondents might be less willing to participate than when the survey is anonymous.

In-Person and Phone Interviews

Surveys of respondents with limited literacy and education might be best handled with a structured interview. This method of data collection is guided by an interview schedule, which consists of the instructions to the interviewer, a standardized set of questions asked of the respondents, and response options for closed-ended questions. The structured interview is somewhat like a conversation, except that the interviewer controls the topic, asks the questions, and does not share opinions or experiences and the respondent is often given relatively few options for responses. In the ideal, the questioning of all respondents is standardized, so that if respondents' answers differ, it can be assumed that the differences are due to real differences among respondents rather than because of differences in the instrument (Denzin, 1989: 104).

In general, structured interviews have reasonably high response rates, result in few skipped questions, allow for some flexibility in administration and make possible the clarification of questions and the use of follow-up questions. On the other hand, interviews are more expensive than questionnaires as interviewers need to be hired and trained. In addition, using an interview adds another factor into the data collection process—the **interviewer effect,** or the change in a respondent's behavior or answers that are the result of being interviewed by a specific interviewer. Interviewers have different personal and social characteristics and might present questions differently. Or, the same interviewer might present questions distinctively on different days or in different settings.

interviewer effect, the change in a respondent's behavior or answers that are the result of being interviewed by a specific interviewer.

in-person interview, an interview conducted face-to-face.

The **in-person interview** was the dominant mode of survey data collection in this country from 1940 to 1970 (Groves, 1989: 501). Depending on the sampling procedure—as, for example, when the researcher has a list of home addresses—and the need for privacy in conducting the interview, interviews can be conducted in respondents' homes or in public places, such as an employee cafeteria or an office. The in-person interview is a good choice for questions involving complex reports of behavior, for groups difficult to reach by phone, for respondents who need to see materials or to consult records as part of the data collection (Bradburn and Sudman, 1988:102) and when high response rates are essential. In-person interviewers tend to have better **rapport**, or sense of inter-personal harmony, connection, or compatibility with respondents than do phone interviewers. Rapport might be why people are more willing to provide information on such topics as income in face-to-face interviews than on the phone (Weisberg, Krosnick, and Bowen, 1989: 100). Furthermore, the in-person interviewer can obtain visual information about respondents and their surroundings and a sense about the honesty of their answers.

rapport, a sense of interpersonal harmony, connection, or compatibility between an interviewer and a respondent.

On the other hand, in-person interviews are the most expensive of the self-report methods because they involve time to locate and contact respondents as well as travel time to the interview. Other disadvantages include the inability of interviewers to make contact with respondents in some buildings, such as in apartment buildings with doormen, and the reluctance of interviewers to go into rough neighborhoods (Bradburn and Sudman, 1988:

102). Furthermore, interviewers introduce a whole new set of factors into the situation, including their own characteristics, expectations, and biases.

In the Chapter 7 focal research, we read about Rachel Filinson's study of senior citizens living in public and publicly subsidized housing, a study that used a structured in-person interview. Carefully trained, middle-aged interviewers achieved a good response rate; interviewing people in their own homes helped to put respondents at ease. The interviewers could re-phrase questions and use follow-ups to help the respondents understand the questions. As a result, the data were of good quality with few questions unanswered by respondents. On the other hand, Filinson's cost for each interview was $20. With a limited budget, her sample included only 50 respondents.

phone interview, an interview conducted over the telephone.

Before the 1960s, the proportion of households with telephones in the United States was too small to use the **phone interview** for national probability samples. By the 1970s, 90 percent of U.S. households had telephones, and the phone interview was commonplace (Lavrakas, 1987: 10). Today it is the most widely used method for large-scale public surveys (Dillman, 1991). It has become a preferred approach because it can yield substantially the same information as an interview conducted face-to-face at about half the cost (Groves, 1989; Herzog and Rogers, 1988) and, using random digit dialing, can be conducted without names and addresses. Phone interviews are especially useful for groups that feel they are too busy for in-person interviews, such as physicians, managers, or sales personnel (Sudman, 1967: 65). The interviewers do not have to travel, can do more interviews in the same amount of time, and require fewer supervisors (Bradburn and Sudman, 1988: 99).

On the other hand, the major disadvantages of phone interviews include the elimination of those without phones from the sample and limitations on the length of the interview and the kinds of questions that can be asked. Phone interviews tend to have a quicker pace and elicit shorter answers than the in-person version (Groves, 1989: 551). It is unlikely for respondents to stay on the phone for more than 20 or 30 minutes, but in-person interviews often last 40 or more minutes. Complicated questions are more difficult to explain over the phone than in person. In-person interviews allow respondents to read or see something (Lavrakas, 1987: 13) and enable interviewers to better judge respondents comprehension. "Sensitive" questions are answered less frequently and perhaps less accurately on the phone (Aquilino, 1994; Frey, 1983: 71; Johnson, Hougland and Clayton, 1989). Finally, although the use of telephone answering machines has not been a major factor in recent surveys, their increasing use by white, nonrural, younger, and better-educated respondents to screen calls (and the bias that this practice introduces) is a potential source of concern (Oldendick and Link, 1994).

Issues for Structured Interviews

Because there is no precontact with sample members in most telephone and many face-to-face survey interviews, the introductory statement made by the interviewer is crucial. The interviewer should be credible and encourage participation. To allow potential respondents to make informed decisions, the interviewer must provide sufficient information about the survey's topic, the sponsor of the study, and the amount of time the interview will take. Interviewers should not feel defensive and guilty for taking up respondents' time and should convey confidence and legitimacy to respondents (Converse and Schuman 1974: 43).

In interviews where an appointment is scheduled, the rest of life is "put on hold" to some extent while the interview takes place. In most surveys, interviewers call or ring doorbells while respondents are in the midst of daily life. Interruptions, like a baby crying or a neighbor stopping over, are often part of the interview process (Converse and Schuman, 1974).

The interviewer must often call back to contact the person specified by sampling decisions. In urban areas, six calls per household are about average to reach respondents (Fowler, 1988: 52). The best times to obtain a good completion rate for both phone and in-person surveys are the evening hours, with 6 to 7 P.M. typically being the most effective (Vidgerhous, 1989).

Interviewers' social characteristics, especially race and gender, have been found in some studies to affect responses for at least some topics (Finkel, Gutterbok, and Borg, 1991; Kane and Macaulay, 1993; Davis, 1997). The majority of paid interviewers in many large surveys are women, and interviewers and respondents frequently differ on important traits such as gender, race, ethnicity, and age. Studies with racial topics as part of the survey usually try to ensure that the interviewer and the respondent are of the same race because of the belief that more valid answers will result. Groves (1989: 401) cautions, however, that such matching can yield overreports of extreme positions on measures of racial attitudes.

Whether they are volunteers or paid staff, interviewers must be trained so that they understand the general purpose of the study, be familiar with the introductory statement and interview schedule, and understand how to record answers. The interviewer's manner should be polite, open, and neutral, neither critical nor overly friendly. In other words, the interviewer should try to establish rapport or a sense of connection with the respondent, although this is more critical for qualitative interviewing, the topic of Chapter 10. In keeping with ethical standards of research, all interviewers should enter into an agreement not to violate the confidentiality of the respondents.

Constructing Questions

Types of Questions

Questions can be asked with or without answer categories. In an interview, **open-ended questions** allow the respondents to answer in their own words; on a questionnaire, one or more blank lines are designated, indicating an approximate length for the respondent's answer. Interviewers can follow up questions with probes that ask the respondent for an expanded or more detailed answer.

open-ended question, question that allows respondents to answer in their own words.

closed-ended questions, questions that include a list of predetermined answers.

exhaustive and mutually exclusive answer categories, closed-ended answer categories that allow respondents to select one and only one answer.

Closed-ended questions ask the respondent to pick from a list of answer categories. The answer categories must be **exhaustive** and, if only one answer is wanted, **mutually exclusive.** Exhaustive answers must have enough answers listed so that every respondent can find an answer that represents or is similar to what he or she would have said had the question been open-ended. (Sometimes "other" is included as one of the answers to make the list exhaustive.) Answer categories should be mutually exclusive so as to allow each respondent to find *only one* appropriate answer. For example, in the questionnaire reproduced in Appendix B, Gray, Palileo, and Johnson used the category 'single' rather than 'never married' in the question on marital status. Some divorced and widowed respondents might define themselves as single as well as divorced or widowed and be tempted to check two categories. Another example is the common error made by beginning researchers to list overlapping categories, such as when age groups are used as closed-ended answers to a question about age. The listing of ages as 15 to 20, 20 to 25, and 25 to 30 makes a category set that is not mutually exclusive.

STOP AND THINK *Gray and her associates used only closed-ended questions in their study of rape victim blame. They included five answer categories from "strongly agree" to "strongly disagree" as responses to each of the rape myth statements; the questions on risk taking had four possible answers, ranging from "often" to "never." Why do you think they used closed-ended as opposed to open-ended questions?*

As the example in Box 9.3 demonstrates, a question can work in either an open- or closed-ended format. Selecting which to use is a complex decision as there are advantages and disadvantages to each kind. Leaving the answers open-ended allows respondents to use their own words and frames of reference and can encourage fuller and more complex answers that closed-ended categories miss. On the other hand, a small percentage of respondents will not answer open-ended questions (Geer, 1988), and other respondents give incomplete or vague answers. Furthermore, open-ended questions are more expensive to record or transcribe, and, for both questionnaires and interviews, all open-ended responses must be coded before doing statistical analyses. Coding, described in more detail in Chapters 15 and 16, is the process of creating a limited number of answer categories for each question and classifying the responses into the categories. Coding is an expensive and time-consuming task that can result in the loss of some of the data's richness.

Offering answer choices with closed-ended questions has advantages. Answers can provide a context for the question, and they can make the completion and coding of questionnaires and interviews easier. On the other hand, respondents might not find the response that best fits what they want to say and possible answer categories (such as "often" or "few") can be interpreted differently by different respondents. In addition, respondents' "presentation of self" might be influenced by using the response alternatives to infer which behavior is "usual," if they assume that the "average" person is represented by the middle category of a response scale (Schwarz and Hippler, 1987: 167).

One common practice is to use open-ended questions in a pre-test and then develop a closed-ended version from the responses for the actual

BOX 9.3

One Question—Two Formats

Open-Ended Version:
Thinking about your job and your spouse's, whose would you say is considered more important in your family?

Probe questions for follow-up:
Could you tell me a little more about what you mean?
Could you give me an example?

Closed-Ended Version:
Whose job is considered more important in your family?

- ☐ my spouse's job is considered more important
- ☐ both jobs are considered equally important
- ☐ my job is considered more important

interview or questionnaire. Another alternative is to use mostly closed-ended questions and include a few open-ended ones to obtain greater detail.

In the focal research, Gray, Palileo, and Johnson restricted their instrument to closed-ended questions. They wanted to minimize the time necessary to complete the instrument to encourage faculty to allow the survey to be conducted. As Dave Johnson told us,

> The most precious commodity in this study was classroom time—time the students had paid for, and time the professors were jealous of. We feared that open-ended questions could drag the time out on at least some occasions. In hindsight, I think I could have asked at least one or perhaps two questions asking for explication of the date rape incidents (personal communication, 1995).

screening question, a question that asks for information before asking the question of interest.

contingency questions, questions that depend on the answers to previous questions.

A useful type of question is the **screening question,** a question that asks for information before asking the question of interest. It's important to ask if someone voted in November, for example, before asking for whom they voted. Screening questions can help reduce the problem of "nonattitudes" (when people with no genuine attitudes respond to questions anyway). By asking "Do you have an opinion on this or not?" or "Have you thought much about this issue?," the interviewer can make it socially acceptable for respondents to say they are unfamiliar with a topic (Asher, 1992 :24). When this type of question is used, some surveys have found sizable numbers say they have no opinion (Asher, 1992). **Contingency questions** are questions that depend on the answers to previous questions. For example, in an interview, those who answer "yes" to the question, "Do you work for pay outside the home?," can then be asked the contingency question, "How many hours per week are you employed?" As the example in Box 9.4 shows, the respondent to a questionnaire can be directed to applicable contingency questions with instructions or arrows. Gray and associates could have used

BOX 9.4

Using a Contingency Question

Have you applied to a graduate or professional program?

☐ no

☐ yes → If yes, please list the program(s)

____

____

____

gender as a contingency question, followed by questions on aggression for the men and questions on victimization for the women. Instead, they chose to give the men and women different questionnaires, which made it easier for respondents to identify the appropriate questions.

Most surveys use direct questions to ask for the information desired, such as "What is your opinion about whether people convicted of murder should be subject to the death penalty?" Occasionally indirect questions, where the link between the information desired and the question is not as obvious, are used. Asking what others think—as in the question, "How do you think your co-workers would feel if a woman manager were appointed to your group?"—as an indicator of the person's own opinion is an example of an indirect question.

How to Ask Questions

Questionnaires and interview schedules should be kept as *short* as possible and ask only the questions that are necessary for the planned analysis. It's useful to start by making a list of the variables that are of interest and then do a literature search[2] to see if another study's measurement scheme can be used or adapted. A draft of a survey should be pre-tested with a small sample of respondents similar to those who will be selected for the actual study. The pre-test helps the researcher to see if the questions and instructions are clear, if the open-ended questions generate the kinds of information wanted, and if the closed-ended questions have sufficient choices and elicit useful responses.

STOP AND THINK *An interviewer was thinking of asking respondents, "Do you think that men's and women's attitudes toward marriage have changed in recent years?" and "Do you feel that women police officers are unfeminine?" Try to answer these questions. Do you see anything wrong with the wording?*

[2] In addition to seeking articles and books reporting the results of research that include discussions of their measuring schemes, there are listings of many questionnaires and interview schedules. For example, *Handbook of Research Design and Social Measurement* (Miller, 1991) and *Handbook of Family Measurement Techniques* (Touliatos, Perlmuter, and Straus, 1990) each include dozens of scales, indexes and inventories.

Guidelines for Question Wording

Survey research is based on language, so it's important to remember that words can have multiple meanings and can be misunderstood by respondents. Questions should be clear and concise and words carefully selected as wording can affect responses. A study of affirmative action, for example, found only 35 percent agreeing that where there has been job discrimination against blacks in the past, "preference in hiring and promotion" should be given, while 55 percent said they favored "special efforts" to help minorities get ahead (Verhovek, 1997: 32). Studies with questions about public spending priorities found much more support for "providing assistance to the poor" than "spending for welfare" (Smith, 1989). Even minor changes in wording (for example, "dealing with drug addiction" versus "drug rehabilitation") can result in large changes in the percentage agreeing with a statement (Rasinki, 1989).

After reading the following suggestions for question wording, review the poorly worded questions and suggested revisions in Box 9.5.

1. Avoid *loaded words,* words that trigger an emotional response or strong association by their use.
2. Avoid *ambiguous* words, words that can be interpreted in more than one way.
3. Don't use *double negative* questions, questions that ask people to disagree with a negative statement.
4. Don't use *leading questions,* questions that encourage the respondent to answer in a certain way, typically by indicating which is the "right" or "correct" answer.
5. Avoid *"threatening"* questions, that is, questions that make respondents afraid or embarrassed to the give an honest answer.
6. Don't use *double-barreled* or *compound* questions—questions that ask two or more questions in one.
7. Ask questions in the language of your respondents, using the idioms and vernacular appropriate to the sample's level of education, the vocabulary of the region, and so on.

STOP AND THINK *Did you identify what was wrong with the questions in the previous Stop and Think? Did you see that the first question is double-barreled in that it asks two questions about marriage, one about men's attitudes, and one about women's? Did you notice that the word "unfeminine" in the question about women police officers is really a loaded word?*

Response Categories Guidelines

When response categories are used, a large enough range of answers is needed. Some common closed-ended responses are listings of answers from strongly agree to strongly disagree, excellent to poor, very satisfied to very dissatisfied, or a numerical rating scale with only the end numbers (such as 1 and 10) given as descriptors. For questions with more than one possible answer, we can use a list and ask the respondent to identify all that apply, by asking, for

BOX 9.5

Examples of Poorly Worded Questions and Suggested Revisions

Ambiguous question (adapted from Fowler, 1988):

How many times in the past year have you seen or talked with a doctor about your health?

Revised:

We are going to ask about visits to doctors and getting medical advice from them. How many office visits have you had this year with all professional personnel who have M.D. degrees or those who work directly for an M.D. in the office, such as a nurse or medical assistant?

Double negative question:

Do you disagree with the view that the President has done a poor job in dealing with foreign policy issues this past year?

Revised:

What do you think of the job that the President has done in dealing with foreign policy issues this past year?

Double-barreled question:

What is your opinion of the state's current economic situation and the measures the governor has taken recently?

Revised:

What is your opinion of the economic situation in the state at present?

What do you think of the measures that the governor has taken in the past month?

Leading question:

Like most former presidents, the President faces opposition as well as support for his economic policies. How good a job do you think that the President is doing in leading the country out of the economic crisis created by the national debt?

Revised:

Are you aware of the President's policies to reduce the national debt?

If so, what is your opinion of them?

Threatening questions: (adapted from Bradburn and Sudman, 1979)

Have you ever smoked marijuana?

Revised:

Different people use different terms for marijuana. What do you think we should call it so you understand us? *(If no response or awkward phrase, use "marijuana" in the following questions. Otherwise use respondent's word(s).)* ________ is commonly used. People smoke ________ in private to relax, with friends at parties, with friends to relax, and in other situations. Have you, yourself, at any time in your life smoked ________?

example, "Which of the following television shows have you watched in the past week? Check all that apply." Arguing against a "mark all that apply" approach, Rasinki, Mingay, and Bradburn (1994) assert that more responses are elicited when the question is broken down into separate questions, each followed by "yes" or "no" answers. Another debate in the research literature is the question of whether a "no opinion," "don't know," or "middle category" should be offered. (For example, "Should we spend less money for defense, more money for defense, or continue at the present level?") Some argue that it is better to offer just two polar alternatives, whereas others claim it is more valid to offer a middle choice. Studies have found people to be much more likely to select a middle response if it is offered to them (Asher, 1992; Schuman and Presser, 1989) and that listing the "middle" option last rather than in the middle makes it more likely to be selected (Bishop, 1987).

Question Order and Context

Responses to questions can be affected by the question order. People will often interpret successive questions as related to the same topic unless they are told otherwise (Clark and Schoeber, 1992). If there are several questions on one topic, for example, question placement can affect the responses to later questions. In the later questions the respondent, consciously or unconsciously, might try to be consistent, might be tired, or might use the information or context provided by a previous question in formulating an opinion (Frey, 1983). For example, King (1994) found that reports of alcohol consumption were higher when questions were included on the Semi-Quantitative Food Frequency Questionnaire than when they were used on a questionnaire exclusively targeting alcohol use.

In ordering questions, it is important to begin with interesting, nonthreatening questions, and to save questions about sensitive topics for the middle or near the end of the survey. If questions about emotionally difficult topics are included, it's preferable to have some "cool down" questions follow them.

Gray, Palileo, and Johnson began their survey with nonthreatening questions about social characteristics (see Appendix B) before going on to questions on sexual victimization and sexual aggression on the women's and men's surveys respectively. A middle section included questions about nonsexual victimization and risk-taking behavior. The final items were the rape myth statements. There are tradeoffs in ordering the questions in any given way. Sociologist Dave Johnson had this to say about the existing question order:

> We decided to ask nonthreatening questions first to ease the respondents into the questionnaire. We asked the sexual victimization questions fairly soon because it was the most important information we wanted, and we did not want to reveal our "agenda" prior to asking the victimization questions. We felt that the rape myth questions might have done this, so we held those until the end (personal communication, 1995).

As a result of the question order, respondents were asked about victimization or aggression before turning to questions on rape myths. Thus the questions on victimization or aggression provided a context for answering questions about blaming victims. Answers to the victim blaming questions might have been different had they been asked first.

Technology and the Self-Report

Technology has made possible some interesting innovations in the self-report method. Computer-assisted data collection is now used in some situations as a substitute for paper and pencil questionnaires. In this modification of a traditional tool, the researcher provides the respondent with a computer programmed to present the questions and register the answers. This approach allows a randomly selected set of questions, a question sequence with the contingency choices appropriate for each respondent, and the possibility of validating answers by repeating questions (Saris, 1991). New technology can also be used to take advantage of the increasing numbers of people using electronic mail to "interview" respondents via the Internet. Although this technique is not yet common, one of us, Emily, was recruited through an Internet mailing list to be part of a nonrandom sample and completed a brief e-mail questionnaire on professional identities.

One especially interesting innovation designed to reduce memory inaccuracy is the experience sampling method, or ESM, used by Larson and Richards (1994) to "sample" experiences. In *Divergent Realities* (1994), they describe employing the technique to study the way that family life was experienced by different members of the same family. The mother, father, and one child in a sample of 55 families each received an electronic pager programmed to beep randomly eight times a day for a week. Using a self-report booklet, each family member recorded privately what he or she was doing, thinking, and feeling just before being paged. As the technology changes, we are likely to continue to see new ways to ask and answer questions.

Summary

Self-report methods are widely used in social research. They can profitably be used to collect information about many different kinds of variables, including attitudes, opinions, levels of knowledge, social characteristics, and some kinds of behavior. Careful wording of introductory statements and instructions and the judicious use of open- and closed-ended questions are critical if questionnaires and interviews are to provide useful results. As we saw in the focal research article on rape victim blame, clearly worded questions and answer choices and respondents willing to participate allowed the researchers to collect a great deal of data that could not be obtained either through documents or by observation.

Questionnaires and structured interviews are less useful when the topic under study is difficult to talk about or when questions about the topic are unlikely to generate honest or accurate responses.

In choosing between the various kinds of questionnaires and interviews, factors like cost, response rate, anonymity, and interviewer effect must be considered. Table 9.1 sums up the advantages and disadvantages of the types of questionnaires and structured interviews.

TABLE 9.1 **Advantages and Disadvantages of Questionnaires and Structured Interviews**

Data Collection Method	*Advantages*	*Disadvantages*
Group administered questionnaire	Inexpensive Time efficient Can be anonymous, which is good for "sensitive" topics Good response rate	Not all populations possible Must have literate sample Must be fairly brief Need permission from setting or organization
Mailed administered questionnaire	Fairly inexpensive Completed at respondent's leisure Time efficient Can be anonymous which is good for "sensitive" topics	Response rate may be low Need names and addresses Must have literate sample Can't explain questions
Individually administered questionnaire	Good response rate Completed at respondent's leisure	Most expensive questionnaire Need names and addresses Must have literate sample
Structured phone interview	Good response rate Little missing data Can explain questions and clarify answers Can use "probes" Don't need names or addresses Good for populations with difficulty reading/writing	Expensive Time consuming Respondents may not feel anonymous Some interviewer effect
Structured in-person interview	Good response rate Little missing data Good for populations with difficulty reading/writing Can assess validity of data Can explain questions and answers and use "probes"	Most expensive Most time consuming Need names and addresses Interviewer effect Confidential, not anonymous

SUGGESTED READINGS

Bradburn, N. M., and S. Sudman. 1988. *Polls & surveys: Understanding what they tell us.* San Francisco: Jossey-Bass.

In addition to introductory material outlining the basics of survey research and a brief history of public opinion polling, Bradburn and Sudman provide extensive methodological coverage for users of survey data. Three major methods (mailed questionnaires, face-to-face and phone interviews) are discussed, along with sampling issues, question construction, and data analysis.

Converse, J. M. 1987. *Survey research in the United States: Roots and emergence 1890–1960.* Berkeley: University of California Press.

Focusing on the historical development of survey research, Converse traces the origins and "ancestors" of surveys, the way in which polls and government funding facilitated their growth in the 1930s and 1940s, and their establishment and struggles within the university setting in more recent decades.

Dillman, D. A. 1991. The design and administration of mail surveys in *Annual Review of Sociology 1991*, edited by W. R. Scott and J. Blake). Palo Alto, CA: Annual Reviews.

Dillman gives a good overview of recent research on ways to improve the design and administration of mail surveys. He emphasizes sampling error, sample selection, nonresponse, and measurement error.

Fowler, F. J. 1988. *Survey research methods.* Newbury Park, CA: Sage.

Fowler provides a comprehensive overview of the issues that confront survey researchers, including ethical concerns, sampling, nonresponse, choice of data collection strategy, question construction, and data analysis.

Marsh, C. 1982. *The survey method.* London: Allen & Unwin.

One of the best overviews of the survey, this book includes a detailed history of its use in Great Britain and a response to the critics of the survey method.

EXERCISE 9.1

Evaluating Choices

Use the Gray, Palileo, and Johnson study in this chapter to answer the following questions.

1. List two of the hypotheses in this study and identify the independent and dependent variables for each.

2. This research used a group-administered questionnaire. What were the advantages and disadvantages of this data collection method for this study?

3. If the researchers had used a structured in-person interview instead, what advantages and disadvantages would have resulted?

4. Pick two of the variables used in the study and, using the questionnaire provided in Appendix B, list the question(s) that were used as indicators for each variable (include answer categories if closed-ended questions were used). If there is information on the scale or index they constructed for the variable, include that information as well.

5. For each of the two variables you selected for question 4, discuss whether or not you think that the questions you listed above are valid indicators.

6. Do you feel that *any* of the questions in the questionnaire were inadequate or problematic in any way? If so, describe the problem and, if possible, rewrite the question to improve it.

EXERCISE 9.2

Open- and Closed-Ended Questions

Check with your instructor about the need to seek approval from the Human Subjects Committee, the Institutional Review Board, or the committee at your college or university concerned with evaluating ethical issues in research. If necessary, obtain approval before completing the rest of the exercise.

1. Conduct a brief interview with five people who are working at least five hours per week for pay. Use the following open-ended questions for your interview.

    1. Are you currently employed?

    2. What is your job?

    3. On average, how many hours per week do you work?

    4. How well do you like the sort of work you are doing?

    5. How would you compare your job with other jobs that you're familiar with?

    6. If it was up to you, what parts of your job would you change?

    7. What parts of your job do you find most satisfying?

2. Before data can be analyzed statistically, it must be **coded.** Coding involves creating a limited number of categories for each question so that each of the responses can be put into one of the categories. Code your

respondents' answers to five of the questions you asked. For each question keep the number of the response categories you create to no more than four.

3. Based on your respondents' answers, your coding scheme, and other possible answers, rewrite two of the questions into closed-ended questions.

EXERCISE 9.3

Asking Questions (or Have I Got a Survey for You!)

Check with your instructor about the need to seek approval from the Human Subjects Committee, the Institutional Review Board, or the committee at your college or university concerned with evaluating ethical issues in research. If necessary, obtain approval before completing the rest of the exercise.

1. Select between two and four concepts or variables to study. Do not choose "sensitive" or threatening topics. (Appropriate topics include opinions about candidates running for office, career aspirations, involvement in community service, marriage and family expectations, social characteristics, and so forth.) List the variables you have selected.

2. Construct between one and five questions to measure each variable. Feel free to use questions from articles you have previously read or those listed in books like *Handbook of Research Design and Social Measurement* (Miller, 1991) or *Handbook of Family Measurement Techniques* (Touliatos, Perlmuter, and Straus, 1990). Develop an interview schedule with at least 5 and no more than 10 questions.

3. Make six copies of the interview schedule, leaving enough space to write down respondents' answers. Select six people who agree to be interviewed. Administer your interview schedule to three people as a face-to-face interviews and to another three people over the phone.

4. Turn the interview schedule into a questionnaire, complete with instructions and a cover sheet explaining the research and its purpose. (Introducing yourself as a student and the survey as a part of the requirements of your course would be appropriate.) Give the questionnaire to three new people to complete.

5. Compare the information you obtained using the three methods of data collection. Comment on the validity and reliability of the information you received using each method. Which did you feel most comfortable using? Which was the most time consuming method? If you were planning an actual study using these questions, which of the three methods would you select? Why?

6. Based on your experience, select one of the methods and revise the questionnaire or interview schedule you feel you would use in an actual study. Turn in a copy of the revised version with this assignment.

7. Attach the six completed interviews and the three completed questionnaires to this exercise as an appendix.

EXERCISE 9.4

Structured Interview Questions from the General Social Survey

The General Social Survey (GSS) uses a structured interview schedule. It has been administered to a national probability sample almost every year since 1972. Some questions have occurred on every survey, other questions have rotated, and some have been used only in a single survey.

The GSS measures a great many variables. The variables included on past surveys cover a wide range of behaviors, among them job responsibilities, drug use, TV viewing, and the practice of religion, as well as numerous attitudes, such as opinions about affirmative action, capital punishment, family roles, and confidence in the military.

Select one behavior and one attitude either from those listed here or one of each of your own choosing, and, using the GSS web site, find two questions on each of these that have been included on the GSS.

Answer the following questions based on the GSS home page (http://www/icpsr.umich.edu/gss/home.htm).

1. What specific behavior did you select for this assignment?

2. List two GSS questions that focus on this behavior.

3. What specific attitude did you select for this assignment?

4. List two GSS questions that focus on this attitude.

10 Qualitative Interviews

Sandra Enos began her research at a state prison for women to meet the requirements of a graduate level course in qualitative methodology. She worked for a time as a volunteer in a parenting program for inmates while collecting research data as a participant observer. She discovered her own previously unrecognized expectations about inmates and realized that she knew far less about their experiences as mothers than she thought she did. Although people often speak about how important it is to be a good parent and how central mothers are to child development, Enos recognized that relatively few opportunities to systematically observe the practice of mothering exist. The prison setting gave Enos access to a group of women with diverse racial, ethnic, and class backgrounds and the opportunity to study mothering under strained circumstances.

Enos's work at the parenting program site helped her to understand the complex nature of families and the differences between biological motherhood and "mothering" —the array of activities that include caretaking, managing caretaking, and assuming ultimate responsibility for children. As a volunteer, she soon became an accepted member of the group that assembled for extended visitations with children. She had many informal conversations with the women, their children, and the program staff. As a student and researcher, she could ask questions, be curious, and have things explained.

Over the course of the semester, Enos began to see patterns in the narratives and to develop an understanding of how women attempt to manage motherhood while they are in prison. She became intrigued by the ways that the options available to the women were related to race and class and to definitions of motherhood. After completing the course requirements, Enos decided to continue the project by adding another method of data collection—the qualitative interview. The focal research in this chapter is the result of those efforts.

Introduction

> What does it all mean, anyway, when the talking is done, the microphone cable is rewound, and the tapes are labeled and transcribed? What is the status of an interview narrative? An interview is not, in any simple sense, the telling of a life as much as it is an incomplete story angled toward my questions and each woman's ever-changing sense of self and how the world works (Frankenberg, 1993: 41).

Qualitative interviewing has been a social science tool for a century, and its uses have been debated for more than sixty years (Lazarsfeld, 1944; Thurlow, 1935). The less structured interview, which has been called "the art of sociological sociability, the game which we play for the pleasure of savoring its subtleties" (Benney and Hughes, 1956: 137), is an important method of data collection. These interviews are opportunities to learn about people's backgrounds and experiences, their attitudes and perceptions of themselves, and their views about the groups of which they are a part and the organizations with which they interact.

STOP AND THINK *We discussed the structured interview in the last chapter. How do you think it differs from the qualitative interview that we'll be discussing in this chapter?*

qualitative interview, a data collection method in which an interviewer adapts and modifies the interview for each interviewee.

structured interview, a data collection method in which an interviewer reads a standardized interview schedule to the respondent and records the answers.

interview schedule, a set of preformulated questions that are used to guide a structured or semi-structured interview.

The **qualitative interview** has much in common with the **structured interview** discussed in the last chapter. Both kinds of interviews rely on self-reports, and anticipate that the interviewer will do more of the asking and the respondent will do more of the answering. Both kinds of interviews, "far from being a kind of snapshot or tape-recording ... in which the interviewer is a neutral agent who simply trips the shutter or triggers the response" (Kuhn, 1962: 194), are opportunities for data to emerge *from the social interaction* between the interviewer and interviewee.

Despite these commonalities, the two kinds of interviews are distinct. In the structured interviews that we discussed in Chapter 9, the interviewer uses a standardized **interview schedule** heavily weighted toward closed-ended questions. The questions and closed-ended answer choices are delivered to each respondent with as little variation as possible to achieve uniformity of questioning. In addition, some kinds of structured interviews, such as phone interviews using random digit dialing, can be done anonymously without identifying the respondent.

In contrast, the qualitative interviewer uses an interview schedule with open-ended questions (either preformulated or constructed during the interview), can modify the order and the wording of the questions, and often asks respondents to elaborate or clarify their answers. Before the interview, the interviewer typically knows the topics the interview will cover and the kind of information that is desired, but has the opportunity to tailor each interview to fit the participant and situation. In qualitative interviews, the answers are usually longer, and the interviewees can phrase answers in their own terms. The interviewer typically knows the identity of the interviewee and, ethically, must keep confidential all information obtained.

observational techniques, methods of collecting data by observing people, most typically in their natural settings.

Qualitative interviewing has been used as the sole data collection technique in a wide variety of studies, including Frankenberg's (1993) study of white women's social constructions of race, Rubin's (1976; 1994) studies of working-class families, and Weiss's (1990) study of occupationally successful men. Informal interviewing has also been used in conjunction with **observational techniques** (collecting data by observing people in their natural settings, which we will discuss in detail in Chapter 11), as in, for example, Rollins' (1985) study of domestic workers and their employers, Hochschild's (1989) research on "the second shift"—the family work that must be negotiated when both the spouses are employed outside the home—and a study by Rogers and Henson (1997) on structural vulnerability and sexual harassment in temporary clerical employment.

This chapter's focal research section is by Sandra Enos, a sociologist whose research interests included both families and corrections. She combined these interests by asking how the social processes of mothering get worked out when mothers are in prison. While she did work as a volunteer in the prison parenting program, observing and talking informally to the participants, in the excerpt that follows, she draws on her qualitative interviews for her analyses of "mothering" while in prison.

FOCAL RESEARCH

Managing Motherhood in Prison[1]

by Sandra Enos

Introduction

Critiques of existing research on families have identified ideological assumptions buried in mainstream sociological work in this area (Baca Zinn, 1992). Scholars have argued that myths about the family not only underlie public discourse, but inform much of the social science work in this area (Baca Zinn and Eitzen, 1994). Traditional sociological work has posed one normative family form and defined other forms as deviant or at least underdeveloped (O'Barr, Pope, and Wyer, 1990).

Some recent work has included reconsideration of "mothering" as a status, role, and practice. The assumptions about effective mothering in much of the social science literature and the popular discourse have been described: Mothers are totally absorbed in mothering, the mother is the only figure who can give the child what s/he needs, the mother's performance is the link

[1]This article is adapted with permission from Sandra Enos's paper of the same name presented at the 1995 Annual Meetings of the Eastern Sociological Society in Philadelphia.

between the child's upbringing and his/her future success, the mother has resources to give the child what he needs, and the mother should be able to manipulate the child's behavior toward successful adulthood (Chodorow and Contratto, 1980; Caplan and Hall-McCorquodale, 1985).

These assumptions do not make room for mothering styles outside the middle class (Maher, 1992; Glenn, 1987; Dill, 1988) or the way motherhood is experienced by women in different race/ethnic categories or life cycle positions. There is only a small body of work (Collins, 1990; Rubin, 1994; Stack, 1974; Young and Wilmott, 1957) that has investigated how mothers under strain and economic stress "do mothering." Few have studied mothering in a way that makes room for the interpretation of the actors—the mothers themselves. It is important to analyze mothering, including race and class differences, in a way that goes beyond overly idealistic images or negative categorizations of mothers as evil or neglectful caretakers.

For women who come into conflict with the state and its agents of social control, specifically criminal justice, correctional, and drug control and treatment organizations, being a mother brings both positive and negative judgments. These women are accepted as "normal" because they have succeeded in fulfilling the gendered expectation that women bear children. But, they suffer denunciation because the behavior which leads to incarceration is not congruent with what is generally considered "motherly."

While the "imprisonment boom" between 1980 and 1992 increased the male population behind bars by 167 percent, the number of sentenced women increased by over 250 percent during the same time (Gilliard and Beck, 1994). Although the number of women in prison remains small (45,9000) in comparison to the number of incarcerated men (769,700) (Gilliard and Beck, 1994), their situation creates special problems. Most women in prison are mothers, with approximately 65 percent to 80 percent of them having children under the age of 18 (Baunach, 1985). Because arrest, pre-trial holding, and court processing are usually unplanned, the placement of children may be haphazard and not controlled by the offender. Imprisonment means that existing family and extended family units must reorganize to adjust to life without the offender. Because women are the caretakers of children and other family members, their imprisonment typically means a larger disruption in the life of a child. Nationally, 90 percent of children who have fathers in prison live with their mothers, while only 25 percent of those with mothers in prison reside with their fathers (Snell, 1994).

In recent years, the rehabilitation of women offenders has come to mean attending to their needs as parents as well. Responding to these needs, some jurisdictions now offer parenting programs with enhanced contact and visitation opportunities for mothers and children. The premise is that the responsibility of parenting may serve as a rehabilitative element. How mothers remain connected to their children, how they maintain a place and position in their families, how they arrange and evaluate child-care options are elements of the social process of "managing motherhood" in prison. They are the research interests which guided this research.

Methodology

The Sample and Method

A women's prison in the Northeast served as the study site. The facility is small in size, holding a maximum of 200 women awaiting trial or serving sentences. Approximately half of the mothers in the prison were involved with the state's child welfare agency as a result of complaints of child abuse, neglect, or the presence of drugs in infants at birth. In other cases, the women sought help from child welfare to place children when they felt they had no other viable option.

The parenting program in this prison is operated off-site and allows considerable freedom from usual security constraints. The program facility is filled with toys and more closely resembles a school than a correctional site. Child welfare authorities view the program as a way to demonstrate a willingness to become rehabilitated (an indication of "proving fitness" as a parent) and to work toward getting custody of children upon release. For these reasons, participation in the program is a welcome opportunity.

To participate in the parenting program, the women had to be mothers or guardians of minor children, have either minimum or work release security status, and be free from work obligations at the time of the program. The participants in the parenting program were serving sentences similar to those of the general population—ranging from six months to a few years for nonviolent offenses such as embezzlement, fraud, soliciting for prostitution, drug offenses, breaking and entering.

I was granted access to the parenting program by the Warden and the Director of the parenting program in 1993. I assumed two roles—observer and volunteer/program aide. The mothers gave me permission to sit in on the parenting group (an hour or more discussion group designed to help mothers work out issues related to reunification with children) and to observe the children's visits. Attendance at the program varied from week to week, attracting from seven to twelve women and between twelve and twenty-five children. There was turnover in the population; approximately twenty women participated during the four months I was an observer.

I attended 16 sessions and held informal conversations with both the women and their children during the parent-child visits. The conversations were typically about the child. Although I was able to interview three women fairly completely during this time, the informal conversations with the women and children were typically limited to a few minutes at each session so that my research interests did not detract from the real purposes of the parenting program—to allow contact and enhance communication between parents and their children.

I made two key observations during this preliminary work: The placement of children and their mothers' paths to prison was different for African-Americans and whites. An interest in the patterns and possible connections among them led to the interview stage of data collection.

I conducted 12 additional interviews, most with women in the parenting program, but several with women identified by staff as serving terms of longer duration for more serious crimes. I identified myself as a researcher and told them the interview was about motherhood. I made it clear that, unlike other conversations about their children with authority figures from child welfare, counseling agencies, drug treatment facilities, school departments, and so on, their conversations with me would not directly affect them. Most were curious about my study and wanted to share their stories hoping it might help others or lead to changes in the parenting program. The women I talked to were white, African-American, or Hispanic, ranged in age from 19 to 40 years old, and were responsible for a total of 22 children.

In the lower security levels, the interviews were held in a hallway or program area. In the higher security level, a correctional official stood outside the room where I was allowed to conduct interviews. I would characterize the exchanges as directed conversations rather than structured interviews. Some of the participants allowed me to tape record the interview and all but one of the rest agreed to my taking notes. One woman insisted on neither notes nor tape, which required reconstructing the interview from memory immediately after leaving the site. Each interview took approximately two hours.

The Emotional Content of Interviewing

I was concerned about how forthcoming the women would be in relating their experiences of mothering. Given the dynamics related to the presentation of self and to the pressures for women to conform to idealized conceptions of mothering and doing motherhood (for example "Mothers love their children unconditionally"; "Mothers put their children first"; "Mothers consider the welfare of their children before acting"; etc.), I was concerned about the fruitfulness of certain lines of questioning. For the most part, I directed conversations along paths initially suggested by women themselves. In later interviews, I occasionally paraphrased the responses from previous interviews as a way of giving the women something to respond to. In very few instances did I ask questions that would result in a simple yes or no answer.

My aims were to strike a balance between protecting each woman's privacy, getting an accurate picture of how each viewed her role and place as a mother, and hearing each tell her story in her own voice. To establish this balance, I asked very broad questions in a semi-structured interview format. For instance, women were asked "How did you get here? What brought you to prison?" and "Where are your children living?" (See the Appendix to this essay for the entire interview schedule.) The answers often required additional clarification. Some women had been to prison several times and were confused about whether I was referring to their first incarceration or their most recent term. The question about where their children were living was too vague, as some women had been caring for children who were not their biological children and others had already lost custody of their children to the state or other caretakers before imprisonment. Many issues not originally in the interview outline were brought into the conversation by the women themselves. For instance, several women, more typically the middle-class women, brought up problems with confinement although I did not ask about them.

In the interviews, the women were direct and forthcoming. Some interviewees stated very simply that they were not emotionally bonded to certain of their children; others expressed overly optimistic views of the possibility of reestablishing good and easy relations with their children; others were not in touch with their children because current caretakers prevented contact; some discussed in detail strategies for maintaining a drug habit while caretaking their children. In other words, while the emotional context of the interviews was rich and the conversation occasionally difficult, the mothers rarely presented an idealized view of themselves as mothers or a view of "mothering" as nonproblematic.

The women expressed an appreciation for the time spent speaking about their children and their hopes for the future. Another rewarding aspect for the mothers was that I offered to photograph their children, giving them both the photographs and negatives. Since many of the women didn't have recent pictures of themselves or their children, this opportunity was appreciated.

Findings

Between 25 percent and 34 percent of imprisoned mothers do not live with their children prior to incarceration (Snell, 1994). For those who do co-reside with their children, child placement becomes a challenge upon imprisonment. Both in the national data and among the women I interviewed, there are important differences in the white, African-American, and Hispanic child placements (see Tables 10.1 and 10.2). White mothers are much more likely to leave their children with husbands or to have them placed in non-relative care (that is, foster care or with friends), while African-American women are more likely to place their children with their mothers or other relatives. While this difference has been noted (Bresler and Lewis, 1986), it has largely been ignored in the literature. Making a decision about placing a child is not easy. A mother may see only problematic choices. That is the situation faced by Stacey,[2] a white woman in her twenties, who wanted to protect her child but also wanted to keep state attention and intrusion away from her family. Her four-year-old son was living with his father, and she had heard from friends that the father had resumed a drug habit and the selling of narcotics. Stacey felt that contacting the child welfare authorities would prompt an investigation of the father but was also likely to bring the family to the attention of an agency that could remove the child from the home. As she noted, "Most of the time, trying to get help means getting into trouble."

Maintaining presence and "force" as a mother is difficult when a woman is absent from the home and is even more problematic when the mother has few resources. Bernice, a pregnant African-American mother of a five-year-old son who had been in many foster homes, echoed Stacey's concern. She said, "I feel like I'm having babies for the state. Once they [the child welfare agency] get in your life, they're always there. They stick with you forever." This mother remarked that foster care was the worst place for an

[2]Stacey and the names of all the other women in this paper are pseudonyms.

TABLE 10.1

Race/Ethnicity of Female Inmates and Residence of Children (N = 15)

	White	*Black*	*Hispanic*
Child living with			
Husband	3	0	1
Mother or sister	1	5	1
Foster care	3	1	0

TABLE 10.2

Residence of Children of Female Inmates in all State Prisons by Race/Ethnicity, 1991*

	White	*Black*	*Hispanic*
Child living with**			
Father	35.2%	18.7%	24.4%
Grandparents	40.6%	56.7%	54.9%
Other relatives	14.7%	23.7%	22.8%
Friends	5.7%	2.7%	4.2%
Foster home	12.6%	5.8%	6.5%
Agency/institution	2.1%	1.8%	2.1%
Alone	1.9%	2.3%	2.3%

* Snell, Tracy L. 1994. *Women in Prison: Survey of State Inmates, 1991.* Bureau of Justice. Washington, D.C.: U.S. Department of Justice.
**Totals might exceed 100 percent because inmates with more than one child may have reported multiple responses in different placements.

African-American child to be. Her child was placed with a foster family only because her family and friends wouldn't (or couldn't) respond to her need for a place for him to live during her incarceration.

Other women were more positive about putting their children in the care of foster parents. Some white women characterized it as a "life saver." Louise talked about her satisfaction with foster care and she was careful to make a distinction between being a child's mother and the caretaking of the child. Her comments indicate that she was able to separate the biological identity of mother and the social and psychological work of "mothering."

Interviewer: So, you had a baby just before you came to jail. Where is he now?

Louise: He's with foster parents who really love him a lot. They are just great with him.

Interviewer:	Does he visit you here?
Louise:	Sure he does, a little, the foster parents bring him to me.
Interviewer:	And does he know you?
Louise:	Well, he knows who I am. But, he doesn't know me know me.
Interviewer:	Can you explain that?
Louise:	Well, he knows I'm his mother, but he doesn't know me that way.

The distinction Louise makes may imply special challenges when the mother resumes caretaking of the child or when she reflects upon how she may handle the tasks associated with caretaking. Pam, a white middle-class woman, noted that her husband had taken up much of the responsibility for their developmentally delayed son. She felt that some activities can be delegated, while others are better for mothers to do. As an example, she described what happened when the father took the son shopping for school shoes. She was dismayed at his choice, saying "I looked at what they picked out and thought to myself, 'My God, I would have never bought such a thing for him. No mother would have.' There is a difference in what a mother would have done."

Unlike other women, Pam had no difficulty maintaining her position in her family. She was still defined as the family's major caretaker. She reported that her husband visited often and that they maintained the close "working relationship" that parenting a special child required.

Family members that assume caretaking responsibilities immediately after a child's birth were viewed differently than those who became involved later in the child's life. Vanessa, an African-American woman told me, "Just after I had the baby, my mom took him. Then, I got a short bid. But ever since he was born, he was her baby. I never had a chance." Vanessa is suggesting she has a marginal role in the child's life as his mother—either because others have asserted ownership in her absence or because of her lack of effort.

Other women noted the same development, but some suggested that this may have turned out for the best. Kate, a white woman said, "I don't know what was wrong with me then. I just never had any feeling for my first. I just had it and then my mom took over. We used to do drugs together and party all the time and then she went straight and was preaching to me all the time. Then, the baby came and she took over." Kate conveyed a simpler delineation of responsibility by putting her child in her mother's hands with little regret.

Significant differences were also revealed as women described their paths to prison. White women were much more likely to attribute blame to their families of origin for their problems, suggesting that they were pushed out of their homes or subjected to early sexual and physical abuse and neglect. Few white women relied on their families to take care of their children and most were wary of incurring "unpayable debts" if they utilized family resources to help them with child care.

On the other hand, African-American women tended to trace their paths to prison to their attraction to the "fast life," arguing that their families had tried unsuccessfully to divert them from a life of crime. Placing children with family and friends was expected and acceptable. In many cases, the mothers had previously prevailed upon relatives to care for their children when they were engaged in criminal activities. These children were taken care of by a variety of adults before the women's imprisonment, and most African-American women expected to continue using these shared child caring arrangements after release.

There were exceptions to these patterns. Not all children of white women were in the care of husbands, the state or other non-kin, and not all children of African-American mothers lived with relatives or other kin. White women with access to deviant life styles had placements similar to the African-American inmates. That is, white women with parents or relatives involved in law-breaking behavior, or who lived in areas where entry into this lifestyle was easy, did place children with their own mothers. On the other hand, African-American women who traced their imprisonment to deficient families of origin and were estranged from them placed their children in foster care.

The African-American women suggested that because their families had an expectation that members would face "trouble" (i.e., attractions of the "street" and conflicts with educational, social and state institutions), the family units were usually prepared to extend resources when needed. They also felt that the expectations for girls in white families were more rigid and punitive, with young women expected to "stay out of trouble," and to follow gendered prescriptions for behavior. Most of the white women did report that their families had expected them to "follow the rules," and handled deviations by constricting the family's resources to protect the other members of the family and to keep the offender isolated. In contrast to the African-American women, the white women did not perceive family patterns as differing by race.

Summary and Conclusion

Imprisonment provides an important vantage point for sociologists to examine family responses to women in trouble and to understand more completely how these responses are affected by larger structural forces. While we generally think that families keep their members out of trouble, we also must also understand that membership in some families means having direct connections to careers involving illegal behaviors.

Two patterns of women's incarceration and race were revealed by the current investigation. The first pattern was that African-American women were found to be more likely than white women to rely on kin to care for their children before, during, and after incarceration and that white mothers tended to depend on foster care placement. The second pattern was that African-American women were more likely than white women to attribute their path to prison as being the result of personal life choices; white women tended to see it as a result of problems within their families of origin. As I've suggested, these two patterns might be related to one another because of differences in family expectations regarding the likelihood of encountering "trouble" and the acceptability of asking for help from kin.

This examination has also focused on the complex nature and meanings of motherhood. Motherhood might mean a biological connection to child, taking care of a child, maintaining overall responsibility for a child's welfare (even if not directly engaged in the care), or sustaining a unique and irreplaceable relationship with a child. We can gain insight into the management of motherhood by seeing how women in prison understand motherhood and meet its challenges.

REFERENCES

Baca Zinn, M. 1992. Reframing the revisions: Inclusive thinking for family sociology. In *The knowledge explosion*, edited by D. Kramarae and D. Spender. New York: Teacher's College Press.

Baca Zinn, M., and D. S. Eitzen. 1993. *Diversity in families*, 3rd ed. New York: HarperCollins College.

Baunach, P. J. 1985. *Mothers in prison*. New Brunswick, NJ: Transaction.

Bresler, L., and D. K. Lewis. 1986. Black and White women prisoners: Differences in family ties and their programmatic implication. *The Prison Journal* 63: 116–122.

Caplan, P. J., and I. Hall-McCorquodale. 1985. Mother-blaming in major clinical journals. *American Journal of Orthopsychiatry* 55 345–353.

Chodorow, N., and S. Contratto. 1982. The fantasy of the perfect mother. In *Rethinking the family*, edited by B. Thorne and M. Yalom. New York: Longman.

Collins, P. Hill. 1990. *Black feminist thought: Knowledge, consciousness, and the politics of empowerment*. New York: Routledge.

Dill, B. T. 1988. Our mothers' grief: Racial ethnic women and the maintenance of families. In *Race, class and gender: An anthology*, pp. 215–237. Newbury Park, CA: Sage.

Glenn, E. N. 1987. Gender and the family. In *Analyzing gender: A handbook of social science research*, edited by B. B. Hess and M. M. Ferree, pp. 348–380. Newbury Park, CA: Sage.

Gilliard, D. K., and A. J. Beck. 1994. *Prisoners in 1993*. Bureau of Justice Statistics: Washington, D.C.: U.S. Department of Justice.

Maher, L. 1992. Punishment and welfare: Crack cocaine and the regulation of mothering. In *The criminalization of a woman's body*, edited by C. Feinman, pp. 157–192. New York: Harrington Park.

O'Barr, J., D. Pope, and M. Wyer. 1990. Introduction. In *Ties that bind: Essays on mothering and patriarchy*, edited by J. O'Barr, D. Pope, and M. Wyer, pp. 1–14. Chicago: University of Illinois Press.

Rubin, L. B. 1994. *Families on the fault line*. New York: HarperCollins.

Snell, T. L. 1994. *Women in prison: Survey of state inmates, 1991*. Bureau of Justice Statistics. Washington, D.C.: U.S. Department of Justice.

Stack, C., 1974. *All our kin: Strategies for survival in the Black community*. New York: Harper & Row.

Young, M., and P. Wilmott. 1957. *Family and kinship in East London*. London: Routledge & Kegan Paul.

Appendix
Interview Guide

Prior to each interview, information was collected from the correctional staff and official records about the women and checked in the course of the interview, as appropriate. This information included the woman's race and ethnicity, living arrangement of child(ren), length of sentence, recidivism status, and involvement of child welfare.

Paths to prison

Can you tell me what brought you to prison?

How did you get involved in drugs/crime?

Did your family and friends provide an entry? Did a boyfriend?

How would you describe your family when you were growing up?

What has been the hardest thing about being in prison?

Children

Where are your children living right now?

Were you living with your children before you came to prison? If not, who was taking care of them?

How old are your children?

Is DCF (child welfare) involved? If so, in what way?

What are your plans after you are released? Will you live with your children? immediately? eventually?

Have any of your children been adopted?

People say that crime and getting into trouble with the law might be passed down through generations. Are you concerned about your children getting into trouble? What can be done, if anything, to prevent this?

Caretaker characteristics

How did you decide where to place your children?

Did you feel you had some good options?

What are the pros and cons of placing your children with your parents?/foster care?/your husband?/relatives?

Do you think your child's caretaker is doing a good job?

Do you think that the caretaker is doing a better job than you can just because of your situation?

Are you involved in making decisions about your children, like where they will go to school? If they need to go to the doctor?

What kind of burden do you think it is for other people to take care of your children? Do you think it is hard for them?

What kinds of obligations do you think families have for each other?

What are family members supposed to do if somebody in the family needs help?

Are you comfortable asking your family for help?

Management of motherhood

Women who are serving a long sentence must have to make lots of arrangements for their children. What are some of these and how is that different from women who are serving a short sentence?

Are there things that only mothers can give and do for their children? If so, what are these?

Are there things that you think your children are missing because someone else is taking care of them?

Do your children understand that you are in prison? What do you think they think about this?

How often do you see your children? How often are you in touch with them?

Do your children understand that you are their mother even though someone else is taking care of them?

Have you heard of instances where children call women other than their mother "Mom?"

What are some things mothers are supposed to do for their children?

How do you think women in prison try to make up for the fact that they are in prison?

What are some of the things that make it hard to be a good mother to your children?

In terms of your family and friends, do you need to prove to them that you are a good mother? How do you do that?

In terms of child welfare, do you need to prove to them that you are a good mother? How do you do that?

Women's understanding of other mothers

There are some racial differences in where children live when their mothers come to prison. African-American and Hispanic children seem to be more likely to live with relatives while white kids seem more likely to go into foster care or live with husbands. Can you explain why you think that happens?

Can you tell when women in prison are ready to make a change in their lives and go straight? What are some of the signs?

There has been a lot of talk about making it easier for the state child welfare agency to terminate parental rights if there has been a child death in the family or if the children have been exposed to drugs. What do you think about these new laws?

Do you think it is possible to tell if someone is a good mother by seeing how she acts with her children in prison?

Do you think the courts are easier on women who have children or harder? Why does this happen?

Qualitative versus Structured Interviews

The standardized techniques of the structured interview are quite appropriate when studying homogeneous populations not too different from the investigator (Benney and Hughes, 1956: 137) and when large samples and cost containment are necessary. As we noted in the last chapter, structured interviews are useful when the researcher wants to collect data on many topics, especially when surveying representative samples.

On the other hand, structure in an interview can limit the researcher's ability to obtain in-depth information on any given issue. Furthermore, using a standardized format implicitly assumes that all respondents understand and interpret questions in the same way. Another concern is that the interaction in such interviews is very asymmetric and hierarchical, as interviewers control the interaction and direct the talk while keeping their own views private (Mishler, 1986). Some critics have gone so far as to argue that the approach used by the structured interview "breaks the living connection with subjects so that it is forever engaged in the dissection of corpses" (Mies, 1991: 66).

For researchers less interested in measuring variables and more interested in understanding how individuals subjectively see the world and how they make sense of their lives, less structured approaches, like the one used by Enos, are quite useful. Qualitative researchers feel that questions should be adapted to the experiences of those questioned to ask questions with fixed *meanings* rather than fixed wordings (Denzin, 1989: 107).

In qualitative interviewing, the interviewer can "break the frame of the interview script" and shift her or his perspective to that of the interviewee. Encouragement to elaborate can increase the interviewee's personal investment in the interview and decrease the sense of being a machine producing acceptable answers to questions (Suchman and Jordan, 1992: 251). The group of women with whom Enos talked were able to tell their stories in ways that were meaningful to them, using their own words rather than responses from sets of answers.

Perhaps because of its engaging nature, the qualitative interview typically has a high response rate. Most of those asked agree to participate and complete an interview. For example, in a study of heroin and methadone users, 90 percent of those approached in two methadone clinics agreed to complete a one- to two-hour interview (Friedman and Alicea 1995: 435).

STOP AND THINK *Although being able to modify each interview to fit the situation clearly has advantages, what do you think some disadvantages are?*

Conducting and transcribing interviews is a time- and labor-intensive task. As a result, qualitative interviewing is an expensive method that can translate into a small sample and limited generalizability. Because of the benefits and the costs, the qualitative interview is most appropriate for exploratory research, for research that seeks an understanding of the interviewees' worlds and meanings, and when the researcher is interested in generating "grounded theory"—that is, theory "developed in intimate relationship with data, with researchers fully aware of themselves as instruments for developing that grounded theory" (Strauss, 1987: 6).

Variations in Qualitative Interviews

Number and Length of Interviews

Qualitative interviews can vary in several ways: the *number of times* each member of the sample is interviewed, the *length of each interview*, the *degree of structure* in the interview, and the *number of interviewees and interviewers* participating in the session. Most researchers feel that it is adequate to interview each member of the sample only once as Enos did. Despite the additional cost, however, some researchers suggest dividing the material to be covered into sequential interviews. Seidman (1991: 11–12), for example, suggests a three-session series, with the first focused on the past and background to provide a context for the participant's experience, the second to cover the concrete details of the participant's present experience in the topic area, and the third directed to the participant's reflections on the meaning of the experience. Other researchers advocate multiple interviews because they feel that the additional contact will give interviewees more confidence in the procedure and increase their willingness to report fully. In one study Robert Weiss (1994) found that only in a fourth interview did a respondent talk about his wife's alcoholism. In another study, women who were single parents were interviewed every 2 weeks for about 5 months but typically did not talk about the emotional ups and downs in relationships with boyfriends until the fifth or sixth interview (Weiss, 1994: 57).

In any given study, there is typically variation, sometimes quite a lot, in how long each interview takes. The interviews conducted for the focal research took between 45 minutes and 4 hours. In a study of women's fears of rape and their coping strategies, most of the interviews lasted 90 to 100 minutes, but some lasted as long as three hours (Gordon and Riger, 1989: 198). In Bart and O'Brien's (1985) study of rape victims and avoiders, women were encouraged to talk as long as they wanted to and the interviews lasted from 1½ to 6 hours. Kaufman found that her interviews of women who had consciously chosen orthodox Judaism took between 2½ to 5½ hours, with an average of slightly over 3 hours and some were interviewed twice for an average of 8 hours of interview time (1985: 546).

STOP AND THINK *Sometimes an interviewer using a structured interview schedule feels foolish using it rigidly, not being able to modify, add, or delete questions. How do you think you'd feel in the opposite situation—without a "script," and needing to spontaneously construct questions as the interview proceeded?*

semi-structured interview, interviews with an interview guide containing primarily open-ended questions that can be modified for each interview.

interview guide, a list of topics to cover and the order in which to cover them that can be used to guide less structured interviews.

Semi-Structured and Unstructured Interviews

Qualitative interviews can vary from totally unstructured to semi-structured interactions. **Semi-structured interviews** are designed ahead of time but are modified as appropriate for each participant. They begin either with an **interview guide,** which is a list of topics to cover and the order in which to cover them, or an interview schedule, a set of specified, preformulated questions that can be modified. Although some semi-structured interviews use at least a few closed-ended questions, more typically there is a core set of open-ended questions, supplemented liberally with probes and questions tailored to the specific person. "This sort of hybrid, flexible approach appears optimum for studies that require some quantifiable data but also a good deal of in-depth interpretative material" (Sinclair and Brady, 1987: 67). Constructing questions ahead of time makes the interviewer's job easier because there is a "script," to ensure coverage of all the topics in each interview. The questions must still be judged appropriate or not and re-ordered and re-worded if necessary, and probes and encouragement must be used when necessary to help the interviewee answer fully.

In her study of college students, described in Chapter 3, Hoffnung chose the semi-structured interview for her first data collection to make the interview more like a conversation, to be able identify new issues as they came up and, given her fully heterogeneous sample, to have the ability to re-word questions for each individual respondent (Hoffnung, personal communication, 1997). Similarly, Enos felt that the semi-structured interview best suited her purposes since pre-formulated questions helped her organize her thoughts and paths of investigation, but left things open enough to pursue "surprises"—unexpected but interesting avenues of investigation (personal communication, 1995).

The semi-structured approach is most useful if you know in advance the kinds of questions to ask, feel fairly sure that you and the interviewees "speak the same language," and want to do an analysis that requires the same information from each participant. A semi-structured interview schedule can be made available to other researchers, allowing them the opportunity to evaluate the questions and to replicate the interview in other settings.

unstructured interview, a data collection method in which the interviewer starts with only a general sense of the topics to be discussed and creates questions as the interaction proceeds.

Another approach is an **unstructured interview.** All researchers doing unstructured interviews start with a sense of what information is needed and formulate questions as the interview unfolds. Some also develop a general interview guide ahead of time. Interviewee "digression" tends to be valued as much as core information. Flexibility in questioning allows the interviewer to probe for the meaning behind an answer (Denzin, 1989: 107), understand the view of the other, and is useful for "doing research with people rather than on them" (Reinharz, 1992: 426).

The unstructured interview lends itself to situations where the researcher is developing hypotheses because as hypotheses are formulated, additional questions can be added to the interview. In their study of women attempting to move into the labor market, Acker, Barry, and Esseveld conducted unstructured interviews, trying not to anticipate what would be important before the interviews:

> Our intention was to let the concepts, explanations, and interpretations of those participating in the study become the data we would analyze . . . In our continual process of analysis we had to confront discrepancies between our ideas and interpretations and those of the women we interviewed. As the interview process proceeded, we decided to bring up certain questions if they did not emerge in the interviews (Acker, Barry, and Esseveld, 1991: 138).

In their work on rape avoidance, Bart and O'Brien used a similar approach. "Because of the exploratory nature of the study, questions were added as patterns began to emerge . . . Only after a woman's story was complete did the interviewers ask more focused questions to ensure that the same information was available from each woman in the study" (Bart and O'Brien, 1985: 134).

Reinharz (1992) labels the unstructured approach an "interviewee-guided" interview because the focus is more on understanding the interviewee than on getting specific questions answered. She notes that this kind of interview "requires great attention on the part of the interviewer and a kind of trust that the interviewee will lead the interviewer in fruitful directions" (Reinharz, 1992: 24). This method is most useful when the researcher does not know in advance which questions are important to ask, when interviewees are expected to find different meanings in questions or possess different vocabularies (Berg, 1989:16), or when the researcher wants to work inductively (as discussed in Chapter 2). The less structured interview can also be used to develop themes to be included in more structured interviews as Pain (1997) did in a study of the elderly and the fear of crime.

STOP AND THINK *So far, all of the interviewing we've discussed has been one-to-one. Are there any situations you can think of when it might be better to have more than one interviewer or more than one interviewee?*

Joint Interviewers

The use of more than one interviewer is fairly uncommon. When used, it works best when the person being interviewed is not easily intimidated, and the perspective of two or more interviewers is important. For example, in the study of the women who served in the Rhode Island legislature (Adler and Lemons, 1990), one of us, Emily, interviewed several of the more than fifty women jointly with her co-researcher, historian J. Stanley Lemons. This enabled the researchers to develop similar interviewing styles and to elicit information that was of interest both sociologically and historically. After the joint interviews, we moved to one-to-one interviewing to be able to complete the interviews (each lasting two to three hours) in a timely fashion.

Group Interviews and Focus Groups

group interview, a data collection method with one interviewer and two or more interviewees.

A more common approach is the **group interview**, where one interviewer or moderator directs the inquiry or interaction in an interview with at least two respondents. The individuals in the group are selected because they have something in common, but they either might know each other (such as a married couple or members of the same church) or might be strangers (such as teachers from schools in different towns or patients in a given hospital). Group interviews can be based on a predetermined set of questions, or can use an unstructured format.

Group interviews can have several benefits. They can release the inhibitions of individuals who are otherwise reluctant to disclose what for them are private matters. Because some "are more willing than the others to speak of personal experiences and responses, they tend to be among the first to take active part in the discussion. As one ventilates his experiences, this can encourage others to ventilate theirs ... and establishes a standard for the rest who progressively report more personalized responses" (Merton, Fiske, and Kendall, 1956: 142–143). In addition, the interaction in the group might remind each individual of details that would otherwise not be recalled (Merton, Fiske and Kendall, 1956: 146). Comparing the responses of spouses given in separate, individual interviews to the information shared in a joint, couple interview, Bennett and McAvity (1994:95) noted that "those interviewees who provided limited answers in the individual session became more forthcoming and involved in the couple interview." When the respondents have had shared experiences, better factual information can be obtained, perhaps because it is harder to forget or misrepresent it when someone else who knows what's occurred is nearby (Aquilino, 1993).

Group sessions have several advantages. They stimulate debate and generate multiple descriptions of reality and explanations of causality (Billson, 1991: 208). They are less time consuming than interviewing the same number of participants individually, and have "the advantages of being inexpensive, data rich, flexible, stimulating to respondents, recall aiding, and cumulative and elaborative over and above individual responses" (Fontana and Frey, 1994: 365). Group interviews can be used as a way to obtain shared perceptions of family life or varying perceptions of family history (Bennett and McAvity, 1994: 88). The interaction between participants can replace the need for the interviewer to extract further information because the naturally unfolding dialog can be highly relevant to the topic at hand (Bennett and McAvity, 1994: 96). In addition to using the joint interview to provide either a shared or disparate view of their experiences, the researcher can watch the interaction between participants while the interview is in progress.

STOP AND THINK *Although interviewing several people at once clearly has advantages, there are drawbacks as well. Can you think of some of them?*

One concern about group interviews is the possible suppression of negative attitudes. Aquilino (1993: 372) found that even though the absolute magnitude of effects was small, the presence of a spouse influenced subjective assessment of marriage in the direction of more positive assessment. Some argue that it is better to interview husbands and wives separately to enable them to talk more freely about their feelings and views of each other (Adler, 1981; Rubin, 1983). In addition, even among those who are strangers, participants who become group leaders can structure the situation for others, monopolize the discussion, or inhibit the comments of others (Merton, Fiske, and Kendall, 1956: 149).

focus group, a type of group interview where participants converse with each other and have minimal interaction with a moderator.

A special kind of group interview is the **focus group,** a research tool that focuses on group interaction on a topic selected by the researcher (Morgan, 1996). Focus groups were used in social sciences in the 1940s and 1950s (Merton, Fiske, and Kendall, 1956), but they became more of a market research tool until the 1980s when they were rediscovered by social scientists (Krueger, 1988; Morgan, 1988). In 1994, for example, more than 100 articles using focus groups appeared in academic journals (Morgan, 1996). Focus groups are used both as a self-contained method and in combination with other methods, as they can provide useful qualitative data and precede or supplement a questionnaire or structured interview. Examples of focus group research include a study of social services use by homeless veterans (Applewhite, 1997); an analysis of interracial marriages (St. Jean, 1998); and research on the cognitive and emotional bases of sexual decisions made by young Black men (Gilmore, DeLamater, and Wagstaff, 1996).

Focus groups are designed to have between 4 and 12 participants, selected because they are homogeneous on the characteristic for which the researcher recruited them, such as people who have been recently widowed or those with specific health concerns. The participants usually don't know each other before the group interaction. Projects vary in the number of focus groups that they use. For example, Applewhite (1997) used only five groups in his study of homeless veterans, and Kitzinger (cited in Morgan, 1996) used 52 in studying attitudes toward AIDS in many different populations.

The interaction in a focus group is directed by a moderator who asks questions and keeps the discussion on the topic (Stewart and Shamdasani, 1990: 10). Some groups are more structured, with the moderator more involved in controlling the group dynamics, and others are more "self-managed" (Morgan, 1996). In comparison with other kinds of interviews, in a focus group the participants' interaction with the interviewer is replaced by interaction within the group, leading to a greater emphasis on participants' points of view (Morgan, 1988). Some argue that the group interaction can highlight issues and concerns that would have been neglected by questionnaires (Powell, Single, and Lloyd, 1996) and provide researchers with a better sense of the everyday reality of participants (St. Jean, 1998).

The use of focus groups, however, is not without problems. One-to-one interviews can be more useful in eliciting yes/no answers about specific behaviors and experiences and can have less of a "polarization" effect, which is what happens when attitudes become more extreme after discussion (Morgan, 1996). In a focus group, "the emerging group culture may interfere with individual expression, the group makes it difficult to research sensitive topics, 'group think' is a possible outcome, and the requirements for interviewer skill are greater because of group dynamics" (Fontana and Frey, 1994: 365). In addition, if small and nonrandom samples are used, focus groups have limited generalizability (Stewart and Shamdasani, 1990). Finally, there is the issue of confidentiality of responses. In all group interviews, including the focus group, the need to keep confidential all the information shared by participants must be stressed and can be included on an informed consent form.

Locating Respondents and Presenting the Project

In all qualitative interviews, once the researcher decides on the population, kind of sample, and specific sample to be interviewed, potential interviewees must be contacted. If the researcher has names, addresses, or phone numbers, the approach could be a letter or a phone call. Hoffnung, whose panel study of women college graduates was discussed in Chapter 3, began with a list of seniors that had been selected at random from each of five colleges. She called each woman, introduced herself, and asked each to participate in a study of women's lives, starting with a one hour in-person interview. In her study of working parents and family life, Hochschild (1989) contacted a random sample of employees at a large corporation and then asked that sample for the names of friends and neighbors. In her study of widowhood, Lopata (1980) discovered that finding widows or a sample of them in a geographic area was not easy, even though there were more than 10 million of them the United States. She tried, with limited success, to locate widows using a modified area sample that assigned blocks to interviewers, but did better when she used lists of beneficiaries provided by the Social Security Administration.

STOP AND THINK *What about groups of people for whom there are no lists? For example, what would you do if you wanted to talk to homeless people or women who avoided being raped? How could you locate samples like these?*

Qualitative researchers are frequently interested in studying groups of people for whom there are no lists—such as mothers in prison, "street women" (Miller, 1986), Jamaican, Mennonite, and Inuit Canadian women (Billson, 1995), or rape victims and avoiders (Bart and O'Brien, 1985)—so selecting a random sample might not be possible. In these cases, researchers have recruited participants through friendship networks, newspaper ads, press releases, posting notices, and giving talks (Bart and O'Brien, 1985; Kelly, 1989) or have contacted them through "gatekeepers," that is, people who control access to others. "Gatekeepers can range from absolutely legitimate (to be respected) to self-declared (to be avoided)" (Seidman, 1991: 34). Legitimate gatekeepers include parents and guardians of children under 18 and, as in Enos's study, the heads of institutions, agencies, or groups whose members you want to contact. However, you might not need to gain access through an authority if you are researching an experience or process that takes place in a number of sites, rather than in one particular site. For example, if you wanted to interview high school teachers working in many schools scattered throughout a region, you could go directly to them without asking permission from their principals (Seidman, 1991: 35).

If you want to interview elites—those in positions of higher status or power—access can be difficult. Being able to interview members of Congress, for example, has become distinctly more difficult in recent decades, and researchers interested in conducting such interviews must be prepared to spend extended periods of time to schedule and complete them (Sinclair and Brady, 1987: 63).

STOP AND THINK *What kind of respondents might be particularly reluctant to be interviewed? What do you think about offering a fee to interviewees? Do you think payment encourages participation?*

The more "political" or "deviant" one's topic is, the more difficult it will be to get access to and agreement from potential respondents. For example, when Frankenberg, a researcher interested in the social construction of race, told people that she was "doing research on white women and race," she was greeted with interest in some circles, but suspicion and hostility in others. She realized that her approach was "closing more doors—and mouths—than it was opening" (Frankenberg, 1993: 32). She was more successful when she told white women that she was interested in whether they interacted with people of different racial or cultural groups and whether they saw themselves as belonging to an ethnic or cultural group (Frankenberg, 1993: 35).

Other difficulties include the concern of potential interviewees that the information they provide will be used against them or that the researcher is not who she says she is. The women whom Enos interviewed did not think that she was either from the prison administration or child welfare, and they believed her pledge of confidentiality of information, including her promise not to use their real names, but other researchers have faced suspicion. Miller, for example, felt that some of the "street women" she approached thought she might be a vice officer or narcotics agent in disguise (1986: 186), and Harkess and Warren (1993: 324) believed that some participants in their studies confused them with investigative reporters, union organizers, or industrial spies.

Some researchers provide interviewees with incentives to encourage participation. Bart and O'Brien for example, offered the participants in their study $25 and any babysitting expenses they incurred to complete the interview (1985: 134). Miller (1986) found that offering $10 to women in pre-release centers and halfway houses increased their willingness to be interviewed.

Most researchers don't offer financial incentives, feeling that payment doesn't seem to have a big influence on the decision to participate. In most studies, the respondent agrees because he or she enjoys the interview experience and believes it can make a contribution to social science or society. For example, in a study of ministers' wives and mothers participating in children's playgroups, Finch (1984) reported that, with guarantees of confidentiality, most respondents welcomed the opportunity to talk about themselves to a friendly interviewer.

Even those talking about difficult subjects can find the experience valuable. The great majority of the women who talked to Kelly (1988: 12) about being sexually assaulted felt that the interview had been a very positive experience and that they had learned things through their participation. In the focal research on people living with HIV in Chapter 4, Weitz reported that many people saw their interviews as legacies; they participated in the project despite the emotional and sometimes physical pain it caused them because they believed their stories would help others. In Enos's study, the women did appreciate her offer of photographs of their children, but participated because they enjoyed being listened to with interest and respect and because they thought the study could help others in their position.

Planning the Interview

Using Consent Forms

informed consent form, a statement that describes the study and the researcher and formally requests participation.

Anonymous surveys like the ones discussed in the previous chapter are often exempted from the requirements of using **informed consent forms.** Some researchers rely on verbal rather than written permission because it is possible in an interview to explain the purpose of a study, offer confidentiality of answers, and assume that the decision to participate implicitly means giving consent. However, many institutional review boards, the groups explicitly designed to consider ethical issues, now require interviewers to use a written informed consent form. In a formal statement, the researcher can provide potential participants with information about the purpose of the study, its sponsor, the benefits to participants and others, the identity and affiliation of the researcher, the nature and likelihood of risks for participants (such as the possibility of raising sensitive issues), and an account of who will have access to the study's records and for what purposes. With such information, the benefits, risks, possible uses of the data, and the study's results can be weighed by individuals when deciding whether to participate.

The informed consent form used by Enos is provided in Box 10.1. The statement specifies both the risks and benefits of participation, including the legal limits to confidentiality. One signed copy was kept by the researcher and another given to the interviewee. (For a comparison, see the informed consent form used by Weitz included in Chapter 4).

Constructing an Interview Guide or Schedule

rapport, a sense of connection between the interviewer and interviewee.

Once the objectives of the research are set, it's time to determine if there is a core of essential information that should be collected for all participants. If the interview is to be unstructured, then a list of topics, an introductory statement, and a general sense of the order of discussion is sufficient. The interviewer typically starts with general questions and follows up participants' comments. Questions can be reframed if the context, chronology, or meaning of the answer is not clear. Usually, it's better to wait until **rapport,** a sense of connection between the participants, is developed before raising difficult or sensitive issues.

For unstructured interviewing, doing practice interviews helps to prepare for the actual experience and allows the interviewer to gain experience covering issues and developing appropriate conversation generators. Questions that might elicit additional information include "Can you tell me more about that?" "When did that happen?" "Who did you go with?" and "What happened next?" If the researcher notices not only what is mentioned, but what is *not* mentioned, then additional questions can be asked.

An interview schedule, including both basic and follow-up questions, must be constructed for a semi-structured interview. Sometimes "filler questions" are needed to provide a transition from one topic to another. A good question makes it easy for the respondent to provide the material needed.

BOX 10.1

Informed Consent Form

I have been asked to take part in a research project by Sandra Enos. The researcher will explain the project to me in detail. I should feel free to ask questions. If I have additional questions later, Sandra Enos, the person mainly responsible for the study, will come here to discuss them with me.

Description of the project: I have been asked to take part in a project that examines how women in prison manage motherhood while being incarcerated. The aim of the project is to learn about the challenges and obstacles that face women who are attempting to maintain relationships with their children while serving time in prison.

What will be done: If I decide to take part in the project, I will be asked about my children, their living arrangements, what brought me to prison and how imprisonment is affecting my relationship with children and other family members. I may also be asked about involvement with the child welfare agency. My part in the study will involve an interview which will last about 1–2 hours and which will be tape recorded. My name will not be on the tape and after the interview is typed, the tape recording will be destroyed.

Risks: The possible risks in the study are small. I may feel some discomfort as I talk about the past. There is a chance that some of the interview questions may result in my feeling uncomfortable or anxious. If that is the case, the interview may be suspended at that point if I wish. The researcher will ask several times during the interview if I want to stop. The decision to participate in the study is up to me. I may terminate the interview at any time. Whatever I decide will not be held against me. I understand that the researcher is not affiliated with the Department of Corrections and that my participation in the interview will not have an impact on my treatment, criminal processing or, any other matter.

Reportable child abuse: If, while talking with Sandra Enos, I tell her about some abuse or neglect of a child that I say has not been reported to DCYF, the researcher will inform me that (1) she is required by law to report the abuse/neglect to DCYF and (2) that we must terminate the interview. The tape recording of our interview will be immediately destroyed. I understand that the purpose of the study is not to track reportable incidents of abuse or neglect.

Benefits: There are no guarantees that my being in this research will provide any direct benefit to me. I understand that taking part in this research will have no effect on my parole, classification status and/or inmate record. My taking part will provide important information for people who are trying to understand the impact of prison on women and their families.

Confidentiality: My participation in the study is confidential to the extent permitted by law. None of the information collected will identify me

(continues)

BOX 10.1 *(continued)*

by name. All information provided by me will be confidential. Within two weeks after the interview, the tape will be transcribed, and the recording will be destroyed. No information that is traceable to me will be on the transcript. Transcripts will be maintained in a locked cabinet in a secure location available only to the researcher. No information collected by the research that identifies me will be given to the Department of Corrections.

Decision to quit: The decision whether or not to take part is up to me. I do not have to be in the study. If I decide to take part in the study, I can quit at any time. Whatever I decide is OK. If I want to quit, I simply tell the researcher.

Rights and complaints: If I have any concerns about the research, I may contact Sandra Enos at 456-____ or ask the warden to contact her for me.

I have read the consent form and understand what is stated. Any questions I have about the research have been answered. By signing the form, I am indicating my willingness to participate in the study. The consent form will be kept in a locked cabinet and will not be attached to any transcripts or other materials.

_______________________ _______________________

Researcher signature Interviewee signature

_______________________ _______________________

Date Date

One especially useful approach is to ask the interviewee to give more detail or fill in a gap. The guidelines for question construction provided in Chapter 9 are useful for qualitative interviewing as well. Advice for less structured approaches includes staying away from double-barreled, double-negative, or threatening questions and wording that is ambiguous or leading unless the interviewer is intentionally trying to provoke the interviewee into a response.

The choice of specific questions is affected by the intended analysis. If the idea is to describe experience, then the questions should help people tell their stories. "If we take seriously the idea that people should make sense of experience and communicate meaning through narration, *then in-depth interviews should become occasions in which we ask for life-stories,* . . . narratives about some life experience that is of deep and abiding interest to the interviewee" (Chase, 1995a: 2). In her study of female school superintendents, Susan Chase found that abstract, sociological questions (for example, asking the women if their experiences fit with sociologists' ideas about women in male-dominated professions) encouraged answers that had little to do with how the women lived their lives. On the other hand, questions about specific experiences—such as asking the subjects to describe their work histories—produced lively, lengthy, and engrossing stories that helped construct narratives about the educators (Chase, 1995a: 8). Using another approach,

Billson (1991) sought to understand relationships and roles in the participants' communities. She typically asked interviewees, often in group settings, not only to share their experiences, but to engage in sociological analysis by answering questions such as, "What is it like for women here?"

Once an interview guide or schedule is constructed, it should be pretested with people similar to those who will be interviewed during the actual data collection. A pre-test can determine the questions most useful for focusing the conversation on desired topics. The pre-test interviewees can be asked which questions they saw as most effective and which were difficult to understand. In qualitative studies, an interview schedule can continue to evolve, especially if the focus of the study changes or becomes sharper.

STOP AND THINK *As a college student, you use words and language in certain ways. Are there any groups of Americans that you think might use language so differently from the way you do that you might initially have difficulty communicating with them?*

Speaking the Same Language

Because language is a cultural artifact, it is important to be familiar with the cultural milieu of your respondents. Given cultural and subcultural differences, it's important to use conceptually equivalent language rather than dictionary translations (Deutscher, 1978: 201). Enos had picked up some of the "prison lingo" during the observation part of her study, so for the most part she knew what the interviewees meant when they reported "getting jammed up" (getting in trouble) or "catching a bid" (getting sentenced). If she didn't understand something, she'd ask for it to be repeated or explained. In the context of her study, it isn't surprising that she sometimes had to ask the women to explain family relationships since some of the children they talked about "mothering" were not their biological children, but those for whom they were informal guardians.

In *Street Woman,* Eleanor Miller noted that many of the black women she interviewed accommodated her by using "White English" rather than the "Black English" they would have normally used (1986: 185). She commented on having problems with the meanings of particular words and phrases. "Although I had admitted my ignorance of their world, I didn't want to appear too square. As a result, when I couldn't determine the meaning of a word or phrase from the context in which it was used, I felt comfortable asking for one or two definitions. Beyond that, I would save my questions about terminology for another time or another informant" (Miller, 1986: 185). Booth and his associates (1993) in studying behavior change among injection drug users, found that the right word was sometimes critical for getting accurate information. For example, drug users who'd purchased syringes with others and then shared them would typically respond negatively to questions about how often they *lent or shared* someone else's syringe (perhaps because of the implied degree of intimacy), but would answer in the affirmative if asked if they'd injected with "works" that had been *used* by someone else (Booth, Koester, Reichardt, and Brewster, 1993: 179).

Conducting the Interview

Where and How to Interview

Interviews can be held in offices with co-workers nearby, or in the interviewee's home, where children and spouses might be present. For example, in her study of married couples, when one of us, Emily, was interviewing wives (while her associate was interviewing the husbands), she found that, on occasion, children would be present for part of the interview—sometimes commenting on the woman's comments! If privacy is needed in an interview, it's important to consider that when scheduling the time and place.

Although the interview might not follow all the conventions of conversation, it *is* a social interaction. The interviewer must be conscious of all that is being communicated both verbally and nonverbally. For the interviewer, it is critical to *be as neutral as possible* in tone of voice, inflection, phrasing of questions, and body language. The point is to communicate openness to whatever the interviewee wants to share, rather than conveying expectations about his or her answers. The interviewer should focus on the answers to ask for details, clarification, and additional information when appropriate. It is especially important to note inconsistencies in the interviewees' verbal and nonverbal messages and changes in them (Gorden, 1975: 97). One useful skill is knowing when to keep quiet because interested and active listening is an important talent in obtaining complete responses.

Recording the Interview

When an interview covers a large number of topics, a tape recording is invaluable; the more unstructured the interview, the more necessary taping becomes. If there is no strict order of questioning and probing is an important part of the process, the interviewer will not be able to attend adequately to what the interviewee is saying if trying to write everything down (Lofland 1984: 60). But the impact of a tape recorder is hard to assess. On the one hand, it can help rapport by allowing the interviewer to make eye contact, nod, and show interest and help the interviewer concentrate on follow-up questions. On the other hand, tape recorders can intimidate interviewees and inhibit the frankness of their responses.

STOP AND THINK *After describing his relationship with his father, the person you are interviewing asks you how you get along with your father. What would you do? Would you tell him that what you think really isn't important because he's the one being interviewed, answer that you'll tell about your father after the interview is over, describe your relationship with your father honestly but briefly, or launch into a detailed description of how you get along with your father?*

Being "Real" in the Interview

In more traditional qualitative interviewing, the interviewer maintains social distance from the interviewee. This means using a style that gives evidence of interest and understanding in what is being said (nodding, smiling, murmuring "uh-huh"), but that prohibits reciprocal sharing or "real conversation." (Even saying things like "Yeah, I know what you mean" is not allowed [Weiss, 1994: 8]). The interviewer is advised not to share opinions or any personal information with respondents because it can increase the chance of leading subjects to say what they think the interviewer wants to hear and shift attention from the interviewee.

Critics of the traditional method dispute the view that the traditional interviewer response is really neutral. Frankenberg (1993: 31) feels that "evasive or vague responses mark one as something specific by interviewees, be it 'close-mouthed,' 'scientific,' 'rude,' 'mainstream,' 'moderate,' or perhaps 'strange'—and many of those are negative characterizations in some or all of the communities in which I was interviewing."

Some go so far as to argue that the interviewer and interviewee should treat each other as full human beings. Fontana and Frey (1994: 374), for example, believe that rather than remaining objective, faceless interviewers, interviewers should disclose themselves as they try to learn about the other. In this alternative view of interviewing, there is mutual sharing, and the interviewer attempts to minimize the status differences of the traditional interviewing hierarchy. Some argue that it is more honest, morally sound, and reliable, to treat the respondent as an equal, and allows him or her to express personal feelings. They believe the results present a more 'realistic' picture than can be uncovered using traditional interview methods (Fontana and Frey, 1994: 371).

Although Enos was only asked an occasional personal question about herself or her opinions, quite a few interviewers report interviewees asking numerous questions—about the study, about their personal lives, or for advice on the topic under study (Acker, Barry, and Esseveld, 1991; Kelly, 1988; Oakley, 1981). Each researcher must decide how to respond. Acker and her associates

> always responded as honestly as we could, talking about aspects of our lives that were similar to the things we had been discussing about the experience of the interviewee—our marriages, our children, our jobs, our parents. Often this meant also that our relationship was defined as something which existed beyond the limits of the interview situation. We formed friendships with many of the women in the study. We were offered hospitality and were asked to meet husbands, friends, and children. Sometimes we would provide help to one or another woman in the study. . . . However, we recognized a usually unarticulated tension between friendships and the goal of research. The researcher's goal is always to gather information; thus the danger always exists of manipulating friendships to that end. Given that the power difference between researcher and researched can not be completely eliminated, attempting to create a more equal relationship can paradoxically become exploitation and use (Acker, Barry, and Esseveld, 1991: 141).

Being "real" as an interviewer can also mean acknowledging that at least some of what you hear is painful, difficult, or upsetting. Although the demands of interviewing on interviewers is occasionally described, many researchers are not prepared nor trained to handle the emotional impact of the interview process (Batchelor and Brigg, 1994). In the focal research in Chapter 4, Weitz describes some of the personal difficulties she faced in interviewing people living with HIV—dealing with death, disability, and fear of contagion as well as having to witness the pain of the participants' lives.

In their study on sexual violence, Gordon and Riger (1989: xiii) said of the interview process, "Constantly reading about and discussing rape and other forms of violence against women often left us anxious and depressed. Staff working with us also found themselves disturbed." Eleanor Miller conducted her interviews in a variety of settings, and at times felt afraid, intimidated, and uncomfortable. Hearing the details of the women's lives sometimes made her angry that their childhoods had been so awful, upset that people could be so brutal to one another, and depressed about the lack of realistic options for the women and the probable futures of their children. Sometimes, however, she found the fieldwork exhilarating and, in the long run, Miller (1986: 189) felt the study was worth the effort both personally and professionally.

Interviewing Across the Great Divides

STOP AND THINK *How effective do you think you would be as an interviewer if you were interviewing someone considerably older than you are? How about if the person was of the opposite gender or a different race? Do you think the topic of the interview would make a difference in interviewing someone whose background was very different from yours?*

interviewer effect, the change in a respondent's behavior or answers that is the result of being interviewed by a specific interviewer.

As we discussed in the previous chapter, the data from all interviews face the problem of the **interviewer effect**, the change in a participant's behavior or the answers given to questions as the result of being interviewed by a specific interviewer. Each interviewer brings unique qualities and characteristics to the interview situation, so an ongoing debate concerns the necessity of matching interviewers and interviewees on social characteristics, such as gender, race, ethnicity, age, and class. Some argue that interviewers and participants of good will who are of different backgrounds can have a good interviewing relationship (Seidman, 1991). In the short piece in Box 10.2, Don Naylor shares some of his thoughts about the way that gender affected a recent study.

The most common perspective is that it's advisable to match the participants in an interview because interviewers who differ from interviewees on class, ethnicity, gender, or race tend to get different responses than when the participants have the same backgrounds (Hyman, 1975; Reisman, 1987; Kane and Macaulay, 1993). It might be best to match interviewer and interviewee on the most important characteristics, if possible, because the shared assumptions that come from common backgrounds can make it easier to build rapport. Enos, for example, believes that her gender was helpful in the interview process. "A man in that setting could have been problematic, especially since the women would have had some trouble understanding why a man would be

BOX 10.2

Reflections on Matching the Gender of the Interviewer and Interviewee

by Donald C. Naylor, Graduate Student, Department of Sociology, University of Southern California[3]

In planning a study in which I'd be interviewing men about gender, I knew that the men in my sample might not be completely honest with me. But, I didn't think I needed to be concerned about the interviewer effect or be worried about getting the polite answers that sometimes result when people are interviewed by someone who is perceived as different and possibly unreceptive to their views. In my study, the interviewees and I would be the same gender and would be similar in other ways — age, class, and sexual orientation. Of the many problems associated with interviewing, interviewer effect was *not* one I thought I would have to consider.

Upon reading transcripts of my interviews though, I noticed that what was said seemed to have been affected by something happening between myself and the men I was interviewing. In one instance, I noticed that the man's answer seemed like something he felt he had to say to appear masculine. I wondered if he believed what he'd said or if it was just what one guy thought he was "supposed" to say to another. While the literature on interviewing says how much better it is when the interviewer and interviewee are similar, I wondered if there were also some problems with similarity.

Reflecting on the interview process, I've had some thoughts about what was occurring. My first is that there is an effect even when the interviewer and interviewee are similar. It is a different dynamic than when the two are different, but it is still there. Similarity can affect the content of the interview, the types of questions asked, the answers given, and the way the interviewer and interviewee interact. In addition, it seems harder to observe critically a familiar interaction.

Second, I began to see that there was more than interviewer effect going on. The interviewee also had an active part in constructing the interview. This was especially true since I was doing an unstructured interview. When the men varied in what they said and how they said it (amount of openness, attempts to control, emotions displayed), I would change my style and demeanor. They were affecting me as much as I was affecting them. I began to think not so much in terms of interviewer effect, but about interviewer-interviewee effect as we were jointly constructing the interview.

[3] Permission granted by Donald C. Naylor.

(continues)

BOX 10.2 *(continued)*

Third, it began to be apparent to me that the interview process is a gendered one. We were "doing" an interview not just as two people but as two men. We were "doing gender." The dialog being created was typical of how men often converse. Of course variations resulted since there are a lot of masculinities (and femininities) and these men (and myself) were not identical. But the more I read the transcripts, the more apparent it became that we had been "doing gender." I noted that some men tried to control the situation (as did I) and most avoided topics with much emotional content (as did I some of the time). They all tried to present themselves as competent, and each wanted to avoid a discussion of problems, perhaps seeing these as failures rather than as normal difficulties (did I collude in this?). The ways I responded to these men and the ways they reacted to me seemed to be typically "male" ways of interacting.

As I thought about the basic structure of the kind of interview I used, it became obvious that this was a method that did not fit easily with the style of many men. Having a stranger walk into their livingrooms and ask personal questions is not what men typically do. Many men are hesitant about disclosing personal information, especially to strangers in unfamiliar situations.

I found it most useful to try not try to press the men for very personal information. I found that when I was accepting, uncritical, and nonthreatening, the men did not clam up or shut down. Oddly, at times, the less I asked, the more they told me. Perhaps unconsciously I was trying to conduct an interview in a way that was "guy like." This is "doing gender." But it also, I believe, allowed me to obtain more information than if I had tried a different style.

Interviewer-interviewee effect and "doing gender" can not be avoided. The question for me was, given the inevitability of this, what do I consider the effects on the information obtained to be? We can acknowledge and understand these effects, and perhaps even profit from them.

interested in mothering. Also a number of the women expressed exasperation with relying on men and others were still looking for Prince Charming" (Enos, 1995, personal communication).

It might not be possible, however, to match interviewer and interviewees on more than one or two characteristics, and the choice might be between leaving important work undone and interviewing across the "great divides." Enos told us that she would have loved to have worked with both African-American and Hispanic co-researchers, but didn't have that option. She believes that despite the differences, she was able to communicate effectively across race, class, and ethnic lines.

Issues of Validity

As with the other self-report methods, there are questions about the validity of the data produced by qualitative interviews. Inaccurate memories, misunderstandings, and miscommunications must be considered in evaluating the information obtained. As early as 1935, researchers debated the "questionable value" of the interview for securing reliable data when people become defensive concerning their private and personal lives (Thurlow, 1935). The fact that in a qualitative interview, the interviewer is not a passive listener, is another validity concern. The way an interviewer questions, responds, and acts can affect the way a participant responds. It's possible that a different interviewer or the same interviewer using different wording would have been told a different account. Denzin (1970: 188) cautions that the tremendous range and variation in interviews in the same study can make comparability a fiction.

The counterpoint to these concerns is that standardizing the interview might be no better at producing comparability because different respondents can hear the same question in different ways. The ability to reword and restate questions can give the researcher a chance to present them in a way that's more understandable by each participant. In addition, a relatively natural interactional style allows the interviewer to better judge the information obtained. Enos and many others who have used the qualitative interview believe that, overall, interviewees tell the truth as they understand it and rarely offer false information knowingly.

After the Interview's Over

STOP AND THINK *Imagine that you're almost at the end of an interview on students' relationships with significant others. After describing how the most recent love relationship ended, the college student you're interviewing looks up and says, "I'm so depressed, I feel like killing myself." What would you do?*

Researchers need to be prepared for requests for help or the emotional aftermath of interviews when they are on emotionally difficult topics. Helena Lopata reporting on her study of widows, said, "over and over, we found the respondents expecting some sort of help as a result of the interview, a solution of problems, and even a complete change in life. They assumed that the interviewer, or at least the university staff, has the power to bring societal resources to them or to change the attitudes or behavior of significant others toward them. It is difficult to be faced with a respondent, caller, or letter writer who is obviously in pain or need whom we are not trained to help" (Lopata, 1980: 78).

At the very least, researchers usually include a series of "cool down" questions at the end of an interview so the interview doesn't end immediately after talking about sensitive subjects. Some researchers prepare something to leave with participants, most typically a list of local organizations that provide services in the area under discussion. A few researchers have offered to locate or provide counseling or therapy sessions for the interviewees after the interviewing process is over (Burgess and Homstrom, 1979; Bart and O'Brien, 1985).

Analyzing Interview Data

STOP AND THINK *Imagine that you've conducted qualitative interviews with a small sample of nurses about work satisfactions and problems. Each interview took about 30 minutes, and, when transcribed, is approximately 20 pages long. How would you go about analyzing the data?*

If interviews have been tape recorded, they are usually transcribed. The interview process generates a great deal of text and the process of analyzing such data is typically more inductive than deductive. That is, although the researcher might come to the data with some tentative hypotheses or ideas from another context, the most common approach is to read with an open mind, while looking for motifs.

At this point we will concentrate on some of the possible approaches to interview data and will focus on the more technical aspects of data reduction and analysis in Chapter 16. One general approach to data analysis is to take the topics that the participants have talked about and to use them as a sorting scheme. Categorizing the material by topic, the researcher can look for patterns and themes in the accounts. Enos's analysis describes the patterns between race and the children's placement and those that exist between race and the women's views of their paths to prison. In their study of rape victims, Bart and O'Brien saw a pattern between the relationships of intended victims and attackers and the woman's likelihood of avoiding rape. For example, the women were more likely to avoid rape if attacked by a stranger and more likely to be raped if attacked by a man they knew (1985: 29).

Another approach is to construct life histories, profiles, or "types" that typify the patterns or "totalities" represented in the sample. General phenomena can be described by focusing on their embodiment in specific life stories (Chase, 1995a: 2). Laurel Richardson uses the term "collective story" to describe this approach. A collective story is one that "displays an individual's story by narrativizing the experiences of the social category to which the individual belongs, rather than by telling the particular individual's story or by simply retelling the cultural story" (1990: 25).

Cases can also be examined one at a time to see if most fit a hypothesis, and then the "deviant cases"—those that don't fit the hypothesis—can be examined more closely (Runcie, 1980: 186). If a description fits a series of respondents, the investigator can propose a more general statement as a hypothesized minitheory. "Each new interview can then be a test, the results of which will support the minitheory, discredit it, or most likely, require that it be augmented or qualified" (Weiss, 1994: 179). This is similar to what Billson (1991) calls progressive verification, and that Glaser and Straus (1967) call theoretical saturation. In these approaches, the researcher determines that he or she is getting an accurate picture when similar things are repeated by successive interviewees. Billson, for example, knows that she has achieved at least a basic understanding when her analysis is consistently verified by subsequent participants. She argues that in qualitative research, stability is similar to traditional tenets regarding statistical reliability in quantitative research (Billson, 1991: 209).

Summary

The qualitative interview is an important and useful tool for researchers interested in understanding the world as others see it. The information provided can be especially useful for exploratory and descriptive research. Less structured interviews can help the researcher develop insights into other people's realities and to work inductively toward theoretical understandings.

The qualitative interview uses either an interview guide or an interview schedule. Using mostly open-ended questions, the interview is modified and adapted for each interview, and participants are encouraged to "tell their stories." Because it is an expensive and time-consuming method of data collection, the qualitative interview is most frequently used with relatively small samples.

There are several choices to be made in using qualitative interviewing as the method of data collection. Among the most important are the degree of structure in the interview, whether to do individual or group interviews, whether to "match" interviewer and interviewee, and the extent to which the interview is more or less like a "real" conversation in which the interviewer shares opinions and information with interviewees during the interview.

When the interviewing process is completed, the creative process of looking for patterns and themes in the accounts begins. Data analysis can be a rewarding task, but is often very time consuming.

Qualitative interviews can't provide comparable information about each member of a large sample as easily as more structured methods can, but they can be important sources of insight into particular realities. In Table 10.1, that follows, we summarize some of the advantages and disadvantages of the types of qualitative interviews described in this chapter.

TABLE 10.1 **Advantages and Disadvantages of Qualitative Interviews**

Method of Data Collection	*Advantages*	*Disadvantages*
Semi-structured interview	Allows interviewer to develop rapport with study participants Good response rate Questions can be explained and modified for each participant Useful for discussing complex topics Can be used for long interviews Useful when the topics to be discussed are known	Time consuming Requires highly skilled interviewer Interviewer effect Data analysis is time consuming
Unstructured interview	Allows interviewer to develop rapport with study participants Good response rate Questions can be explained and modified for each participant Useful for exploratory research on new topics Useful for discussing complex topics Can be used for long interviews	Time consuming Expensive Requires highly skilled interviewer Will not have comparable data on all participants Interviewer effect Data coding and analysis is time consuming
Focus group interview	Allows participants to have minimal interaction with interviewer and keep the emphasis on participants' points of view Typically generates a great deal of response to questions Often precedes or supplements other methods of data collection	Attitudes can become more extreme in discussion Possibility of "group think" outcome Requires highly skilled interviewer Data less useful for analysis of individuals

SUGGESTED READING

Mishler, E. G. 1986. *Research interviewing context and narrative*. Cambridge, MA: Harvard University Press.

Mishler critiques survey interview practices and argues for interviews that allow the interview to be a form of discourse that empowers the interviewee. His presentation of methods of analyzing interviews as narrative accounts is especially interesting.

Morgan, D. L. 1988. *Focus groups as qualitative research*. Newbury Park, CA: Sage.

A good overview of focus groups in social science research, including discussions of planning, conducting, and analyzing the results of this kind of group interview.

Seidman, I. E. 1991. *Interviewing as qualitative research*. New York: Teachers College Press.

In a thorough discussion of in-depth interviewing, Seidman connects method and technique with broader issues of qualitative research. The chapter on interviewing technique is especially useful.

Weiss, R. S. 1994. *Learning from strangers: The art and method of qualitative interview studies*. New York: Free Press.

This is a comprehensive presentation of interviewing as a way of getting access to the observations of others. Weiss draws on his extensive experience as a qualitative interviewer to advise researchers on recruiting respondents, preparing for interviewing, conducting the interview, analyzing the data, and writing the report.

EXERCISE 10.1

Evaluating Methods

In her research with incarcerated mothers, Sandra Enos used an in-person semi-structured interview. Two other self-report methods are the in-person structured interview and the mailed questionnaire (both discussed in Chapter 9). What advantages and disadvantages would Enos have had with each of these compared to the method she used?

1. Mailed Questionnaire:

   List the advantages of this method for this research.

   List the disadvantages of this method for this research.

2. In-person structured interview:

   List the advantages of this method for this research.

   List the disadvantages of this method for this research.

EXERCISE 10.2

Doing an Unstructured Interview

Pick an occupation that you know something, but not a great deal, about (such as police officer, waiter, veterinarian, letter carrier, high school principal, and so on). Find someone who is currently employed in that occupation and is willing to be interviewed about her or his work.

Conduct an unstructured interview of about ½ hour in length, finding out how the person trained or prepared for her or his work, the kinds of activities the person does on the job, approximately how much time is spent on each activity, which activities are enjoyed and which are not, what the person's satisfactions and dissatisfactions with the job are, and whether or not he or she would like to continue doing this work for the next ten years.

Transcribe at least 15 minutes of the interview, including your questions and comments as well the interviewee's replies. Include the transcription with your exercise.

Consider this a "case study" and write a brief analysis of the person you talked to. If you can, draw a tentative conclusions or construct a hypothesis that could be tested if you were to do additional interviewing. Include a paragraph describing your reactions to using this method of data collection.

EXERCISE 10.3

Constructing Questions

Construct a semi-structured interview guide for a research project on work in an occupation of your choice.

Write an introductory statement that describes the research project, and write at least 12 interview questions. Include both basic and follow-up questions. Focus on the same kinds of information called for in exercise 10.2, including how the person trained or prepared for her or his work, the kinds of activities the person does on the job, the amount of time spent on each activity, which activities are enjoyed and which are not, what the person's satisfactions and dissatisfactions with the job are, and so on.

11 Observation Techniques

Adam's now ten. When he was four, we signed him up for T-ball at the local Jewish Community Center. What a hoot! Adam, like many of his teammates, would stand in the field on defense, studying earthworms, cloud patterns, and the scatological humor of other players (for example, "What a poopie hit that was." Immoderate laughter.) Anything but paying attention to the game. His mother and I pretty much decided that spring, implicitly at least, that we wouldn't push Adam into organized activities until he was old enough to want them.

Adam's never shown interest in watching sports on TV and so has only come to his interest in basketball through shooting with me or with friends. He still rarely plays even pick-up games, much preferring the solo pleasures of shooting alone, sometimes for hours in the backyard or in the gym provided by his school's afterschool program. He went to basketball camp this summer, where he seemed to develop some appreciation for the team-oriented aspects of the game.

After he saw all three of the then-extant versions of "Karate Kid," Adam expressed great enthusiasm for the martial arts. Ever since, his mother and I have spent hours every week transporting him to a school for Taekwondo. He's been at that for almost two years, has won several medals in tournaments, and should earn his black belt next spring.

Watching the different activities in which Adam engages, I've often wondered if the specific choices we've made will have significant consequences for his future. I found it interesting that two other sociologists and parents had a similar question.

—Roger Clark, Adam's dad, 1996

Introduction

The relatively new social institution of adult-controlled, afterschool activities in which Adam's participated merits research investigation. Unlike subjects such as attitudes toward rape victims (studied by Gray, Palileo, and Johnson in Chapter 9), housing for the elderly (examined by Filinson in Chapter 7), and TV violence (looked at by Messner in Chapter 12), organized afterschool activities haven't received much public attention. Consequently, it's difficult to come up with many educated guesses (or hypotheses) about them. This was especially true when two parents and sociologists, Patricia A. Adler and Peter Adler, became so interested in their own children's afterschool activities that they produced the focal research of this chapter. It was also true when Arlie Hochschild (1983) became interested in how people act on feelings, or stop acting on them, or even stop feeling them; when Jay MacLeod (1987) wanted to find out how young male inhabitants of low-income housing come to feel trapped in a position of inherited poverty; and when Elliot Liebow (1993) became intrigued by the apparently inexplicable humor and courage of homeless women he met. All these authors practiced a brand of research that we broadly call **observational techniques,** and under which rubric we include methods that are sometimes called participant and nonparticipant observation. **Participant observation** is performed by observers who take part in the activities of the people they are studying; **nonparticipant observation** is conducted by those who remain as aloof as possible. Adler and Adler call their own research participant observation because each participated as "parent, friend, counselor, coach, volunteer, and carpooler" in the activities of the children they studied.

observational techniques, techniques that include participant and nonparticipant observation.
participant observation, observation performed by observers who take part in the activities they observe.
nonparticipant observation, observation made by an observer who remains as aloof as possible from those observed.

FOCAL RESEARCH

Social Reproduction and the Corporate Other: The Institutionalization of Afterschool Activities[1]

by Patricia A. Adler and Peter Adler

Introduction

Since the early 1970s we have witnessed the rise of a new phenomenon: the broad expansion and institutionalization of the adult-controlled "afterschool"

[1] This article is adapted from Patricia and Peter Adler's (1994) "Social Reproduction and The Corporate Other: The Institutionalization of Afterschool Activities," *The Sociological Quarterly*, Vol. 35, No. 2, 309–328, copyrighted by the Midwest Sociology Society, by permission.

period (Berlage, 1982, Eitzen and Sage, 1989). Instead of merely coming home and playing in the house, neighborhood, or schoolyard, elementary, middle school, and junior high youths are likely to be registered in some extracurricular activity organized through a local YMCA, recreation center, community center, or private association founded for this purpose. As a result, afterschool activities have become one of the most salient features of many families' childrearing experience and are instrumental in defining the developing identities of participating youth.

The rise of the institutionalized afterschool phenomenon may be rooted in two cultural conditions that have occurred over the last generation. First, with the massive entry of women (especially middle-class women) into the labor force, a need arose for childcare and/or child supervision following the school day. While we have seen the development of the "latchkey" kid (see Rodman, 1990), we have more often seen activities where children could be taken care of, entertained, and enriched. Afterschool activities thus derived reinforcement from the prevalence of dual-career families and a social consciousness that demanded the cultural edification of children apart from traditional instruction offered in the classroom. Second, there have been rising concerns about leaving children unsupervised in public, outdoor places (Cahill, 1990). Afterschool activities represented a safe place where children could spend recreational time.

Our work builds on previous research that has focused primarily on organized youth sport, some of which has extolled adult-run leisure (for example, Webb, 1969), while some has been less sanguine (for example, Devereaux, 1976). First, we describe more fully the characteristics of available afterschool activities. Second, we illustrate a model of developmental progression young people frequently follow in advancing their extracurricular participation. Finally, we analyze the socializing effects of these experiences through the culture that is generated and the effect this has on the reproduction of social structure and the development of the self.

Methods

We draw on data gathered through participant observation with students at elementary and junior high schools. Over the course of six years (1987–1992) we observed and interacted with children both inside and outside of their schools. The children we studied attended public schools drawing predominantly on middle- and upper-middle-class neighborhoods in a large, mostly white university community. While doing our research, we occupied several roles: parent, friend, counselor, coach, volunteer, and carpooler (Fine and Sandstrom, 1988). We undertook these diverse roles both as they naturally presented themselves and as deliberate research strategies, sometimes combining them as opportunities for interacting with children became available through familial obligations or work/school requirements.

In interacting with children we varied our behavior; there were times when we acted naturally, expressing ourselves fully as responsible adults, yet there were times when we cast these attitudes and demeanors aside and tried to hang out with the children, getting into their gossip and (mis)adventures. Through these varied approaches, and through the often irrepressible candor of children, we were able to gain an insider's access to their thoughts, beliefs, and assessments.

Conducting most of our research outside of school settings, we tried to develop the parameters of the "parental" research roles by observing, casually conversing with, and interviewing children, children's friends, other parents, and teachers. We followed our daughter, son, and their neighbors and friends through their school experiences, gathering data on them as they developed. The children we befriended relished the role of research subjects because it raised their status in the eyes of adults to "experts," whose lives were important and who were seriously consulted about matters ranging from "chasing and kissing" games to the characteristics of "nerds."

In addition we conducted informal interviews with a range of young people from a variety of ages and types of activities. We selected interview subjects on several bases: age, gender, broad interests, or degree of specialization in particular activities. In addition, new subjects were referred to us by people we had already talked to who had specific knowledge of or participation in something on a different level than our previous subjects. We continued this snowball referral (Biernaki and Waldorf, 1981) until we felt that we had adequately covered the range of available afterschool activities. The topics we focused on in discussing extracurricular participation with parents and children included the range and type they had experienced, perceptions and feelings about them, identification and commitment to them, organizations and individuals associated with them, and decisions to continue or disaffiliate with various activities.

The Extracurricular Career

A broad range of adult-organized afterschool activities are available to youth. Three categories emerge, varying in their degrees of organization, rationalization, competition, commitment, and professionalization: recreational, competitive, and elite. These categories represent a developmental model, where young people progress through various kinds of activities, increasing their depth, fervor, and skill of involvement as they escalate their participation. Not all young people follow the progression of the extracurricular career; some remain at more recreational levels or retreat from intense types of afterschool participation to more moderate and less demanding activities. However these are exceptions to the norm, and the ensuing depiction represents the path followed by most young people.

Spontaneous Play

Traditionally, afterschool activities were those planned and directed by children themselves. Beginning in earliest childhood, children would come home after school, play in each other's houses, backyards, outside in the neighborhood, at the school playground or athletic field, and in child-organized games at any of these locations.

There are myriad organizational skills children have to master to accomplish spontaneous play. For instance, they have to plan what to do, decide on a location, establish rules and roles for participants, and set handicaps to ensure equitable and enjoyable play (see Coakley, 1990). Negotiation is another skill learned through spontaneous play. Children must routinely resolve competing desires, settle different interpretations of what has occurred and what it means, select among competing plans, and make adjustments when things are not going well. Finally, spontaneous play involves a considerable amount of problem solving. This involves significant power dynamics, compromise, and communication. For example, we observed one child who would often quit the game, without warning, when things were not going as he desired. The other children learned ways to deal with him, ranging from ignoring him to begging him to return, depending on their needs and his mood. Invariably, after this grandstanding behavior, he would reluctantly return to play.

Recreational

Recreationally organized afterschool activities vary widely in scope and character. They can be fundamentally centered on the social functions of companionship and play, such as those offered by established bureaucratic organizations (for example, YMCA, local Rec, or Community Centers). Second, these recreational activities can be centered around fitness. For instance, the Kidsport Fun and Fitness Club promotes their "Fun to be Fit" program for 8- to 12-year olds, emphasizing the health, nutrition, grooming, and fitness advantages of Participation (Kidsport flyer). Third, recreational activities focus on extracurricular *learning*. Schools, seeking to capitalize on the afterschool market, offer a variety of (paid) programs designed to keep children on the grounds after the traditional day has concluded. These programs offer cooking, art, creative writing, and computer classes. Finally, the greatest majority of afterschool activities involve *skill development* in the arts (dance, acting, singing, music), crafts (cooking, sewing, needlework), and individual sports (horseback riding, skiing, martial arts, tennis, skating).

In recreationally organized team sports, scoring is de-emphasized. The YMCA organizes its youngest "T-ball" league so that players bat through the lineup once on each team per inning, until the designated time is reached. One eight-year-old boy expressed his feeling about this orientation:

> Yeah, it's fun. Me and my friends, we all play on the same team. Jack's dad is the coach and he's pretty nice. But sometimes it's not so fun, because everybody has to get an equal chance to play every position, no matter how good or bad they are. And some of those guys out there can't even throw or catch the ball. And he puts me in outfield and them at shortstop, and if I catch a fly ball, I can't throw it in to second base for the double play, because they can't catch it.

Past the youngest ages, recreational teams may keep score and acknowledge a winner and loser. However, there are no league standings or playoffs at the end of the season. Children, especially boys, who value displays of domination (Adler, Kless, and Adler, 1993; Best, 1983; Thorne, 1986), may still undercut this by infusing hierarchy. In a football league for nine-year-old boys, one team kept their own record of all the teams' victories so that they could track league standings, and taunted players on losing teams. Still, the emphasis at this level is on children forming broad interests in many fitness and extracurricular activities, without becoming particularly enmeshed in any specific activity.

These types of programs have several defining characteristics. First, they are adult-organized and supervised. This brings with them a set of rules and regulations for how, when, and where things should be done, introduces a teacher-student style relationship, and establishes a clear situation of authority and hierarchy (compare Coakley, 1990; Eitzen and Sage, 1989). Second, they are located outside of the home or neighborhood/yard, in institutional settings, where adults are responsible for decision-making and safety. Third, in contrast to child-directed play, where participation may be exclusive, these programs are geared toward democratic acceptance of interested parties. Finally, they are guided by a philosophy of noncompetitive structure (although participants may compare themselves informally to each other). Young participants are thus socialized to value acceptance and fairness, team spirit and camaraderie, knowledge acquisition and skill development, and submission to adult rules and authority.

Competitive

Succeeding this recreational base are afterschool activities (such as music, gymnastics, and team sports) organized on a more competitive plane. Although universal participation is initially held as an ideal, this progressively diminishes as competition escalates.

Along with the formalization of the activity's structure, children are ranked and placed in skill levels. A hierarchy is established, with children who do not measure up progressively excluded, ostracized, or denigrated. A continuum can be outlined, from an initial philosophy of democracy to one of increasing meritocracy, accompanying the increasingly competitive nature of children's play. All of the accompanying factors such as seriousness, commitment, bureaucratization, and professionalism shift as one moves from the lower end of this continuum toward the top.

Democratic At the democratic end of the competitive continuum, activities are structured hierarchically, yet the rhetoric of recreation is still espoused. Organizations responsible for coordinating and offering such afterschool activities are usually specialized and tied to a single enterprise. Many of them hold democratic principles resembling this one stated in a letter to the coaches of a girls' softball association:

> The key message is that the program or activity must be fun if the participants are going to get something out of it and continue to play. We have tried to structure the program on that basis and our #1 goal is that the girls have fun.

Yet sometimes young participants get mixed messages about goals from adults. In a soccer league office at the beginning of the season, as mothers thronged to switch their children onto teams with their friends, one administrator sardonically remarked to another, "What do they think this is, a social club? This is a soccer club." After overhearing this, another mother whispered, "They may not think it's a social club, but I sure do."

Other times it is the parents, more than the administrators, who take the activity seriously. Attending a soccer game of a friend's nine-year-old son in another state, we observed

> The parents were all running up and down the sidelines screaming and shouting at the players and coach. While this looked fairly typical to us, several of the parents felt the need to approach us during the game and offer accounts for their behavior. "Oh, you'll have to excuse us, we get kind of carried away."

Their words suggested that they perceived their intensity, involvement, and competitiveness as deviant relative to the league.

While more talented players may find the increase in competition satisfying, those with weaker skills often find themselves unhappy. In one democratically geared competitive league, a nine-year-old baseball player described his dissatisfaction with his coach's philosophy and practice:

> My coach wants to win too much. It's not fun for me. I'm not getting enough playing time . . . . I never get to start . . . He only puts me in to play the outfield. I can play other positions, it's not fair. I want to get a chance to pitch. I can do it.

At this level, differences become more apparent between the stated ideals of the program and the actual practices of the participants. Adults usually espouse the commonly held notion of democratic play and skill development, but in practice often emphasize winning. Children have no such ambiguities. Despite the admonitions that "It's only a game," they are quite aware of performance differentials. They easily note that, when faced with crucial situations, the rhetoric of democracy is abandoned and unmitigated competition comes to the fore. The values communicated to youngsters through this level of afterschool participation include a beginning awareness of inequality in talent and the legitimization of how this affects opportunity as well as the introduction of focus on competition and winning.

Meritocratic Organizations sponsoring afterschool activities progressively detach from the vestiges of recreation and democracy as they become more competitive and professionalized. Skill, performance, and talent criteria associated with a meritocracy, become increasingly rewarded, while effort, fairness, and democratic decision-making diminish in centrality. As one 12-year-old girl explained,

> I really like being in the performing company. Having tryouts cut out all of the people who can't really dance, who aren't coordinated, and who just don't pay attention. For so many years my dance classes were filled with those kids and they dragged the class down, made us go slowly when they couldn't get the steps or they forgot them. Now we learn

> much more because we can move faster, the classes are tough, and the rehearsals are serious. The people with the solos are the ones who are the best and who work the hardest, and you don't have someone's mother complaining . . . that their daughter should have a solo, because they know it's based on who's best.

The shift in character associated with the move from democratic to meritocratic deters some participants from continuing to advance in the activity. One boy dropped out of his quiz bowl team when the level of competition exceeded his abilities. The funneling process also led young people to abandon some activities while concentrating on others. As one 10-year-old boy commented,

> I'm not going out for soccer this year. It was boring last year because I didn't get to play enough, and I also had a problem because I was in a play and there was an overlap between the play and soccer ... [S]ometimes rehearsals were on Saturday mornings when we had soccer games, and I had to leave games to go to rehearsals, and my coach yelled at me. So I think I'll just stick with acting this year. I don't have any friends in it, but I like it more and I'm better at it. It's more *me.*

The funneling effect of youngsters dropping out of extracurricular activities is thus associated with the rise in commitment these activities require, so that the youngsters have to choose the one to which they will dedicate themselves. These choices then represent a deliberate move on the youngsters' part (as opposed to their parents signing them up) and usually reflect their developing identities.

Beyond whole leagues organized at the more competitive level, coaches or parents occasionally push individual squads into a more competitive posture. Most youngsters enjoyed participating in some of the more competitive afterschool activities. They gave greater concentration and effort, drew a greater sense of self-involvement from them, and received greater attention from the surrounding adult world. Youngsters also noted that the more competitive activities were for older and more talented participants, and appreciated the greater status they derived. Part of growing up for these middle-class youngsters involved being exposed to a variety of diverse extracurricular experiences and acquiring many skills. But they also learned from their recognition of the escalating ladder of extracurricular participation that adults and society value depth involvement. Recreation and breadth, then, were less notable than competition and depth. Other values associated with advancing to this level of afterschool included the stratification of hierarchy and ranking, exclusion, seriousness of purpose, and identification with participation in the activity.

Elite Almost all competitive afterschool activities can also be performed on an elite level. School plays or local dance troupes can be replaced by participation in adult theater or dance performances, band players can move up to junior symphony, local competitions can give way to regional or national meets, and competitively oriented leagues can be succeeded by elite leagues.

The decision to pursue an elite activity means allocating it a greater amount of time, making it a priority ahead of other activities, and eliminating other projects. Not everyone wants to make this commitment, but for those who do, not only their own but their families' time may have to be shifted and activities rescheduled around what was once an afterschool activity. One 14-year-old ballet enthusiast regularly took a lighter scholastic load so she could spend several hours every day in the studio, while another student left school during the day to attend art classes at the local university. Many families plan their vacations around the demands of their children's activities. Parents have to transport their children to and from these activities daily, adjusting their work, mealtime, and family schedules. One mother expressed her concerns:

> Ever since my son started to play hockey seriously my days have been ruined. He has to be on the ice at 5 A.M. and then he practices after school. My husband and I haven't had a free weekend in months.

The accoutrements associated with elite afterschool activities are superior to those at the organized and competitive levels, featuring enhanced, often personalized, uniforms, better equipment, better arenas, and upgraded transportation. All of this costs more money, frequently a substantial burden for both students and parents. Concomitant with these changes, exclusiveness becomes salient in elite activities. All remaining vestiges of democratic participation are replaced by a philosophy of selectivity. The unconcealed exclusiveness may hurt young people who do not qualify for participation. Others are more sanguine about their failures, putting them aside and turning to other activities.

Once selected for the squad, participants soon discover that the adults treat elite activities seriously and expect adolescents and pre-adolescents will make the commitment. Practices may move from weekly or twice weekly to almost daily and require strenuous concentration. Many participants and parents regard elite activities as a track into some future enterprise. As one 15-year-old girl remarked,

> My goal is to get more speed on my fastball and to stick with softball through the teenage years, which are the tough ones. Then I want to get a scholarship to college. My coach has already told me that I can get one, it's just a matter of how much I improve that will make the difference of if I go to, say, Arizona State or end up at Western Kansas. And then after that, there's no professional softball for women right now, but maybe by the time I get out, there will be.

As the youngsters stay on the professionalization path (Vaz, 1982), the funnel continues to narrow.

Although elite participants make sacrifices for the team, they never abandon their focus on developing their own careers. Individuals accord the coach authority but never abdicate their own decision-making completely. Decisions are made by the coach for the competitive advancement of the group, overruling parents' authority. Yet participants will only stay with a coach as long as it suits their best interests.

Decisions are no longer made on the basis of fairness or ethical values, but on merit, pragmatics, and outcomes (Berlage, 1982). Coaches and leaders foster a mentality of winning. Young people become introduced to some of the other less idealistic and "fair" aspects of society such as politics, favoritism, and financial considerations.

Berlage (1982) has suggested that the meanings and values embedded in elite afterschool activities are those of the corporate world. These values socialize children to the attitudes and behavior they will encounter in corporate jobs. They are predominantly transmitted through the specialization of roles, the subjugation of individualism and family for the team, the emphasis on sacrifice and self-discipline, the structuring into well-defined units, and the processual model of advancement. Other values associated with the elite level include subordination, interdependence, deferred gratification, professionalism, commercialism, and a strong focus on winning. While these characteristics define the elite level, as time progresses, the characteristics of each tier of afterschool activity tends to diffuse down to the developmentally preceding stage.

Conclusion

In considering the benefits or harm generated by the infusion of adult control and the associated adult-oriented structures and values into the leisure of children and youth, we see elements of both. Our observations suggest that in progressing through the stages of afterschool activities children learn several important norms and values about the nature of adult society. They discover the importance placed by adults on rules, regulations, and order. Creativity is encouraged, but acknowledged within the boundaries of certain well-defined parameters. Obedience, discipline, sacrifice, seriousness, and focused attention are valued; deviance, dabbling, and self-indulgence are not. Coordination with others, the organic model of working toward challenging and complex goals, stands as the ultimate model toward which young people are directed. This anticipatory socialization to the organized, competitive world used to be the more exclusive domain of young boys, giving them what Borman and Frankel (1984) have claimed is a competitive advantage in the adult corporate world. The recent massive growth of institutionalized competitive play has extended into the female realm, however, and this may portend greater gender equalization for this generation's future adults. In fact, as we noted in recent research (Adler, Kless, and Adler, 1993), girls' gender role images appear to be expanding to a much greater extent than boys'. Girls, then, are already beginning to change the world by moving into boys' realms.

At the same time, the earlier imposition of adult norms and values onto childhood may rob children of developmentally valuable play, unchanneled and pursued for merely expressive rather than instrumental purposes. By participating in increasing amounts of adult-organized activities, children are steered away from goal-setting, negotiation, improvisation, and self-reliance toward the acceptance of adult authority and adult pre-set goals.

While these afterschool activities may prepare children for the formally rational, hierarchical, and, in Foucault's (1977) words, disciplined adult world, children's spontaneous play may teach different but important social lessons. Adult-organized activities may prepare children for passively accepting the adult world as given; the activities that children organize themselves may prepare them for creatively constructing alternative worlds.

Moreover, entering into these adult-organized activities draws children into junior versions of the existing social order. Not only are the norms and values of adult culture embodied in organized afterschool activities, but the structural inequalities of race, class, and, to a lesser extent, gender are inherent as well. It takes money to support a child going "up the ladder" of afterschool activities, and while families from lower socioeconomic groups participate in these endeavors, they cannot afford them to the same extent as more affluent households. If these experiences prepare youngsters for the corporate work world—partly through their enhanced "cultural capital" (Bourdieu, 1977a) of additional knowledge, skills, and disposition and partly through the "habitus" (Bourdieu, 1977b), the attitude and experience of achievement they acquire—then afterschool activities are yet another route to reproducing social inequities. The channeling of low-income, racial group members into inexpensive, segregated activities may also become exacerbated as more extracurricular activities are moved from the realm of public education into the private domain. Finally, while girls' activities have made significant strides toward parity with boys' over the last decade, gender segregation and stereotyping still remain, with patriarchal structures and attitudes continuing to dominate significant portions of the field.

The afterschool phenomenon thus offers a doorway into the youth subculture for adults. In exploiting this opening, adults have seized the opportunity to transform the character of children's play. What they have created, in doing so, is an ironic juxtaposition of work and play: play has become used as the vehicle for infusing adult work values into children's lives.

REFERENCES

Adler, P. A., S. J. Kless, and P. Adler. 1993. Socialization to gender roles: Popularity among elementary school boys and girls. *Sociology of Education* 65: 169–187.

Berlage, G. 1982. Are children's competitive team sports corporate values? *ARENA Review* 6: 15–21.

Best, R. 1983. *We've all got scars.* Bloomington: Indiana University Press.

Biernaki, P., and D. Waldorf. 1981. Snowball sampling. *Sociological Research and Methods* 10: 141–163.

Borman, K. M., and J. Frankel. 1984. Gender inequalities in childhood social life and adult work life. Pp. 55–63 in *Women in the workplace,* edited by S. Gideonse. Norwood, NJ: Ablex.

Bourdieu, P. 1977a. Cultural reproduction and social reproduction. Pp. 487–511 in *Power and ideology in education,* edited by J. Katabel and A. H. Halsey. New York: Oxford University Press.

Bourdieu, P. 1977b. *Outline of a theory of practice*. Cambridge: Cambridge University Press.

Cahill, S. 1990. Childhood and public life: Reaffirming biographical divisions. *Social problems* 37: 390–402.

Coakley, J. J. 1990. *Sports in Society*, 4th ed. St. Louis: Mosby.

Devereaux, E. 1976. Backyard versus Little League baseball: The impoverishment of children's games. Pp. 37–56 in *Social problems in athletics*, edited by D. M. Landers. Urbana: University of Illinois Press.

Eitzen, D. S., and G. Sage. 1989. *Sociology of North American sports*, 4th ed. Dubuque, IA: William C. Brown.

Fine, G. A., and K. L. Sandstrom. 1988. *Knowing children*. Newbury Park, CA: Sage.

Foucault, M. 1977. *Discipline and punishment*. New York: Pantheon.

Rodman, H. 1990. The social construction of the latchkey children problem. Pp. 163–174 in *Sociological Studies of Child Development*. Vol. 3, edited by N. Mandell. Greenwich, CT: JAI.

Thorne, B. 1986. Girls and boys together. But mostly apart: Gender arrangements in elementary schools. Pp. 167–184 in *Relationships and Development*, edited by W. Hartup and Z. Rubin. Hillsdale, NJ: Lawrence Erlbaum.

Webb, H. 1969. Professionalization of attitude toward play among adolescents. Pp. 161–178 in *Aspects of Contemporary Sport Sociology*, edited by G. Kenyon. Chicago: Athletic Institute.

STOP AND THINK *What did you do for afterschool activities in elementary school? In junior high school and high school? Does the Adlers' developmental model of afterschool activities jibe with your experiences of afterschool activities as you grew up? Does the "funneling" they describe ring true or was your experience of afterschool activities somewhat different?*

Observational Techniques Defined

The term "observational techniques" is somewhat misleading for what we're discussing in this chapter, though it's slightly less so, in our opinion, than the main alternatives: "qualitative methods" and "field research." Both "qualitative methods" and "field research" connote a variety of methods—the former, qualitative interviewing (Chapter 10); the latter, qualitative interviewing, finding available data (Chapter 12), and others—that we cover elsewhere in this book, as well as "observations," the topic of this chapter. Adler and Adler have clearly asked a lot of questions in the course of their work (hence all the participants' recountings and remarks), but they've also made a lot of observations, including having observed nine-year-old boys in one football league keeping track of their own record for comparative purposes and seen mothers throng, at the beginning of a soccer season, to a league office to switch their children onto teams with friends. We'd like to stress that a variety of methods *are* and *should be* used in the field, but here we want to focus on those observational methods that are used in the field but have not been covered in other chapters.

Of course, the data from questionnaires, interviews, content analysis, experiments, and so on, *are* all based on observations of one kind or another. For questionnaires and interviews, the observations are about responses to questions. For content analysis, the observations are about communication of one sort or another. In experiments, however, observations are frequently of the sort that closely resemble those used by participant and nonparticipant observers. In most cases, experimental observations are of a more controlled sort than are those used in the field. **Controlled observations** involve clear decisions about just what kinds of things are to be observed. In the "Coins in the Kettle" experiment by Bryan and Test (1982), described in Chapter 8, observers focused solely on whether people would or wouldn't contribute coins to a Salvation Army kettle after they'd seen, or not seen, another person do so (the stimulus). But such experimental observations are not the focus of this chapter.

controlled observations, observations that involve clear decisions about what is to be observed.

What Adler and Adler do that distinguishes their work from these other approaches is to immerse themselves in the natural setting of their subjects. Rather than merely read about or ask about "afterschool activities," they've gone to and observed those activities in person (though they've obviously read about and asked about the activities as well). Moreover, although they have questioned participants and made their presence felt in one way or another (as when they've acted as "parent, friend, counselor, coach, volunteer, and carpooler"), their intervention has been minimal. They have, as they've said elsewhere, tried to create a situation in which "behavior and interaction continue as they would without the presence of the researcher, uninterrupted by intrusion" (1994: 378). We here confine our attention to observational techniques, then, that are used to observe actors who are, at least in the perception of the observer, acting naturally in natural settings.

STOP AND THINK *Could Adler and Adler have used only questionnaires to study afterschool activities? What kinds of insights might have been "lost" if they had?*

The origins of modern observational techniques are the subject of some current debate. Conventional histories attribute their founding to anthropologists like Bronislaw Malinowski, with his (1922) *Argonauts of the Pacific*, and sociologists like Robert Park, who, around the same time, began to inspire a generation of students to study city life as it occurred naturally (see, for example, Wax, 1971 and Hughes, 1960). More recently, however, Reinharz (1992) has argued that Harriet Alice Fletcher, studying the Omaha Indians in the 1880s, and Harriet Martineau, publishing *Society in America* in the 1830s, promoted essentially the same techniques that Malinowski and Park received fame for in the 1920s, suggesting the earliest "social science" observers might have been American women.

Uses of Observations

When are observational techniques desirable? As we've suggested, observation can be useful when you don't know much about the subject under investigation. A classic technique in anthropology, the ethnography, or study of culture, is primarily an observational method that presumes that the culture under study is unknown to or poorly known by the observer. Margaret Mead's (1933) classic investigation, *Sex and Temperament*, of gendered behavior among three tribes in New Guinea, for instance, found evidence that sex-role definitions were more malleable than convention held at the time. Similarly, many well-known sociological investigations have been of things less than well known beforehand. Laud Humphreys' famous (1975)—some would say infamous (see Chapter 4)—study of impersonal sex among men in public restrooms described practices that were not broadly known. Adler and Adler justify their investigation of adult-controlled afterschool activities, in part, by how new those activities are to the American cultural landscape.

But observational techniques also make sense when one wants to understand experience from the point of view of those who are living it or from the context in which it is lived. Denzin makes a distinction between **thin** and **thick description**, the former merely reporting on an act, the latter providing a sense of "intentions, motives, meanings, context, situations, and circumstances of action" (1989: 39). Elliot Liebow (1993), for instance, knew that homeless women existed before he embarked on his research. Moreover, as a volunteer in a soup kitchen, he'd observed that many of them coped remarkably well with their homelessness. His study, however, is an attempt to get behind the coping behavior, to understand it through a description of those meanings, contexts, and situations that make it possible.

thin description, bare-bone description of acts.

thick description, reports about behavior that provide a sense of things like the intentions, motives, and meanings behind the behavior.

The goal of understanding experience is clearly not antithetical to the goal of studying a subject that is previously unknown. The two goals frequently go hand in hand. On the one hand, the Adlers' progressive typology of afterschool activities (from recreational to competitive to elite) gives us an interesting new way of understanding those activities. On the other hand, this and other new insights come from focusing on a subject (afterschool activities) that had received very little attention before.

Observational techniques are also useful when you want to study quickly changing social situations. Kai Erikson largely based his account in *Everything in Its Path* (1976) of the trauma visited on an Appalachian community by a disastrous flood in 1972 on descriptions gleaned from legal transcripts and personal interviews. He, as a professional sociologist, had been asked by a law firm to assess the nature and extent of the damage to the Buffalo Creek community by a flood that had been caused, in part, by the neglect of a coal company. His book presents the legal case he made then, using content analyses and unstructured interviews. But when he attempted to present the effects of the favorable judgment on the community he studied, he described a gathering "in a local school auditorium to hear a few announcements and to mark the end of the long ordeal." He interprets the gathering as "a kind of graduation . . . that ended a period of uncertainty, vindicated a decision to enter litigation, and furnished people with sufficient funds to realize whatever plans they were ready to make. But it was also a graduation in the sense that it propelled people into the future at the very moment it was placing a final seal on a portion of the past" (1976: 248–249). Erikson then used direct observation to mark a momentous occasion: the moment (two years after the flood) when Buffalo Creek began, as a community, to heal itself after its calamitous ordeal. If he hadn't been there when he was, this moment of significant change would necessarily have gone unnoticed by himself and his readers.

Another reason for using observational techniques is that they offer a relatively unfiltered view of human behavior. You can, of course, ask people about what they've done or would do under given conditions, but the responses you'd get might mislead you for any number of reasons. People do, for instance, occasionally forget, withhold, or embellish in ways that are self-serving. But if you see parents "running up and down the sidelines screaming and shouting at the players and coach," as Adler and Adler did, you've got data about behavior that might be difficult to come by otherwise.

There are other reasons why one might use observational techniques rather than others described in this text. Questionnaire surveys can be very expensive, but participant or nonparticipant observation, although time-consuming, can cost as little as the price of paper and a pencil. And ethically, studying homeless people in their natural settings, as Liebow did, raises fewer questions than would studying, say, the effects of homelessness by creating an experimental condition of homelessness for a group of subjects. We'd like to emphasize again, that a researcher can be moved by various combinations of the previously mentioned motives. Judith Stacey's (1990) comparative case study of two women's lives in *Brave New Families* was a self-conscious effort to study "postmodern" families that were created in one American city in the late 1980s. Stacey spent two years studying the public and private lives of these two (extraordinarily generous!) women to discern how and why their family lives changed (and deviated from one another) to get a handle on the more general emergence of alternative family forms (what Stacey calls divorce-extended and matrifocal family forms) in America, and to discern what that emergence meant to the individuals experiencing it.

complete participant role, being, or pretending to be, a genuine participant in a situation one observes.

participant-as-observer role, being primarily a participant, while admitting an observer status.

observer-as-participant role, being primarily a self-professed observer, while occasionally participating in the situation.

complete observer role, being an observer of a situation without becoming part of it.

Observer Roles

We suggested earlier that one of the defining characteristics of observational techniques is their relative inobtrusiveness. But this inobtrusiveness varies with the role played by the observers and with the degree to which they are open about their research purposes. Raymond Gold's (1958) typology of researcher roles suggests a continuum of possible roles: the **complete participant**, the **participant-as-observer**, the **observer-as-participant**, and the **complete observer**. It also suggests a continuum in the degree to which the researcher is open about his research purpose. As a complete participant, the researcher might become a genuine participant in the setting under study, or might simply pretend to be a genuine participant. When a genuine participant, the complete participant not only risks the fear of being found out but also entertains the danger of **going native,** of over-identifying with other participants, which in some cases can lead to abandoning the research for the joys of participation. One of us, Emily, intended, for instance, to report on a feminist consciousness-raising group in the early 1970s until her participation altered her consciousness, as it were, and she decided that her membership in the group precluded her using it as an object of study. The fate of Emily's project (its abandonment) is just one the of reasons conventional wisdom has warned against "going native"; another concern has been the bias that tends to be associated with "over-rapport" (see, for example, Hammersley and Atkinson, 1992). Current "critical" (for example, Morrow, 1994) and "feminist" (for example, Reinharz, 1992) perspectives have questioned this concern, sometimes even claiming that understanding other people requires becoming close to them.

going native,
a danger of complete participation, the condition of over-identifying with other participants.

In any case, the complete participant, in Gold's typology, is defined by his withholding of the fact that he is, in fact, doing research. Thus Humphreys assumed the role of the "watchqueen," the look-out who kept an eye out for unwanted intruders in his study of impersonal sex in public restrooms:

> The very fear and suspicion encountered in the restrooms produces a participant role, the sexuality of which is optional. This is the role of lookout ("watchqueen" in the argot), a man who is situated at the door or windows from which he may observe the means of access to the restroom. When someone approaches, he coughs. He nods when the coast is clear . . . (1975: 27).

Humphreys described his discovery of the "watchqueen" role as the "real methodological breakthrough" (27) in his research. It was a breakthrough that involved withholding the true purpose of his presence to those he observed. But, to the degree that he successfully "passed" as a watchqueen, his participation was unobtrusive or unlikely to affect what was going on in the restrooms.

STOP AND THINK *Some have suggested that it is unethical for a researcher to withhold his research purposes from his subjects (see Chapter 4). Do you think Humphreys was unethical in disguising his research purpose from the people he studied? Why or why not?*

The opposite of the complete participant role is that of the complete observer: one who observes a situation without becoming a part of it. Lyn Lofland (1972, 1973) studied how people in cities bring order to their relations with strangers. She anonymously watched people in public places like libraries, stores, and restaurants and recorded as much of what she could see as possible. But unlike Humphreys, who played a well-defined social role (watchqueen), she remained socially invisible to those whom she observed. She didn't tell those she watched that she was observing them and in doing so she was able to discern patterns of behavior, such as how people size up the situation as they enter public settings.

Hammersley and Atkinson (1992: 96) note the paradoxical fact that complete observation shares many of the disadvantages and advantages of complete participation. Neither is terribly obtrusive: Neither Lofland nor Humphreys was likely to affect the behavior of those they watched. On the other hand, neither permits the kind of probing questions that one can ask if one admits one is doing research. In addition to the question of (not being able to ask scientifically interesting) questions, the lack of openness (about one's research purposes) implicit in pure participation or pure observation has raised, for some, the ethical issue of whether deceit is ever an appropriate means to a scientific end, an issue, you will recall, that has come up in our discussion of other methods as well.

Both the participant-as-observer and the observer-as-participant make it clear that they are doing research; the difference is the degree to which they emphasize participation or observation. One senses that Adler and Adler were sometimes primarily participants, as when they undertook the diverse roles of "parent, friend, counselor, coach, volunteer, and carpooler." At times, they seem to have been and, perhaps more important, been perceived to be observers, as when they attended a friend's son's soccer game and observed parents shouting at players and coaches but then offering accounts for their behavior. Liebow is more explicit about his alternation between being primarily participant and primarily observer in the homeless women's shelter called The Refuge:

> I remained a volunteer at the Refuge [throughout the study], and was careful to make a distinction between the one night [a week] I was an official volunteer and the other nights when I was Doing Research (that is, hanging around). (1993: x)

Adler and Adler divulged their research purposes to their respondents (and report that child respondents enjoyed playing "expert"), especially, of course, when they began to conduct interviews. But,

> While we did not go out of our way to hide the fact that we were researchers (as we did in one of our previous books, *Wheeling and Dealing*), in some cases we didn't go out of our way to divulge it either. Due to the characteristics of the children (i.e., young), we did not think it necessary to announce our intentions at every event (personal communication, 1995).

Their research for *Wheeling and Dealing* (Adler, 1985), incidentally, had been of drug dealers and smugglers and might therefore have been compromised by a frank disclosure of their research interest.

Liebow clearly announced his interest and even records the announcement of his purpose, and his request for permission:

> "Listen," I said at the dinner table one evening, after getting permission to do a study from the shelter director. "I want your permission to take notes. I want to go home at night and write down what I can remember about the things you say and do. Maybe I'll write a book about homeless women" (1993: ix).

One danger of both the participant-as-observer and observer-as-participant roles is that they are more obtrusive than those of the pure participant or the pure observer and that those being studied might shift their focus to the research project from the activity being studied, much as they might in responding to a questionnaire or an experimental situation. This might have happened occasionally to Adler and Adler, but the verisimilitude, what they elsewhere (1994) call **vraisemblance,** of their writing style tends to dispel one's doubts about the overall validity of their reports.

vraisemblance, the verisimilitude, or sense of truth, that a reporter can convey through his or her writing.

STOP AND THINK *Recall that within their section on "democratic" activities, Adler and Adler claim that "sometimes young participants get mixed messages about goals from adults." What about their supporting evidence in this section gives you the feeling this claim is essentially true?*

The danger of obtrusiveness aside, there are times when announcing one's research intention offers the only real chance of access to particular sites. The only other way participant observers, Becker, Geer, Hughes, and Strauss (1961), could have gained access to the student culture in medical school for their classic study *Boys in White* would have been to apply and gain admission to medical school themselves. Think of all the science courses they would have had to take!

One thing Adler and Adler's study makes clear is that within any given research project, researchers can play several different observational roles. There were times when they were primarily participants, and times when they were primarily observers. In fact, as they've recently suggested (Adler and Adler, 1996), capitalizing on their parent role meant that their engagement in both of these capacities was, depending on the situation, more nuanced still; at times, they were just *parents-in-the-home*, at times, *parents-in-the school*, at times, *parents-in the community*, and sometimes, more explicitly, *parents-as-researchers*. It's also reasonably clear that different research purposes require different roles. It's hard to imagine, for instance, that Humphreys' study of impersonal sex in public places or that the Adlers' of drug dealers and smugglers would have worked if they'd announced their research intentions. It is hard to decide which role to adopt in a given situation. You should obviously be guided by methodological and ethical considerations, but there are, unfortunately, no definite guidelines.

Getting Ready for Observation

We've seen how researchers using other techniques (for example, experiments or surveys) can spend much time and energy preparing for data gathering, in formulating research hypotheses and developing measurement and sampling strategies, for example. Such lengthy preparations are usually not as vital for studies involving observational techniques, where design elements can frequently be worked out on an *ad hoc* basis. The exception to this rule, however, is **systematic observations,** which are defined (not unlike the controlled observations of experiments, mentioned earlier) by their use of explicit plans for selecting, recording, and coding data (McCall, 1984). Goodsell's (1984) study of welfare waiting rooms, for instance, was a systematic study of 28 waiting rooms in seven states and the District of

systematic observations, observations that employ explicit plans for selecting, recording, and coding data. Similar to controlled observations.

Columbia, intended to discern whether all such waiting rooms were as dismal in appearance and condition as the research literature suggested. Because Goodsell began with a more or less well-defined research question (are welfare waiting rooms dismal?), he could structure much of his observation with a checklist of items: things like "room configuration, nature and arrangement of furniture, decor and decoration, audible sounds, comportment of clerks, clients, and others, the wording of visible signs, and the presence of amenities such as furnished reading matter and vending machines" (1984: 470). One can imagine Goodsell filling out a codesheet that looked something like the one used Clark, Lennon, and Morris (1993) for their content analysis of children's books (see Chapter 13) after each of his 28 waiting room visits.

Far more often, however, observers begin their studies with less clearly defined research questions and considerably more flexible research plans. Adler and Adler (1994) suggest that an early order of business for all observational studies is selecting a setting, a selection that can be guided by opportunistic or theoretical criteria. The selection is made opportunistically if, as the term suggests, the opportunity for observation arises from the researcher's life's circumstances (Riemer, 1977). Adler and Adler chose to study afterschool activities because their own children were engaged in those activities. In fact, they have confided that their general preference is for opportunistic settings:

> We try to research in our own "backyard" (in this case [that is, in the after school activities study], literally) so that we can spend as much time in our research setting as our life will allow . . . We got interested in our children's afterschool activities, and then developed a theoretical interest in them (personal communication, 1995).

Liebow studied shelters for homeless women because, after he'd been diagnosed with cancer and quit his regular job in 1984, he decided to work in soup kitchens and shelters, only to find their clients extraordinarily interesting. Riemer (1977: 474) advocates opportunistic settings for ease of entry and of developing rapport with one's subjects, as well as for the relatively accurate levels of interpretation they can afford.

On the other hand, Liebow's other book, the *Tally's Corner (*1967), was based on a 12-month study of African American men in Washington, D.C., an area that was chosen because of the light it might cast on theories of poverty that were prevalent at the time. Similarly, Hochschild studied airline flight attendants and bill collectors for her book *The Managed Heart* (1983) because they "illustrate[d] two extremes of occupational demand on feeling" (16), the one demanding that one "feel sympathy, trust, and good will," the other, "distrust, and sometimes positive bad will" (137). The settings for *Tally's Corner* and *The Managed Heart*, then, were chosen primarily for their theoretical relevance rather than for their ready access.

STOP AND THINK *Do you imagine Humphreys' decision to study public restrooms was informed primarily opportunistically or theoretically? What makes you think so? (Humphreys' answer appears in the next Stop and Think paragraph.)*

The suggestion that setting selection can be made on either opportunistic or theoretical grounds might be slightly simplistic, however, insofar as it implies that one chooses only one site, once and for all, and that it is primarily opportunity or theory that informs the choice. In fact, Adler and Adler are not so very unusual in their decision, initially, to immerse themselves in settings that were available, but then to seek out interviews that spoke to ideas they developed on their own:

> [W]e derived our ideas about what to write *about* from our observations of and casual interaction with kids, ... but to get quotes from kids themselves and to see how they felt or if their experiences jibed or differed from our perspectives, we conducted interviews ... These interviews were selected by theoretical sampling to get a range of different types of kids in different activities, age ranges, and social groups (personal communication, 1995).

Especially in seeking their interviews, then, Adler and Adler were guided by the effort to test their own theoretical notions against those of other participants, an approach they recommend elsewhere (Adler and Adler, 1987). By emphasizing their efforts to "get a range of different types of kids," in fact, they suggest they attempted to approximate the ideal of **theoretical saturation** (Glaser and Strauss, 1967, Adler and Adler, 1994)—that is, the point where new interviewees or settings looked a lot like interviewees or settings they'd observed before. Opportunity, then, might have inspired their early choices; some mix of opportunity and theory influenced their later choices.

theoretical saturation, the point where new interviewees or settings look a lot like interviewees or settings one has observed before.

STOP AND LEARN *Humphreys says he embarked on his study of sex in public restrooms after he'd completed a research paper on homosexuality. He states that his instructor asked about where the average guy went for homosexual action, and that he (Humphreys) articulated a "hunch" that it might be public restrooms. His further comment, "We decided that this area of covert deviance, tangential to the subculture, was one that needed study" (1970: 16), suggests that his decision to study public restrooms was theoretically, rather than opportunistically, motivated.*

Moreover, Adler and Adler's study suggests what seems to be true in general of sampling in studies using observational techniques: Observations are most often done in a nonrandom fashion. Devising a sampling frame of the kinds of things studied by participant and nonparticipant observers (for example, afterschool activities, occupations involving emotion work, homeless women, bathrooms used for homosexual encounters) is often impossible, though again, this might be less true of the (relatively infrequent) studies involving systematic observations. Goodsell (1984) could conceivably have found a sampling frame of welfare waiting rooms in the states he studied, even though he didn't do so. What he did do was to pick ten offices from large cities, nine from medium-sized cities, and nine from small towns. In other words, he collected a **purposive sample**, or one based on his intuitive sense of what would offer a reasonably full understanding of his subject matter. Purposive and other nonrandom sampling seem to be the most commonly used sampling approaches in observational studies.

purposive sampling, sampling based on one's intuitive sense of what will offer a reasonably full understanding of the subject matter. A form of nonprobability sampling (see Chapter 5).

Gaining Access

Although choosing topics and selecting sites are important intellectual and personal tasks, gaining access to selected sites is, as Lofland and Lofland (1995: 31) point out, an important *social* task. At this stage a researcher must frequently use all the social skills or resources and ethical sensibilities she has at her command.

Lofland and Lofland (1995) suggest that the balance between the need for social skills or resources and ethical sensibilities, however, can depend, as much as anything, on whether the observer plans to announce her presence as observer or not. If the observer plans not to reveal the intention to observe, the major issues entailed in gaining (or using) access are often ethical ones (see Chapter 4): especially because of the deceit involved in all covert research. This can be a particularly knotty problem when one seeks "deep cover" in a site that is neither public nor well known, as was the case for Patricia Adler in her study of drug dealers and smugglers (1985), but it exists nonetheless even when one plans to study public places (as Humphreys did while watching impersonal sex in public restrooms) or to study familiar places (as you might do if you decided to study your research methods class today while you continued playing your role as student). According to Lofland and Lofland (1995), the decision to engage in covert research and thereafter to establish access, however, is ethically acceptable, if other concerns (such as ensuring lack of harm to those observed and pursuing worthwhile topics in settings that cannot be studied openly) neutralize or overwhelm concerns about deception.

Gaining access to, and then getting along in, a site becomes much more a matter of social skill or resources when the observer intends to announce her observer role. The degree to which such skills or resources are taxed, however, is generally a function of one's current participation in the setting. Furthermore, for Adler and Adler, who were already participants in many afterschool settings, at least early in their study, their pre-study participation not only made access easy, but also made it relatively easy to decide whom they needed to tell about their research, whom they needed to ask permission from, and whom they needed to consult for clarifications about meanings, intentions, contexts, and so on. Ease of access, then, is certainly one major reason why the Adlers advocate "starting where you are."

STOP AND THINK *Suppose you wanted to study a musical rock group ("The Easy Aces") on your campus, to which you did not belong. How might you gain access to the group for the explicit purpose of studying the things that keep them together?*

Access particularly taxes social skills or resources when you want to do announced research in a setting outside your usual range of accustomed settings. In such cases, Lofland and Lofland (1995) emphasize the importance of four social skills or resources: connections, knowledge, accounts, and courtesy. If, for instance, you want to study a rock group ("The Easy Aces") comprising students on your campus, you might first discuss your project with several of your friends and acquaintances, in the hopes that one of them knew one of the "Aces" well enough to introduce you. In so doing, you'd be employing a tried and true convention for making contacts—using "who you know" (*connections*) to make contact. In general it's a good idea to play the role of "learner" when doing observations, but such role-playing can be overdone on first contact, when it's often useful to indicate some *knowledge* of your subject. Consequently, before meeting "The Easy Aces," it would be a good idea to learn what you can about them—either from common acquaintances or from available public-access information. (If, for instance, "Aces" had made a CD, you might want to listen to a copy and refer to their songs, perhaps even admiringly, at your first meeting.) You'd also want to have a ready, plausible, and appealing explanation (an *account*) of your interest in the group for your first encounter. The account needn't be terribly clever, long-winded, or high-brow (something reasonably direct like "I'd like to get an idea of what keeps a successful rock group together" might do), but it might require some thought in advance (blurting out something like "I'm doing a research project and you guys seem like good subjects" might not do). Finally, and perhaps most commonsensically, a bit of common *courtesy* (such as the aforementioned expressions of appreciation for the "Aces'" music) can go a long way toward easing access to a setting.

All the aforementioned skills and resources—connections, knowledge, accounts, and courtesy—can be more difficult to muster effectively when there are gender, class, language, or ethnic differences between the researcher and other actors. Shulamit Reinharz (1992) cites numerous reports from female observers about the kinds of "interactional shitwork" they sometimes feel compelled to do when encountering one form of sexual harassment or another in mixed-gender or predominantly male settings. One example is from Arlene Kaplan Daniels (1967), who found that studying military men required unusual deference on her part.

> The military officers resented the introduction of a sociologist, a civilian, and a female into their midst ... In the view of the liaison chief and his officers, my attitude was overbearing, demanding, and all too aggressive . . . Interpersonal problems with the officers were exacerbated at each contact by my initial obtuseness about and resistance to the demeanor they thought I should exhibit. (Daniels, 1967: 267)

Still, the eventual success of Daniels' project, and that of other female observers in mixed-gender settings, as well as that of white-male observers, like Elliot Liebow who, for *Tally's Corner* (1967) studied poor, urban, African American males, suggests that even gender, race, and class divides are not necessarily insurmountable.

Participant and nonparticipant observations, perhaps more than any other techniques we've dealt with in this book, lend themselves poorly to "cookbook" preparations. (We can't tell you precisely what to do when you come up against bureaucratic, legal, or political obstacles to your setting, but we can tell you that others have encountered them and that authors like Lofland and Lofland [1995] provide some general strategies for surmounting them.) We can, however, make two general recommendations. First, we think potential observers (and, perhaps, especially if they are novices) should review as much literature in advance of their observations as possible. Not only are such reviews likely to sensitize the researchers to the kinds of things they might want to look for in the field; the reviews can also suggest new settings for study. You should know, however, that our advice on this matter is not universally accepted. In fact, the Adlers tell us that they now try to follow the advice of Jack Douglas (1976), who advises researchers to immerse themselves in their settings first, and avoid looking at whatever literature is available until one's own impressions are, well, less impressionable (personal communication, 1995). They suggest it is only once the researcher begins to "grasp the field as members do" that they should begin to "compare their typologizations to those in the literature." Second, whatever decision you make about when to review the literature on your own topic, we strongly recommend spending some time leafing through earlier examples of participant or nonparticipant observation, just to see what others have done. Particularly rich sources of such examples are the *Journal of Contemporary Ethnography* (which was, in earlier avatars, *Urban Life* and *Urban Life and Culture*) and *Qualitative Sociology*.

Gathering the Data

Perhaps the hardest thing about any kind of observation is recording the data. But how does one do it?

STOP AND THINK *Quick. We're about to list three ways of recording observations. Can you guess what they are?*

Three conventional techniques for recording observations are writing them down, recording them mechanically, as on a tape recorder or camera, and recording them in one's memory, to be written down later (hopefully only shortly later). Given the notorious untrustworthiness of memory, you might be surprised to learn that the third technique, trusting to memory for short periods of time, is probably the most common. Even Goodsell's (1984) relatively systematic observations of welfare waiting rooms, many of which observations obviously could have been made in the waiting rooms themselves, were recorded only after Goodsell had left the premises. And despite gaining the explicit permission of his subjects, Liebow (1993) waited until he got home at night to write down what he'd observed about homeless women. So did Herbert Gans during his (1962) study of working-class men on Boston's West Side. All were cognizant that even their relatively well-trained powers of observation and memory were subject to later distortion (whose aren't?) and so were at pains to record their observations as soon after they'd made them as possible. Their goal, the goal of all observers, was to record observations, including quotations, as precisely as possible, as soon as possible, before the inevitably distorting effects of time and distance were amplified unnecessarily.

Adler and Adler admit to being not "as thorough about this phase of research as others" (personal communication, 1995). Occasionally, they would come back home and write "jotted fieldnotes." But one of the advantages of their "team" approach was that when the time came "to make outlines and cull the data for examples," they could brainstorm "together to try to recall examples and talk to each other about what happened and what was said."

But why not record them on the spot? Joy Browne, in her (1976) reflections on her study of used car dealers, gives one view when she advises prospective observers,

> [D]o not take notes; listen instead. A panicky thought, perhaps, but unless you are in a classroom where everyone is taking notes, note taking can be as intrusive as a tape recorder and can have the same result: paranoia or influencing the action. (1976: 77)

The intrusiveness of note-taking, tape-recording, and picture-taking, then, make them all potentially troublesome for recording data.

On the other hand, short-hand notes and tape-recordings are certainly appropriate when they can be taken without substantially interrupting the natural flow. Maher and Tetreault (1994), for instance, taped classes of professors to discern varieties of feminist teaching styles. A problem of tape-recording, however, is that one can gather daunting quantities of data: it can take a skilled transcriber three hours to type one hour of tape-recorded conversation (Singleton, Straits, Straits, and McAllister, 1988: 315). And even if you're able to take notes or tape-record, it is still wise to transcribe and elaborate those recordings as soon as possible, while various unrecorded impressions remain fresh enough that you can record them as well.

Participant and nonparticipant observers commonly supplement their observations with interviews and available data. Adler and Adler refer to informal interviews with "a range of young people," a Kidsport flyer, and a letter to coaches. Interviewing other participants, known as **informants,** frequently provides the kind of in-depth understanding of what's going on that motivates researchers to use observation techniques in the first place. One concern raised by feminist approaches to observation, in fact, is how to retain as much of the personal perspective of informants as possible (for example, Reinharz, 1992: 52). How, in effect, to give "voice" to those informants? Some observers (for example, Maher and Tetreault, 1994) have even asked permission of their respondents to report their names, and have thus abandoned the usual practice of offering confidentiality or anonymity. We recommend, however, that confidentiality or anonymity, offered to all interviewees (see Chapters 4, 9, and 10), be given to interviewees in the field as well. Informants who later see their words used to substantiate points they themselves would not have made, for instance, may regret having given permission to divulge their identities in print, no matter how generous the intention of the researcher.

informants, participants in a study situation who are interviewed for an in-depth understanding of the situation.

Analyzing the Data

The distinction between data collection and data analysis is generally not as great for participant and nonparticipant observation as it is for other methods (for example, questionnaires, experiments) we've discussed. (See Chapter 16, where we discuss qualitative data analysis in greater depth than here.) For, although these other techniques tend to focus on theory verification, observational techniques are more often concerned with theory discovery or generation. Thus, Adler and Adler have generated a theory of afterschool activities that speaks to their functions in American society (for example, the preparation of characteristics that are amenable to corporate labor) as well as their classification (for example, as recreational, competitive, and elite). This is not to say that observational studies can't be used for theory verification or falsification. Perhaps the primary theoretical upshot of Goodsell's (1984) study of welfare waiting rooms was to falsify the hypothesis that such rooms are necessarily dismal: Some rooms (such as those he described as "dog kennels") did, in fact, look pretty degrading to Goodsell; but some (like the ones he described as "bank lobbies" and "circus tents") looked much less so.

But more studies based on observation techniques are like Adler and Adler's than like Goodsell's, more concerned, that is, with theory generation or discovery than with theory verification or falsification. Important questions are these: When does the theory-building begin and how does it proceed? The answer to the first question—when does theory building begin?—seems to be soon after you record your first observation. Let's say, for instance, that one of Adler and Adler's early observations was of the soccer game in which parents screamed and shouted on the sidelines. The very act of recording that observation, rather than whatever else they might have recorded (for example, meteorological conditions, the presence or absence of hot-dog concessions, the kind of uniform worn by players, and so on), implies that the observer attaches some importance or relevance to the observation. This choice is likely to be based on some preconceived notion of what's (theoretically) salient, a notion that the sensitive researcher will, sometime during or after the initial observation, try to articulate to himself. (Could the screaming of parents on the sidelines of a soccer game be typical of adult behavior that might give children a mixed message about fun-oriented activities?)

Once these notions do become articulated by the researcher, they become concepts or hypotheses, the building blocks of theory. At this point, the researcher can begin to look for behaviors that differ from or are similar to the ones she's observed before.

STOP AND THINK

Suppose an early observation led Adler and Adler to the suspicion that parents (and coaches) of children who are in activities described as primarily "for fun" will occasionally help create a competitive atmosphere. What other kinds of observations might they make that would tend to confirm this suspicion? What kinds of observations might they make that would tend to disconfirm it?

grounded theory, theory derived from data in the course of a particular study.

The pursuit of similarities can lead to the kind of generalizations on which **grounded theory** (Glaser and Strauss, 1967), or theory derived from the data, is based. Hence, the observation, repeated many times, that older children tend to get involved in one activity, shunning the multiple activities of earlier childhood, led Adler and Adler to the concept of a continually narrowing funnel of activities. But this "observation" is really founded as much on differences as similarities, isn't it? For instance, it depends on the observation of differences between what younger and older children tend to do when they participate in afterschool activities. Moreover, apparently disconfirming observations (like "not everyone wants to make this commitment") are acknowledged to suggest the probabilistic, rather than absolute, nature of the generalizations that emerge.

The process of making comparisons, of finding similarities and differences among one's observations, is the essential process in analyzing data collected during observations. It is also the essential ingredient for making theoretical sense of those observations. Through it, Humphreys (1970) came up with his typology of those who used public restrooms for impersonal sex and Goodsell, his of welfare waiting rooms. Through it, Liebow (1993) discovered the processes that created group solidarity among homeless women in shelters; Lofland (1973) conceived the "entrance sequence" used by most people on approaching an urban "public space"; Maher and Tetreault (1994) found differences among self-described feminist teachers in how much they encourage, for instance, the input of students of different races; Erikson (1976) came up with the concept of "collective trauma"; Hochschild (1983) developed the concept of "emotive dissonance" (the strain that develops when one's occupation makes one feign what one is not feeling); and Enos (Chapter 10) saw, for the first time, that there were differences in the ways that black and white women mothered from prison.

By now it might have occurred to you that the analysis of observed data is not so very different from what you do in everyday life to make sense of your world. But the metaphor of everyday thinking, and especially of the simple pursuit of similarities and differences, surely understates the complexities involved in analyzing observed data. The Adlers stress this point when they speak of the varied sources of theoretical insight in their larger opus, as well as in their afterschool piece:

> On the issue of analysis and theory-building, we have taken different tacks in the various papers we have written. For some, the patterns and their implications jumped out at us easily and in a clear-cut manner. For others, we see contradictory patterns and have to mull things over for months and years to try to grasp the underlying themes explaining how and why things work the way they do. We interview all types of participants, alone and in groups to try to reconcile conflicting interpretations into one that incorporates all the data. This is not always smooth, as life is contradictory and complex. Usually, we spend 18 months to 2 years on a given article, and begin to mull over ideas at the beginning, working on them throughout the whole time. Anything in our life can pop up and make us think about an issue and analyze or re-analyze our thoughts on it ... In the case of the afterschool paper, people kept challenging us to take a side on the issue, determining if it was good or bad. There was a lot of pressure placed on us to condemn afterschool activities (by scholars and reporters), so this made us think about these issues more and rehash them between ourselves ... No matter what, we kept seeing both good and bad elements in these activities. Finally, we had to go with this dual approach, despite the pressure to morally condemn

> these programs. Ultimately we asked, "Why do they all do it if it's so bad?" (personal communication, 1995).

STOP AND THINK *Do you think the Adlers achieved the even-handedness they aspired to in their analysis? Do you think they successfully avoided "condemning afterschool activities?"*

As with other techniques we will discuss), there is a qualitative difference between understanding the basics of analyzing observed data and using them effectively. We encourage you to try the exercises at the end of the chapter to hone your own analytic skills while you make observations.

Advantages and Disadvantages of Observational Techniques

The advantages of observational techniques mirror the purposes for using them: getting a handle on the relatively unknown, obtaining an understanding of how others experience life, studying quickly changing situations, studying behavior, saving money (see the subsection labeled "Use of Observations" earlier). We'd like to emphasize, once again, the possibility of gaining an in-depth understanding of a natural situation or social context, including a sense of what might be called an "insider's view" of that situation and context. We reiterate that Adler and Adler might well have used a survey, or even an experiment, to study afterschool activities, but it is very unlikely that they would have been able to tell us such a plausible and detailed story about the meaning of such activities unless they'd actually gone and observed the activities as they naturally occurred.

Adler and Adler (1994) suggest that one of the chief criticisms of observational research has to do with its validity. They claim that observations of natural settings can be susceptible "to bias from [the researcher's] subjective interpretation of situations" (Adler and Adler, 1994: 381). They are surely right. But compared with measurements taken with other methods (for example, surveys and interviews), it strikes us that measures used by observers are especially likely to measure what they are supposed to be measuring, and therefore be particularly valid. We can imagine few combinations of questions on a questionnaire that would elicit as clear a picture of the funneling process that leads some young people to abandon some activities while concentrating on others as the statement by the ten-year-old boy quoted by Adler and Adler:

> I'm not going out for soccer this year. It was boring last year because I didn't get to play enough, and I had a problem because I was in a play and there was an overlap between the play and soccer ... So I think I'll just stick with acting this year. I don't have any friends in it, but I like it more and I'm better at it. It's more *me*. [From the Focal Research earlier in this chapter.]

A final advantage of observation is its relative flexibility, at least compared with other methods of gathering social data. Observation is unusually flexible for at least two reasons: one, because a researcher can, and often does, shift the focus of research as events unfold and, two, because the primary use of observation does not preclude, and is often assisted by, the use of other methods of data gathering (for example, surveys, available data, interviews). Thus, as their study progressed, the Adlers' focus shifted from providing an outline of the characteristics of available afterschool activities to one of assessing a model of developmental progression that young people frequently follow in the course of pursuing extracurricular interests, even as their primary method of data acquisition alternated among observation, interview, and using available data.

On the other hand, generalizability is a perpetual question for participant and nonparticipant observation alike. This question arises for two main reasons. The first has to do with the reliability, or dependability, of observations and measurements. One example will stand for the rest. In Erikson's (1976) study of the effects on a flood on one community, he concludes that the community experienced "collective trauma" because all the individual presentations given by survivors were "so bleakly alike" in their accounts of "anxiety, depression, apathy, insomnia, phobic reactions, and a pervasive feeling of depletion and loneliness" (1976: 198). But it remains a question whether another observer of the same accounts would have detected the bleak sameness that Erikson did, much less come to the conclusion that that sameness was an indicator of "collective trauma," rather than of many individual traumas.

The second reason that questions of generalizability arise is a direct result of the nonrandom sampling procedures used by most observers. The statement of the ten-year-old boy quoted by Adler and Adler is, as far as we're concerned, compelling evidence that he was funneling his activities as he grew older. We also have no doubt that the bulk of the information that Adler and Adler collected about children of his age and older was consistent with a similar funneling process. In short, we think funneling occurred within the sample they examined, but whether it occurs among some larger population of children in the United States depends on how generalizable their sample was. And because that sample was not randomly selected (how could it have been?), the question of its generalizability is one that, of necessity, must wait another day (and replication [by one of you?] or, perhaps, the use of another method).

STOP AND THINK *Can you imagine how the generalizability of the "funneling" concept of Adler and Adler might be studied with another method? Would you be inclined to use a survey? A cross-sectional or longitudinal design? A probability or nonprobability sample?*

demand characteristics, characteristics that the observed assume simply as a result of being observed.

A third problem associated with observational methods, especially when they are announced, is the bias caused by **demand characteristics**, or the distortion that can occur when people know (or think) they are being observed. One wonderful example of demand characteristics in action comes from Rosenhan's classic (1971) study entitled "On Being Sane in Insane Places." Rosenhan was concerned about the accuracy of psychiatric diagnoses and so, with seven accomplices, gained admission as a pseudopatient to eight psychiatric hospitals. In this first instance, all pseudopatients took the role of complete participants and, although all were as sane as you and us, none was diagnosed as sane once he or she was admitted as a patient. In a second instance, however, Rosenhan told the staff in one hospital that, during a three-month period, one or more pseudopatients would be admitted. Of the 193 patients who were admitted in the next three months, forty-one were alleged by at least one member of the staff to be pseudopatients; twenty-three of them by attending psychiatrists. Because no pseudopatients, to Rosenhan's knowledge, had attempted to be admitted, the results of this second "experiment" suggest that even the threatened presence of observers can distort what goes on in a setting. Thus, despite the effort of most observers to create a situation in which action goes on as if they weren't there, there's little reason to believe that these efforts can be totally successful—especially when their presence is announced.

Particularly when that presence is unannounced, of course, the ethical problem of deceit comes into play. Should Rosenhan and his accomplices, in the first instance mentioned, have disguised their intention to observe when they had themselves admitted as pseudopatients (while, incidentally, claiming initially to possess symptoms none of them actually had)? The range of responses to the ethical dilemma of deceit is almost as wide as there are authors who have written about it, but there can be little doubt that observation, even when it *is* announced, frequently involves the researcher in some measure of deceit.

Three other problems of observational methods are worth emphasizing. First, in comparison with virtually every other technique we present in this book, observation can be extremely time-consuming. The Adlers speak of spending from 18 months to 2 years on most articles; Elliot Liebow saw about nine years pass between the conception and publication of his book on homeless women. Second, and clearly related to the issue of time consumed, is the problem of waiting around for events "of interest" to occur. In almost every other form of research (administering a questionnaire, an interview, an experiment, or collecting available data), the researcher exercises a lot of control over the flow of significant events. In the case of field observations, control is handed over to those being observed—and they might be in no hurry to help events unfold in an intellectually interesting way for the observer. Finally, although observation can be fun and enlightening, it can also be terribly demanding and frustrating. Every stage—from choosing appropriate settings, to gaining access to settings, to staying in settings, to leaving settings, to analyzing the data—can be fraught with unexpected (as well as expected) difficulties. Observation is not for the unenergetic or the easily frustrated.

Summary

Although observation is often used in experiments, the observational techniques we've focused on here—those used by participant and nonparticipant observers—are those used to observe actors who are, in the perception of the observer, acting naturally in settings that are natural for them. Adler and Adler, for instance, studied children, parents, and other adults as they participated in the wide variety of natural settings that have been created for adult-organized, afterschool activities. Observational techniques are frequently used for several purposes: to get a handle on relatively unknown social phenomena (surely a major aim of the Adlers), to obtain a relatively deep understanding of others' experiences (surely another aim of the Adlers), to study quickly changing situations, to study behavior, to save money, and to avoid ethical problems associated with other techniques.

A sense of the variety of roles a participant or nonparticipant observer can take is suggested by Gold's (1958) classic typology. A researcher can become a complete participant or a complete observer, either of which roles implies that the researcher withholds his or her research purposes from those being observed. Although both of these roles entails the advantage of relative inobtrusiveness, they also entail the disadvantages of incomplete frankness: for example, of not being able to ask probing questions and of deceiving one's subjects. Both the participant-as-observer and the observer-as-participant do reveal their research purposes, purposes that can prove a distraction for other participants. On the other hand, by divulging those purposes, they earn the advantage, among others, of being able to ask other participants detailed questions about their participation. The Adlers were, alternately, participants-as-observers and observers-as-participants in their study of afterschool activities, and thereby gained the methodological advantage, among others, of being able to ask other participants about the meaning of events for them.

Observers can choose their research settings for their theoretical salience or their natural accessibility, or for some combination of theory and opportunity. The Adlers emphasize the natural accessibility in their own account of why they chose to study afterschool activities. In some instances, it might be possible to sample multiple settings randomly, and thereby maximize their representativeness, but usually the selection of such settings is done in a nonrandom fashion (as it was by the Adlers). Generally, participant and nonparticipant observation begin without well-defined research questions, measurement strategies or sampling procedures, though systematic observations can involve some combination of these elements.

Once in a setting, researchers can record observations by writing them down, by using some mechanical device (for example, tape recorder or camera), or by trusting to memory for short periods of time. Most often observers, like the Adlers, prefer trusting to memory because alternative techniques involve some element of intrusiveness, but in all cases, the effort to record or transcribe "notes" should be made soon after observations are made, improving the chances for precise description and for recalling less precise impressions. The analysis of the data collected during observations involves comparing observations with one another and searching for important differences and similarities.

The advantages of participant and nonparticipant observation mirror the purposes for using them and include the likelihood of valid measurement, as well as considerable flexibility. Both types of observation are particularly useful for gaining an in-depth understanding of particular settings or situations. On the other hand, disadvantages of observational techniques include almost unavoidable suspicions that they might yield unreliable and nongeneralizable results. Observation can also affect the behavior of those observed, especially when the observation is announced, and contains an element of deceit, especially when the observation is not announced. Observation can also be a particularly time-consuming and sometimes frustrating path to social scientific insight.

SUGGESTED READINGS

Denzin, N. K. and Y. S. Lincoln, editors. 1994. *Handbook of qualitative research.* Thousand Oaks, CA: Sage.

This collection offers insight into cutting-edge issues in qualitative research generally, and observation techniques particularly. In addition to an article by Adler and Adler, in which they provide an alternative to Gold's typology of researcher roles, there are splendid pieces on grounded theory, feminist approaches to qualitative research, and many others.

Hammersley, M., and P. Atkinson. 1992. *Ethnography: Principles in practice.* London: Routledge.

Extraordinarily readable and accessible, this book provides a systematic and elegant account of ethnographic techniques, of which observational techniques constitute an important subset.

Lofland, J. and L. Lofland. 1995. *Analyzing social situations.* Belmont, CA: Wadsworth.

A wonderful manual that covers key topics in field research, A to Z.

Reinharz, S. 1992. *Feminist methods in social research.* New York: Oxford University Press.

An extraordinary resource for understanding feminist methods generally, the book's chapter on feminist "ethnography" and "case studies" are great resources for would-be participant and nonparticipant observers.

EXERCISE 11.1

Social Reproduction and the Corporate Other Revisited

For this exercise, you'll need to locate two similar settings in which children engage in afterschool activities. (Hint: if you don't know where such settings are, you might try to find informants who can help. You could call a local YMCA or recreational center, for instance, and ask about organized activities for children near you.) You should try to play the role of a pure observer in one—that is, you should not announce your research intention. (You might pretend to be an interested observer, which, of course, you are.) In the other, you should try to play the role of participant-observer—that is, you should try primarily to observe, but only after announcing your intention to do so to as many of the participants as you can—perhaps, even to the assembled group before it starts the activity. Tell them you are a student, collecting information for a homework assignment. Tell them what school you're from. Tell the truth. Use the same data recording technique in both settings. If you choose to take notes in the one, take notes in the other. If you choose to tape-record in one, tape-record in the other. If you choose to trust to your memory, short-term, in one, do the same in the other. Your research question in each setting is this: To what extent is competitiveness encouraged in this setting? You should spend at least a half- hour in each setting. Once you leave, you should write up your data: Don't summarize what the participants have done or said; try to record what they've done and said as precisely as you can. Then summarize your overall impressions and respond to the research question. Finally, answer the following questions:

1. Did you notice any differences in the ways your two groups acted toward you or toward one another that you might attribute to your announcement of your research purpose in one and your lack of such an announcement in the other?

2. Which setting did you prefer to observe? Why?

3. Discuss the extent to which you think competitiveness was encouraged in each afterschool setting by children and adults. Give support for your conclusion by summarizing some portions of your field notes.

4. Compare your research findings with those of the Adlers.

EXERCISE 11.2

Participant Observation

Try to replicate what you've done for exercise 11.1, only this time with two groups you're already a member of (for example, your family [at dinner?], [a meeting of?] a school organization, [a meeting of?] a class at school). (For best results, try to make the settings somewhat comparable. You probably don't want to study your own family at dinner, one time, and a dinner in which you meet your fiancé[e]'s parents at another time.) In one of the groups, do not announce your research purpose; in the other, make a general announcement. Again, for the second group, tell them of your assignment and your plan to take notes on what goes on. Your research question, again, is this: How competitive is this group? Use the same note-taking regime you used in exercise 11.1 for both of these observations. Write up your notes, again trying carefully to make precise accounts of what went on and what was said, and then answer these questions:

1. Did you notice any differences in the ways these two groups acted toward you or toward one another that you might attribute to the announcement of your research plan in one and the absence of that announcement in the other?

2. Did you feel any different in these two settings?

3. Compare your experiences in the two exercises, 11.1 and 11.2. Which of the four observations felt most comfortable to you? Which was the least comfortable? Why? In which do you feel you gained the greatest understanding of the setting? In which did you gain the least? Why? In which did you learn the most about the setting (a different question)? In which did you learn the least? Why?

4. Which of the groups examined for 11.1 and 11.2 seems to have been most competitive? Give support for your conclusion by summarizing some portions of your field notes.

12 Using Available Data

The following article made recent front-page news:

RESEARCH FINDINGS USED TO JUSTIFY THE AMERICAN MEDICAL ASSOCIATION'S DECISION TO ENCOURAGE DOCTORS TO EDUCATE THEIR PATIENTS ABOUT THE POTENTIALLY HARMFUL EFFECTS OF TELEVISION

- The average American child will witness more that 200,000 acts of violence on television, including 16,000 murders before age 18.
- The average child spends about 28 hours a week watching television, which is considerably more time annually than they spend in school.
- A study of population data for various countries showed homicide rates doubling within 10 to 15 years after the introduction of television, even though television was introduced at different times in each country.

Providence Journal, September 10, 1996
Page 1

Research findings such as these, especially when presented as part of a case against the viewing of TV violence by the AMA, seem to support the conclusion that viewing TV violence contributes to the rise of violent crime in our society. At least that's what Steven Messner thought before doing the research presented in this chapter.

Introduction

You've probably heard television viewing blamed for a lot of what ails us as a nation: our alleged family problems, political apathy, and violent criminality, to name but three. Suppose you wanted to test one of these assertions. How might you do it? Let's say you wanted to focus on the connection between viewing violent television programming and criminal violence. You might employ an experimental design (see Chapter 8) and determine whether subjects exposed to violent television were more likely to do something nasty than were subjects exposed to prosocial television (as Milgram and Shotland did in 1973). You might use a questionnaire (see Chapter 9) to determine whether young people who admit to watching violent programming describe their behavior as more aggressive than those who don't (as Eron, cited by Kolbert [1994], has done several times). Both of these approaches would permit "micro-level" assessments (that is, assessments based on samplings of individuals) of your hypothesized relationship. Both would require the use of data about individuals.

But no micro-level approach, whatever its interest, can address a central question: the *degree to* which violent television programming actually contributes to crime in our society (Comstock, 1980). This question moved Steven Messner, the author of this chapter's focal research. Messner's paper illustrates the use of **available data** in the social sciences. Available data are, as the term suggests, data that are available to the researcher. They are frequently, but not always, data that have been collected by someone else—other researchers, governments (as when they collect census data), or businesses (as when they prepare for annual reports). Data that have been collected by someone else are sometimes called secondary data instead of available data. Although **primary data** are data that are collected and used by the same researcher, **secondary data** are data that have been collected by someone else. Secondary analysis, in fact, is the name given to the increasingly popular form of research that involves the reanalysis by a researcher (or researchers) of data that were originally collected, often for another purpose, by another researcher (or researchers). The increasing popularity is suggested by the fact that, although Roger found that only about 12 percent of a probability sample of sociology articles published in 1956 involved the analysis of data that had been collected by governments, organizations, or private researchers, 36 percent did so in 1976 and almost 50 percent did in 1996 (Clark, 1999 [forthcoming]).

available data,
data that are easily accessible to the researcher.

primary data,
data that the same researcher collects and uses.

secondary data,
data that have been collected by someone else.

A particularly accessible form of available data (for research purposes) is that which is presented in published reports of governments, businesses, and organizations. Messner, you will see, makes revealing use of such data in the following essay.

Messner's essay assumes some knowledge of an idea you might not be familiar with: the idea of correlation. We'll discuss **correlation** in greater detail in Chapter 15, but for now you should know that it (and the related idea of regression) refers to a technique for assessing the connection or association between two variables—say, the crime rate of a city and the percentage

correlation,
a statistical technique for assessing the strength and direction of the association between two variables.

of its population watching violent programming. What you need to know here is that a correlation (or standardized regression) coefficient can vary between -1 and 1. When the coefficient is 1, it means that as one of the variables (say, the percentage of the population watching violent television programming) goes up, the other variable (say, the crime rate of a city) goes up in a very predictable way. When it is -1, it means that as one of the variables goes up, the other goes down in a very predictable way. When the coefficient is zero, it indicates that the two variables are unrelated to each other. In a study of self-esteem among junior-college female students, for instance, Clifford and Clark (1995) found that the correlation of student self-esteem with grade point average was .25. This was higher than the correlation of .10 for the relationship between self-esteem and family income, indicating that, in this sample, self-esteem was more strongly associated with grade point average than with family income. (Can you imagine why that might be?)

Whether positive or negative, the farther away a coefficient is from zero, the "stronger" the relationship it corresponds to; the closer to zero, the "weaker" the corresponding relationship.

STOP AND THINK *If you had to compare two correlation coefficients, one of which was .09 and the other, -.19, which would indicate the* stronger *relationship? (Hint: which coefficient is "farther away" from zero?)*

FOCAL RESEARCH

Television Violence and Violent Crime: An Aggregate Analysis[1]

by Steven F. Messner

Introduction

Television is one of the most important media of mass communication in the United States. It is not surprising therefore, that this medium has been at the center of controversy and public debate. Perhaps the most intense and enduring controversy concerns the viewing of television violence.

The purpose of this paper is to address an aspect of the television controversy that has received comparatively little attention. This is the claim that violence on television is an important cause not simply of "aggression" but of "criminal violence" in society at large. The key question guiding the analysis can be stated quite simply. Do population aggregates with high levels of exposure to violent television content also exhibit high rates of criminal violence?

[1] This article is adapted with permission from Steven Messner's (1986) "Television Violence and Violent Crime: An Aggregate Analysis," *Social Problems*, Vol. 33, No. 3, 218–235, copyright by the Society for the Study of Social Problems.

Previous Research

There is an extensive body of literature dealing with the potential effects of television viewing on human behavior. A National Institute of Mental Health review of over 2,500 publications between 1972 and 1982 finds that "recent research confirms the earlier findings of a causal relationship between viewing television violence and later aggressive behavior" (NIMH, 1989, 89). Even though some consensus seems to be emerging with respect to the existence of a causal connection between television violence and aggression, the larger social significance of this relationship has not been firmly established. The question arises as to whether the findings for the kinds of aggression typically studied by social scientists can be extended to more serious forms of aggression such as criminal violence.

Most previous research has focused on mild forms of aggression because of the preference of social psychologists for "micro-level" research designs. These designs, although strong in many respects, are not well suited for studying a dependent variable such as criminal violence. Ethical considerations preclude the use of an experimental design in the study of violent crime because it would be unethical to introduce a stimulus, either in the laboratory or in the field, suspected of provoking harmful behavior. Correlational studies, in contrast, are not subject to the same ethical constraints, but the "micro-level" correlational design (that is, a sampling of individuals) encounters problems of a different sort. Because serious crimes of violence are extremely rare, the probability of selecting an appreciable number of individuals who have engaged in such acts is highly remote unless samples are unrealistically large.

More macro-level attempts to examine the relationship between television content and rates of violent crime (for example, Clark and Blackenburg, 1972, Henningan, Del Rosaria, Cook, Wharton, and Calder, 1982, and Phillips, 1983) have yielded much less conclusive findings. Consequently, George Comstock has observed that the research to date does not tell us with any certainty "the extent or degree to which television programming may contribute to crime or serious antisocial behavior" (1980: 109).

The purpose of the present study is to introduce an original source of evidence that bears upon the relationship between television violence and criminally violent behavior. The evidence derives from a cross-sectional analysis of levels of exposure to violent television content and rates of criminal violence for samples of population aggregates. There are two noteworthy advantages associated with this type of research design. First, the use of population aggregates rather than individuals facilitates the investigation of the extremely serious forms of criminal aggression that are at the center of public concern about media violence. Second, the selection of a cross-section of population aggregates at a single point in time results in a fairly large sample of observations. The availability of a large sample makes it possible to conduct multivariate analyses and to assess the importance of indicators of television violence in comparison with other independent variables.

Theoretical Rationale

The central hypothesis is that there will be a significant positive relationship between levels of exposure to television violence and rates of violent crime. In essence, this hypothesis is based on theories and empirical findings concerning the determinants of mild forms of aggression at the individual level which I extend to account for more serious forms of aggression at the aggregate level.

Social psychologists have identified several relevant processes: (1) "modeling," or the process through which individuals learn to behave aggressively by observing the behavior of others (Bandura, 1973); (2) "desensitization," or the process by which children who are exposed to portrayals of violence become more accepting of it (Eyzeneck and Nias, 1978); (3) "disinhibition," or the process by which previous socialized inhibitions against aggressive impulses are broken down through exposure to a steady diet of media violence (NIMH, 1982: 39). All of these theorized processes provide grounds for anticipating that individuals who are exposed to violence on television will be especially prone to violent behavior.

Methods

A crucial requirement in this study is a measure of the level of exposure to television violence. Such a measure must reflect two components: the number of persons viewing television programs, that is, audience size, and the content of these programs scaled along the dimension of violence. The ideal procedure would be to estimate the viewing audience for all television programs broadcast in a given geographical area for some specified time interval, weight these viewing estimates by evaluations of the violent content of the respective programs, and then combine the weighted estimates into a composite index. However, this approach would be highly impractical because it would require detailed information on the content of all shows broadcast during the specified time interval. Conducting the content analysis required for such a measure would be a monumental task.

Instead, I used an alternative approach which involved identification of the "most violent" regular television series and measurement of the typical audience size attracted to these violent programs. Selection of the "violent" television series was based on the content analyses performed by the National Coalition on Television Violence (NCTV, 1981). The NCTV regularly monitors the amount of violence portrayed in prime time network telecasts and records the number of "violent acts per hour." A list of the shows that are the most violent according to the "violent acts" criterion is then compiled. I selected the five most violent regular prime-time series as a "strategic sample" to be used in the construction of the index of exposure to television violence.

The violent content ratings of the NCTV are evidently quite reliable. The NCTV newsletter reports a score-by-score agreement of .81 for its monitoring team (NCTV, 1981:4).

Estimates of audience size for these violent television series were obtained from the A. C. Nielsen Company. During four months of the year—November, February, May, and June—the Nielsen Company measures viewing behavior at the local level. The estimates for November were selected for the present analysis because this time interval corresponds most closely with the time period covered by the NCTV content analyses.

After identifying the five most violent television series and gathering information on the proportion of households regularly tuned to these shows over the course of a one-month interval, I computed an index of exposure to television violence by simply averaging the proportions across these five violent shows. The resulting score measures the average relative size of the viewing audience attracted to programs with characteristically violent content. High values on this index indicate that large numbers of residents are regularly viewing television programs with violent content; low values imply the reverse.

The computation of the television exposure index involved the use of a rather unusual unit of analysis. The Nielsen data on audience size are aggregated on the basis of "Designated Market Areas" (DMAs). DMAs are geographical units comprising the counties served by local television stations. DMAs are larger than the more common governmental units of "Standard Metropolitan Statistical Areas" (SMSAs). To be able to use the Nielsen data on audience size in an SMSA analysis, it was necessary to locate each SMSA within an encompassing DMA. The DMA value on the television index was assigned to the "residing" SMSA.

The dependent variables are the SMSA rates (per 100,000 population) of violent crime. The offenses include criminal homicide, forcible rape, robbery, and aggravated assault. Crime rates are measured for 1981 since the television index refers to a period near the very end of 1980. The source for the crime data is the Federal Bureau of Investigation's (1982) *Uniform Crime Reports* (UCR). The offenses are considered singly, and in combination as a composite index of violent crime. The composite index is simply the sum of the individual rates.

The UCR data are based on reports to the police. As such, they are subject to all of the criticisms of official crime statistics that have been discussed at great length in the literature (see among others Hindelang, 1974; Savitz, 1978). The problem of reporting biases is especially worrisome in the present analysis because a plausible case could be made that such biases would be systematically related to the independent variable of primary interest—exposure to television violence. The reporting behavior of residents in communities with low levels of exposure to television violence might very well differ from that of residents in communities with low levels of exposure. This would be expected, for example, if "desensitization" is produced by exposure to television violence. The possibility of such biases will be considered in the context of the actual results.

In an effort to minimize the possibility of observing "spurious" associations, data were also gathered for a fairly comprehensive list of controls. The selection of the control variables was guided by previous research indicating relationships between these variables and either television viewing or levels of violent crime. On the basis of the research revealing that television viewing is associated with low socioeconomic status, low academic achievement, and minority status (Comstock, 1975), measures of the following socio-demographic characteristics of SMSAs were included: the percentage of the population below the poverty line, the percentage with less than a high school education, and the percentage black. A measure of the level of economic inequality, the Gini coefficient of household income concentrations, was also included. Blau and Blau (1982) reported that the level of inequality across SMSAs is related to rates of criminal violence. Another economic control variable for the analysis was the average monthly payment to recipients of Aid to Families with Dependent Children (AFDC). Significant relationships between AFDC payments and rates of violent crime were reported by De Fronzo (1983).

Four population variables were also measured: total population; population per square mile; the percentage of the population 18-34; and the number of males per 100 females. Finally, given the well-known thesis of a "regional culture of violence" (Gastil, 1971; Hackney, 1969; but see Loftin and Hill, 1974), a dummy variable to represent regional location (South = 1; non-South = 0) was also included in the list of controls.

Information for the percentage below the poverty line, percentage with less than high school education, the Gini coefficient, and males per females was taken from the 1980 U.S. Census tapes (U.S. Bureau of the Census, 1982b). The regional dummy variable has been scored in accordance with Census Bureau classifications. The source for all other control variables is the *State and Metropolitan Area Data Book, 1982* (U.S. Bureau of the Census, 1982b).

Results

Regression equations for the violent crime rates of SMSAs are reported in Table 12.1. The results are quite surprising. For each measure of violent crime, the estimate for the level of exposure to television violence is negative, and, with the exception of assault, all coefficients are statistically significant. In other words, SMSAs in which large audiences are attracted to violent television programming tend to exhibit *low* rates of violent crime.

There are several additional findings of interest in Table 12.1. The percentage black has a fairly strong, positive effect on each measure of violent crime. This is consistent with previous research detecting a relationship between racial composition and rates of violent crime for SMSAs (Blau and Blau, 1982). This relationship, many criminologists agree, is a function not only of group differences between blacks and whites in age, income, occupation, education, family background, and other social characteristics, but also of differences in the experience of historical and contemporary patterns of discrimination. The measure of economic inequality (the Gini coefficient) also yields consistently positive effects—SMSAs with high levels of income inequality tend to have high rates of violent crime. Sex composition and population size are both positively related to violent crime rates—that is, large SMSAs and SMSAs with

TABLE 12.1

Regression Results for Violent Crime Rates, TV Violence Exposure, and Additional Characteristics of SMSAs (N = 281)[a]

	Dependent Variables				
Independent Variables	*Criminal Homicide*	*Forcible Rape*	*Robbery*	*Aggravated Assault*	*Violent Crime Index*
Exposure to TV Violence	–.17*	–.17*	–.15*	–.12	–.17*
Percent Black	.43*	.42*	.39*	.32*	.43*
Percent 18–34	–.08	–.18*	–.18*	–.17	–.20*
Males/Females	.14*	.16*	.13*	.19*	.19*
Percent less than HS education	.21*	–.37*	–.06	–.05	–.09
Population (ln)	.29*	.28*	.41*	.09	.30*
Gini Coefficient	.41*	.28*	.36*	.36*	.42*
AFDC Payments	–.18*	–.01	–.03	.03	–.01
Population per sq mi. (ln)	–.09	–.15*	.20*	–.03	.08
Percent Poor	–.12	.05	–.05	.00	–.03
South	–.08	.05	–.23*	.05	–.09
Adj. R^2	.58	.39	.65	.27	.52

Notes:

a. Standardized regression coefficients are reported.

* $p < .05$

large numbers of males relative to females tend to exhibit high rates of violent crime. Finally, contrary to a "regional culture of violence thesis," the only significant effect of region is on robbery, and the effect indicates that robbery rates tend to be lower in Southern than in non-Southern SMSAs.

Given the highly unexpected nature of the findings for the television exposure measure in Table 12.1, I performed several checks (not reported in this version of the paper) to ascertain whether the observed patterns might reflect certain methodological features of the analysis. One concern, already mentioned, was that residents in communities with high levels of exposure to television violence might be less likely than those in communities with lower levels to report criminal incidents to the police because exposure had "desensitized" them to crime. However, evidence in these data diminishes the plausibility of this "underreporting" hypothesis. Criminologists generally agree that the official statistics for homicide are not susceptible to appreciable reporting biases. Therefore, if the underreporting argument were correct, the results for homicide would diverge appreciably from those for the other offenses which are subject to these reporting biases. But, since the findings for homicide are very consistent with those for other types of crime, it is unlikely that the surprising inverse associations are merely "artifacts" for any of the types of crime. And, suffice it to say here, none of the other possible methodological problems I considered were found to "explain away" the findings either.

Discussion

The results of the analysis clearly raise questions about generalizing directly from previous research on television violence and mild forms of aggression at the micro-level to more serious forms of aggression such as criminal violence at the level of population aggregates. Contrary to what might be expected on the basis of previous research, there is no evidence in the aggregate data to support the claim that high levels of exposure to television violence are related to high rates of violent crime. In fact, the relationship that does emerge is in the opposite direction. The data consistently indicate that high levels of exposure to violent television content are accompanied by relatively low rates of violent crime.

Certain limitations of the present study require that these empirical findings be interpreted cautiously. Most importantly, perhaps, the analysis presented above is cross-sectional in nature. A particularly promising extension of the research design would be to collect data on audience size for population aggregates at multiple points in time. Longitudinal data of this sort would permit the use of statistical techniques such as panel regression analysis and would also allow for closer examination of the direction of causal influences. However, it is interesting to note that the finding of an *inverse* relationship between television violence and violent crime is rather surprising irrespective of the nature of the causal ordering.

Despite limitations, the consistency of the empirical findings is quite impressive. Negative associations are observed for each of the four offenses taken singly and for the combined index of violent crime. The question that naturally arises is whether or not these anomalous findings can be placed within a meaningful theoretical context. Why should high levels of exposure to television violence be associated with *low* rates of violent crime?

One plausible theoretical account is the subcultural argument proposed by Howitt and colleagues (Howitt and Cumberbach, 1965; Howitt and Dembo, 1974). This argument begins with an assertion that criminal behavior results from socialization into delinquent or criminal subcultures. Further, it suggests that this socialization occurs primarily by means of associations with criminally inclined peers. Therefore, the effect of a given mass medium on the probability of criminal behavior will depend on whether or not that medium promotes associations with criminally inclined peers at the expense of associations with more "conventional others."

Howitt and Cumberbach observe that television is not the kind of medium that encourages extensive associations with delinquent or criminal peers. This argument implies that television viewing, by keeping youths at home, decreases opportunities for socialization into criminal subcultures, and in so doing, reduces the likelihood of criminal behavior. Moreover, since total television viewing and the viewing of television violence are highly intercorrelated, the same sort of relationship with indicators of crime is expected for both of these measures of television viewing behavior.

A slightly different theoretical interpretation can be formulated by means of the "routine activities approach" developed by Cohen and Felson (1979). Cohen and Felson argue that aggregate crime rates reflect the probability that motivated offenders, suitable targets, and the absence of capable guardians converge in space and time. This probability will depend on the "routine activities" in which people are engaged. Cohen and Felson observed that "routine activities" centered around the household and family involve a relatively low risk of criminal victimization (1979:594). The reason for this low victimization risk is that family and household activities are unlikely to promote the convergence in space and time of the three requisites for criminal incident—offender, target, no guardian.

Television viewing is by its very nature "an activity largely confined to the home." The logic of the routine activities approach implies that crime rates are low in population aggregates characterized by high levels of exposure to television programming because activities in these areas are concentrated around the household, and because the concentration of activities around the household reduces the available opportunities for criminal victimization.

Admittedly, I have applied the subcultural theory and the routine activities account in an "ex post facto" manner. Consequently, neither has been tested in any formal sense. Nevertheless, these two theories provide plausible explanations for results that are very surprising yet remarkably consistent. Irrespective of the cogency of these theoretical arguments, the evidence presented here clearly raises serious questions about the claim that high rates of urban crime can be attributed in any simple and direct way to heavy exposure to television violence.

REFERENCES

Bandura, A. 1973. *Aggression: A social learning analysis.* Englewood Cliffs, NJ: Prentice-Hall.

Clark, D., and W. Blackenburg. 1972. Trends in violent content in selected mass media. Pp. 188–243 in G. Comstock and E. Rubinstein (eds.), *Television and social behavior,* Vol. 1. Washington, DC: U.S. Government Printing Office.

Cohen, L., and M. Felson. 1979. Social change and crime rate trends: A routine activity approach. *American Sociological Review* 44: 588–608.

Comstock, G. 1975. *Television and human behavior. The Key Studies.* Santa Monica, CA: Rand.

Comstock, G. 1980. *Television in America.* Beverly Hills, CA: Sage.

De Fronzo, J. 1983. Economic assistance to impoverished Americans: Relationships to incidence of crime. *Criminology* 21:119–136.

Eyzeneck, H., and D. K. B. Nias. 1978. *Sex, violence, and the media.* New York: Harper & Row.

Federal Bureau of Investigation. 1982. *Uniform crime reports for the United States—1981.* Washington, DC: U.S. Government Printing Office.

Gastil, R. 1971. Homicide and a regional culture of violence. *American Sociological Review* 36: 412–427.

Hackney, S. 1969. Southern violence. *American Historical Review* 74:906–925.

Henningan, K., M. Del Rosario, L. Heath, T. Cook, J. D. Wharton, and B. Calder. 1982. Impact of television on crime in the United States; empirical findings and theoretical implications. *Journal of personality and social psychology* 42: 461–477.

Hindelang, M. 1974. The uniform crime reports revisited. *Journal of Criminal Justice* 2: 1–17.

Howitt, D., and G. Cumberbach. 1975. *Mass media violence and society.* New York: Wiley.

Howitt, D. and R. Dembo. 1974. A subcultural account of media effects. *Human Relations* 27:25–41.

Loftin, C. and R. Hill. 1974. Regional subculture and homicide: An examination of the Gastil-Hackney thesis. *American Sociological Review* 39:714–724.

National Coalition on Television Violence. 1981. NCTV News. 2 (January).

National Institute of Mental Health. 1982. *Television and behavior: Ten years of scientific progress and implications for the eighties.* Washington, DC: U.S. Government Printing Office.

Phillips, D. 1983. The impact of mass media violence on U.S. homicides. *American Sociological Review* 48: 560–568.

Savitz, L. 1978. Official police statistics and their limitations. Pp. 69–81 in L. Savitz and N. Johnston (eds.) *Crime and society.* New York: Wiley.

U.S. Bureau of the Census. 1982a. *State and metropolitan area data book, 1982.* Washington, DC: U.S. Government Printing Office.

U.S. Bureau of the Census. 1982b. *Census of the population and housing, 1980: Summary Tape File #3C.* Washington, DC: U.S. Bureau of the Census.

STOP AND THINK *A good question to ask after reading a piece of research is "What was the unit of analysis, or type of subject, studied by the researcher?" What was Messner's unit of analysis?*

STOP AND THINK AGAIN *Other good questions are "What were his major independent and dependent variables?" "How did he measure those variables?" "How did he expect those variables to be related?" "How were they related?" Try to answer as many of these questions about Messner's article as you can.*

Sources of Available Data

Messner takes advantage of four sources of available data in his study: the National Coalition on Television Violence on the most violent television programs during the period of his study; the Nielsen Company, from which he obtained estimates about the audience for these television programs in various SMSAs; the FBI, from which he obtained estimates of the incidence of violent crime in his SMSAs; and the U.S. Bureau of the Census, from which he took information about other variable characteristics of these SMSAs (for example, socio-demographic characteristics such as the percentage of the population below the poverty line). The imaginative combination of a variety of data sources is quite common in analyses involving available data, and it's what can make them a great deal of fun, as well as illuminating. In fact, part of the art of such analyses is finding sources that complement one another the way Messner's did in the pursuit of a given research question or hypothesis. It is, however, possible to "round up the usual suspects" and to introduce you to some of the most frequently used sources of available data (so you'd recognize them in a line up).

Governments

Governments, national and local, are an extraordinarily prolific source of available data in modern society. Messner employs two of the most frequently employed sources in his paper: the U.S. Bureau of the Census and the FBI. He's used two of the available census products: a printed report and a computer tape. (There are also microfiche and, now, CD-ROM laser disk formats.) The FBI's *Uniform Crime Reports* has become a standard source for crime research in the United States.

STOP AND THINK *Let's examine data about two states on two variables. A 1993 Census Report shows that the population of New York in 1993 was just over 18 million, while the population of Wyoming was just over 400 thousand. FBI* Uniform Crime Reports *data show the overall violent crime rate of New York for 1992 was 1,221.1 per 100,000 inhabitants, while the overall crime rate for Wyoming was 319.5 per 100,000 inhabitants. What general finding from Messner (see, for instance, Table 12.1) about SMSAs is illustrated at the state level by these available data?*

The Census Bureau and the FBI produce great tomes that might be available in the reference section of your library. In addition to the census volumes, the Census Bureau produces such regular publications as the *County and City Data Book* and such irregular ones as *Historical Statistics of the United States*. But you don't really have to dig deeply to obtain fascinating data derived from these sources. Almost any almanac contains large amounts of information about cities and states in the United States ready for your use. A stunning distillation of data about cities and states in the United States, however, is to be found in *Statistical Abstract of the United States,* produced each year by the United States Bureau of the Census.

Especially since the United Nations started setting standards for data collection and measurement in the late 1940s, the amount of comparable available data from nation to nation has become voluminous. Check out your library's reference section, and you're bound to be impressed by what's there. The U.N.'s *Demographic Yearbook* and the *Statistical Yearbook*, and The International Labor Office's *Labor Force Estimates and Projections,* for instance, are extraordinary sources for data to test hypotheses about "development," "modernization," or "international dependency." (Again, almanacs alone can be a fruitful source for cross-national research.) We teamed up with a colleague (Clark, Ramsbey, and Adler, 1991), for example, to use data from these sources in an investigation of whether regional culture, vigor of the national economy, and the investment of multinational corporations affected access of women to formal labor forces around the world between 1965 and 1985. In our analysis, we found evidence that all three had independent effects—that, for instance, women in "richer" nations made the greatest inroads into the formal labor force; women in nations whose economies were least dominated by multinational corporations made greater inroads than others; and women in nations that were predominantly Islamic were less likely to have made labor force inroads than others, while women in Marxist regimes were more likely to have made such inroads.

We should emphasize that most types of data mentioned here so far (for example, census data, FBI data) are really available statistics—statistics that summarize at the aggregate level data that have been collected at the individual level. Thus, when the census provides information about the percentage of the population of a given SMSA that is below the poverty line, that percentage represents a summary of information elicited from a question about income on a census questionnaire, filled out by many respondents. We'd like to point out, however, that some of these data, as well as much other data, are available in their original, non-aggregated form. Thus, although it's extremely easy to obtain aggregated census data, it's also possible to obtain Public-Use Microdata Sample Files, which contain all the census data collected about each person in a small sample of unidentified households, through special requests to the Census Bureau.

Similarly, although it is relatively easy to obtain aggregated data (through, among other things, the *Monthly Vital Statistics Report,* published by the National Center for Health Statistics) on births, deaths, marriages, and divorces, many researchers have also used non-aggregated records of births, deaths, marriages, and divorce to significant effect. Webb, Campbell, Schwartz, and Sechrest (1966) report that Winston, for instance, used birth records to document the preference for male offspring among upper-class families (where a preference for males was measured by the male-female ratio of the last born). Sometimes such data are used in conjunction with data from other sources (say, questionnaire or interview surveys) to produce results that neither could have achieved alone. Thus, Karney and associates (1995) used information about jobs and education on marriage certificates to determine that higher-status couples were more likely to respond to mailed questionnaire surveys than were other couples.

Of course, the use of government statistics constitutes one of the richest research traditions in the social sciences. It provided grist for Émile Durkheim's classic ([1897] 1964) study, *Suicide,* in which he found that Protestant countries had higher suicide rates than Catholic countries, as well as for Karl Marx's argument for economic determinism in his ([1867] 1967) *Das Capital.* It is hard to imagine what these early social scientists would make of the ever increasing amounts of government statistics available today—and, frankly, there's no way we could do justice to them here (though a brief discussion with a reference librarian might give you a greater appreciation)—but it's almost a certainty they would have done a lot.

STOP AND THINK *Where, again, did Messner get his information on violent crime rates? In which source for available data mentioned earlier would you expect to find cross-national data on the labor force participation rates for men and women? In which might you look for the age and religion of recently divorced individuals?*

Nongovernmental Agencies or Researchers

If we've given you only a small taste of the kinds of data that are available to you from government sources, we'll do even greater injustice to the panoply of nongovernmental sources. Messner's discovery of both the Nielsen and the NCTV data is an interesting story. Messner was beginning his career at Columbia University when Murray Straus, from the University of New Hampshire, came as a visiting scholar. The two compared notes on their research: Straus on constructing a state-level index of "legitimate violence," Messner on the effects of TV violence. Straus, according to Messner's memory, mentioned the NCTV data, but neither of them had given much thought to the Nielsen possibility, until Messner serendipitously mentioned their interests to his father, who was employed at an advertising agency. Messner's father, in turn, led Messner and Straus to William Behanna, of the A. C. Nielsen Company, who was kind enough to explain technical features of the Nielsen data and provide them with relevant publications. According to Messner,

> Looking back now, I realize that my research in this area probably would not have unfolded the way it did without several serendipitous conditions: a visiting colleague with similar interests, a relative with valuable connections, and an unusually helpful contact in private industry (personal correspondence, 1995).

Commonly used sources of data for cross-national investigations are the World Bank's *World Development Report, World Tables, Social Indicators of Development, and the World Debt Tables*. Interested in testing something about world-system or economic dependency theory? Take a look at Ballmer-Cao and colleagues (1979) *Compendium of Data for World-System Analysis* or Bornschier and Chase-Dunn's (1985) *Transnational Corporations and Underdevelopment*. But frequently the "discovery" of appropriate available data involves the same kind of serendipity it did in Messner's case. Not long ago, one of us (Roger) and a student (Jeffrey Carvalho) read a fascinating article (by Collins, Chafetz, Blumberg, Coltrane, and Turner, 1993), purporting to provide "A General Theory of Gender Stratification." We became particularly intrigued by a portion of the argument that dealt with the causes of women's movements in different countries and by the fact that earlier tests of models involving "second wave" (or post-1960s) movements had been limited to 16 or so cases—the number of nations for which two researchers

(Chafetz and Dworkin, 1986) had been able to find enough information about women's movements from disparate sources. It seemed to us that there ought to be some more central repository of observations of women's movements—perhaps for an even greater number of countries. Then it occurred to Roger that his sister had given him, several years before, a copy of Robin Morgan's (1984) *Sisterhood is Global: The International Women's Movement Anthology*, an almanac-like volume on women's movements throughout the world. Sure enough, Roger and Jeff were able, through the content analysis of 66 entries in Morgan's book, to abstract enough information to test models about the occurrence and intensity of women's movements using 50 more (a little over 300 percent more) cases than had ever been used in such tests before (see Clark and Carvalho, 1996).

The Nielsen data, the NCTV data, the World Bank data, and the data from Morgan's collaborative effort with women around the world are all examples of data taken from organizations whose specialized interests in collecting the data differed from those of the researcher or researchers who reused them. These examples are merely illustrative of the wide variety of data sources from nongovernmental agencies "out there," if you can find them. Again, your best bet for finding those that are appropriate for any given research topic is your college or university library's reference section (and, within that, your reference librarian).

STOP AND THINK *Suppose you wanted to do a quick check of the hypothesis that cartoon TV programs were more likely to appear on nonschool days than on school days. What available data source might permit such a check?*

We don't want to leave you with the impression that nongovernmental sources are only good for aggregated data. On the contrary, it has become increasingly common since the 1970s for researchers who have done large-scale surveys to share their individual-level data with other researchers for secondary analyses. By 1996, nearly two-fifths of all articles published in sociology were based on reanalyses (or secondary analyses) of survey data that had been collected by other researchers (Clark, 1999 [forthcoming]).

The first large-scale social survey conducted explicitly for the use of secondary analysts was the 1972 General Social Survey. The General Social Survey is administered every year by the National Opinion Research Center to a sample of about 2,000 persons (Davis and Smith, 1992: 54), and asks respondents about present and past experiences, behavior, and opinions. Its questions cover many topics and include those that were asked on older surveys during the 1940s, 1950s, and 1960s.

One of the largest clearing houses for data available for secondary analyses is the Inter-University Consortium for Political and Social Research (ICPSR) at the University of Michigan. A blind grab at the *Social Science Quarterlys* behind Roger's desk located one issue (March, 1989) in which five of the ten major research articles used data sets the authors had acquired from *ICPSR:* a *Panel Study of Income Dynamics, a Study of Time Use,* the *General Social Survey* (twice), and an *American National Election Survey.* The topics of these articles ranged from the "economic risks of gender roles" and the "fragmentation of health care" to the question of whether there is a "black gender gap." The point is this: If you are planning a major research project, you might want to consult your professor about whether your college or university already has one of these data sets for your use or whether there might be another convenient way for you to gain access to such data.

The World at Large

physical traces, physical evidence left by humans in the course of their everyday lives.

erosion measures, indicators of a population's activities created by its selective wear on its physical environment.

accretion measures, indicators of a population's activities created by its deposits of materials.

You don't need to depend on outside agencies to make your "available data" available. In their now classic list of alternative data sources, Eugene Webb and his associates (1966) included **physical traces** and private records. Physical traces include all kinds of physical evidence left by humans in the course of their everyday lives and can generally be divided into what Webb and his associates call **erosion measures** and **accretion measures.** An erosion measure is one created by the selective wear of a population on its physical environment. Webb and associates cite Mosteller's (1955) study of the wear and tear on separate sections of the *International Encyclopedia of the Social Sciences* in various libraries as a quintessential example of the use of erosion measures, in this case to study which sections of those tomes received the most frequent examination. An accretion measure is one created by a population's remnants of its past behavior. The "Garbage Project" at the University of Arizona has studied garbage since 1973. By sorting and coding the contents of randomly selected samples of household garbage, William Rathje and his colleagues have been able to learn a great deal, including what percentage of newspapers, cans, bottles, and other items aren't recycled and how much edible food is thrown away (Rathje and Williams, 1992: 24). Because there are "certain patterns in which people wrongly self-report their dietary habits" (Rathje and Williams, 1992: 24), the proponents of "garbology" argue that this type of available data can sometimes be more useful than asking people questions about their behavior.

STOP AND THINK *Suppose that, like Kinsey and associates (1953), you wanted to study the difference between men's and women's rest rooms in the incidence of erotic inscriptions, either writings or drawings. Would you be taking erosion or accretion measures?*

private records, records of private lives, such as biographies, letters, diaries, and essays.

Private records include autobiographies, letters, diaries, and essays. With the exception of autobiographies, these records tend to be more difficult to gain access to than do the public records we've already discussed. Frequently, however, private records can be used to augment other forms of data collection. Roger has used autobiographies, for instance, to augment information from biographies and questionnaire surveys to examine the family backgrounds of famous scientists, writers, athletes, and other exceptional achievers (Clark and Rice, 1982; Clark, 1982; Clark and Ramsbey, 1986). Personal diaries, when they are accessible, hold unusually rich possibilities for providing insight into personal experience. And letters can be an especially useful source of information about the communication between individuals. Lori Kenschaft (forthcoming), for instance, has studied the ways in which two important nineteenth-century educators, Alice Freeman Palmer and George Herbert Palmer, who happened to be married to each other, affected each other's thinking and

American education by carefully studying their written correspondence (available at the Harvard University library). Similarly Karen Lystra's (1989) *Searching the Heart: Women, Men and Romantic Love in Nineteenth-Century America* uses letters almost exclusively to build an argument that nineteenth century conventions of romantic love required people to cultivate an individual, unique, interesting self that they could then present to the beloved other—that is, that romance is an historical and psychological root of individualism.

Kenschaft's and Lystra's works point to one of the great advantages of available data: the potential for permitting insight into historical moments that are inaccessible through more conventional social sciences techniques (such as interviews, questionnaires, and observation). Other sources of such historical data are newspapers and magazines, as well as organizational and governmental documents. Beverly Clark (forthcoming) has scrutinized nineteenth- and early twentieth-century newspapers and magazines for insight into the popular reception of (children's) books by, among others, Louisa May Alcott. Clippings from the *Cleveland Leader* and the *Albany Daily Press and Knickerbocker* indicate that the Cleveland public library required 325 copies of *Little Women* in 1912 to satisfy constant demand; the New York City branch libraries acquired more than a thousand copies.

The real point is, however, this: You really don't want to be guided in your search for available data by any of the "usual suspects" we've suggested. A much sounder approach is to immerse yourself in a topic by reading the available literature on it. In the course of your immersion, take a look at the sources other researchers have used to study the topic. If you discover that they've typically used the FBI *Uniform Crime Reports,* you might want to use that too. If they've used the U.N.'s *Demographic Yearbook,* you might want to look into it as well. Very often, using earlier research as a resource in this way is all you'll need to find the data you'll want to use. On the occasions when you still need to find that extra source that will set your research apart, it will still be useful to know where you can get the data that will enable you to stand on the shoulders of those who have gone before.

The Internet

These days a significant new source of available data is the Internet. Want to know how the populations of special racial or ethnic groups changed in the United States between 1980 and 1990? Why not check out the Statistical Abstracts Web site (http://www.census.gov/stat_abstract/)? That's right, an abridged version of the *Statistical Abstracts of the United States* we mentioned is actually available on the World Wide Web. Want to get some idea of what cutting-edge issues are in the field of gender research? Check out a academic discussion group called SEX-LINK (listserv@cunyvm.cuny.edu). Want access to the latest nationally representative survey data on alcohol, drug abuse, violence, or HIV/AIDS or many other health-related topics? Why not contact Gerry Hendershot of the National Center for Health Statistics (e-mail: geh2@nch08a.em.dcdc.gov)? For about $20, you can purchase information (on a CD-ROM) on about 100,000 sample persons.

You can use the Internet in many ways to acquire access to available data, but all three of the Internet's major functions—remote access to other computers, communication, and transfer of files from one computer to another (Hacker, 1998: 6)—can be useful at one time or another. If you need assistance with how to use the Internet for gaining access to available data, seek help from your college computer center or one of the growing number of books on the topic (for example, Hacker, 1998; Kurland and John, 1997). We'll just mention a few general resources that are available:

- *Research Engines in the Social Sciences* (http://www.carleton.ca/~cmckie/research.html) provides links to Internet sites around the world, including databases of bibliographic and statistical information.
- *The American Sociological Association's Home Page* (http://www.asanet.org) includes, among other things, a list of publicly available data sources, such as the annual survey of health-related topics kept by the National Center for Health Statistics, mentioned earlier, and e-mail and home page addresses for the Inter-University Consortium for Political and Social Research.

  (A word of warning: Although these data sets are described as "publicly available," they're not necessarily cheap, especially for an undergraduate student. But some, like the survey of health-related topics, are [at about $20] comparative bargains.)
- *The Bureau of the Census Home Page* (http://www.census.gov) offers, among other things, census data on population available in a computer searchable way. In particular, the bureau's *Statistical Abstract* (http://www.census.gov/stat_abstract/) provides useful information on crime, sex and gender, social stratification, and numerous other topics.
- *Demography and Population* (http://coombs.anu.edu.au/ResFacilities/DemographyPage.html) has very a complete guide to worldwide demographic statistics available on the Internet.
- *Population Division Home Page* (http://www.census.gov/ftp/pub/population/www/) contains population information about the U.S., particularly say, about the number of single parent families in the United States.
- *Bureau of Labor Statistics Data Home Page* (http://stats.bls.gov:80/datahome.htm) has, among other things, data on men's and women's labor force participation.
- *Social Documentation and Analysis Home Page* (http://socrates.berkeley.edu:7502/archive.htm) is an incredible site that not only provides codebooks for several social survey data sets (The General Social Survey is one) but also permits users to analyze the data while they visit the site.

We've barely scratched the surface here. For those of you interested in a good sense of social science data on the web, we recommend looking at the *Social Science Data Archives* (http://www.socsciresearch.com/r6.html). For those interested in asking experts particular questions, for instance, you can use professional discussion groups to tune into the social sciences. For a list of these in sociology try http://www.w3.org/vl/Sociology/listserv.html. (Although professional discussion groups welcome the participation of students, they especially welcome the participation of students who have already done some research, have focused questions, and are prepared to share preliminary findings. It's often a good idea to "lurk" for a week or so first, to get an idea of how a particular group interacts.) But scratching the surface is about all we can do here, because opportunities for accessing data on the Internet are increasing daily and will be much greater by the time you read this than they are as we write it. We recommend Kurland and John's (1997) book for those who seek more information about using the Internet.

Advantages and Disadvantages of Using Available Data

Available data offer many distinct advantages over data collected through questionnaires, observation, or experiments. Most striking among these are the savings in time, money, and effort they afford a researcher. A properly administered random survey of 1,000 to 1,500 individuals can easily cost more than $100,000 and years of preparation and execution. The secondary analysis of such data might cost the price of a phone call (though, depending on your institutional affiliation, there can be a charge of several hundred dollars—still relatively little—for access to and the right to use such data) and the time it takes to figure out how you want to analyze them. Each national census might cost a government millions of dollars, but comparing 150 nations based on the results of those collections might cost you only the price of a library card.

A related advantage of using available data is the possibility of studying social change over time as well as social differences over space. Using data from the 1968 and 1980 American National Election Surveys, Ellison and Gay (1989) were able to show that although blacks were significantly more likely than whites to vote and participate in American political campaigns in 1968, there was a significant convergence in such activity during the 1970s and 1980s. And using U.S. census data collected since 1790, gerontologists have been able to discern that the proportion of the American population that is 65 years or older has grown from 2 percent in 1790 to 4 percent in 1890 to 12.5 percent in 1990. Most individual researchers don't live long enough to observe such trends directly, but, because of available data, and the imagination to use them, they don't have to.

STOP AND THINK *Imagine you want to find out how much the size of the "average" American family has changed over the last two centuries. Would you be more inclined to use a survey, participant observation, an experiment, or U.S. census reports as your data source? Why?*

Available data also mean that retiring and shy people, people who don't like the prospect of persuading others to answer a questionnaire or to participate in a study, can make substantial research contributions. Moreover, the researcher can work pretty much where he wants and at whatever speed is convenient. Roger once coded data from a source he'd borrowed from the library while waiting (along with 150 other people who could do no more than knit or play cards or read) in a courthouse to be impaneled on a jury. (At other times, when unable to sleep, he's collected data in the wee hours of the morning—a time when researchers committed to other forms of data collection pretty much have to shut down.)

These examples suggest a final advantage of using available data: The collection itself is unobtrusive and extremely unlikely to affect the nature of the data collected. Unlike researchers who use questionnaires, interviews, observations, or experiments, those of us who use available data rarely have to worry that the process of data collection will affect our findings—only that it affected the original collection, if we're using secondary data.

On the other hand, the unobtrusiveness of using available data is bought at the cost of not knowing the limitations of one's data as well as those who collect their data firsthand. One of the major disadvantages of available data is that one is never quite sure about its reliability. You may recall that in Chapter 6, Rockett and Smith questioned the reliability of data on suicide for several reasons: for example, because different jurisdictions might employ different personnel (for example, coroners or medical examiners) or because certain cultures are more or less amenable to acknowledging suicide. The same is obviously true for crime data. Some jurisdictions might be more or less ready to recognize and categorize crime. Messner was concerned, for instance, that SMSAs where the viewing of television violence was particularly common might, because of the desensitization of people to violence brought about by such viewing, be less apt to report violent crime when it occurred. Similarly, cross-national research involving female labor force participation has been plagued by an awareness that, in many countries, especially in those where women's work outside the home is frowned upon, there is considerable undercounting of that work.

STOP AND THINK *What kinds of nations, do you think, would be most likely to undercount women's work outside the home? Why?*

Sometimes the best you can do is be explicit about your suspicion that some of your data are not reliable. You can at least try to clarify the kinds of biases the sponsors of the original data collection might have had, as we did when we used available data on female labor force participation. This kind of explicitness provides readers with a chance to evaluate the research critically. At other times, checks on reliability are possible, even without the capacity to perform the kinds of formal tests of reliability (for example, interrater tests—see Chapter 6) that you could use while collecting primary data. Messner was concerned, for instance, that the "underreporting" of violent crimes in SMSAs where many people were desensitized to crime by the viewing of violent television programming might have been responsible for his unexpected finding that such viewing was negatively associated with the presence of such crimes. But he knew that one kind of violent crime, homicide, was very unlikely to be underreported under any circumstances. So, when he found that the unexpected negative association between viewing violent programs and violent crime rates persisted even for homicide rates, he concluded that it was less plausible that his other unexpected findings were only the result of systematic underreporting.

STOP AND THINK *Why is homicide less likely than, say, shop-lifting to be under-reported, whatever the television viewing habits of inhabitants?*

A second disadvantage of using available data is that you might not be able to find what you want. In some cases, there just might not have been anything collected that's remotely appropriate for your purposes. Modern censuses are designed to enumerate the entire population of a nation, so you can reasonably expect to be able to use their data to calculate such things as sex ratios (the number of males per 100 females, used, for instance, by Messner as a control variable in his analysis). But ancient Roman censuses focused on the population of adult male citizens (the only group, after all, owning property and eligible for military service and therefore worth knowing about for a government interested primarily in taxation and military recruitment [Willcox, 1930, cited in Starr, 1987]), and so the data necessary for the computation of, say, sex ratios, not to mention population density or the percentage of the population that's a certain age, might simply not be available.

Sometimes, however, you can locate data that very nearly (if still imperfectly) meet your needs, so the issue of validity arises. Messner, for instance, is self-conscious about the validity of his measure of the level of exposure to television violence: the proportion of households regularly tuned to the five most violent shows (as determined by the National Coalition on Television Violence) in a one-month interval. His ideal would have been "to estimate the viewing audience for all television programs broadcast in a given geographical area for some specified time interval, weight these viewing estimates by evaluations of the violent content of the respective programs, and then combine the weighted estimates into a composite index," a measurement project that he plausibly claims would have been a "monumental task." So he settled for the perhaps less valid, but much more accessible data he used.

STOP AND THINK *Can you see why Messner's actual measure may not perfectly tap his ideal?*

Admittedly you really have to stop and think to see the validity problem associated with Messner's measure. But this is not always the case. Numerous cross-national studies (for example, Ward, 1984, Marshall, 1986, Clark, Ramsbey, and Adler, 1991, Clark, 1992) have used the women's labor force participation rate as a rough measure of women's economic status in a nation. A women's labor force participation rate refers to the proportion of women in an area (in this case, a country) who are employed in the formal workforce. The rationale behind its use as a measure of women's economic status is the notion that, if women don't work, they won't have the power and prestige within the family and in the community that access to an independent income can buy. But what if women and men don't always (as we know they don't) receive equal monetary returns for their work? Wouldn't the economic status of women then depend on, among other things, how closely their average income approximated that of men's—something that their labor force participation rate doesn't begin to measure? We're sure you can think of other problems with labor force participation as a valid measure of economic status, but you probably get the idea: Sometimes researchers who employ available data are stuck using imperfectly valid indicators of the concepts they want to measure because more perfect indicators are simply not available.

One peculiar danger of using available data is the possibility of becoming confused about your unit of analysis (or subjects). This is not likely to happen when you're doing a secondary analysis of survey data collected by another researcher. If, for instance, you use General Social Survey data to show that as the education of parents increases, the education of your respondents also increases, it's unlikely you'll forget that people are your units of analysis. Nonetheless, Messner's study is, from a methodological point of view, at bottom a cautionary tale. It warns against the temptation to infer from the many individual-level ("micro-level") studies that have found a connection between viewing television violence and personal aggression that aggregate viewing of television violence will necessarily lead to aggregate-level violent crime. In fact, as Messner suggests in his conclusion, aggregate viewing of violent television (and television in general) might actually protect a population aggregate from crime by reducing its exposure to situations in which crime is most likely to occur. In this sense, it warns against making inferences about one unit of analysis (in this case, aggregates) based on information collected for another unit of analysis (in this case, individuals).

Perhaps the more famous problem, however, is that of making inferences about individuals based on information about groups. This problem is so famous, in fact, that it has a special name: the **ecological fallacy.** Messner wanted to see whether places where people tend to watch more television violence had higher crime rates than did places where people tend to watch less television violence, and so using SMSAs as a unit of analysis made a great deal of sense. So far so good. But can you conclude from Messner's finding that SMSAs where people watch more television violence have lower crime rates than those where they watch less that individuals who are exposed to more TV violence are less likely to commit crimes? Such a conclusion would risk the ecological fallacy because it might be those individuals who are watching violent TV in SMSAs where relatively little violent TV watching goes on (and elsewhere) who are committing crime. The problem is that we have looked at *SMSAs* as our units of analysis and then tried to draw conclusions about *individual people.*

ecological fallacy, the fallacy a researcher commits when making inferences about individuals from information about groups or aggregates.

STOP AND THINK *You might remember that Émile Durkheim (in* Suicide*) found that Protestant countries and regions had higher suicide rates than Catholic countries and regions did. He inferred from this the conclusion that Protestants are more likely to commit suicides than Catholics. Is this a case of the ecological fallacy? Explain.*

Summary

In contrast with primary data, or data that the researcher collects herself, available data exist before you begin your research. Common sources of available data are governmental agencies, like the U.S. Bureau of the Census, or supra-governmental agencies, like the United Nations, or private organizations, like the Nielsen company. These data tend really to be available statistics, or summaries at some aggregate level of information that has been collected at the individual level.

Increasingly, however, social scientists are using survey data that have been collected by other researchers and have been made accessible to them for secondary analysis. One major repository for such data is the ICPSR at the University of Michigan. Even more recently, social scientists have begun to use the Internet as a source of much available data.

The main advantages of using available data are that the data are cheap and convenient. They also allow the study of social change over long periods of time and of social difference over considerable space. Finally, their collection is **unobtrusive** (or unlikely to affect the interactions, events, or behavior under consideration) even if their original collection may not have been unobtrusive.

unobtrusive measures, indicators of interactions, events, or behaviors whose creation does not affect the actual data collected.

On the other hand, because available data have been collected by someone else, their reliability is more likely to be unknown or unknowable by their user. Moreover, one frequently has to settle for indicators that are known to be imperfectly valid. Finally, especially when using available statistics (or summaries at some aggregate level of information that has been collected at the individual level), the researcher has to be especially cognizant of the temptation to commit the ecological fallacy, of making inferences about individuals from data about aggregates.

SUGGESTED READINGS

Kurland, D. J., and D. John. 1997. *Internet guide for sociology.* Belmont, CA : Wadsworth.

This book is designed to introduce students to the Internet. You'll want to read and keep this book by your computer for ready reference.

United Nations. 1995. *Statistical yearbook.* New York: United Nations.

An exceptional source of social indicators for many nations.

U.S. Bureau of the Census. 1998. *Statistical abstract of the United States, 1998, National Data Book & Guide to Sources.* Washington, DC: U.S. Government Printing Office.

With the possible exception of any almanac you can buy, this is conceivably the best buy you can make as a social scientist. It puts all sorts of data at your fingertips.

Webb, E., D. T. Campbell, R. D. Schwartz, and L. Sechrest. 1966. *Unobtrusive measures: Nonreactive research in the social sciences.* Chicago: Rand McNally.

An especially well-written book on various ways of collecting unobtrusive measures.

EXERCISE 12.1

Urbanization and Violent Crime: A Cross-Sectional Analysis

This exercise gives you the opportunity to examine one assumption that seems to lie behind Messner's analysis: Cities, or more particularly SMSAs, are especially good places to study violent crime. We'd like you to see whether, in fact, population aggregates that are more highly urbanized (that is, have a greater proportion of their population in urban areas) have higher violent crime rates than those that are less highly urbanized.

Let's examine a population aggregate that can be more or less urbanized: states in the United States. Find a recent resource, perhaps in your library, that has data on two variable characteristics of states: their levels of urbanization, and their overall violent crime rate. Collect data on these variables for 10 states. (Hint: *The Statistical Abstract of the United States* is a good source of the data you'll want. Most college libraries will have this in their reference section.)

1. Prepare ten code sheets like this (actually, you can arrange all data on one page):

   State: Total Violent Crime Rate: Percent Urban:

2. Go to your library and find the data on crime and urbanization. Using the code sheets you've prepared, collect data on 10 states of your choice.

3. Once you've collected the data, analyze them in the following ways:

   a. Mark the five states with the lowest percent urban as **LU** (low urban) states. Mark the five states with the highest percent urban as **HU** (high urban) states.

   b. Mark the five states with the lowest total violent crime rates as **LC** (low crime) states. Mark the five states with the highest total violent crime rate as **HC** (high crime) states.

   c. Calculate the percentage of **LU** states that you have also labeled **HC**. Calculate the percentage of **HU** states that have been labeled as **HC**.

4. Compare your percentages. Which kinds of states tend to have the highest violent crime rates: highly urbanized or not so highly urbanized states? Interpret your comparison relative to the research question that sparked this exercise.

EXERCISE 12.2

Life Expectancy and Literacy: A Cross-National Investigation

This exercise asks you to investigate the connection between education and health by examining the research question: Do more educated populations have the longest life expectancies? Investigate this hypothesis using data from 10 countries (your choice) about which you obtain data on the percentage of the population that's literate and either female *or* male life expectancy. (Female and male life expectancy are so highly correlated that either can be used as a pretty reliable indicator of national life expectancy.) We encourage you to seek data in an almanac. (Hint: Your library will have several of these. *The World Almanac* has data on literacy and life expectancy almost every year in its "Nations of the World" section.)

1. Prepare 10 code sheets like this (again, you can arrange all data on one page):

   Country: Literacy: Life Expectancy:

2. Go to your library and find the almanac that offers appropriate data. Using the code sheets you've prepared, collect data on 10 countries of your choice.

3. Once you've collected the data, analyze them using the technique introduced in Exercise 12.1.

   a. Mark the five countries with the lowest literacy rate as **LL** (low literacy) nations. Mark the five countries with the highest literacy rates as **HL** (high literacy) nations.

   b. Mark the five countries with the lowest life expectancy as **LLE** (low life expectancy) nations. Mark the five countries with the highest life expectancy as **HLE** (high life expectancy) nations.

   c. Calculate the percentage of **LL** nations that you have labeled **HLE**. Calculate the percentage of **HL** nations that are labeled **HLE**.

4. Compare your percentages. Which kinds of nations tend to have the highest life expectancy: those with high literacy rates or those with not-so-high literacy rates? Interpret your comparison relative to the research question that sparked this exercise.

5. What is the **ecological fallacy**? Do you think you succumbed to this fallacy in your interpretation of your analysis in this exercise? Why or why not?

EXERCISE 12.3

For Baseball Fanatics

This exercise gives you the chance to examine an age-old social science dispute: whether hitting or pitching wins ball games. Help shed light on this dispute by, again, turning to a recent almanac, this time to collect winning percentages, team batting averages, and team-earned run averages for the 14 teams in either the American or the National League for one year. Use data collection techniques similar to those you used for the last two exercises, this time collecting data on three variables, rather than two. Moreover, you should use the data analysis techniques of each exercise, this time examining the relationship between winning percentage and team batting average, on the one hand, and winning percentage and team earned run average, on the other. Are both batting average and earned run average associated with winning percentage? Which, would you say, is more *strongly* associated with it? Why? Write a brief essay in which you state how your analysis bears on this research question.

EXERCISE 12.4

What's Today's Population? Using the Internet.

This exercise asks you to use the Internet, particularly the World Wide Web, to investigate some timely questions about population. Your first task will be, perhaps with the assistance of another student or someone at your computer center, the *Bureau of Labor Statistics Data Home Page*. (See this chapter's section on the Internet for the address.) Then see if you can find the answers to the following questions:

1. What is the approximate population of the United States as you investigate the home page?
2. What is the approximate population of the world as you investigate the home page?
3. What was the population of the state in which you reside in 1900? In 1950? In 1990?
4. How many people living in the United States last year were born in

   a) Mexico?

   b) Canada?

   c) Germany?

EXERCISE 12.5

Exploring Social Science Data Archives on the Web

This exercise is meant to give you a sense of the variety of available data sources on the Web. We're not as concerned with getting the right answers as we are with exploring what's out there.

On the Web, go to a site entitled "Social Sciences Archives" (http://www.socsciresearch.com/r6.html). See what's available at this site. See if, during your exploration, you can find a likely title for data on each of the following topics. Write the name of an actual title you could use to find data about:

1. U.S. education:
2. The holocaust:
3. International trade:

Now click on the topic "statistical resources on Web" and repeat the kinds of explorations you just did. See if you can find an actual title that is a likely source of data on each of the following subjects and write it down.

1. Health:
2. Labor:
3. Demography:
4. The elderly:
5. Crime:

Now click on the topic "sociology." Under that, click on "crime." Then "hate crime." Now find the name of your state and write down the number of hate crimes committed in your state in the most recent period for which data are available.

13 Content Analysis

Let's look at illustrations from two award-winning children's books, published about 22 years apart. Figure 13.1 is from Evaline Ness's (1966) *Sam, Bangs & Moonshine* and shows Samantha, the book's female protagonist, tearfully begging her father to rescue a friend whose life she realizes she might have inadvertently jeopardized by sending the friend on a fanciful mission. Figure 13.2 is from Patricia McKissack's and Jerry Pinckney's (1988) *Mirandy and Brother Wind* and shows Mirandy helping her friend Ezel restack logs on a pile of kindling he has just inadvertently knocked over. Which of these illustrations depicts the stronger, more independent female character? Which image, do you think, would be more likely to plant the seed of suspicion in a child reading or hearing the story that girls can't take care of themselves (or that they can), that they tend to screw things up (or that they fix them), that they're basically weak (or basically strong)? Which would you prefer to share with your children?

FIGURE 13.1

Source: From *Sam, Bangs, and Moonshine,* written and illustrated by Evalene Moss, © 1966 by Evalene Ness. Reprinted by permission of Henry Holt and Company, Inc.

FIGURE 13.2

Source: From *Mirandy and Brother Wind,* by Patricia McKissack, illustrated by Brian Pinckney. Illustrations copyright © 1988 by Brian Pinckney. Reprinted by permission of Alfred A. Knopf, Inc.

Introduction

These are the kinds of questions that led one of us, Roger—the father of two young children—to question whether the messages that children's books sent about sex roles had changed over time. So he enlisted the aid of two students, Rachel Lennon and Leanna Morris, and together they produced the article from which this chapter's focal research is drawn. This report illustrates the technique of **content analysis**. Content analysis is a method of data collection in which some form of communication (speeches, TV programs, newspaper articles, films, advertisements, even children's books) is studied systematically. Content analysis is one form of available data analysis.

content analysis, a method of data collection in which some form of communication is studied systematically.

FOCAL RESEARCH

Engendering Junior: Changing Images in Children's Books[1]

by Roger Clark, Rachel Lennon, and Leanna Morris

Introduction

A suitable way to mark the twentieth anniversary of the classic study, "Sex-Role Socialization in Picture Books for Preschool Children," by Weitzman, Eifler, Hokada, and Ross (1972) is to replicate and expand this work with updated samples. That study showed what some parents already suspected: American picture books for preschoolers depicted male and female characters in stereotyped ways, often to the point of making female characters literally invisible. Weitzman's primary focus was on winners and runners-up of the Caldecott Medal (the major national award for children's picture books), especially those selected during the five years from 1967 to 1971. In this paper we examine Caldecott winners and runners-up from the 1987–1991 period to determine whether the social action that Weitzman's work was part of had a significant "equalizing" effect on the depiction of male and female characters in prize-winning books.

Method

One reason for lavishing research attention on Caldecott winners is the unusual influence these books have on tastes for children's literature. The Caldecott is not only the most prestigious award for preschool literature, but it also guarantees its winners phenomenal sales (Clark 1992: 6). For the purposes of

[1]This article is adapted from Clark, Lennon, and Morris' (1993) "Of Caldecotts and Kings: Gendered Images in Recent American Children's Books by Black and Non-Black Illustrators," *Gender & Society*, Vol. 7, 227–245.

comparison, we have closely studied the 18 Caldecott winners and runners-up from 1967 to 1971 (Weitzman et al.'s sample) and the 16 Caldecott winners and runners-up between 1987 and 1991.

Davis (1984) and Williams, Vernon, Williams, and Malecha (1987) have aptly criticized Weitzman et al. (1972) for failing to specify their units of analysis and their operational definitions for many variables. We specify our units of analysis as we report our findings, but admit that one of our variables—gender itself—is not always easy to determine. In *Owl Moon*, for instance, writer Jane Yolen (1987) and illustrator John Schoenherr successfully obscure the gender of the young child who finally gets to go owling with her father. With this exception, however, we found that, once we used the text to supplement information available in illustrations, we were able to make confident gender assignments.

Davis (1984) refined a set of variables dealing with various aspects of gender-related behavior. We also employ this set of variables (see Appendix to this report for variable definitions). Given these variable traits, we observe that mainstream gender stereotyping entails expectations that female characters will be more dependent, cooperative, submissive, imitative, nurturant, emotional, and passive and that they will be less independent, competitive, directive, persistent, explorative, aggressive, and active than male characters. Both male and female characters are expected to be creative, but usually in different ways, with male characters often being stereotyped as creative problem solvers (as in getting themselves out of difficult situations) and female characters as being more naturally or domestically creative (as in giving birth to a child or, perhaps, devising a clever cleaning technique).

We decided to code whether each of four possible characters in each book possessed a given trait (say, dependence). These were the book's central character, the most important other character of the same sex, and the two most important characters of the opposite sex, where supporting characters were deemed sufficiently visible for analysis. Overall intercoder or interrater reliability was 76 percent, although it was particularly low for two traits (cooperativeness [63 percent] and passive activeness [62 percent]), suggesting that the results for these traits should be interpreted with particular care.

Results

Weitzman et al. (1972) demonstrated that females were invisible in children's books of the late 1960s by focusing, alternately, on pictures with humans, pictures with animals, and books as a whole as units of analysis. Table 13.1 depicts indicators of male and female visibility as they were reported by Weitzman and as we calculated them for the most recent Caldecotts.

Following Weitzman,[2] we defined human single-gender illustrations as ones depicting only male or only female characters. The percentage of such

[2]In trying to replicate Weitzman et al.'s (1972) study, we were not always able to do so perfectly. For instance, we counted 20 more single-gendered illustrations of both male and female characters than did Weitzman et al. We are consoled that our essential findings (for example, the ratio of male to female illustrations) were always practically the same as those of Weitzman et al.

TABLE 13.1 **Comparisons of Gender Visibility**

	1967–1971	*1987–1991*
Human single-gender illustrations		
Total number	188	293
Percentage female	11.7	26.6 sig.
Male/female ratio	7.5:1	2.8:1
Nonhuman single-gender		
Total number	96	68
Percentage female	1.0	22.1 sig.
Male/female ratio	95.0:1	3.5:1
Books		
Total number	18	16
Percentage with no female character	33.3	25.0 ns.
Percentage with central female character	11.1	43.2 sig.

Notes: sig. indicates that difference is significant at .05 level; ns. indicates that difference is not significant at this level.

illustrations depicting human females, which was only 11.7 percent of all single-gender illustrations in the Caldecotts of the late 1960s, had grown to 26.6 percent by the late 1980s. This growth would have been more impressive if it hadn't been for Bill Peet's (1989) autobiography in the most recent sample, a book with an excess of 99 single-gender male illustrations.[3] If *Bill Peet* were dropped from the late-1980s sample, the percentage of human single-gendered illustrations that were female would be almost 40 percent. In any case, the overall trend since the late 1960s is in the direction of greater female visibility.

Weitzman et al. (1972) also considered nonhuman single-gendered illustrations to demonstrate the male bias of the late-1960s Caldecotts. They found that females appeared in a strikingly low 1 percent of all such images in their sample. We found that 22.1 percent of single-gendered animal illustrations in the late 1980s Caldecotts depicted female characters, again suggesting a trend toward greater female visibility. Similarly, while Weitzman found 33.3 percent of the Caldecotts she examined lacked any female character at all, we found only 25 percent like this in the late 1980s sample.

On the pivotal issue of the gender of central character, Weitzman reported that only 2 of the 18 Caldecotts (11 percent) of the later 1960s had females in the central role. We found that for the Caldecotts of the later

[3]The late-1960s sample contained a similar troubling outlier—Sendak's (1970) *In the Night Kitchen*, which contained 52 illustrations of male characters only—although its removal from that sample would still have left the 1960s Caldecotts with a small proportion (16.5 percent) of single-gendered female illustrations.

TABLE 13-2

Comparisons of Behavioral Traits
Percentage of Characters Exhibiting Traits

Traits	Late-1960s Caldecotts Females N = 7	Late-1960s Caldecotts Males N = 28	Late-1980s Caldecotts Females N = 14	Late-1980s Caldecotts Males N = 20
Dependent	43	25	36	20
Independent	14	42	43	35 (b)
Cooperative	43	25	29	25
Competitive	0	11	14	0(b)
Directive	0	21	14	25
Submissive	28	14	14	10
Persistent	14	21	29	20
Explorative	0	14	14	10(b)
Creative	28	21	14	30
Imitative	0	0	0	0
Nurturant	57	18(a)	36	30
Aggressive	0	18	21	5(b)
Emotional	43	28	21	45(b)
Active	28	57	43	40
Passively Active	14	14	21	20

(a) Indicates the difference between males and females in Caldecott sample is significant at .05 level.
(b) Indicates the difference of proportions between the late-1960s and the late-1980s Caldecotts is significant at .10 level.

1980s, 7 of 16 (44 percent) showed females in this role, a marked changed toward greater female visibility.

The general pattern, in terms of visibility, then, is that among Caldecotts, female characters made gains on all dimensions mentioned by Weitzman et al. But this only raises the question of how those who are visible are portrayed. Weitzman et al. (1972) found a fair amount of stereotyping in the few female characters that they found. We have used the more systematic procedures outlined by Davis (1984) for discerning gender stereotyping, with two goals in mind: (1) to see if we, using more systematic procedures, although slightly different indicators, would come to the same conclusions that Weitzman et al. did about the late-1960s Caldecotts, and (2) to see if we could detect differences in the amount of stereotyping appearing between that sample and the later 1980s sample.

Table 13.2 provides considerable support for the kinds of conclusions drawn by Weitzman et al. (1972). Here one can see that female characters in the late-1960s Caldecotts were substantially more likely to be depicted as dependent, cooperative, submissive, nurturant, and emotional than male characters, whereas they were substantially less likely to be shown as independent, competitive, directive, persistent, explorative, aggressive, and active. Because female representation in the late 1960s was so small, the only difference that proves to be significant at the .05 level is the one for nurturance. Nevertheless, no fewer than 11 of 15 characteristics show substantial differences, and in the stereotyped direction. Thus our more systematic analysis, using slightly different variables, leads us to concur with the conclusion of Weitzman et al. that gender stereotyping did exist in the late-1960s Caldecotts.

Table 13.2 also indicates substantial change in the presentation of gender among Caldecotts between the late 1960s and the late 1980s. For although female characters of the 1980s are still more likely to be depicted as dependent, they are only slightly more likely to be depicted as cooperative and nurturant, and slightly less likely to be depicted as directive. Moreover, they now *more* likely to be depicted as independent, competitive, persistent, aggressive, and active. In all cases, change has been in the less stereotyped direction, and, in the cases of independence, competitiveness, explorativeness, aggressiveness, and emotionality, the change is statistically significant. Those differences that remain in the predicted direction in terms of traditional stereotypes of males and females in American culture, then, are smaller in the late-1980s sample than in the late-1960s one, but few of the differences actually remain in that direction.

A comparison between two prominent female characters in the two groups of Caldecotts illustrate the point. Shang in Young's (1989) *Lon Po Po* is an extremely independent, explorative, creative, and aggressive little girl. Her curiosity about her "grandmother's" true identity leads to an ingenious and ruthless scheme to save herself and her two sisters from the intruding wolf. Compare her to Sam in Ness's (1967) *Sam, Bangs, and Moonshine*, who induces her loyal friend Thomas to take a perilous journey on her behalf. When she realizes the danger she has put Thomas in, Sam sits at home and cries until her father rescues the boy. Thus, whereas Shang actively shapes events around her, Sam shapes them in a much more passive fashion and observes what happens from the safety of her home.

Conclusion

The visibility of females in Caldecott medal winners and honor books increased considerably between the late 1960s and the late 1980s. The depiction of gendered characters who are visible has also changed. The change is evident for almost all behavioral traits measured in this study.

We conclude that the egalitarian values of Weitzman et al. have found embodiment in recent award-winning children's books. The Caldecott committee is today clearly acknowledging books that give more visibility to female characters than it did in the 1960s. One may hope that the legitimacy bestowed upon award-winning books may help create a society, via their young readers, in which the aforementioned values become embodied as well.

Appendix
Behavioral Definitions from Davis (1984)

Dependent: seeking or relying on others for help, protection, or reassurance; maintaining close proximity to others.

Independent: self-initiated and self-contained behavior, autonomous functioning, resistance to externally imposed constraints.

Cooperative: working together or in a joint effort toward a common goal, complementary division of labor in a given activity.

Competitive: striving against another in an activity or game for a particular goal, position, or reward; desire to be first, best, winner.

Directive: guiding, leading, impelling others toward an action or goal; controlling behaviors of others.

Submissive: yielding to the direction of others; deference to wishes of others.

Persistent: maintenance of goal-directed activity despite obstacles, setbacks, or adverse conditions.

Explorative: seeking knowledge or information through careful examination or investigation; inquisitive and curious.

Creative: producing novel idea or product; unique solution to a problem; engaging in fantasy or imaginative play.

Imitative: duplicating, mimicking, or modeling behavior (activity or verbalization) of others.

Nurturant: giving physical or emotional aid, support, or comfort to another; demonstrating affection or compassion for another.

Aggressive: physically or emotionally hurting someone; verbal aggression; destroying property.

Emotional: affective display of feelings; manifestation of pleasure, fear, anger, sorrow, and so on via laughing, cowering, crying, frowning, violent outbursts, and so on.

Active: gross motor (large muscle) physical activity, work, play.

Passively active: fine motor (small muscle) activity; alert, attentive, activity but with minimal or no physical movements (for instance, reading, talking, thinking, daydreaming, watching TV).

REFERENCES

Clark, B. L. 1992. American children's literature: Background and bibliography. *American Studies International* 30: 4–40.

Davis, A. J. 1984. Sex-differentiated behaviors in non-sexist picture books. *Sex Roles* 11:1–15.

Ness, E. 1967. *Sam, bangs, and moonshine*. New York: Holt, Rinehart & Winston.

Peet, B. 1989. *Bill Peet*. Boston: Houghton Mifflin.

Sendak, M. 1970. *In the night kitchen*. New York: Harper & Row.

Weitzman, L., D. Eifler, E. Hokada, and C. Ross. 1972. Sex-role socialization in picture books for preschool children. *American Journal of Sociology* 77:1125–1150.

Williams, J., J. Vernon, M. Williams, and K. Malecha. 1987. Sex role socialization in picture books: An update. *Social Science Quarterly* 68: 148–156.

Yolen, J. 1987. *Owl Moon.* New York: Philomel.

Young, E. 1989. *Lon Po Po.* New York: Philomel.

Appropriate Topics for Content Analysis

Although Clark, Lennon, and Morris applied content analysis to children's books, content analytic methods can be applied to any form of communication,[4] including movies, speeches, letters, obituaries, editorials, and song lyrics. In the last decade, Naisbitt and Aburdene (1990) examined local newspapers to determine major trends in American life, as did Manoff and Schudson (1987) to outline the role of journalists and journalism in American society. Much earlier, the freed slave Ida Wells (1892, cited in Reinharz [1992]) and G. J. Speed (1893, cited in Krippendorff [1980]) analyzed newspaper articles to show, respectively, the extent to which black men were being lynched in the South and how gossipy newspapers in New York were becoming during the 1880s. More recently, Fissell (1995) looked at popular childbearing guides and conduct books from seventeenth and early-eighteenth century Great Britain to examine the cultural construction of female bodies; Giordano (1995) studied handwritten messages in junior and senior high school yearbooks from throughout the twentieth century to see how adolescents learn to navigate their social world; and Matcha (1995) examined newspaper obituary notices from early twentieth-century Ohio to assess marriage and family patterns there. Studies have focused on suicide notes (Osgood and Walker, 1959, Bourgoin, 1995, the latter to amend Durkheimian suicide theory), trial transcripts (Lasswell, 1949 and others), magazines (Friedan, 1963, Ho, 1984 and many others), wills (Finch and Wallis, 1995), textbooks (Walworth, 1938, Gordy and Pritchard, 1995, and many others), TV programs (Brouwer, Clark, Gerbner, and Krippendorff, 1969, and many others), radio programs (Albig, 1938, Washburn, 1995), want ads (Baize and Schroeder, 1995, and others), speeches (Umennachi, 1995), verbal exchanges (Bales, 1950), diaries (for example, Reinharz, 1987), cookbooks (Stricker, 1979), films, both conventional (Asheim, 1950, Deegan, 1983, Denzin, 1995) and pornographic (Cowan and Campbell, 1995), and most recently e-mail messages (Schleef, 1995). The list is endless. Systematic content analysis seems to date back to the late 1600s, when

[4]In fact, while content analysis is most often applied to communications, it may, in principle, be used to analyze any *content*. Thus, for instance, Rathje and Williams' (1992) "garbology" project mentioned in Chapter 12 employed content analysis to study the contents of garbage cans.

Swedish authorities and dissidents counted the words in religious hymns and sermons to prove and disprove heresy (Dovring, 1973, Krippendorff, 1980, Rosengren, 1981).

Generally the questions asked by content analyzers are the same as those asked in all communications research: "Who says what, to whom, why, how, and with what effect?" (Lasswell, 1965, 12). Weitzman, Eifler, Hokada, and Ross (1972) in the study the focal research replicated, had been interested in what children's books were saying about gender to young children, and holding that content up against liberal feminist standards that advocate that males and females be shown in equal numbers and with similar capabilities and tendencies. Against these standards, award-winning books of the late 1960s looked pretty bad. Clark, Lennon, and Morris (1993), on the other hand, were interested in whether changes in the content of award-winning children's books since the late-sixties constituted progress, regress, or no gress at all by the same standards—and found evidence of progress. Evaluating communication content up against some standard and describing trends in communication content are two of the general purposes for which content analysis has been used (see Holtsi: 1969, 43).

STOP AND THINK *Another frequent use of content analysis is hypothesis testing. For instance, do you think local, regional, or national newspapers are more likely to "headline" international stories on the top of the front page? What is your hypothesis? Why? (You might test this hypothesis in exercise 13.2)*

Units of Analysis

Sampling for content analysis involves many of the same problems that sampling for any type of social research involves. First you need to determine the units of analysis (or elements), the kinds of subjects you want to focus on. In the focal research, this task was simplified because the authors were replicating an earlier study (Weitzman et al.'s), whose units of analysis they obviously wanted to use again, if they could. And when it came to questions of female visibility, such as how many books had female main characters, or even how many human single-gender illustrations were of females, determining units of analysis was a snap: They were clearly books, in the first instance, and human single-gender illustrations, in the second. These were the units the Weitzman study had used.

When it came to questions of characterization, however, the earlier study was less helpful. Weitzman and colleagues had concluded, for instance, that in the world of picture books "the females are passive while the males are active" (1972, p. 1139). But, rather than say which females and males they'd looked at (were they main characters? main characters and supporting characters? all characters?), the researchers suddenly became curiously anecdotal, referring to a main character here, a secondary character there, a barely visible character in the next instance. What were Clark, Lennon, and Morris to do?

What they actually did was to stumble (as a result of a systematic literature review) on Albert Davis' (1984) article in which he reported his content analysis of books that had been recommended by liberal feminist organizations. Davis had identified and operationalized the 15 behavioral traits (for example, dependence, independence, and so on) listed in the Appendix to the focal research and used these variables to content-analyze all illustrations and text-messages in those books. Clark, Lennon, and Morris decided that Davis' behavioral traits (his variables) were appropriate for their purposes, but that his units of analysis (all illustrations and text messages) were less so. In particular, they thought that the overall presentation of notable characters in books would be more likely to impress young readers than the details of individual illustrations or text messages.

So Clark, Lennon, and Morris decided to focus (somewhat arbitrarily) on the two major characters of each gender in each book. (If there weren't two of each, the researchers didn't make them up. They evaluated those characters that were available). Clark, Lennon, and Morris then decided, after they'd read the book, whether each of the four characters embodied any of the behavioral traits developed by Davis.

units of analysis: the units about which information is collected.
units of observation: the units from which information is collected.

Earl Babbie (1995) offers another useful distinction: between **units of analysis** (the unit about which information is collected) and **units of observation** (or the unit from which information is collected). This, of course, is a distinction without a difference when the unit of analysis and observation are the same. The units of observation in the focal research were always children's books (that is, the authors *always* got information from the books) and in some parts of the analysis (when the authors were determining the gender of the main character of each *book*), the units of analysis were also books. But when the units of analysis and observation are different, it is important to remember that the constituents of one's sample are the units of analysis, not the units of observation. Thus, when Clark, Lennon, and Morris shifted their focus (their units of analysis) to illustrations or characters, they needed to remember that they were then interested in variable characteristics of illustrations and characters (within books) and not in the books themselves.

STOP AND THINK *Suppose you were interested in whether commercials during children's TV programs were more utopian (that is, focused on improvement rather than on things as they are) than were those for adult programs (see exercise 13.3, below). What would your unit of analysis be? What would your unit of observation be?*

If you were interested in the difference between commercials during children's and adults' TV programs, your units of observation (or the units from which you collected information) would be children's and adults' TV programs (because commercials from other types, say, adolescent TV programs, would be irrelevant), but the units of analysis (or the units about which you'd get information) would be the commercials attached to those programs.

Sampling

In content analysis, units of analysis can be words, phrases, sentences, paragraphs, themes, photographs, illustrations, chapters, books, characters, authors, audiences, or almost anything you want to study. Once you've chosen the units of analysis, you can sample them with any conventional sampling technique (for example, simple random, systematic, stratified, or cluster—see Chapter 5) to save time or effort. In the analysis of human single-gender illustrations, for instance, Clark, Lennon, and Morris might have numbered each one and systematically sampled 25 percent of them by random sampling the first one and picking every fourth illustration after that. As things turned out, however, they decided that, because Weitzman and colleagues had examined the whole population of human single-gender illustrations in Caldecotts between 1967 and 1971, and because none of the populations were extraordinarily large, Clark and his colleagues should do the same for the period 1987 to 1991. And, in general, they used this kind of saturation sampling in their research.

STOP AND THINK *How might Clark, Lennon, and Morris have collected a stratified sample of picture books? (Hint: Can you think of any characteristic of authors they might have used to divide their sample?) How might they have cluster sampled illustrations?*

Creating Meaningful Variables

Content analysis is a technique that depends on the researcher's capacity to create and record meaningful variables for classifying units of analysis. Thus, for instance, in classifying books, Clark, Lennon, and Morris were interested in knowing whether they had female characters or not and whether they had female central characters or not; in classifying human and animal single-gender illustrations, whether the depicted character was female or not; in classifying main characters, whether they appeared dependent or not, independent or not, and so on. In each of these cases, variables were of *nominal* scale, but this wasn't necessary. In analyzing books, for instance, the authors could have asked "how many" major male or female characters appeared, in which case they would have created a *ratio* scale variable.

In fact, their coding sheet, depicted in Figure 13.3, makes it look as though Clark, Lennon, and Morris did collect ratio-scale information about single-gender illustrations after all.

STOP AND THINK *Can you think of a question Clark, Lennon, and Morris might have asked about children's books that would have elicited ordinal-scale data?*

Question 3 in Figure 13.3 asks each coder, for instance, about the number of illustrations that depict characters of clear and singular gender (as opposed, for example, to those that show characters of both genders or characters of ambiguous gender). But the appearance of interval categories is a slight-of-hand, made possible because the unit of observation (picture books), in most cases,

FIGURE 13.3

Sample Coding Sheet From "Gender and Junior" Study

Book: Goggles Author: E. J. Keats Year of Publication: 1969
Publisher: MacMillan: New York

1. Does the book have any female characters?: No
2. Does the book have a central female character?: No
3. Number of human illustrations that depict a single and clearly determinate gender: 16
4. Number of these that depict females: 0
5. Number of animal illustrations that depict a single and clearly determinate gender: 0
6. Number of these that depict females: 0

BEHAVIORS OF CENTRAL CHARACTERS

	Main Character	Supporting Character of Same Gender	Most Important Character of Opposite Gender	Supporting Character of Opposite Gender
ROLE:	PETER	ARCHIE		
Dependent				
Independent	X	X		
Cooperative	X	X		
Competitive	X	X		
Directive	X			
Submissive		X		
Persistent	X	X		
Explorative				
Creative	X	X		
Imitative				
Nurturant				
Aggressive				
Emotional	X	X		
Active	X	X		
Passively Active				

contains several units of analysis (single-gender illustrations) and because all the authors were really interested in, after all, was the proportion of all single-gender illustrations in a given sample of books that depicted females.

STOP AND THINK *How do you think Clark, Lennon, and Morris used the data from their coding sheets to calculate the proportion of all single-gender illustrations that showed females?*

Clark, Lennon, and Morris calculated the proportion of all single-gender illustrations that showed females by dividing the response to item 3 (number of human illustrations that depict a single and clearly determinate gender) into the response to item 4 (number of these that depict females). The unit of analysis here was the human single-gender illustration and the variable was gender (a nominal, not an interval, variable).

As in any measurement operation, the scheme left room for unreliable observations, or the possibility, in this case, that parallel observations of the same phenomenon would yield inconsistent data. Thus, two of the three coders (in this case, the three authors) disagreed about several books even on the number of unambiguous, single-gender illustrations. The authors handled these disagreements (amicably, you'll be glad to know) by chatting about them. In most cases, discussion soon made it obvious that one or more of them had simply under- or over-counted cases they could all quickly agree about. In the case of *Owl Moon* mentioned in the article, however, two coders spotted what the third did not: all illustrations involving the main character, a child who was out owl-spotting with his/her father, were ambiguous because the child was not clearly gendered either in the illusfrations or the text. (The third observer had perceived the child to be a girl.) In this case, discussion once again resolved the issue. But this time the issue was more clearly a matter of genuine perceptual differences than of arithmetic miscalculation.

STOP AND THINK *If Clark, Lennon, and Morris hadn't used the discussion method described in this paragraph, what could they have done with their three estimates of the number of single-gender illustrations in a book to arrive at a final, single estimate for that book?*

If the authors hadn't used the discussion method they could have taken the arithmetic average of their three estimates of the number of single-gender illustrations as their final, single estimate for each book.

Clark, Lennon, and Morris were tempted to use the discussion method to resolve differences in the classification of behavioral traits as well. But early attempts suggested that many differences here were not only perceptual (Morris might honestly disagree with Clark that a certain character was persistent, despite a common agreement about what Davis [1984] meant by his definition of persistence [see Appendix in the focal research]) but also not easily reconcilable. So when it came to the final classification of characters by whether they possessed a particular behavioral trait, the authors agreed that they'd finally classify a character as possessing such a trait (for example, persistence) only if all three coders independently (and without discussion) coded him or her that way.

Quantitative or Qualitative Content Analysis?

We have discussed the piece by Clark, Lennon, and Morris to give you some appreciation for the potential of a relatively quantitative brand of content analysis: namely, a brand that was designed with statistical analysis in mind. Quantitative content analysis is the most common kind in the social sciences. But we wouldn't want to leave you with the impression that content analysis must be quantitative to be effective (see also Reinharz, 1992). In fact, we think content analysis is well employed when qualitative social scientists aim for primarily verbal, rather than, statistical analysis of various kinds of communication. Roger did, with another student, Heidi Kulkin, a qualitative content analysis of 16 recent young-adult novels about non-white, non-American, or non-heterosexual characters (Clark and Kulkin, 1996). Roger and Heidi came to these books with few preconceived notions of what they would "see" (that is, their major purpose was exploration) and so did not try to find ways to "quantify the data," as the previous section suggests they might have. Instead they found themselves "listening" to themes of oppression and resistance and noting patterns.

Compared with the relatively well-accepted techniques (for example, for quantifying and ensuring the reliability of data) for doing quantitative content analysis, however, guidelines for doing qualitative content analysis are few and far between. (For those interested in such guidelines, however, we recommend Reinharz, 1992, and Strauss, 1987, 28ff.) Heidi and Roger tried to discern theme patterns by a process of comparison and contrast that would be congenial to the advocates of grounded theory (see also Glaser and Strauss, 1967). Thus, when they noticed in novels about American slavery (for example, Gary Paulsen's [1993] *Nightjohn*), that authors seemed to put a premium on collective (for example, the participation in highly subversive "night" schools), rather than individual, resistance to oppression, they began to think about an empirical generalization that seemed to be borne out throughout their sample: Authors who focus on oppression as community-wide phenomena have a "higher regard for communal efforts to subvert oppression" than do authors who focus on it as an individual's problem.

Generally speaking, then, quantitative content analyses tend to be of the deductive sort described in Chapter 2, analyses that begin, perhaps, with explicit or implicit hypotheses that the researcher wants to test with data. Implicit in Clark and colleagues' analysis, for instance, is the hypothesis that Caldecott-winning books of the late 1980s are more likely to make female characters visible than their counterparts of the late 1960s were. Qualitative content analyses, on the other hand, tend to be of the inductive sort, analyses that might begin with a research question, but are then likely to involve observations about individual texts (or portions of texts) and build to empirical generalizations about texts (or portions of texts) in general. Clark and Kulkin, for instance, began with the general research question "What kinds of themes are typical of young-adult novels about non-white, non-American, or non-heterosexual characters?" and then saw what themes emerged as they read such novels.

Advantages and Disadvantages of Content Analysis

One great advantage of content analysis is that it can be done with relatively little expenditure of time, money, or person power. Clark, Lennon, and Morris worked as a team of three on their examination of children's books, all of which they borrowed from local libraries. They did so partly for the fun of working with one another and partly for the enhanced data reliability that comes from cross-checking one another's observations in various ways. In principle, however, their project could have been done by one of them, with no special equipment other than a library card.

Another related plus is that content analysis doesn't require that you get everything right the first time. Frankly, despite appearances to the contrary, the coding sheet shown in Figure 13.1 is really a compilation (and distillation) of several separate efforts. Indeed, the authors had made a first pass at all the books in their sample to discern the visibility of female characters (in illustrations, and so on) *before* they discovered Davis' scheme for coding behavior. They then returned to their subjects (the books) to code behavior. This kind of recovery from early oversights is much less feasible in survey or experimental research.

STOP AND THINK *Compare this to having surveyed 500 people and then deciding you had another question you wanted to ask them.*

Yet another advantage is that content analysis is unobtrusive; our content analysis itself can hardly be accused of having affected the content of the books we read. This is not to say that content analysis might not affect the content of subsequent communications. In fact, we would guess that the findings of Weitzman and colleagues (1972) did contribute to the kinds of attitude shifts (in award-givers, in publishers, and in writers) that can account for some of the changes Clark, Lennon, and Morris documented.

Like using other kinds of available data, content analysis permits the examination of times (and peoples) past—something surveys, experiments, and interviews don't. Fissell's (1995) analysis of seventeenth and eighteenth century British childbearing guides and conduct books is one good example, as are Olson's (1976) use of the Code of Hammurabi, a list of laws and punishments from ancient Babylonia.

STOP AND THINK *The Code of Hammurabi consists of 282 paragraphs, each specifying an offense and its punishment. One could code each of the paragraphs by the severity of the punishment associated with an offense, for instance, to gain a sense of what kinds of behavior the Babylonians most abhorred. Based on the punishments associated with stealing from a temple and striking another man, mentioned in the paragraphs below, which of these offenses was held in greater contempt?*

Paragraph 6. If a man steal the property of a god (temple) or palace, that man shall be put to death . . .

Paragraph 204. If a freeman strike a freeman, he shall pay ten shekels of silver.

One of Emily's future projects is an analysis of life expectancy, family structure and gender roles in previous centuries by doing content analysis of the inscriptions on head stones in New England cemeteries.

Yet another advantage of content analysis is that, because it typically permits one to avoid interaction with human subjects, it also permits the avoidance of the usual ethical dilemmas associated with research involving human subjects and with the problems of dealing with Human Subjects Committees, unless one plans to use private documents.

Some disadvantages of content analysis might have occurred to you by now. Because content analysis is usually only applied to recorded communication, it can't very well be used to study communities that don't leave (or haven't left) records. Thus, we can study the legal codes of ancient Babylonia because we've discovered the code of Hammurabi, but we can't study the codes of societies for which no such discoveries have been made. And Clark, Lennon, and Morris can content-analyze children's books with some sense that what children read (or heard) might have predictable consequences for their attitudes. But to tap those attitudes directly, it would doubtless be more valid to analyze literature produced *by*, not *for*, children. Peirce and Edwards (1988) actually did content-analyze stories written by 11-year-olds, but this would be pretty hard to do for pre-literate, and certainly pre-verbal, children, the not infrequent audience for picture books. Content analysis can also involve social class biases, because communications are more likely to be formulated by educated persons.

There are also questions about the validity of content-analytic measures. For example, is the invisibility of female characters in award-winning children's books of the late 1960s really a valid indicator of sexist attitudes? Perhaps, but maybe not. And even if one were pretty sure that the invisibility was a valid measure, whose attitudes are being measured: those of the authors, the publishers, the award-givers, or the adult purchasers and readers?

STOP AND THINK

One of us, Emily, and a colleague thought they'd found evidence of significant social change when they content-analyzed 20 years of the "Confidential Chat" column (which published reader contributions) in the Boston Globe *and found the column increasingly dealing with women's work outside the home between 1950 and 1970. They were discouraged when a reviewer of their research report for a scholarly journal suggested that their results might reflect changes in the editorial climate at the newspaper, rather than changes in the attitudes of the readership. When they interviewed the editor of the column, they found, indeed, he and the former editor published only about 10 percent of the letters he received and that changes in content were likely to reflect his sense of what his readership (mostly women) wanted to read. Emily and her co-author never resubmitted their paper because they now felt it failed to say anything significant about social change. What do you think of this decision?*

Finally, content analysis doesn't always encourage a sensitivity to context that, say, literary criticism often does. A male character, for instance, might be portrayed as aggressive, but the point of the story might be to show how foolhardy that aggressiveness can be. Is it the aggressiveness or the author's implicit condemnation of the aggressiveness that will affect a young listener? Questions of validity plague those who do content analysis just as much as they plague those who employ other methods of data collection.

STOP AND THINK

Suppose your main concern was whether authors endorsed, rather than simply portrayed, male aggressiveness? What questions might you ask of your books to code such an endorsement? Must content analysis be as insensitive to context as the previous paragraph suggests?

Summary

Content analysis is an unobtrusive, inexpensive, relatively easy, and flexible method of data collection. It involves the systematic and typically, but not necessarily, quantitative study of some form of communication. It can be used to test hypotheses about communication, to compare the content of such communication with some standard, or to describe trends in communication, among other things. Clark, Lennon, and Morris simultaneously looked at trends in the depiction of gender in children's picture books and measured these trends against an egalitarian standard established by Weitzman and colleagues in their 1972 study.

An early step in content analysis is to identify a unit of analysis. This is the unit about which information is collected and can be words, phrases, or paragraphs; stories or chapters; themes; photographs; illustrations; authors; audiences; or almost anything associated with communication. In Clark, Lennon, and Morris's study, the unit of analysis was sometimes whole books, sometimes illustrations, and sometimes characters.

Once a unit of analysis has been identified, any conventional sampling technique can be employed to ensure representativeness. Sometimes, as in Clark, Lennon, and Morris's study, investigating a whole population is also a feasible route to this goal.

The quality of a content analysis depends on the creation and meaningful usage of variables for classifying units of analysis. In studying main characters, for instance, Clark, Lennon, and Morris used a set of variables, created by Davis (1984), to examine various aspects of gender-related behavior. Explicit and reliable coding schemes for recoding variation are most important. Checks on measurement reliability can be pretty simple, but are always appropriate. Clark, Lennon, and Morris used discussion to check the reliability of some of their classification efforts and statistics to check others.

On the down side, content analysis cannot tell us much about communities that have left no recorded communications. Moreover, content analysis is frequently criticized on validity grounds, not only because those criticizing aren't sure whether the content analyzers are measuring what they think they are, but also (sometimes) because those criticizing aren't sure the analyzers know what community they're describing. Such critiques are frequently warranted, but should not deter you from trying your hand at content analysis and making your own judgment about it.

SUGGESTED READING

Holsti, O. R. 1969. *Content analysis for the social sciences and the humanities*. Reading, MA: Addison-Wesley.

Probably the most referred-to manual on content analysis, Holsti's fifth chapter on coding and sixth chapter on sampling, reliability, and validity are where to turn first if you want to move beyond the treatment of these topics in the current chapter.

Reinharz, S. 1992. *Feminist methods in social research*. New York: Oxford University Press.

Chapter 8 on "Feminist Content Analysis" provides some insight into a more qualitative brand of content analysis than is emphasized in this chapter.

Weber, R. P. 1990. *Basic content analysis*. Newbury Park, CA: Sage.

A fine primer on content analysis.

EXERCISE 13.1

Of Caldecotts and Kings: Children's Books

This exercise gives you a chance to replicate the chapter's focal research, only with books that have won the Coretta Scott King award (between 1987 and 1991). The King award is given to black illustrators. Your research question: Are female characters more or less visible in books that have won the King award than in their Caldecott counterparts? This is an exercise that might require about 1 hour in your local public library. We recommend that you collect all your data there, and leave the books for other students to find and use.

Here is a list of King award winners and honor books from 1987 to 1991:

Bryan, Ashley. 1986. *Lion and the Ostrich Chicks*. New York: Atheneum.

Dragonwagon, Cresent. 1986. *Half a Moon and One Whole Star*. New York: Macmillan.

Cummings, Pat. 1986. *C.L.O.U.D.S.* New York: Lothrop, Lee & Shepard.

Greenfield, Eloise. 1988a. *Nathaniel Talking*. New York: Black Butterfly Children's Books.

Greenfield, Eloise. 1988b. *Under the Sunday Tree*. New York: Harper & Lee.

McKissack, Patricia. 1988. *Mirandy and Brother Wind*. New York: Knopf.

Price, Leontyne, 1990. *Aida*. San Diego: Gulliver.

Rohmer, Harriet. 1987. *The Invisible Hunters*. San Francisco: Children's Book Press.

San Souci, Robert. 1989. *The Talking Eggs*. New York: Dial Books for Young Readers.

Steptoe, John. 1987. *Mufaro's Beautiful Daughters*. New York: Lothrop, Lee & Shepard.

Stolz, Mary. 1988. *Storm in the Night*. New York: Harper & Row.

1. Prepare five code sheets like this:

   Book: Author: Publication date:

   a. Does the book have any female characters?

   b. Does the book have a central female character?

2. Go to your local public library and find five of the books listed. Using the code sheets you've prepared, collect data on each of the books.

3. Once you've collected the data, analyze them in the following ways:

   a. Calculate the percentage of books you've studied that have a female character.

   b. Calculate the percentage of books that have a central female character.

4. Compare your findings with those about late-1960s and late-1980s Caldecotts reported in Table 13.3 of the article by Clark, Lennon, and Morris. Interpret your comparison relative to the research question.

EXERCISE 13.2

How Newsworthy is the News?

This exercise offers a chance to observe, and speculate about, differences in the treatment of a given news story in daily newspapers. Your research question in this exercise is this: Does the treatment of a given news story vary from one newspaper to another and, if so, what might account for this variation?

1. Select a current national or international news story. What story did you choose?

2. Select four national, regional, or local newspapers (either from a newsstand or your library). Get copies of these papers on the same day. List the names of the newspapers and the date.

   Date:

   Newspaper 1:

   Newspaper 2:

   Newspaper 3:

   Newspaper 4:

3. Decide how you will measure the importance given a story in a newspaper. (Hint: You might consider one of the following options, or combinations thereof: What page does the story appear on? Where does the story appear on the page? How big is the typeface used in the headline? How many words or paragraphs are devoted to the story? Is it illustrated? What percentage of the total words or paragraphs in the paper are devoted to it? Does the newspaper devote an editorial to the story?) Describe the measure you will use to determine the relative importance given your story in each of the papers.

4. Report your findings. Which paper gave the story greatest play, by your measures? Which paper, the least? Interpret your findings by trying to account for differences in terms of characteristics of the papers you've examined.

EXERCISE 13.3

The Utopianism of TV Commercials

It has been said that commercials for children's TV programs are utopian—that they tap into utopian sentiments for something different, something better; that they express utopian values of energy, abundance, and community (see, for example, Ellen Seiter: 1993, 133)—rather than extolling the virtues of things as they are. But are commercials for children's programs really all that different in this regard from those for adult programs? This is your research question for this exercise.

1. Select a children's TV show and an adults' TV show on commercial stations. (Hint: Saturday mornings are particularly fruitful times to look for children's commercial TV programs.) Pick shows of at least 1/2 hour in duration (preferably one hour).

2. Indicate the shows you've chosen and the times you watched them.

   Children's Show:

   Time watched:

   Adults' Show:

   Time watched:

3. Describe the measures you will use to determine whether a commercial appealed to utopian sentiments or values. (For example, you might ask whether the commercial makes an explicit appeal to the viewer's desire for a better life [for instance, by showing an initially depressed-looking character fulfilled by the discovery of a soft drink or an initially bored breakfaster enlivened by the acquisition of a new cereal] or if it makes implicit appeals to, say, the viewer's wish for energy [is there dancing or singing associated with a product?], abundance [is anyone shown to be constrained by considerations of price?] or community [do folks seem to get along better after acquiring brand X?].) Show an example of the code sheet you will use, including a space for the product advertised in the commercial and a place for you to check if you observe the presence of what it is you're using to measure an appeal to utopian sentiments or values.

4. Use your coding scheme to analyze the first five commercials you see in both the children's and the adult's show.

5. Report your findings. In which kind of programming (for children or adults) did you find a higher percentage of commercials appealing to utopian sentiments or values? Interpret your findings.

EXERCISE 13.4

Gender Among the Newberys: An Internet Exercise

The American Library Association (ALA) awards the Caldecott medal, as well as the Coretta Scott King Award (Exercise 13.1) and the Newbery Award, for adolescent fiction. For this exercise, you'll want to use the ALA web site (www.ala.org) to answer questions about the gender of characters in Newbery winners and honor books over the last decade. (Hint: one way of finding such a list, after getting on the ALA site, is to select the "search Web Site" option, type "Newbery" at the request line, request information about one of the recent winners, and request the list of other winners and honor books at the bottom of that page.)

1. Find a list of Newbery award winners and honor books for the last 10 years.
2. Begin your content analysis of the titles of these books by listing the titles that suggest to you that there is a female character in the book.
3. Now list those books whose titles suggest to you that there is a male character in the book. (The two lists might overlap. For instance, the title *Belle Prater's Boy* suggests to us that there are at least two characters in the book, one male and one female. It should, therefore, make both your lists.)
4. Does your brief content analysis of recent Newbery award and honor books suggest that more attention has been paid to males or females?

14 Evaluation Research

One afternoon in May, John's[1] mother asked him what had been the best part of school that day. The sixth grader had a ready response—the class taught by Officer Jackson, the local DARE police officer. John wanted to become a police officer someday, so he had paid particular attention in each DARE class. By the end of the school year, John had learned a great deal about tobacco, alcohol, and drugs. He felt he knew what using them would do to his body and his mind. He was sure he'd be able to stay away from them, and he hoped his friends would do the same.

John takes part in an anti-drug program that receives the support of many school and police departments. Numerous local officials believe that there are many students like John—youngsters with very positive feelings about DARE and DARE officers. But, locally and nationally, questions have been raised about the long-term effect of programs like DARE. What impact does it have on students as they move from elementary to secondary schools? What will students like John say about drugs, alcohol, and tobacco five or six years from now? Will there be a difference between students who took part in DARE and those who did not?

More than a decade ago, Earl Wysong, Richard Aniskiewicz, and David Wright started with questions like these. They wanted to evaluate a drug education program and its long-term effects. In this chapter, we'll discuss the research strategies they selected to answer their questions and the political context of their work.

[1]John is a composite character, aggregated from the experiences of several students.

Introduction

By now we've covered most of the essential parts of social research. Starting with concepts and variables, we've explored a variety of topics: research questions and hypotheses, study designs and sampling strategies, methods of data collection, and ethical and practical research concerns. Before going on to discuss data analysis and sum up the research process, we want to discuss an especially useful kind of research, applied research, and in so doing, provide the details of an exemplary research project. Now that you're familiar with issues involved in the research process, we'll describe the specific set of research decisions made by one set of researchers to answer a specific research question.

Kinds of Research

All social research is designed to increase our understanding of human behavior, and all research can help society or specific groups and individuals. Although we've already described many of the ways that research projects differ from each other, we'll now discuss a final distinction. Some research is basic research, and other work is applied or evaluation research. We can distinguish between the kinds of research by examining the specific research questions they ask or hypotheses they pose.

basic research, research designed to add to our fundamental understanding and knowledge of the social world regardless of practical or immediate implications.

Basic research is designed to add to our knowledge and understanding of the social world *for the sake of knowledge and understanding*. Basic research can focus on anything within the realm of social reality, even if the understanding provided by the study has no immediate or practical use. Basic research can originate from a number of sources, including the desire to satisfy scientific curiosity, and can focus on any topic of interest. Most of the focal research projects we've presented in this text are basic research. The Adlers' work on children's after-school activities; the content analysis of children's books by Clark, Lennon, and Morris; the interviews of mothers in prison by Enos; the Gray, Palileo, and Johnson survey on rape victim blaming; Rockett and Smith's work on classifying deaths as suicide; and Weitz's study of people living with HIV are examples of basic research. They are research projects designed to provide us with additional understandings of our social world, although it is clear some practical applications can be derived from them.

applied research, research that is intended to be practical and useful in the immediate future.

evaluation research, research specifically designed to assess the impact of a specific program, policy, or legal change.

In contrast, **applied research** is work that is intended to be useful in the immediate future and to have specific practical implications. Schools, legislatures, government and social service agencies, health care institutions, corporations, and the like all have specific purposes and ways of "doing business." Applied research is designed to solve social problems in organizations like these and to provide information that is of immediate use to those participating in institutions or programs. One of the most common forms of applied social science is **evaluation research**, research that is designed to assess the impact of programs, policies, or legal changes. Often the focus of an evaluation is whether the program, policy, or law has succeeded in effecting intentional or planned change. An example of this kind of work is the Adler and Foster experiment (in Chapter 8) on the effect of the school reading project on caring.

Evaluation research is not a unique research *method*. Rather, it is research with a specific *purpose*. Rossi and Freeman (1993: 15) call evaluation research "a political and managerial activity, an input into the complex mosaic from which emerge policy decisions and allocations for the planning, design, implementation, and continuance of programs to better the human condition."

The Need for Applied Research

In the United States, programs directed toward solving social problems became increasingly common in the early twentieth century. The efforts before World War I focused on literacy, occupational training, and controlling infectious diseases (Rossi and Freeman, 1993: 9). Evaluations were conducted to determine if funding initiatives in these areas were worthwhile. Through World War II and into the 1960s, as federal human service projects in health, education, housing, income maintenance, and criminal justice became widespread, so too did the efforts to demonstrate their effectiveness.

By the end of the 1970s, many interested parties started questioning the continued expansion of government programs, partly because of changes in ideology, but also because of disenchantment with the outcomes and implementation of many of the programs advocated by public officials, planners, and politicians (Rossi and Freeman, 1993: 24). Despite disappointments with earlier efforts, the obvious need to deal with both old and newly identified social problems remained, and programs continued to be created and implemented. With millions of program dollars spent on local, state, and federal levels, we can see the unending need to make critical decisions about programs. Are they effective? Should they be continued as is? Modified in some way? Eliminated entirely? Because of limited funding resources and the necessity to make choices among programs, there is an ongoing need for applied research.

STOP AND THINK *What do you think are the most serious social problems that we face as a society? Poverty? Racism? AIDS? Teenage pregnancy? Homelessness? High crime rates? Are you aware of a program in your local community that is designed to work toward solving one of these problems? If you are familiar with such a program, do you think it is effective? Are those who are served by such programs involved in deciding if they are useful?*

Social Problems and Social Solutions

Unemployment, high crime rates, homelessness, illiteracy, child abuse, teen suicide, or the distribution of health care and its costs can be social issues that you believe warrant immediate attention. We sometimes learn of new programs and policies that deal with some of these problems from newspaper headlines ("New Heath Care Policy Developed," "Alternative Curriculum Implemented"). However, many policies and programs designed to deal with social problems receive little press. And we rarely learn about the *effectiveness* of programs and policies designed to address social problems. Even more rarely do we learn how those affected by the programs and policies feel about the situation and what their research concerns would be.

In the past two decades, one social concern that's received widespread public attention is the use of drugs by American youth. Over the years, public policies defined this as a critical area for action, and a variety of programs have been proposed to provide at least partial solutions. Drug Abuse Resistance Education (DARE) was one of the American anti-drug education programs developed in the mid-1980s. Implemented in thousands of communities, DARE reached millions of students in a matter of years. Central to its continued use is the *assumption* that it is effective in helping students resist pressures to use tobacco, alcohol, cocaine, heroin, and similar drugs. Considering the widespread implementation of DARE and the amount of money spent on this program, it is clearly important to test the assumption about its effectiveness.

STOP AND THINK *Did you take part in an anti-drug education program through your elementary or secondary school or through a religious or community youth group? If so, was it a DARE program? If not, what did your program involve?*

Sociologists on the faculty of Indiana University, Kokomo, were given the chance to ask and answer an important research question: Does DARE work? As in other studies, the social and political climate, chance factors, and the professional training and interests of the social scientists provided the impetus for an interesting and important project.

In the mid-1980s, the first police officer recruited and trained to teach the DARE curriculum in the Kokomo schools had been a student in one of Professor Richard Aniskiewicz's sociology courses. DARE had program evaluation as one of its funding requirements, so the officer recommended Professor Aniskiewicz to the local police and schools as a possible evaluator. Aniskiewicz recruited Earl Wysong, and in 1987, after receiving approval and some funding from the local school system, they began an evaluation of the local DARE program that lasted more than seven years.

One article that resulted from their work is the basis for the focal research selection that follows. Because it is evaluation rather than basic research, the story Wysong, Aniskiewicz, and Wright tell not only tries to answer a research question, but also considers the study's practical and political implications. The authors focus on the "real world" context of evaluation research—a context that can make evaluation research simultaneously interesting and frustrating. The decisions that Professors Wysong, Aniskiewicz and Wright faced, and the choices that they made tell us as much about the evaluation process as they do about the success of DARE.

FOCAL RESEARCH

Truth and DARE: Tracking Drug Education to Graduation and as Symbolic Politics[2]

by Earl Wysong, Richard Aniskiewicz, and David Wright

Introduction

The period following former President Reagan's 1986 "War on Drugs" campaign witnessed a rapid expansion of anti-drug programs at federal, state, and local levels with active support from corporations, the mass media, and community agencies (Jensen, Gerber, and Babcock, 1991; Males, 1992). Federal drug-related expenditures for law enforcement, treatment, and education rose from $2.6 billion in 1985 to nearly $13 billion in 1993 (U.S. Office of Management and Budget, 1990; 1992). To coordinate public and corporate policies aimed at reducing both the supply of and demand for drugs, a "National Drug Control Strategy" was initiated (National Drug Control Strategy, 1990). Schools quickly became the locus of such efforts and by 1990 more than 100 school-based drug education programs were being promoted (Miley, 1992: Pereira, 1992) with wide variations reported in their approaches, duration, costs, and reputed effectiveness (Bangert-Drowns, 1988; Pellow and Jengeleski, 1991).

Within the diverse array of drug prevention curricula in the United States, DARE (Drug Abuse Resistance Education) has become the largest and best known school-based drug education program (Rosenbaum, Flewelling, Bailey, Ringwalt, and Wilkinson, 1994). In 1990, DARE programs were in place in more than 3,000 communities in all 50 states and were reaching an estimated 20 million students (U.S. Congress: House, 1990). The DARE curriculum involves the presentation of anti-drug lessons by uniformed police officers who have taken 80 hours of specialized training (U.S. Department of Justice, 1991). With training costs of more than $2,000 per officer and program costs reaching as much as $90,600 per year for each full-time officer-instructor, DARE is expensive to implement and maintain (Miley, 1992; Pope, 1992). In 1993, total nationwide expenditures for DARE programs from all sources were estimated to be $700 million (Cauchon, 1993). The costs of adopting DARE are typically covered by federal, state, and foundation grants, by corporate support, and by local education and law enforcement agencies (U.S. Department of Justice, 1991). However, at this time no institutionalized funding base exists to sustain the program on a permanent basis (Congressional Record, 1992a).

[2] This report is published with permission as an abbreviated version of the article by Wysong, Aniskiewicz, and Wright in *Social Problems* 41 (August, 1994): 448–472.

Despite its high cost and the absence of laws or formal policies mandating its implementation, DARE has gained widespread acceptance and support. Popularity not withstanding, DARE's size, claims, and expense have generated some questions about its effectiveness. Short-term evaluations conducted within weeks or months following completion of the DARE program indicate it does produce "anti-drug" effects among students, especially in terms of creating and/or reinforcing anti-drug attitudes and drug-coping skills (Aniskiewicz and Wysong, 1990; Clayton, Cattarello, Day, and Walden, 1991; Harmon, 1993). However, claims for the program extend beyond the short-term. DARE is now being represented as a "long term solution" to the problem of drug use (U.S. Department of Justice, 1991: i), but studies of the program's long-term effectiveness are virtually nonexistent. Longitudinal studies are underway at DARE Regional Training Centers (U.S. Department of Justice, 1991) and DARE AMERICA, a nonprofit agency in Los Angeles, is now funding an eight-year study in six U.S. cities (DARE AMERICA, 1991). In addition to the issue of efficacy are the less frequently addressed, but equally important, questions concerning DARE's emergence as a potent political force and the implications of this for program evaluation.

Assessment Framework

We view the issues of DARE's efficacy and political potency as interrelated. The present study analyzes data collected from 1987 to 1992 as part of a multiyear DARE assessment in Kokomo, Indiana (a medium-sized midwestern city). We explore DARE's effects on high school seniors five years after their initial exposure to the program using quantitative measures to determine if students' exposure to DARE produces lasting effects related to the program's objectives.

As an extension of our previous work (Aniskiewicz and Wysong, 1990), we also utilize a macro-level process perspective to explore DARE's political potency, organizational support structure, and the implications of these factors for program evaluation and implementation. This approach views DARE as a socially constructed form of symbolic politics growing out of and embedded within the larger Drug War of the late 1980s and early 1990s. It calls attention to the role played by stakeholders in developing, legitimating, and expanding the program as well as the nature and extent of DARE's organizational support structure.

Project DARE

DARE originated in 1983 as a joint venture between the Los Angeles Police Department (LAPD) and the Los Angeles United School District (LAUSD). It is the best known of the "new generation" of drug prevention programs and uses a combination of factual/informational, cognitive/affective decision-making, and psychosocial life skills approaches. Uniformed officers teach a standardized anti-drug curriculum that attempts to shape students' attitudes and social skills to help them resist peer and media pressures to try drugs including tobacco and alcohol.[3] The DARE "core" curriculum is designed for fifth and sixth graders. It typically consists of 17 in-class weekly lessons, each approximately 45 to 60 minutes long, with active student involvement, including questions and answers, role playing, and group discussion (U.S. Department of Justice, 1991:7).

Kokomo was the first city in Indiana to implement the DARE program (Miley, 1992). During 1987–1988, all Kokomo fifth graders received the full DARE core curriculum, while all seventh graders received a shortened 11 week version of the curriculum. At that time, we conducted short-term program assessments for both grades using only the "DARE SCALE," an instrument developed in Los Angeles (Nyre, 1984; 1985), which supposedly provides a general measure of DARE's effectiveness in the areas of drug-related attitudes, knowledge, and anti-drug coping skills.[4] In a pre- and post-test design, we found significant increases in anti-drug attitudes among the fifth but not the seventh graders. While both groups are part of a larger study, the former seventh graders (as 1992 seniors) are used as the basis for evaluating DARE's long-term effectiveness in this inquiry.

Impact Evaluation: Objectives, Sample, Design

The impact dimension of the study uses data from a multipart questionnaire completed by Kokomo High School (KHS) seniors in 1991 (non-DARE group, no DARE exposure) and in 1992 (DARE group, exposed to DARE as 7th graders). Several comparisons are made to assess whether and the extent to which DARE exposure is associated five years later with DARE's primary objectives of preventing/reducing/delaying drug use, including measures of self-reported drug use rates and of drug attitudes, drug knowledge, and drug-resisting coping skills. If DARE exposure in the seventh grade produces lasting anti-drug effects, then measures of the DARE group's characteristics in these areas should be significantly different from the non-DARE group.

The questionnaire was completed by samples of 1991 and 1992 KHS seniors nearly evenly divided by gender. Subjects were chosen each year by distributing questionnaires to all students in a random selection of classes (for example, English, science, and so on) and asking them to participate on a voluntary basis. Asking students not to put their names on the survey and using ballot-box style procedures to ensure confidentiality and anonymity, responses were obtained for 331 1991 seniors (class size = 511) and 334 1992

[3]DARE's objectives regarding student drug use embrace the same "zero tolerance/no-drug/drug free" orientations that have guided national drug policies for the past several years. DARE makes no distinctions between legal and illegal drugs and advocates total abstinence as the only acceptable approach to all types and categories of drugs. The program appears to assume that drug experimentation (with any drug at any level) equals drug abuse or will inevitably lead to it. DARE's objectives are not only unrealistic, but possibly counter-productive because they are obviously unattainable.

[4]See the appendix to this article for a more complete description of the scale and examples of questions.

seniors (class size = 474). Of the later group, 288 were identified as having completed DARE as seventh graders. This sub-population is referred to as the 1992 "DARE group" in the comparisons developed throughout the study.

The questionnaire included both the Drug Use Scale and the DARE Scale.[5] The Drug Use Scale[6] is used to compare the 1991 and 1992 KHS senior groups on five measures of drug use:

1. *Lifetime Prevalence.* Percentages of students reporting use (any amount) of each type/category of drug during their entire lives.
2. *Recency of Use (Past Year).* Percentages of students reporting use (any amount) of each type/category during the past 12 months.
3. *Recency of Use (Past month).* Percentages of students reporting use (any amount) of each type/category during the past 30 days.
4. *Grade Level at First Drug Use.* The mean grade at first use for each type/category of drug.
5. *Frequency of Use.* Percentages of students reporting drug use by number of occasions for three time periods—lifetime, past year, and past 30 days.

The DARE Scale consists of 19 questions with five response categories for each item (strongly agree to strongly disagree). Scores were computed by calculating the percentage of students responding with "appropriate" answers (disagree or strongly disagree) to each item along with the overall mean percentage of "appropriate" responses for all 19 items. According to this coding and scoring approach, the higher the individual item percentages and overall mean score, the more the results are supposedly indicative of anti-drug attitudes, greater drug knowledge, and enhanced drug-resistant coping skills.

Results: Impact Evaluation

Drug Use Scale

Comparisons of data from the Drug Use Scale for 1991 non-DARE and 1992 DARE seniors reveal that self-reported drug use rates among both groups are very similar for (1) Lifetime Prevalence (see Figure 14.1), (2) Recency of Use (Figure 14.2), (3) Grade Level at First Drug Use (Figure 14.3) and (4) Frequency of Use (data not shown). For some drugs and periods, drug use rates are higher for the 1992 DARE group than for the 1991 non-DARE group. However, for other drugs and periods, the rates are reversed. Figure 14.3 illustrates the similarities between the two groups regarding Mean

[5]See the appendix of this article for a more complete description of the instrument and examples of questions from both scales.

[6]The Drug Use Scale categories of hallucinogens, cocaine, sedatives/tranquilizers, and narcotics are used to report use rates among KHS seniors for combinations of specific drugs reported separately in the national senior survey. For example, our questions on hallucinogens asked students to respond within that category if they had used any related drugs (such as LSD, PCP, and so on). By contrast, the national senior survey instrument from which this scale was adapted asks for use rates on each specific drug (Johnston, 1992).

FIGURE 14.1

Drug Use Scale: Lifetime Prevalence

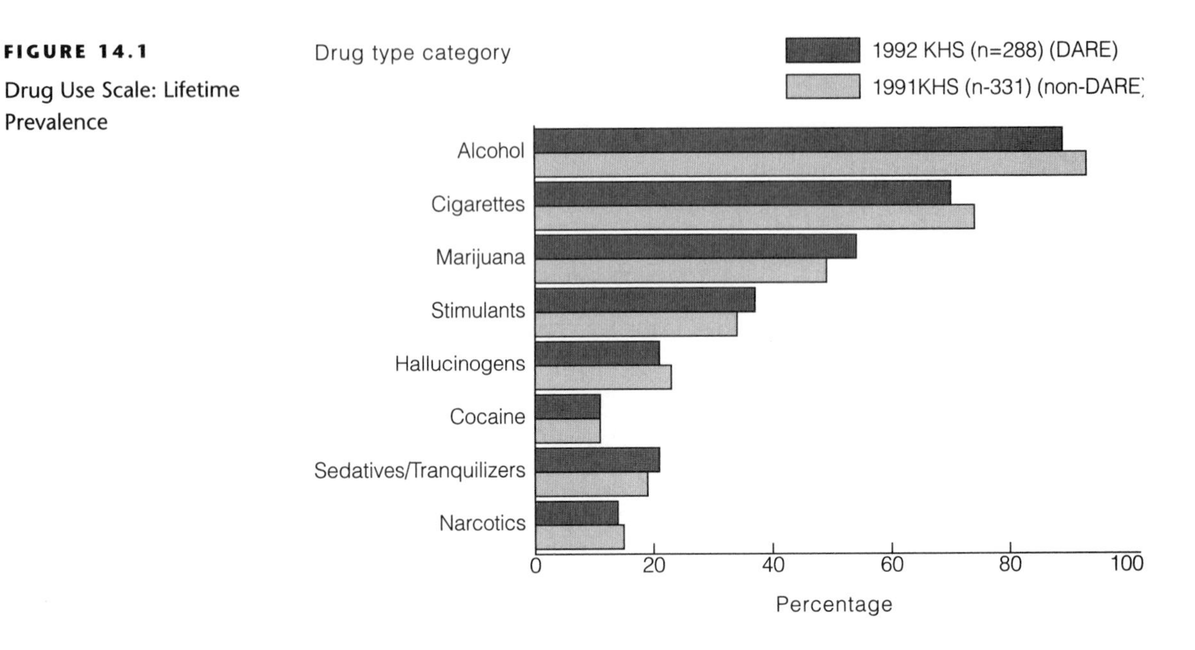

Grade Level at First Drug Use.[7] A separate set of eight t-tests revealed no significant differences between the two groups on this measure for each drug type/category.

Prevalence and recency data were further analyzed using chi-square tests comparing drug use among DARE and non-DARE seniors for all drug types/categories, time periods, and use levels. While too lengthy to report in detail, the results reveal no significant differences in drug use rates or patterns between the two groups in each area (prevalence, recency, frequency) for each drug type/category—with two exceptions. The results were significant for comparisons of the 1991 and 1992 groups' hallucinogen use over the last 30 days in terms of use/no use categories (sig. = .0004) and levels of use (sig. = .02) In both instances *higher* use rates were recorded for the 1992 DARE group.

DARE Scale

Table 14.1 summarizes the DARE Scale results and reinforces the basic findings of the previous section: DARE exposure appears to produce no significant long-term effects in areas related to the program's primary objectives. As Table 14.1 indicates, the mean percentages of "appropriate" responses for all 19 scale items were similar for all three groups with no statistically significant differences apparent in the groups' mean scores.[8]

[7]The "numbers reporting first use" sub-figure illustrates that aside from alcohol, the mean grade level figures reflect drug use among relatively small groups of students compared to the total samples.

[8]While too lengthy to report here, in a separate analysis of "appropriate" mean response percentages for each DARE scale question by each of the three groups identified in Table 14.1, we found similar results for most of the 19 items.

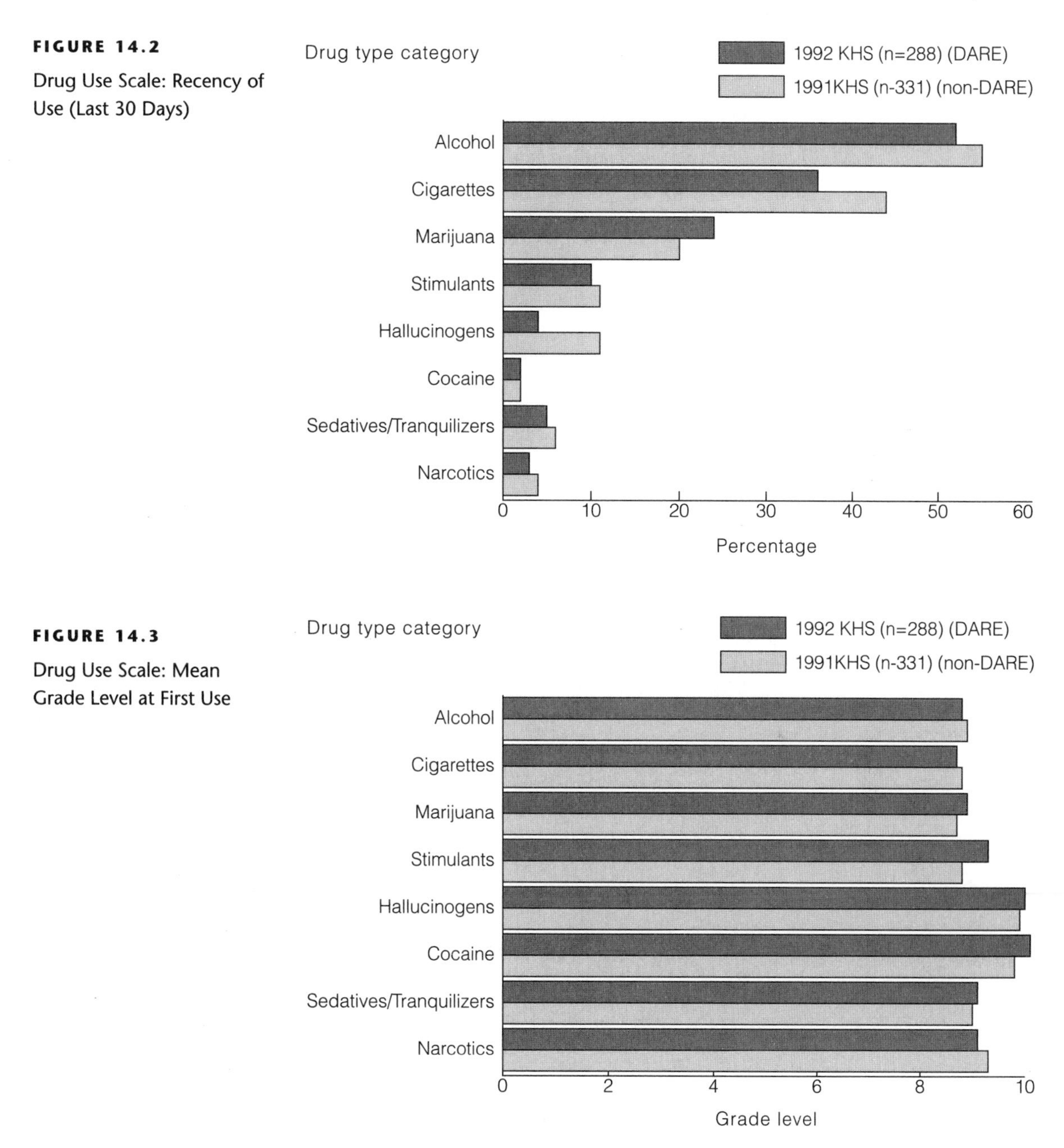

FIGURE 14.2

Drug Use Scale: Recency of Use (Last 30 Days)

FIGURE 14.3

Drug Use Scale: Mean Grade Level at First Use

Number Reporting First Use by Drug Type/Category

	Alcohol	Cigarettes	Marijuana	Stimulants	Hallucinogens	Cocaine	Sedatives/ Tranquilizers	Narcotics
1991 KHS	279	143	175	88	71	36	63	41
1992 KHS	259	137	142	75	64	61	54	29

Notes: The Alcohol category includes those who tried more than a few sips; cigarettes includes those who smoked on a daily basis; and the other categories include those who first tried (any level) the substances. See footnote 7 in text.

TABLE 14.1 **DARE Scale Results: ANOVA of "Appropriate" Response Means (1991 Seniors, 1992 Seniors, and 1992 Seniors as 7th Graders)**

DARE SCALE	*1991 KHS (non-DARE Group)*	*1992 KHS (DARE Group)*	*1992 KHS (as 7th Graders)*	*F-value*	*Sig. F*
"Appropriate" Response Means	82	82	79		
St. Dev:	12.86	12.21	11.10	.313	.732*
Cases:	n = 331	n = 288	n = 596		

Note: * = nonsignificant

Despite the similarities, there are some interesting differences on individual items. For example, there was a sharp decline in positive attitudes toward the police and a growing unwillingness to condemn peer's consumption of alcohol. However, on the other items the "Appropriate" response results were higher for the 1992 DARE seniors than the results recorded as seventh graders. Despite these potentially confusing trends, the DARE and non-DARE senior groups had virtually identical response percentages for all 19 items, indicating that the differences between student responses as seventh graders and later as seniors were not due to DARE exposure.

Finally, the comparisons of results of the DARE Scale with those from the Drug Use Scale call into question the utility of the DARE Scale as a meaningful predictor of drug use. For example, despite the fact that high percentages of both senior groups gave "Appropriate" responses for most items on the scale, substantial majorities in these groups reported using various drugs (especially alcohol and cigarettes). These results are consistent with a number of findings in the drug education literature showing that while anti-drug information is easily imparted, producing changes in drug-related attitudes and behavior is much more difficult and problematic (Bangert-Drowns, 1988; Pellow and Jengeleski, 1991). Moreover, the results suggest that rather than measuring any meaningful attitudinal or behavioral changes, the DARE Scale simply assesses the extent to which DARE encourages students to uncritically recall and repeat information. In this sense, the scale appears to be a very self-serving measure which by the design of its questions and coding procedures elicits the *appearance* of positive results thereby overstating the efficacy of the DARE program.

Micro-Level Process

A micro-level process evaluation of a focus group interview with six KHS seniors, evenly divided by gender and recruited from informal community contacts, supported the quantitative findings. The first part of the interview concentrated on recollections of their DARE experiences as seventh graders; the second part addressed student perceptions of DARE's long-term impact. DARE was judged by all group members as having had no lasting influence on their peers' drug-related attitudes or behaviors, but they did not see it as a waste of time *in the lower grades.* Students thought that DARE had a short-term anti-drug effect and could be seen as giving students the message that adults cared about them.

Symbolic Politics and Organizational Support for DARE

Programs are never implemented in a vacuum nor under ideal conditions (Weisheit, 1983). Our approach views social problems and related public policies as linked to political, social, and cultural processes (Gamson, Croteau, Hoynes, and Sasson, 1992). In the case of DARE, this orientation directs us to explore (1) the links between the media and the politically constructed Drug War and DARE's emergence as a form of symbolic politics, (2) the role of various stakeholders in promoting, legitimating, and expanding the program, and (3) the program evaluation implications of DARE's evolution into a major political force.

Edelman's (1964) concepts of "symbolic politics/action" are useful in understanding the links between the Drug War and DARE's emergence as a popular programmatic policy response. They focus attention on the role and interests of political elites and the media in constructing the Drug War and in promoting policies to deal with it. They alert us to the importance of the public reassurance features of ameliorative programs in generating political and public support (which might be more important than their actual substantive effects). The reassurance value of such programs is linked to the extent to which the programs are grounded in widely respected and legitimate institutions and cultural traditions. Programs imbued with potent symbolic qualities (such as DARE's links to schools and police) are virtually assured of widespread public acceptance (regardless of actual effectiveness) which in turn advances the interest of political leaders who benefit from being associated with highly visible, popular symbolic programs.

In the mid-1980s illicit drug use was defined as a "drug crisis" and a major social problem by the mass media and national political leaders (Jensen, Gerber, and Babcock, 1991). President Reagan officially declared a national War on Drugs (Boyd, 1986), and as the 1980s ended, the Bush administration extended and systematized the Drug War in the "National Drug Control Strategy."

Public attitudes became more supportive of anti-drug programs and political interest in drug education increased. With a positive track record in Los Angeles and its powerful symbolic affirmation of traditional values and institutions, DARE was well positioned to take advantage of this emerging context by offering a convenient individual-level programmatic "solution" to the drug threat. Since expanding the DARE program offered public reassurance in the face of the constructed drug threat, DARE can be seen as a clear example of symbolic politics and action.

As political support for DARE increased, the program shifted from being a local success story to being a national symbol in the war on drugs. This resulted in mutual support and reinforcement among direct stakeholders (such as staff and directly supportive organizations) and *indirect stakeholders* (such as political supporters). Both groups benefited from the reflected approval, support, and legitimacy associated with a program linking a popular cause to traditional authority structures.

The stakeholders collaborated to embed the program within a complex organizational support structure to ensure its continued existence and growth. One important feature of this structure is DARE AMERICA, a nonprofit corporation with a large annual budget ($1.3 million in 1990) (Pope, 1992). Another dimension of the support structure is an interrelated network of organizational ties linking DARE programs to federal, state, and local agencies (such as the U.S. Department of Justice [U.S. Department of Justice, 1991], the National Institute on Drug Abuse [Clayton et al., 1991], the U.S. Department of Education [Rogers, 1990]), and to private corporations, including McDonald's, Security Pacific National Bank, and Kentucky Fried Chicken (Pope, 1992).

Implications for Program Evaluation

DARE's existence as a potent form of symbolic politics has important implications for evaluation. It increases the prospects for tension as negative evaluations threaten the interests of powerful stakeholders. Furthermore, it underscores the importance of framing results in a process perspective offering alternative interpretations for observed outcomes.

While our negative results alone could create the potential for conflict with DARE stakeholders, this prospect is magnified by recent criticisms (Pereira, 1992) and concerns about the expense of the program (Pope, 1992). Our results could make the case for cutting the program's budget, constituting a potential threat to stakeholders and eliciting their wrath. This does not mean evaluators should anticipate stakeholder pressures by discounting the validity or importance of their results. However, when the issue under consideration is important and the stakes are high, program evaluators must be the first to call attention to the complex and inexact nature of the evaluation process. Our obligation is to provide the DARE program and its supporters not only with our findings, but also with an interpretation of factors that may have influenced our results.

Alternative interpretations of our "no effects" findings essentially cluster around three main issues. The first is the influence of contextual factors. While the National Drug Control Strategy was built around the complementary goals of reducing both the demand for and the supply of drugs, the exposure of the local seventh graders to DARE occurred before the implementation of several important facets of the strategy, such as the Partnership for Drug-Free America and the passage of major federal anti-drug laws. Therefore, our DARE group did not get the reinforcing effects of coordinated mass media campaigns, community programs, and federally mandated anti-drug policies until after their initial exposure to DARE.

The second issue is the disjunction between key DARE program assumptions/ideals and the manner in which the program was implemented. DARE views drug use as an individual decision and choice that is heavily dependent upon knowledge, attitudes, and social skills. Consequently, DARE assumes it is possible to intervene in these areas via systematically organized anti-drug lessons that will produce long-lasting attitudinal and behavioral effects. DARE implementation includes the ideas that anti-drug effects are maximized by early intervention and the uniform presentation of a standardized curriculum.

The DARE implementation we evaluated differs from the assumptions in several ways. The group was exposed to the program two years later than recommended, they did not receive the full 17 week program, and uniform presentation and "curriculum fidelity" did not occur in at least some sessions for some students. While these disjunctions could be a threat to DARE's long-term effectiveness, they may also be trivial and perhaps inconsequential factors.

The third set of issues to consider in conjunction with our findings involves the program design limitations related to areas not explicitly considered by the DARE curriculum. DARE's mandatory nature in participating schools means that resistant students will have DARE imposed upon them. Other studies confirm that "when socially deviant youths are required to participate in the school setting in peer-led denunciation of activities they value, they are more likely to become alienated than converted" (Baumrind, 1987:32). "Classroom climate" (including pedagogical styles and classroom teachers' level of enthusiasm for DARE) can also be influential in conditioning student receptiveness to DARE (Aniskiewicz and Wysong, 1990). DARE's implicit embrace of "free will/user accountability" principles means limiting the scope and emphasis to the context within which students' choice are made and ignoring important structural factors such as student' socioeconomic background, family influences, and employment prospects.

Discussion

After tracking DARE for five years, our data point in the direction of no long-term program effects for preventing or reducing adolescent drug use. Despite the massive scope of the Drug War, the existence of powerful DARE stakeholders and the potential controversies that our findings might generate, we believe our evaluation of DARE represents a fair and accurate assessment of the program's long-term effects *for the group we studied.* We believe our results must be viewed as suggestive rather than definitive. We explicitly recognize the limitations of our findings given the local nature of our sample and the complexities of conducting longitudinal evaluation on a popular program addressing adolescent drug use. However, because our efforts have gone far beyond a one-dimension impact evaluation approach, we believe the study deserves widespread attention.

Our work should help stimulate a reconsideration of several issues related to the etiology of drug use among young adults. For example, should drug prevention efforts be expanded beyond those programs aimed at changing attitudes or improving social skills through sophisticated persuasive communication programs? We need a more complete consideration of the appropriate balance between approaches focusing primarily on the psychosocial dimension of drug use and those addressing cultural and structural factors related to it (for example, those that work towards reducing inequalities, alienation, and social isolation among adolescents). To the extent that our evaluation helps to stimulate a more multifaceted discourse on the nature and merits of various psychological and structural drug prevention policies and programs, then DARE, regardless of its long-term effects, will have made an important contribution.

REFERENCES

Aniskiewicz, R., and E. Wysong. 1990. Evaluating DARE: Drug education and the multiple meanings of success. *Policy Studies Review* 9: 727–747.

Bangert-Drowns, R. L. 1988. The effects of school based substance abuse education: A meta-analysis. *Journal of Drug Education* 18: 243–264.

Baumrind, D. 1987. Familial antecedents of adolescent drug use: A developmental perspective. In *Etiology of drug abuse: Implications for prevention,* eds. Coryl L. Jones and Robert J. Battjes, 13–44. Rockville, MD: National Institute for Drug Abuse.

Boyd, G. M. 1986. Reagan's advocate 'crusade' on drugs *New York Times* September 15: A1, B10.

Cauchon, D. 1993. Studies find drug program not effective. *USA Today,* October 11: 1–2.

Clayton, R. R., A. Cattarello, L. E. Day, and K. P. Walden. 1991. Persuasive communication and drug prevention: An evaluation of the DARE Program. In *Persuasive communication and drug abuse prevention,* eds. L. Donohew, H. I. Sypher, and W. J. Bukowski, 295–313. Hillsdale, NJ: Lawrence Erlbaum.

Congressional Record. 1992. Senate Joint Resolution 295. Joint Resolution designating September 10, 1992, as National D.A.R.E. Day. April 30: S5915-S5916.

D.A.R.E. AMERICA. 1991. *Request for proposal.* D.A.R.E. America, Inc. 606 South Olive, Suite 1206, Los Angeles, California.

DeJong, W. 1987. A short term evaluation of project DARE (Drug Abuse Resistance Education): Preliminary indications of effectiveness. *Journal of Drug Education* 17: 279–294.

Edelman, M. 1964. *The Symbolic Uses of Politics.* Chicago: University of Illinois Press.

Gamson, W., D. Croteau, W. Hoynes, and T. Sasson. 1992. Media images and the social construction of reality. In *Annual Review of Sociology.* Eds. J. Blake and J. Hagan, 373–393. Palo Alto, Calif.: Annual Reviews.

Harmon, M. A. 1993. Reducing the risk of drug involvement among early adolescents. *Evaluation Review* 17: 221–239.

Jensen, E. J., J. Gerber, and G. M. Babcock. 1991. The new war on drugs: Grass roots movement or political construction? *Journal of Drug Issues* 21: 651–667.

Johnston, L. D., P. M. O'Malley, and J. G. Bachman. 1992. *Smoking, Drinking, and Illicit Drug Use Among American Secondary School Students, College Students and Young Adults, 1975–1991*. Volume 1. Rockville Md.: National Institute on Drug Abuse.

Males, M. 1992. Drug deaths rise as the war continues. *In these Times,* May 20–26: 17.

Miley, S. L. 1992. DARE beats drugs; truth is, funding's in a 'crisis'. *Indianapolis Star,* May 18: D-1.

National Drug Control Strategy. 1990. *The White House, January 1990.* Washington DC: U.S. Government Printing Office.

National Drug Control Strategy. 1992. *The White House, January 1992.* Washington DC: U.S. Government Printing Office.

Nyre, G. F. 1984. An evaluation of project DARE (Drug Abuse Resistance Education). Los Angeles. Evaluation and Training Institute.

Nyre, G. F. 1985. Final evaluation report, 1984–1985: Project DARE (Drug Abuse Resistance Education). Los Angeles. Evaluation and Training Institute.

Pellow, R. A., and J. L. Jengeleski. 1991. A survey of current research studies on drug education programs in America. *Journal of Drug Education* 21: 203–210.

Pereira, J. 1992. The informants: In a drug program, some kids turn in their own parents. *Wall Street Journal,* April 20: 1, A4.

Pope, L. 1992. DARE-ing program at risk. *Los Angeles Daily News,* September 20: G-6, G-7.

Rogers, E. M. 1990. Cops, kids, and drugs: Organizational factors in the spontaneous diffusion of Project D.A.R.E. Paper presented at the Conference on Organizational Factors in Drug Abuse Prevention Campaigns. Bethesda, Maryland.

Rosenbaum, D. P., R. L. Flewelling, S. L. Bailey, C. L. Ringwalt, and D. L. Wilkinson. 1994. Cops in the classroom: A longitudinal evaluation of Drug Abuse Resistance Education (DARE). *Journal of Research in Crime and Delinquency* 31: 3–31.

U.S. Congress: House of Representatives. 1990. Oversight hearing on drug abuse education programs. Committee on Education and Labor. Subcommittee on Elementary, Secondary and Vocational Education. 101st Congress. 1st session. Serial No. 101–129. Washington DC Government Printing Office.

U.S. Department of Justice Bureau of Justice Assistance. 1991. *PROGRAM BRIEF: An introduction to DARE* (2nd edition). Washington DC: Bureau of Justice Assistance.

U.S. Office of Management and Budget. 1990. *Budget of the U.S. government, Fiscal Year 1991.* Washington DC: U.S. Government Printing Office.

U.S. Office of Management and Budget. 1992. *Budget of the U.S. government, Fiscal Year 1993.* Washington DC: U.S. Government Printing Office.

Weisheit, R. A. 1983. The social context of alcohol and drug education: Implications for program evaluations. *Journal of Alcohol and Drug Education* 29: 72–81.

Appendix: The Questionnaire

Students were given a 15-page, voluntary, and anonymous questionnaire. The instructions stated that they would be asked about their opinions on a number of topics, and that they should view their answers as "votes" on a wide range of important issues. The specified interest was in group results, not individual scores and students were assured that there were no right or wrong answers.

The DARE Scale consisted on 19 questions with five answer categories. The following are examples of the questions:

1. There is nothing wrong with smoking cigarettes as long as you do not smoke too many.

5	4	3	2	1
strongly disagree	disagree	unsure	agree	strongly agree

2. Students who drink alcohol are more grown up.

5	4	3	2	1
strongly disagree	disagree	unsure	agree	strongly agree

3. It is safe to take medicine that a doctor orders for another person.

5	4	3	2	1
strongly disagree	disagree	unsure	agree	strongly agree

4. Police officers would rather catch you doing something wrong than try to help you.

5	4	3	2	1
strongly disagree	disagree	unsure	agree	strongly agree

5. If your best friend offers you a drug, you have to take it.

5	4	3	2	1
strongly disagree	disagree	unsure	agree	strongly agree

6. It is okay to drink alcohol or use drugs at a party if everyone else is.

5	4	3	2	1
strongly disagree	disagree	unsure	agree	strongly agree

The Drug Use Scale consisted of 12 multipart questions. Questions focused on tobacco, alcohol, marijuana, hashish, psychedelics, cocaine, crack, tranquilizers, barbiturates, amphetamines, and narcotics (heroin, morphine, and the like). The following are some examples of questions:

1. Have you ever smoked cigarettes?

____ 1. Never—**Go to Question 2.**
____ 2. Once or twice
____ 3. Occasionally, but not regularly
____ 4. Regularly in the past
____ 5. Regularly now

1A. How frequently have you smoked cigarettes during the past 30 days?

____ 1. Not at all
____ 2. Less than 1 cigarette per day
____ 3. 1 to 5 cigarettes per day
____ 4. About ½ pack per day
____ 5. About 1 pack per day
____ 6. About 1½ packs per day
____ 7. 2 or more packs per day

2. Have you ever used smokeless tobacco?

____ 1. Never—**Go to Question 3.**
____ 2. Once or twice
____ 3. Occasionally, but not regularly
____ 4. Regularly in the past
____ 5. Regularly now

3. Have you ever had any beer, wine, wine coolers, or liquor to drink?

____ 1. No—**Go to Question 4.**
____ 2. Yes

4. On how many occasions have you had alcoholic beverages to drink?

4a. in your lifetime?	**4b. during the last 12 months**	**4c. during the last 30 days?**
____ Zero occasions	____ Zero occasions	____ Zero occasions
____ 1–2 occasions	____ 1–2 occasions	____ 1–2 occasions
____ 3–5 occasions	____ 3–5 occasions	____ 3–5 occasions
____ 6–9 occasions	____ 6–9 occasions	____ 6–9 occasions
____ 10–19 occasions	____ 10–19 occasions	____ 10–19 occasions
____ 20–39 occasions	____ 20–39 occasions	____ 20–39 occasions
____ 40 or more	____ 40 or more	____ 40 or more

5. On how many occasions (if any) have you used marijuana (grass, pot) or hashish (hash, hash oil)?

5a. in your lifetime?	**5b. during the last 12 months**	**5c. during the last 30 days?**
____ Zero occasions	____ Zero occasions	____ Zero occasions
____ 1–2 occasions	____ 1–2 occasions	____ 1–2 occasions
____ 3–5 occasions	____ 3–5 occasions	____ 3–5 occasions
____ 6–9 occasions	____ 6–9 occasions	____ 6–9 occasions
____ 10–19 occasions	____ 10–19 occasions	____ 10–19 occasions
____ 20–39 occasions	____ 20–39 occasions	____ 20–39 occasions
____ 40 or more	____ 40 or more	____ 40 or more

6. On how many occasions (if any) have you used LSD ("acid") or other psychedelics (like mescaline, peyote, psilocybin, or PCP)?

6a. in your lifetime?	**6b. during the last 12 months**	**6c. during the last 30 days?**
____ Zero occasions	____ Zero occasions	____ Zero occasions
____ 1–2 occasions	____ 1–2 occasions	____ 1–2 occasions
____ 3–5 occasions	____ 3–5 occasions	____ 3–5 occasions
____ 6–9 occasions	____ 6–9 occasions	____ 6–9 occasions
____ 10–19 occasions	____ 10–19 occasions	____ 10–19 occasions
____ 20–39 occasions	____ 20–39 occasions	____ 20–39 occasions
____ 40 or more	____ 40 or more	____ 40 or more

7. Amphetamines are sometimes prescribed by doctors to help people lose weight or to give people more energy. They are sometimes call uppers, speed, bennies, dexies, pep pills, and diet pills. On how many occasions (if any) have you taken amphetamines on your own—that is, without a doctor telling you to take them?

7a. in your lifetime?	**7b. during the last 12 months**	**7c. during the last 30 days?**
___ Zero occasions	___ Zero occasions	___ Zero occasions
___ 1–2 occasions	___ 1–2 occasions	___ 1–2 occasions
___ 3–5 occasions	___ 3–5 occasions	___ 3–5 occasions
___ 6–9 occasions	___ 6–9 occasions	___ 6–9 occasions
___ 10–19 occasions	___ 10–19 occasions	___ 10–19 occasions
___ 20–39 occasions	___ 20–39 occasions	___ 20–39 occasions
___ 40 or more	___ 40 or more	___ 40 or more

8. How difficult do you think it would be for you to get each of the following types of drugs, if you wanted some? (please circle your answer for each question)

8a. cigarettes?	1 probably impossible	2 very difficult	3 fairly difficult	4 fairly easy	5 very easy
8b. alcohol (beer, wine, liquor)?	1 probably impossible	2 very difficult	3 fairly difficult	4 fairly easy	5 very easy
8c. cocaine or "crack"?	1 probably impossible	2 very difficult	3 fairly difficult	4 fairly easy	5 very easy

Types of Evaluation and Research Questions

Outcome Evaluation

outcome evaluation, research that is designed to "sum up" the effects of program, policy, or law in accomplishing the goal or intent of the program, policy, or law.

Wysong, Aniskiewicz, and Wright's work is a good example of an outcome evaluation, the most common kind of evaluation research. **Outcome evaluation,** also called impact or summative analysis, seeks to estimate the effects of a treatment, program, law, or policy and thereby determine its utility.

As in the research by Wysong and his colleagues, outcome evaluations typically begin with the research question, "Does the program accomplish its goals?" The study is designed to test the hypothesis that the independent variable (the program) has a positive effect on the dependent variable (the program's objectives). The researcher first determines *what* the goal of the program is and *how* the program was implemented, then conducts a study of use to practitioners and policymakers who need definitive answers to specific questions. At the end of outcome evaluation, the researcher should be able to answer questions about the program's success and to speculate on how to improve it.

Wysong and colleagues evaluated one local version of a large national program, but evaluation research is not limited to this kind of program. Programs can vary in the number of people they serve, their geographic target area, and the extensiveness of their goals. Evaluations can be done when programs serve only a few or a few million people, when they last for days, weeks, or years, and whether they have broad or narrow goals. Examples of the many hundreds of research questions that have guided outcome evaluations in the past decade include these: What is the impact of high school restructuring on student achievement? (Lee and Smith, 1995); does retraining help workers who are dismissed due to work-site closing or significant downsizings find new jobs or receive comparable pay? (Koppel and Hoffman, 1996); do family-based intervention programs designed to prevent the removal of abused and neglected children from their families actually work? (Schuerman, Rzepnicki, and Littell, 1994). One big question for the next decade is this: What is the effect of welfare reform? As of 1997, there were nearly 100 studies designed to analyze the many different versions of welfare reform that states have recently implemented. Together the studies of welfare reform will receive more than $120 million in private and government funding (Jacobs, 1997).

Cost-Benefit Analysis

In the focal research, Wysong and his colleagues focus on whether DARE accomplished its goals and, in so doing, imply that it might be better to expend the resources consumed by the program in other ways. Evaluations can be more explicitly a **cost-benefit analysis**, which is a study designed to weigh all expenses of a program (its costs) against the monetary estimates of the program's benefits (even putting dollar values on intangible benefits), or a **cost-effectiveness analysis**, which estimates the approach that will deliver a desired benefit most effectively (at the lowest cost) without considering the outcome in economic terms. Comparing program costs and benefits can be helpful to policy makers, funding agencies, and program administrators in allocating resources. With large public expenditures on health care, correctional facilities, social welfare programs, and education in the United States, saving even a small percentage of costs is significant.

cost-benefit analysis, research that compares a program's costs to its benefits.

cost-effectiveness analysis, comparisons of program costs in delivering desired benefits based on the assumption that the outcome is desirable.

Typical questions asked in cost-benefit and effectiveness analyses are these: How effective is the program? How expensive is it? Is it worth doing? How does this program compare with alternative programs? These analyses can compare outcomes for program participants with those not in programs or with those in alternative programs.

Collection and interpretation of data is often a complex task when evaluating social programs. For example, a cost-benefit analysis of a drug treatment program, will have both direct expenditures (such as electricity, rent, and salaries) and indirect costs (including transportation expenses, donated time and equipment, and lost work time) (French, 1995). Moreover, the variety of direct benefits of effective drug abuse treatment include abstinence or reduced drug use and lower health care services. Indirect benefits are more difficult to measure because they include quality-of-life improvements, such as increases in productive activities (employment and earnings), improved living conditions, reduced criminal activity, and improvement in the health status of drug users and their future children (French, 1995: 113).

Evaluations for Other Purposes

stakeholders, people or groups that participate in or are affected by a program or its evaluation, such as funding agencies, policymakers, sponsors, program staff, and program participants.

needs assessment, an analysis of whether a problem exists, its severity, and an estimate of essential services.

The tasks an evaluation sets out to accomplish are determined by the stage of the program (whether it is a new, innovative program or an ongoing one) and the needs and interests of the **stakeholders** (those involved in or affected by the program in some way). In some instances, before a program or project is designed, a needs assessment is conducted to determine the need for various forms of service. A **needs assessment** evaluates the existence of a problem, its severity and the number of people affected. It can also include an analysis of the services or programs that are essential and estimate the required duration. For example, if school administrators, elected officials, or parents were concerned about math or reading skills of middle school students, they could initiate a needs assessment. Data would be collected to document the current skills of students. If the analyses and subsequent decision-making process were supportive of a programmatic solution, then a new curriculum might be designed.

formative analysis, evaluation research focused on the design or early implementation stages of a program or policy.

If a new program is funded and there is time in the early stages of design or implementation to make improvements, program staff and developers can benefit from a **formative analysis**. This kind of analysis occurs when an evaluator involved in the beginning stages of program development and implementation provides program staff with an evaluation of the current program and suggestions for improving the program design and service delivery. Suppose, for example, a high school principal asked for assistance with a new AIDS education program. A formative analysis would carefully review the program's goals and its instructional materials (including comparisons of the program to guidelines from state agencies and with programs and materials used in other school systems) and collect data on the ongoing program. Administrators, teachers, and students might be interviewed, teacher training and program sessions might be observed, and the first set of students taking the course might be asked to complete a questionnaire about it or to participate in focus groups. After analysis of the data, the evaluators would be able to offer advice, including suggestions for modifications of instructional materials or teaching techniques or for adjustments in program management or staff development.

process evaluation, research that monitors a program or policy to determine if it is implemented as designed.

A related kind of evaluation is **process evaluation** or an implementation study, which is research to determine if a program was implemented as designed. Variations in program delivery occur because of differences in program deliverers, recipients, sites, and changes over time (Sheirer, 1994: 42). It's especially important for the program staff responsible for delivering subsequent versions of a program to know exactly how the initial version was implemented. For instance, if a state implemented a new victim assistance program, a process evaluation could ascertain the actual staffing patterns, administrative arrangements, and primary program activities. The evaluators would also be able to determine the actual cost of the program, staff responsibilities, typical services provided to the victims of crime, number of participants, and to what extent the program differed from site to site. Monitoring a program's quality allows evaluators to see whether a program included any unplanned aspects or substandard procedures. In process evaluation, the goal is not to see whether there is an effect or how big the effect is but, rather, to try to pinpoint how something works and what variations in the program contributed to whatever effect there was.

Designing an Evaluation

Selecting research strategies, including choices in study design, measurement techniques, sample selection, and data analysis, depends on several aspects of the evaluation. The study's intended audience, the kinds of information needed, the resources available for the project, and the evaluation project's time frame will each affect the selection of research strategies.

Who Is the Study for?

Typically, many parties will have an interest in an evaluation, including legislators, client groups, interest groups, and the general public. An all-purpose evaluation is rarely possible, so each study should be designed to address the information needs of a primary client (Rutman, 1984: 16). Sometimes it is possible to have more than one client, but the information needed by the various constituencies will not always be in agreement. In our focal research on DARE, Wysong and associates were not paid project consultants, but did see the Kokomo Center School Corporation as a client because the organization had contributed resources and had requested the evaluation. The school department provided the researchers with access to the schools for data collection and contributed $1,500 toward the expenses associated with the early phases of the research. However, Wysong and associates also saw the study as having additional clients: the students, families and community of Kokomo, professional colleagues, and the social sciences.

Selecting and Measuring Goals

The objective of most evaluation research is to determine if a program, policy, or law accomplished what it set out to do, so one of the study's concepts or variables must be the goal or desired outcome of the program, policy, or law. A typical evaluation research question asks whether the program has

participatory research, research that involves people in the community as co-participants in the entire research process and produces work that has implications for social change.

BOX 14.1

Another Point of View: Participatory Research

> The explicit aim of **participatory research** is to bring about a more just society in which no groups or classes of people suffer from the deprivation of life's essentials, such as food, clothing, shelter, and health, and in which all enjoy basic human freedoms and dignity . . . Its aim is to help the downtrodden be self-reliant, self-assertive, and self-determinative, as well as self-sufficient (Park, 1993: 2)

In response to the growing awareness of the need for socially relevant social research, new research strategies have been developed. Some newer kinds of research include participatory research, participatory action research, and collaborative research. These approaches ask research questions that have implications for social change and that are relevant to the group being studied; they reject the view of research as a value-free endeavor (Small, 1995: 948). Researchers that value action do not assume that behavioral scientists should discover the basic facts and relationships and leave to others the decisions about the uses of social science discoveries. Instead, those doing action-oriented research believe it is important to devise strategies in which research and action are closely linked so that both science and social welfare are improved (Whyte, 1991: 8).

Some researchers work in partnership with those in the community being studied to obtain and use the knowledge that is generated to empower the community. Although researchers in traditional research are sometimes seen by community organizations and social service agencies as coming in to judge how well they have been performing their jobs, more participatory kinds of research projects are "done *with* the community and not *to* the community" (Nyden, Figert, Shibley, and Burrows, 1997: 7). Rather than treating members of organizations and communities being studied as passive subjects, the researcher doing more participatory or action-oriented research works collaboratively with people in the organization or community throughout the research process—from selecting the problem to discuss the action implications of the study's findings (Whyte, Greenwood, and Lazes, 1991: 20).

One example of participatory research is a project on community accessibility conducted by Mary Brydon-Miller and the Community Accessibility Committee in western Massachusetts. The initial objectives of the project included encouraging participants to see themselves and each other as legitimate experts in the disability field, to develop their potential for advocacy efforts to achieve social change, and to develop a sense of community and ownership of the research process (Brydon-Miller, 1993). Brydon-Miller conducted lengthy, individual interviews with members of the local disabled community and organized a workshop for the participants to identify important accessibility-related problems and issues. One outcome of the workshop was their decision to focus on a local shopping mall as an accessibility target. Over a five-year period, the group evaluated the mall's accessibility, communicated problems to the mall management,

(continues)

BOX 14.1 (Continued)

filed a complaint with the Architectural Barriers Board, and followed through after the initial decision against the mall was appealed and re-appealed. The outcome was successful: The mall was renovated after the case became the first accessibility-related case to be heard by the Massachusetts Supreme Court (Brydon-Miller, 1993).

This innovative research project focused on community concerns, provided a forum for individuals to identify common objectives and possible actions, allowed individuals with specialized training and community members to structure the project, and resulted in a fundamental structural transformation that improved the lives of the people involved (Brydon-Miller, 1993: 136).

accomplished its goal. In our focal research example, Wysong and his colleagues asked the question "Is DARE effective?" That is, did the DARE curriculum produce "anti-drug" effects among students, especially by creating or reinforcing anti-drug attitudes and drug-coping skills?

If a researcher constructs a hypothesis, typically it will have the program, policy, or law as the independent variable and the goal(s) or outcome(s) as the dependent variable(s). Although Wysong, Aniskiewicz, and Wright did not state an explicit hypotheses, the implied hypothesis was that students who participated in DARE would have stronger anti-drug attitudes and drug-coping skills than those who did not participate in DARE.

If a program has broad or amorphous goals, it can be difficult to select one or more specific dependent variables. Therefore, a critical challenge for evaluators is to articulate appropriate dependent variables and select valid measurement techniques. Even when program personnel help define a specific goal, it might still be hard to determine the desired outcome precisely and how to measure it. Some issues that researchers might have to confront are the following:

1. Is the outcome to be short- or long-term?
2. Are attitudinal changes sufficient, or is it essential to study behavioral change as well?
3. Should all aspects of the program be studied or only certain parts?
4. Should all the targets of an intervention be studied or only some of them?

In an earlier study, Aniskiewicz and Wysong (1990) focused on short-term attitudinal changes. Evaluating DARE during the initial implementation year in Kokomo, they found significant increases in anti-drug attitudes and drug information and knowledge during the course of the school year among fifth grade students (Aniskiewicz and Wysong 1990). However, because the researchers felt the ultimate goal of DARE was to prevent or delay student use of drugs (both legal and illegal), they also wanted to conduct a long-term study focusing on *behavioral outcomes*. For that reason, five years after the first students in Kokomo participated in DARE, Wysong and his colleagues conducted the research presented in this chapter using "preventing/reducing/delaying drug use" as the dependent or outcome variable.

STOP AND THINK *In the focal research, Wysong and his associates used two different self-report methods to collect data on the dependent variable (preventing/reducing/delaying drug use). The primary data collection method was an anonymous, group administered questionnaire, and a secondary method was a group interview with a six-person focus group. What do you think are the advantages and disadvantages of the two data collection methods?*

As we discussed in the chapters on data collection, each method has advantages and disadvantages. Wysong and colleagues wanted to identify differences in both attitude and behavior between two groups of people. Attitudes are best studied by asking people questions, whereas measuring behavior is typically most valid through observation. However, because drug use is not a behavior that lends itself easily to observation, the researchers used self-report methods for the data collection. After considering the self-report options, Wysong and colleagues selected a group-administered questionnaire as the primary data collection method. The group-administered questionnaire is a good choice for a large, literate population, that, like theirs, can be found in a group setting to which they had access. This kind of questionnaire typically has a good response rate and allows anonymous collection of data, a "plus" when researchers ask about "sensitive" topics like drug use. The questionnaire provided quantitative data that were augmented by the qualitative data of the focus group interview of six Kokomo High School seniors. The focus group has the advantage of being flexible and providing elaborative answers with the interaction among participants typically replacing many interviewer questions.

STOP AND THINK *Look at the appendix of the focal research for examples of the questions included on the questionnaire. What do you think of the validity of the DARE scale and the DRUG USE scale?*

STOP AND THINK AGAIN *A recent evaluation of a DARE program in Ohio did a survey of teachers and principals (RMBSI, 1995). The researchers found that personnel in schools with DARE gave the program "high marks." (For example, 90 percent of the 400 staff completing the survey reported that the program had made a positive difference in students' attitudes about drugs; 75 percent said it delayed students' use of illegal substances (RMBSI, 1995: 2). What do you think of these data as indicators of DARE's impact?*

Wysong and associates did not ask teachers or administrators to comment on student behaviors. Asking school personnel to report on student attitudes and drug use is too indirect and makes an enormous compromise in validity. Instead, they asked students to report on their own attitudes and behaviors, a technique that we believe was appropriate. It is important to recognize, however, that using self-report data about attitudes and behavior depends on accurate memory and honest reporting. Asking students about behavior that occurred as much as five years earlier can be problematic because events and activities can be remembered incorrectly. In addition, even with anonymity, respondents might not always be truthful about their attitudes or drug use—perhaps under-estimating for fear of discovery or overstating to "look cool."

Selecting a Study Design

Experimental and Quasi-Experimental Designs

causal hypothesis, a statement that hypothesizes that the independent variable causes or affects the dependent variable.

Evaluation researchers almost always have an explanatory purpose because they usually specify a **causal hypothesis** or a testable expectation about the effect of an independent variable (the program, law, or policy) on a dependent variable (the desired outcome).

STOP AND THINK *Can you remember some of the study design issues discussed in Chapters 7 and 8? If you have a study with an explanatory purpose (that is, you want to test a causal hypothesis), what is a particularly useful study design?*

controlled experiment, an experimental design with two or more randomly selected groups (an experimental and control group) in which the researcher controls or introduces the independent variable and measures the dependent variable at least two times (the pre-and post-test measurement).

We talked about **controlled experiments** in Chapter 8 as especially good for testing a causal hypotheses. Built into the experimental design are ways to encourage equivalence in the experimental and control groups before the introduction of the program or policy. If we know that the experimental and control groups were alike before a program or policy was introduced, then we'll have more confidence that observed differences, after receiving program services or policy implementation, are not simply the result of other factors. In addition, because the controlled experiment requires that the program be offered to the members of the experimental group, and only to the members of the experimental group, the researcher can control the introduction of the independent variables.

The Adler and Foster research on the school reading project discussed in Chapter 8 is an example of evaluation research using an experimental design. The hypothesis tested was that participating in a 10-week program of reading and discussing books with themes of caring increased students' support for the value of caring for others. A large, team-taught seventh-grade class was selected, students were assigned to experimental and control groups, the independent variable was "introduced," the dependent variable was measured both before and after the program, and the pre- and post-program results were compared. The experimental design handled many of the possible challenges to **internal validity**. That is, there was support for the assumption that the control group was exposed to all the other factors that the experimental group was exposed to *except* for the program and that the small differences in support for the value of caring were likely to have been the result of the independent variable (participation or nonparticipation in the innovative program).

internal validity, agreement between a study's conclusions of causal connections and what is actually true.

STOP AND THINK *The controlled experiment assumes that assigning people into experimental and control groups is possible. Can you think of situations when a researcher either couldn't or wouldn't want to put people in a control group? That is, when might an experiment be an inappropriate study design choice?*

Not all situations lend themselves to using the classic experimental model. Outside a laboratory, the evaluator might not be able to control all aspects of the design. A well-known field study of the impact of police intervention on wife abuse in Minneapolis (Berk, Smyth, and Sherman, 1988) was supposed to be a controlled experiment. Police were to randomly assign domestic violence calls to one of three treatments (mediating the dispute, ordering the offender to leave the premises for several hours, or arresting the abuser). However, despite the intended design, in about one-third of the calls, officers used individual discretion, not random assignment, to determine treatment. With no pre-test measurement and without random assignment, there was no way to estimate prior differences between the experimental and control groups.

Sometimes practical concerns of time or money preclude using the controlled experiments. Evaluation research field experiments can take five or more years from design to the final report—often too much time for administrators and policy makers (Rossi and Wright, 1984). In other situations, researchers know from the beginning that assigning participants to experimental or control groups is not possible or desirable. For example, when studying the impact of high school restructuring on student achievement, Lee and Smith (1995) could not determine which school a student attended, nor could they control when a school system would make significant changes. In studying the impact of "no fault" divorce laws on husbands' and wives' post-divorce economic statuses, Weitzman (1985) could not regulate where a couple would live when filing for divorce nor when couples would want to end their marriages. And clearly, in situations where there is an ethical responsibility to offer services to the most needy—such as AIDS treatment studies, mental health therapies, and child abuse programs—using random selection as the criterion for assignment to treatment or no-treatment groups is not acceptable. Instead, it is ethical to give treatment and services to the whole sample and have no control group.

STOP AND THINK *The DARE research was not a controlled experiment. What aspects of the controlled experiment design were not a part of the research?*

When the true experiment is not feasible or possible, researchers can eliminate aspects of the experimental design. In the focal research, in 1987, the Kokomo school administration made the decision to implement DARE for *all* fifth and seventh graders. Therefore random assignment of students into control and experimental groups was impossible. In addition, because, as we've already noted, the outcomes being studied—attitudes toward drugs and self-reports of drug use—are best handled with anonymous questionnaires, no names or other identifications were used. Without a way to identify individual responses at different times, individual changes in attitudes and behavior over time could not be tracked. With no possibility of doing a controlled experiment, the researchers selected alternate strategies.

quasi-experiment, an experimental design that is missing one or more aspects of the classic controlled experiment.

In 1987 to 1988, the researchers compared the aggregated before and after scores for all of the fifth and seventh graders on the DARE Scale. In the focal research study, the researchers used a version of a **quasi-experimental design**. The two groups were the 1991 and 1992 seniors at Kokomo High School. The "experimental" group was composed of the former seventh graders who took the DARE program, and the comparison group was the former sixth graders who did not participate in DARE.[9] Using a "post-test only" analysis, the answers of these two groups of students on the DARE Scale and Drug Use Scale were compared.[10] Although it is likely that the former sixth and seventh graders were similar to begin with and, aside from DARE, had similar school experiences, the researchers cannot really know if this is true. Although their design raises concerns about internal validity, it was a "do-able" and ethical option in this "real world" setting.

In the focal research, all the students in one grade were enrolled in the DARE program, and all the students in another grade were not, regardless of personal preferences. When members of a sample are able to make choices about their participation in a program, however, the issue of self-selection leads to other validity concerns. A well-known study, the Coleman Report (National Center for Education Statistics, 1982), which compared the academic achievement of students in public and private high schools, has been vulnerable to the criticism that parental, student, and school selection procedure differences confound any simple comparisons between groups of students in different kinds of schools (Rossi and Wright, 1984).

Longitudinal, Cross-Sectional, and Case Study Designs

panel study, a study design in which data are collected about one sample at least two times and all variables are measured, not controlled by the researcher.

Another common design for evaluation is the **panel study**—a design that uses before and after comparisons of a group receiving program services or that is affected by policy changes. If, for example, a school intended to change a discipline code by increasing the use of in-school suspension and decreasing out-of-school suspension, student behavior for the months preceding the change could be compared with the behavior afterwards. If student behavior improved after the policy was introduced, school personnel might decide to keep the new policy, even if they aren't sure about the extent to which other factors (time of year, a new principal, a new course scheduling pattern, and so on) contributed to the differences in behavior. Panel studies can be useful if a comparison group is not available, even though they are better at documenting a change in the group under study than in identifying the cause of changes.

cross-sectional study, a study design in which data are collected once for all the variables of interest using one sample.

Cross-sectional studies are sometimes used in evaluating programs or policies, especially if other designs are problematic. In a cross-sectional design,

[9] Only students who had not moved from Kokomo in the years between 1987 and 1990 or 1991 were in the sample. The former sixth and seventh graders were asked as seniors to identify the particular school they attended during the 1987–1988 school year and to identify the particular months of that year that they attended if they were not in the school for the entire school year. The indicator of DARE participation was having been in seventh grade in the specified schools for the months of the DARE program.

[10] Some of you might be wondering what happened to the fifth graders who participated in DARE in 1987-1988. After all, they, not the seventh graders were the ones who had experienced short-term changes in the anti-drug attitudes immediately after DARE intervention. Two years after the focal research study, Wysong and Wright (1995) studied these former fifth graders when they were seniors. Once again, they found that the attitudes toward drugs and self-reported drug use among DARE-exposed seniors were very similar to those of non-DARE seniors.

data about the independent and dependent variables are collected at the same time. Say, for example, that a city that has been using a series of public service announcements about AIDS prevention in all buses for the past six months. Officials might commission a survey of city residents to determine the effectiveness of the ads. Interviewers could ask a random sample of respondents if they had seen the ads and their opinions of the ads as well as a series of questions about AIDS and AIDS prevention. If after doing such a survey, the results show that a higher percentage of those who remember seeing and liking the ads are more knowledgeable about AIDS and its prevention than those who either don't remember or didn't like the ads, the city officials could conclude that the ad campaign was successful.

STOP AND THINK *Why would we need to be cautious in concluding that the ad campaign was successful?*

In our hypothetical city with our imagined ad campaign, with this relatively inexpensive and easy-to-do study, we could determine that a *correlation* exists between two variables. Although the connection between noticing ads and knowledge is suggestive of causation, we don't have evidence of the time order. For example, it's possible that those already most interested in and best informed about AIDS prevention would be the people to pay the most attention to the ads. And those who are least informed about AIDS to begin with could be the ones who didn't read the public service announcements on the topic. In other words, the proposed independent variable could be the dependent variable!

case study, a research strategy that focuses on one case (an individual, a group, an organization, and so on) within its social context at one point in time, even if that *one time* spans months or years.

When doing exploratory evaluation work, and in situations where limited generalizability is acceptable, researchers might select a **case study** approach. Focusing on a single policy, event, organization, group, or individual over time can be useful for generating a theory or developing tentative conclusions. For example, if a social service agency developed a "trauma curriculum" for use after a crisis in the local schools (for example, after a death of a student or an assault on a teacher), a case study approach could be used to evaluate it. Each time school personnel requested the agency to implement the curriculum, evaluators could also be on site. Collecting data through observation and asking questions of students, agency staff, and school personnel will provide tentative answers about *if, how,* and *why* the curriculum works. Watching reactions to each crisis and the subsequent interventions will allow researchers to describe how the trauma curriculum was implemented and to suggest reasons for the resulting outcome.

Populations and Samples

In addition to decisions about using one or more groups as part of the study design, researchers must think about the generalizability of the data. As we discussed in Chapter 5, random and nonrandom samples have different implications for the kinds of analyses that are appropriate and the conclusions that can be drawn. The DARE study used all the students in several grades in one school system over the course of several years. The results of this evaluation might be quite different in other cities or in other years. Evaluation research can point up the "one size fits all" fallacy as policies and programs useful for specific geographic or historical contexts might be inappropriate in other settings. For greater generalizability, evaluations should be replicated in a variety of settings and contexts.

Politics and Evaluation Research

The Political Nature of Research

It's important to recognize the political nature of all research and the implications for evaluation research. All research is affected, to some degree, by the social, political, and economic climates that surround the research community. Because knowledge is socially constructed, the choice of research topics and questions and all methodological decisions are related to the social and political context.

In evaluation research, the specific choice of research projects is affected not only by societal values and the priorities of funding agencies but also by the perspectives of various constituencies and program stakeholders. Another important factor is that evaluation is typically conducted in field settings, within ongoing organizations, such as schools systems, police departments, health care facilities, social service agencies, and the like.

> Organizations are very political: There is "turf" to protect, loyalties and other long-standing personal relationships to look after, and there are always careers and economic survival to think about. These are not the ordinary, newsworthy politics of the legislative variety—intellectual issues, public morals, votes and the next election. Rather, these politics are the kind that are internal to organizations, politics spelled with a lower case "p," the politics of authority, sexual relationships, and small groups (Chambers et al., 1992: 11).

Organizations are not "neutral territory"; the researcher is working in someone else's sphere and must understand the circumstances and conditions. The results of an evaluation have the potential to affect the organization and the individuals under study, so organizational participants might try to help or to hinder the research process.

Trust and Mistrust

A special challenge in evaluation research is to obtain the cooperation of program staff for access to data. Sometimes this is especially challenging. When one of us, Emily, was hired to help with the evaluation of special needs educational policies in several communities, feasibility became a concern. The research plan was for Emily to observe school meetings at which individual education plans for special-needs students were designed by school personnel in consultation with parents. There was no difficulty in getting cooperation at three of the schools selected, but, at the fourth school, the principal kept "forgetting" to communicate when meetings were canceled or had been rescheduled because of "unavoidable problems and crises." Several times, on arriving for a meeting at this particular school, Emily was told that the meeting had already been held—the day or even the hour before. The message was clear: The principal did not want an observer. Because the schools selected were a purposive, rather than random sample, the meetings at another school were substituted and observed.

The sentiment that evaluators aren't welcome isn't unusual. As the following excerpt from an editorial in a magazine for administrators of youth programs makes clear, evaluators and their work are often distrusted by program personnel:

> As part of the price of doing business, youth service managers put up with a stream of over- schooled but often under-educated "evaluators" typically drawn from academia or a Beltway Bandit consulting firm. Much of the supposed evaluation time is spent by the staff teaching the evaluators the realities of the rough and tumble world of youth work. When evaluation results are made available years later they speak to a staff, funding mix and business climate that no longer exist. In evaluating the actual helpfulness of evaluators in steering the nation toward better, more cost effective programs for youth, it is reasonable to wonder if this whole mini-industry isn't just another example of white collar welfare masquerading as help for the disadvantaged. (Vanneman, 1995: 2)

Researchers must realize that those who work in programs being evaluated might feel that the time and money spent on evaluation would be better spent on programs. Researchers must be sensitive to the fact that program staff have jobs to do that might be made more difficult by ongoing evaluation.

Having an Impact

In 1845, the Board of Education of the city of Boston initiated the use of printed tests to assess student achievement. The resulting test scores were low, but rather than analyze the causes of the poor performance (Traver, cited in Chambers et al, 1992: 2), the board decided to discontinue the test! Unfortunately, this example, more than a century and a half old, still has relevance today.

Even if evaluators are trusted, allowed to collect data, and do a credible job of evaluating a program or policy, their conclusions may have little impact. Ideological and political interests can sometimes have a greater influence on decisions about the future of social interventions than evaluative feedback. Even if a program is shown to be ineffective, it might be kept if it fits with prevailing values, satisfies voters, or pays off political debts. Often changes in social programs are very gradual because frequently no single authority can institute radical change (Shadish, Cook, and Leviton, 1991: 39). Evaluations might be conducted to satisfy political demands or to demonstrate the program effects of decision-makers' positions (Weiss, cited by Shadish, Cook, and Leviton, 1991: 57).

The DARE case provides a good illustration of the process by which programs that receive negative evaluations continue to flourish. Even though evaluations in Kokomo and elsewhere (Ennett, Tobler, Ringwalt, and Flewelling, 1994) have indicated that the DARE program is less than successful, the stakeholders (organizations with direct and indirect involvement) have accumulated sufficient resources and legitimacy to protect it. In Kokomo, the reaction of local policymakers to the "no effects" evaluation has ranged from silence to indifference to hostility (Wysong and Wright, 1995); most stakeholders simply ignored the results. Other DARE evaluations have met similar fates (Wysong and Wright, 1995; Vanneman, 1995:2). Wysong and Wright (1995) conclude that DARE is likely to continue to survive as America's largest drug program in the immediate future because having such a program at least shows the public that schools and police departments are trying to combat drug abuse among students.

At best, evaluators have an advisory role, "closest to that of an expert witness, furnishing the best information possible under the circumstances; it is not the role of judge and jury" (Rossi and Freeman, 1993: 454). Evaluators can provide reports that have policymaking significance (Newman and Tejeda, 1996), discuss the larger context of their work, argue forcefully for their positions and work toward disseminating their findings widely, but they rarely have the power to institute changes in programs or policies.

When evaluations are bought and paid for by agencies that are also stakeholders, an even more difficult political situation can arise. For example, at least one completed analysis of DARE (one commissioned by the National Institute of Justice and conducted by the North Carolina Research Triangle Institute) has not been released by the funding agency, perhaps to restrict or discredit its negative findings (Wysong and Wright, 1995). The DARE evaluation in the focal research was conducted by researchers working as interested scholars rather than as paid consultants to the DARE program or the school department; the bulk of the funding for the study's expenses ($11,000) came from university faculty grants. These factors made possible the widespread dissemination of the study's findings, regardless of the reaction of local or national stakeholders.[11] Not all evaluation researchers are so fortunate.

Summary

Evaluation research is research with a practical purpose. To evaluate a new or ongoing program or policy, the purpose of the research and the specific research question must first be decided. Typically, a causal hypothesis is

[11] One unexpected outcome of this DARE study was the decision by the Kokomo Center School Board to approve random drug testing for students in grades 8 through 12. The superintendent of schools cited as a major factor in deciding to pursue the policy, the estimates in Wysong and colleagues' data that some kinds of drug use were twice as high among Kokomo High School students as the national average (Hubbard, 1996: 1).

constructed with a program or its absence as the independent variable and the goal of the program as the dependent variable.

As in any research project, a research strategy must be selected. Measurement techniques must be developed or selected, a study design chosen, a sample identified, and data collected and analyzed. In making methodological decisions in evaluation research, it is important to consider the implications of each choice on internal validity.

Evaluative efforts also mean the need to negotiate the complexities of doing "real world" research within political and social contexts. Barriers that will make the process more difficult might be erected by program personnel or other stakeholders. At the end of a study, the analyses and their practical implications should be disseminated to the program staff, other stakeholders, other researchers, and the general public. Evaluation research is a demanding enterprise, but it can help create, modify, and implement programs that make a difference in people's lives.

Evaluation research can provide reasonably reliable, reasonably valid information about the merits and results of particular programs operating in particular circumstances. Necessary compromises in programs and research can mean that the users of the information will be less than fully certain of the validity of the findings. In a world full of uncertainties and hazards, however, it is far better to be approximately accurate than to remain totally ignorant because a "conclusive" evaluation was unaffordable (Hatry, Newcomer, and Wholey, 1994: 601).

SUGGESTED READINGS

P. Park, M. Brydon-Miller, B. Hall, and T. Jackson, editors. 1993. *Voices of Change, Participatory Research in the United States and Canada*. Westport, CT: Bergin & Garvey.

An important collection of articles that focus on the practical, theoretical, and institutional experiences of social scientists doing participatory research in North America.

Rossi, P. H., and H. E. Freeman. 1993. *Evaluation: A systematic approach* (5th ed.). Newbury Park, CA: Sage.

The "gold standard" in the field of evaluation research, this book is a must for anyone planning to do an evaluative study. Rossi and Freeman cover the diagnosis of social problems, the development of designs and strategies for outcome assessment, methods of measuring program efficiency, and the social context of evaluation.

Shadish, Jr., W., R., T, D. Cook, and L. C. Leviton. 1991. *Foundations of program evaluation*. Newbury Park, CA: Sage.

The authors focus on social programs (their origins, operations, and evolution), how to develop and use evaluative information, and methods of conducting and disseminating studies in ways that are sensitive to practical, real-life issues. One of the book's strengths is the connection that the authors draw between evaluation theory and empirical research.

EXERCISE 14.1

Designing Evaluation Research

Find a description of a social program in the local newspaper or use the following article of a hypothetical program. (If you select a newspaper article, attach it to the exercise.)

"Pets are Welcome Guests"

Residents of the Pondview Nursing and Rehabilitation Center have a series of unusual guests once a week. The VIPs (Volunteers Interested in Pondview) have organized a "Meet the Pets Day" at the local facility. Each week, one or more volunteers brings a friendly pet for short one-to-one visits with residents. On a typical day, a dozen or so owners will bring dogs, cats, and bunnies to Pondview, but sometimes companions include hamsters and gerbils.

Last week, Buffy, a spirited golden retriever with a wildly wagging tail, made her debut at Pondview. In a 15 minute visit with Mrs. Rita Williams, an 85-year-old widow recovering from pneumonia, Buffy managed to bestow at least several "kisses" on the woman's face and hands. Mrs. Williams said she has seen more sedate pets, but wasn't at all displeased with today's visit.

Margaret Collins, facility administrator, said that the program had been adapted from one she had read about in a nearby city. She was glad that the VIPs had organized the new program. "It gives the residents something to look forward to," she said. "I think it makes them more alert and attentive. If it really does aid residents' recoveries and results in improved mental health, we'll expand it to several days a week next year."

Design a research project to evaluate either "Meet the Pets Day" or the program described in your local newspaper by answering the questions that follow.

1. What is an appropriate research question that your evaluation should seek to answer or hypothesis that you would test?

2. Describe the social program that is being offered.

3. Who are the programs' participants?

4. In addition to the participants, who are the other "stakeholders"?

5. What is the goal or intended outcome of the program?

6. Describe how you would decide if the program was successful in meeting its goal by designing an evaluation study. Be sure to describe the study design you would use, who your sample would include, and how you would measure the dependent variable.

7. What are the ethical considerations you would need to consider if you were interested in doing this study?

8. What are the practical issues (time, money, and access) that you would need to consider if you were interested in doing this study?

EXERCISE 14.2

Drug Use and Attitudes

Work with another student to complete a survey of attitudes toward drugs with a quota sample of 20 college students.

WHAT TO DO

1. Review the appendix in the focal research on DARE and select at least five of the questions from the DARE Scale or the Drug Use Scale to include on an anonymous questionnaire. Construct an additional question that asks the respondent if he or she participated in a DARE or other drug-education program while in elementary or secondary school. Construct several questions asking about background characteristics (age, gender, and so on) that you think are important. Write an introduction to the questionnaire, instructions for completion, and place the questions in an appropriate order. Make at least 20 copies of the questionnaire to distribute.

2. Check with your instructor about the need to seek approval from the Human Subjects Committee, the Institutional Review Board, or the committee at your college or university concerned with evaluating ethical issues in research. If necessary, obtain approval before completing the rest of the exercise.

3. Approach students for your sample in person or by phone. Tell them that you are conducting a short, anonymous survey of attitudes toward drugs for your research methods class.

   If a student is willing to be in the study, first ask if he/she remembers having participated in an anti-drug education program (like DARE) while in elementary or secondary school. You will be aiming for a sample that has about equal numbers of people who have and have not been exposed to an anti-drug education program.

4. Select a method of returning the questionnaire so that the members of the sample can be anonymous. (For example, students can return it to you at an on-campus address or mail it to you using a self-addressed, stamped envelope that you provide).

5. You and your partner should each find at least 10 people to complete the questionnaire. Keep track of how many "turn downs" you get. Aim for 5 students who have participated in an anti-drug program and 5 students who have not. (This way your combined sample of 20 respondents will have approximately equal numbers in each group.)

6. Distribute as many questionnaires as necessary to get approximately 20 questionnaires returned.

7. Tabulate your results using frequency tables as follows:

   a. Separate the questionnaires of respondents who participated in an anti-drug education program from those who did not.

   b. Construct separate frequency tables for the two groups of respondents for at least four of the questions on the drug use and behavior questions.

   c. Compare the answers for the two groups.

   d. Compare your results to those reported in the focal research article.

WHAT TO WRITE

Write a report of your research that includes the following:

1. Describe the approach you used for finding and encouraging students to be part of your survey.

2. How many did you approach? How many agreed? How many actually responded? Calculate your response rate.

3. Describe how the questionnaires were returned so that you could guarantee anonymity to sample members.

4. Present the results of your survey and how the results compare to those of Wysong and colleagues. (#7 b, c, and d from "what to do.")

5. What are some possible concerns about the validity of your results?

6. If you were to ever repeat this survey again, based on your experience, is there anything you'd do differently?

7. Attach all 20 completed questionnaires to your report as an appendix.

EXERCISE 14.3

Evaluating Your Course

You can think of this course or some aspect of it as an educational program designed to help students understand and apply social research methods. Select one aspect of your course (such as the textbook, the instructional style of the teacher, the frequency of class meetings per week, number of minutes per class, the use of lectures, group work or exercises, and so on) and design an evaluation research project that could test the effectiveness of this aspect of the course on the dependent variable, student learning.

Answer the following questions.

1. What aspect of the course are you focusing on?
2. How could you operationalize the dependent variable, student learning of social research methods?
3. Focusing on your study design and sample, what could you do to determine the effectiveness of the aspect of the course that you are interested in?
4. Using your imagination, what are some results that you might get from your evaluation?
5. Comment on the practical and ethical considerations you would need to consider if you wanted to do this study.

15 Quantitative Data Analysis

Introduction

We'd now like to introduce you to the most exciting part of the research process: analyzing the data. After all the work you've done collecting your information, there's nothing quite like putting it together in preparation for sharing it with others—a little like finally cooking the bouillabaisse after shopping for all the individual ingredients. But like the excitement of cooking bouillabaisse, the excitement of data analysis needs to be balanced by having a few guidelines (or recipes) to follow. In the next two chapters we'll introduce you to quantitative and qualitative data analyses. Rather like the *Impoverished Students' Book of Cookery, Drinkery, & Housekeepery*, which once provided Roger with a rudimentary knowledge of his kitchen and what to do in it, these chapters offer elementary data-analytic recipes. (Based on this chapter you should be able to produce the tuna-noodle casseroles of quantitative data analyses, if not the bouillabaisses.) We will, nonetheless, direct you to books dealing with more sophisticated recipes along the way. Bon appetit!

Quantitative or Qualitative?

What is the difference between quantitative and qualitative data? One answer is that quantitative data results from quantitative research and qualitative data results from qualitative research. This sounds like question begging, and of course it is, but it carries with it an important implication: The distinction drawn between quantitative and qualitative data isn't as important (for data-analytic purposes) as the distinction between the strategies driving their collections. Sure, you can observe, as Dabbs (1982: 32) and Berg (1989: 2) have, that the notion of qualitative refers to the essential nature of things and the notion of quantitative refers to their amounts. But, as we've hinted before (for example, in chapters on qualitative interviewing and observation techniques), there's really nothing about qualitative research that precludes quantitative representations (see Morrow, 1994: 207). Adler and Adler (see Chapter 11) did sometimes count the number of times children attended different kinds of afterschool activities. Similarly, research that might seem primarily quantitative never totally ignores essences. Messner (in Chapter 12) used qualitative judgment to identify violent television before he used quantitative techniques to judge which populations were exposed to greater or lesser amounts of it. Indeed, although the mix of quantitative and qualitative approaches does vary in social science research, almost all published work contains elements of each. Consequently, the distinction we draw (and have used to organize this text) is not one of mutually exclusive kinds of analysis but of kinds that, in the real world, stand side by side, and, in an ideal one, would always be used to complement one another.

An Overview of Quantitative Data Analysis

quantitative data analysis, analysis that tends to be based on the statistical summary of data.

However artificial the distinction between quantitative and qualitative research and data might be, there's no denying that the motives driving quantitative and qualitative researchers are distinguishable. Although there are exceptions, **quantitative** researchers normally focus on the relationships between or among variables, with a natural science-like view of social science in the backs of their minds. Thus, in the fashion of the physicist who asserts a relationship between the gravitational attraction and distance between two bodies, Adler and Foster (in Chapter 8) focused on whether students exposed to literature with caring themes became more caring themselves, Messner (in Chapter 12) studied whether violent crime increased as populations are exposed to violent television programming, and Gray, Palileo, and Johnson (in Chapter 9) reviewed whether females were less likely to attribute rape responsibility to victims than males. That these projects pursued natural science-like hypotheses meant that they engaged in other natural science-like activities as well: for example, looking at aggregates of units (whether these were seventh graders, SMSAs, or college students), more or less representative samples (see Chapter 5 on sampling) of these units, and the measurement of key variables (see Chapter 6 on measurement). For data-analysis purposes, the distinguishing characteristic of all these studies is attention to whether there are associations among the variables of concern: Whether, in general, as one of the variables (say, exposure to themes of caring) changes, the other one (students display caring attitudes) does as well. To demonstrate such an association, social science researchers generally employ *statistical* analyses. Thus, we will aim toward a basic understanding of statistical analyses in this chapter on quantitative data analysis.

STOP AND THINK *Look back at Chapter 12 and identify the major independent variable in Messner's study of crime rates in SMSAs? Do you recall the unexpected relationship he found between this variable and violent crime rates? Do you see that the focus of his article, like that of others using quantitative data, is on* ***variables*** *and not really on people, groups, or social organizations?*

An Overview of Qualitative Analyses

qualitative data analysis, analysis that tends to result in the interpretation of action or representations of meanings in the researcher's own words.

The strategic concerns of qualitative researchers differ from those of quantitative researchers and so do their characteristic forms of data analysis. As a general rule, **qualitative** researchers tend to be concerned with the interpretation of action and the representation of meanings (Morrow, 1994: 206). Rather than pursuing natural science-like hypotheses (as Durkheim advocated in his *Rules of Sociological Method*), qualitative researchers are moved by the pursuit of empathic understanding, or Weber's *Verstehen* (as described in his *Theory of Social and Economic Organization*), or an in-depth description (for example, Geertz' [1983] *thick description*). Enos (in her article in Chapter 10) aspired to an empathic understanding of the meaning of motherhood for women inmates and of the differences in those meanings for different inmates. Adler and Adler (in their article in Chapter 11) provide a thick description of what it means to participate in afterschool activities. Qualitative researchers pursue single cases or a limited number of cases, with much less concern than quantitative researchers about how well those cases "represent" some larger universe of cases. And qualitative researchers look for interpretations that can be captured in words rather than in variables and statistical language. In the next chapter we'll seek a basic understanding of how qualitative researchers reach their linguistic interpretations of data.

Quantitative Data Analysis

Quantitative data analysis presumes one has collected data about a reasonably large, and sometimes representative, group of subjects, whether these subjects (or units of analysis) are individuals, groups, organizations, social artifacts, and so on. Ironically, the data themselves don't always come in numerical form (as one might expect from the term *quantitative* data).

A brief word to clarify this irony: You'll recall that the focus of quantitative data analysts isn't really the individual subjects being studied (for example, school children or SMSAs) but, rather, the variable characteristics of those subjects (for example, whether they were exposed to caring literature or whether they have high crime rates). Remember that variables come in essentially three levels of measurement: nominal, ordinal, and interval-ratio. The most common level of measurement is the nominal level (all variables are at least nominal), because all variables (even gender) have categories with names (like "male" and "female"). Thus, an individual child might be a male or a female, and the information you'd obtain about that child's gender might, for instance, be "female." But "female" is obviously not a number. Nonetheless, it is information about gender that when, taken together with data about gender from other subjects, can be subjected to numerical or quantitative manipulation, as when we count the number of females in our sample.

This is just a long way of saying that the descriptor "quantitative" in "quantitative data analysis" should really be thought of as an adjective describing the kind of analysis one plans to do, rather than as an adverb modifying the word "data." When you do a quantitative analysis, the data might or might not be in the form of numbers.

Coding

Because the goal of quantitative data analysis is to perform statistical operations on relatively large sets of data, and because computers are particularly adept at performing such operations, you might want to use a computer with statistical software. Because we don't know just what software you might be using, we speak here in fairly generic terms.

coding, the process by which raw data are given a standardized form. In quantitative analyses, this frequently means making data computer usable.

One common feature of all computer programs is that they want your raw data, whatever form they come in, to be prepared just so (similar, if you will, to the way you might dice your vegetables before stir-frying them). The process of making data computer usable is called **coding**. In most cases, coding involves assigning a number to each observation or datum. Thus, when coding gender, you might decide to assign a value of "1" for each "female," and a "2" for each male. The assignment of a number is often pretty arbitrary (there is no reason, for instance, that you couldn't assign a "1" for females and a "0" for males[1]), but it should be consistent within your data set.

We'd like to introduce you to coding with a reasonably uncomplicated example. When she was coding her data from housing projects, Filinson (see Chapter 7), collected observations on an interview schedule like the one that is partly reproduced in Box 15.1. She then coded each of 10 items in the following way: 1. case number; 2. gender of respondent (1=female; 2=male); 3. the type of project (mixed-aged project=1; housing for the elderly=2); 4. how many confidantes the respondent has in the building (coded as the number reported); 5. whether respondent knows of residents who take too much medicine (1=no; 2=yes); 6. whether respondent knows of residents who persist in being drunk (1=no; 2=yes); 7. years of school completed (coded as number reported); 8. former occupation (coded as unskilled=1; blue collar=2; clerical sales=3; semi-professional=4; manager=5; professional = 6); 9. whether respondent receives Medicaid (1 = no; 2 = yes); 10. whether respondent receives Supplemental Security Income (1=no; 2=yes). Filinson's observations about the case shown in Box 15.1 could then be reduced to the following line of numbers:

16 1 2 0 1 1 08 1 1 1

STOP AND THINK *What does the "2" mean in this line describing case 16? What does the "0" to its right indicate? What does the "08" mean?*

Filinson used the same procedure for each of the 50 respondents in her sample. Following a very common coding convention, she then lined up the information about each case in rows, as we do here for 20 of her cases (note: we have left spaces between each item for ease of reading).

```
01 1 1 1 1 2 09 2 2 2
02 1 1 1 2 2 09 2 2 1
03 1 1 1 2 2 02 2 2 2
04 1 1 0 2 2 09 2 1 1
05 2 1 0 2 2 09 1 1 1
06 1 1 1 1 1 07 1 2 1
```

[1]One of our reviewers, Dan Cervi, recommends, in fact, that students always use "0" and "1" when coding two-category variables. He suggests that this helps reduce the chances of misinterpretation later. Moreover, "1" can be used to refer to the presence, "0" to the absence of a measured characteristic. This can be especially useful when dealing with "yes/no" or "true/false" responses on questionnaire surveys.

BOX 15.1

Sample Interview Schedule with Respondent's Answers

Case: 016
Gender: Female
Building: Elderly Housing

1. How close are you with other residents living in this building?: [no answer]

   Probe: Are any friends?: Yes
   How many?: Several friends
   Any confidantes?: None really that I would tell personal things to.
   How many?: [not asked]

2. I'm going to give you a list of behaviors some residents might exhibit. After each behavior, I would like you to tell me if there are any residents who persist in this. (Probe for description)

   a. Intentionally taking too much medicine: No
   b. Being drunk: No

3. How many years of school did you complete?: 8th grade
4. What was your major occupation?: Hospital worker
5. Do you currently receive Medicaid?: No
6. Do you currently receive Supplemental Security Income?: No

```
07 1 1 1 2 2 09 2 2 2
08 1 1 0 2 2 12 4 2 2
09 1 1 0 1 2 09 2 2 2
10 2 1 0 1 1 08 2 2 2
11 1 2 2 1 1 09 1 2 1
12 1 2 1 1 1 08 2 2 1
13 1 2 1 1 1 09 1 1 1
14 1 2 1 1 2 08 1 1 1
15 1 2 3 1 1 08 3 1 1
16 1 2 0 1 1 08 1 1 1
17 2 2 2 2 2 12 2 1 1
18 1 2 2 2 1 08 3 2 1
19 1 2 2 1 1 12 4 1 1
20 1 2 1 1 1 09 2 1 1
```

The preceding 20 lines are the kind of things computers eat with relish. Notice that when the information from all cases is piled in this fashion the data form a rectangle, whose contents are said to be a rectangular data matrix.[2] Each horizontal row in the matrix represents the observations from one case; each vertical column, or set of columns, represents observations, from different cases, about a variable. Thus, for instance, the first two "columns" of data consist of the two-digit numbers 01 through 20, each of which stands for the case number of a particular respondent.

STOP AND THINK *The second column (with a gray bar behind it) in the matrix has a bunch of 1s and 2s in it. The 1s indicate that the respondent was female; the 2s, that the respondent was male. What was the gender of respondent "20?"*

To achieve the rectangular appearance of the whole matrix, Filinson followed another common coding convention: She made sure each variable was allocated the number of digits of its largest value. Thus, because cases 08, 17, and 20 had 12 years education, and 12 is a two-digit number, Filinson made sure that every value of this variable had two digits, even if this meant giving a 02 (circled) to the respondent who had only two years of education. A word of caution: However desirable it might be to allocate each variable the number of digits of its largest value, you must still be sure that the resulting values are themselves accurate. Thus, you can accurately represent 2 (years of education) as 02 (a two-digit number), but not as 20 (also a two-digit number). In other words, be careful, when coding, where you place your zeroes.

codebook, a guide for translating computer-ready data back into human-understandable language.

To remember how she'd coded her data, Filinson followed yet another coding convention: She created a **codebook**, or a guide for translating computer-ready data back into some human-understandable language.[3] All good codebooks contain information about every variable's name, where it is located, what each code stands for, and how missing data have been dealt with. A partial codebook for the Filinson's data is on page 400.

STOP AND THINK *In which column(s) did we place the number of confidantes each respondent had in the building? How many of confidantes did respondent number 09 report having?*

Complications That Occur in Coding

The housing data provide a relatively simple example of coding in action, but even they entailed complications. Although there are no examples in the sample of 20 cases we've shown, Filinson, with 50 cases and many more variables than we've shown, had oodles of missing data—data about particular variables that just weren't there. Some people, for instance, didn't know the answers to certain questions (such as the question about whether other residents overused their medication), and didn't want to hazard a guess.

[2] We thank two reviewers, Kevin Thompson of North Dakota State University and Cynthia Beall of Case Western Reserve University, for reminding us that the necessity of coding data into rectangular matrixes is obviated with some of the newer computer statistical software, such as SPSS for Windows—software that accepts data "as they come." For students who have access to such software, the following couple of paragraphs will be superfluous.

[3] Researchers frequently create their codebooks before they code their data for reference during the coding process.

A Codebook For Housing Data

Variable Name	*Longer Name and Codes*	*Column Numbers*
CASENO	Case identification number (01-20)	1-2
GENDER	Gender of Respondent 1 = female; 2 = male	4
BUILD	Type of housing 1 = Mixed-aged housing 2 = Housing for elderly	6
CONFID	Number of confidantes in the building? Code as actual number	8
MEDIC	Are there residents who take too much medicine? 1 = No; 2 = Yes	10
DRUNK	Are there residents who persist in being drunk? 1 = No; 2 = Yes	12
YRSCL	Number of years in school Code as actual number	14-15
OCCU	Major former occupation 1 = No; 2 = Yes	17
MECAID	Does respondent receive Medicaid? 1 = No; 2 = Yes	19
SSI	Does respondent receive Supplemental Security Income? 1 = No; 2 = Yes	21

Filinson followed yet another common coding convention for missing data by leaving blank spaces where the data might have been, and by telling the computer that blank spaces meant that data are missing. You're actually more likely to find data missing when you employ group-administered questionnaires (as Gray and her colleagues did in the focal research in Chapter 9) or mailed questionnaires, where there's no interviewer to prompt respondents, than in the kind of structured interview used by Filinson. In any of these cases, however, a respondent might not know the answer to a question. The coding issue then becomes one of creating a special "don't know" category or treating such responses as missing data.

Other practical complexities of coding real data can be handled fairly readily as long as you remember one or two guidelines. Perhaps the most important of these derives from the fact that, whatever else you do while coding, you are striving to create indicators of key variables. And so, harking back to common standards of measurement mentioned in Chapter 6 and Chapter 9, we want to stress the importance of developing categories that are simultaneously *exhaustive* (so that every case has a coding category it fits into) and *mutually exclusive* (so that all cases can be placed in one and only one category). You might, for instance, want to use information about occupation as a rough measure of social class (or perhaps part of a more complex index of social class). This is basically what Filinson had in mind when she asked her respondents about their occupations before retirement, using the open-ended question, "What was your major occupation?" When she received responses like "hospital worker," "secretary," "railroad worker," "jewelry worker," and "waitress," Filinson was faced with a coding decision. To reduce the variety of these responses to a manageable but meaningful number, Filinson coded her raw occupational data (waitress, housekeeper, and so on) according to a conventional occupational typology (unskilled, blue collar, clerical sales, semi-professional, manager, and professional) to facilitate later analysis.

Coding can be a fairly complicated, time-consuming process. We recommend that you consult other texts (such as Frankfort-Nachmias and Nachmias, 1992, Chapter 14, or Norusis, 1990, pp. 23–26, mentioned at the end of this chapter) for more complete discussions, if your coding adventures take you beyond the scope of this introduction.

Elementary Quantitative Analyses

Perhaps the defining characteristic of quantitative analyses is their effort to summarize data by using statistics. Statistics, being a branch of mathematics, is an area of research methods that some social science students would just as soon avoid. But the advantages of a little bit of statistical knowledge are enormous for those who would practice or read social science research, and the costs of acquiring that knowledge, we think you'll agree, aren't terribly formidable, after all.

descriptive statistics, statistics used to describe and interpret sample data.
inferential statistics, statistics used to make inferences about the population from which the sample was drawn.
univariate analyses, analyses that tell us something about one variable.
bivariate analyses, data analyses that focus on the association between two variables.
multivariate analyses, analyses that permit researchers to examine the relationship between variables while investigating the role of other variables.

To organize and limit our discussion of social statistics, however, we'd like to begin with two basic sets of distinctions: one distinction, between **descriptive** and **inferential** statistics relates to the generalizability of one's results; the other distinction among **univariate**, **bivariate**, and **multivariate** statistics relates to how many variables you focus on at a time. Descriptive statistics are used to describe and interpret data that we have from our sample. Filinson reports that one-half of her sample was widowed, and this meant that the modal (most common) marital status of her sample was "widowed." The mode is a descriptive statistic because it describes only the data one has in hand (in this case, data on about 50 respondents). Inferential statistics, on the other hand, are used to make estimates about characteristics of a larger body of data (population data—see Chapter 5 on sampling). Filinson reports that mixed-aged-housing respondents were not only more likely than were their elderly-housing residents to be subjected to "all kinds of disturbing behavior by other residents" (such as drunken behavior), but they were *significantly* more likely to be so. The word "significantly" in this context means that there's a good chance that, in the larger population from which Filinson drew her sample, mixed-aged-housing residents are subjected to more disturbing behaviors than their elderly-housing counterparts. She used an inferential statistic to draw this conclusion. In the rest of this chapter, we'll focus on descriptive statistics, but we'll give you a brief introduction to inferential statistics as well.

Much powerful quantitative research can be done with an understanding of relatively simple univariate (one variable) or bivariate (two variable) statistics. Univariate statistics, like the mode (the most frequently occurring category), tell us about one characteristic. The fact that the modal marital status was "widowed" in Filinson's sample tells us about the typical marital status of that sample. Bivariate statistics, on the other hand, tell us something about the association between two variables. When Filinson wanted to show that mixed-aged-housing residents were more likely than their elderly-housing counterparts to be exposed to disturbing behaviors by other residents—that is, that there is a relationship between "residence" and "exposure to disturbing behaviors," she used bivariate statistics. Multivariate statistics permit us to examine associations while we investigate the role of additional variables. Filinson hypothesized that the association between where residents lived and their exposure to disturbing behaviors might have been affected by their financial resources because adequate financial resources give people additional housing options, including housing in more protected environments. To show that this was not the reason for the original relationship, she employed multivariate statistical methods. In what follows, we'll focus on univariate and bivariate statistical analyses, while also giving you a taste of multivariate analysis.

Univariate Analyses

frequency distribution, a way of showing the number of times each category of a variable occurs in a sample.

A lot of interesting analysis can be accomplished by studying one variable at a time. One especially important kind of univariate analysis describes the distribution of a sample over various categories of a variable. One of the most commonly used techniques is a **frequency distribution**, which shows the frequency (or number) of cases in each category of a variable. To display such a distribution, the categories of the variable are listed, and then the number of times each category occurs is counted and recorded. Take a look, for instance, at the data about the gender of respondent used by Filinson. The categories, coded as 1 and 2, stand for "females" and "males," respectively. If you count, you'll find 17 of number "1s" and 3 "2s." The frequency distribution for this variable, then, is displayed in Table 15.1.

One thing this frequency distribution demonstrates is that the overwhelming majority of Filinson's respondents were female. The table has three rows. The first two display categories of the variable (gender). The third row displays the total of number of cases appearing in the table.

The middle column shows the number of cases in each category (17 and 3). The number is called the *frequency* and is often referred to by the letter f. The total of all frequencies, often referred to by the letter N, is equal to the total number of cases in the subsample examined (in this case, 20). The third column shows the percentage of the total number of cases that appears in each category of the variable. Because 17 of the 20 cases are female, the percentage of cases in this category is 85 percent.

TABLE 15.1

Frequency Distribution of Gender of Respondent

Gender	*Frequency (f)*	*Percentages*
Female	17	85
Male	3	15
Total	N = 20	100

TABLE 15.2

Frequency Distribution of Number of Confidantes

Number of Confidantes	*Frequency (f)*	*Percentages*
0		
1		
2		
3		
Total	N = 20	100

The variable in this example, gender, is a nominal level variable, but frequency distributions can be created for variables of ordinal and interval level as well. Let's confirm this by working with an interval-level variable from Filinson's subsample: the number of respondent's confidantes who are residents of the respondent's building.

Create the outline of an appropriate table on a separate piece of paper, using Table 15.2 as a guide.

Now complete the distribution. This will require you to do a bit of detective work. To start, find the column in the earlier data matrix that contains information about the number of confidantes. Now count the number of respondents who report having "0" confidantes in the building in which they live and put that number (frequency) in the appropriate spot in the table. Do the same for categories "1," "2," and "3." Do the frequencies in each of these categories add up to 20, as they should? If not, try again. Once they do, you can calculate the percentage of cases falling into each category. (Calculate percentages by dividing each *f* by the *N* and multiplying by 100. We count 5 respondents with 0 confidantes, so the percentage of respondents with 0 confidantes is 5/20 x 100 = 25.) Into which category does the greatest number of respondents fall? (We think it's category "1," with 9 respondents.) In which category does the smallest number of respondents fall? What percentage of the respondents said they had 3 confidantes?

Measures of Central Tendency

Although a frequency distribution is a thorough way of describing the distribution of any variable, it is often too cumbersome for many kinds of reports. Presenting a frequency distribution of the number of years of education among Filinson's respondents, for instance, wouldn't necessarily be the quickest way to provide a sense of their overall educational level. You could, instead, tell readers of the *average* educational level attained by them instead. In most distributions, values hover around an average, or central, value. The statistics that are used to depict this average are called measures of central tendency. The most common of these are the **mode**, the **median**, and the **mean**.

mode, the measure of central tendency designed for nominal level variables. The value that occurs most frequently.

In Chapter 6, we observed that the mode, median, and mean are designed for nominal, ordinal, and interval level variables, respectively. The mode, designed for nominal level variables, is that value or category that occurs most frequently. A look at Table 15.1 tells us that more of Filinson's respondents (in the subsample of 20) were female than male, so the modal gender for the subsample is "female."

STOP AND THINK *Look at Table 15.2, which you completed earlier, and decide what the mode is for "number of confidantes" in this subsample.*

Although the mode has been designed for nominal level variables, like gender, it can be calculated for ordinal and interval level variables (like number of confidantes) because ordinal and interval level variables, whatever else they are, are also nominal. As a result, the mode is an extremely versatile measure of central tendency: It can be computed for any variable.

STOP AND THINK *We say a variable is unimodal when, like "gender" and "number of confidantes" in Filinson's subsample, it has only one category that occurs most frequently. We say it is bimodal when it has two categories that occur most frequently. Check out the data for the variable "does respondent know of residents who persist in being drunk?" Is it unimodal or bimodal?*

median, the measure of central tendency designed for ordinal level variables. The middle value when all values are arranged in order.

The **median**, the measure of central tendency designed for ordinal level variables, is the "middle" case in a rank-ordered set of cases. Though it is designed for ordinal level variables, the median can be used with interval variables as well, because, interval variables, whatever else they are, are also ordinal variables. Thus, the first seven respondents in Filinson's subsample reported having the following years of education: 9, 9, 2, 9, 9, 7, 9. If you arrange these in order, from lowest to highest, they become 2, 7, 9, 9, 9, 9, 9. The fourth, or middle case, in this series is 9, so the median is 9.

STOP AND THINK *Why wouldn't you want to try to calculate the median "gender" for our sample?*

The median is pretty easy to calculate (especially for a computer) when you've got an odd number of cases (provided, of course, you've got at least ordinal level information). What you do with an even number of cases is locate the median halfway between the two middle cases. Thus, because the years of education of Filinson's first eight cases in the subsample are reported as 9, 9, 2, 9, 9, 7, 9, 12, and because, once they are arranged in order, they become: 2, 7, 9, 9, 9, 9, 9, 12, the median for the first eight cases is (9 + 9)/2 = 9, again.

STOP AND THINK *What is the median number of confidantes for cases 01 through 05 in Filinson's subsample? For cases 01 through 06?*

mean, the measure of central tendency designed for interval level variables. The sum of all values divided by the number of values.

The **mean**, the measure of central tendency designed for interval level variables, is the sum of all values divided by the number of values. The mean years of education for Filinson's first seven cases is

$$\frac{9 + 9 + 2 + 9 + 9 + 7 + 9}{7} = 7.7 \text{ years.}$$

STOP AND THINK *What is the mean number of confidantes of the first five cases in Filinson's subsample?*

How does a researcher know which measure of central tendency (mode, median, or mean) to use to describe a given variable? Beyond advising you not to use a measure that is inappropriate for a given level of measurement (such as a mean or a median for a nominal level variable like gender), we can't give you hard and fast rules. In general, though, when you are dealing with interval level variables (such as age, years in school, or number of confidantes), variables that *could* be described by all three averages, the relative familiarity of reading audiences with the mean makes it a pretty sound choice. Thus, although you could report that the modal number of years of education among Filinson's first seven cases is 9 or that their median is 9, we'd be inclined to report that the mean is 7.7 years, and leave it at that.

Measures of Dispersion

Measures of central tendency offer the advantage of reducing all the values of a particular variable to a single representative value. A problem with such measures, however, is that they can hide a great deal about the variable's *spread* or *dispersion*. Thus, a mean of 3 could describe both of the following samples:

Sample A: 1, 1, 5, 5
Sample B: 3, 3, 3, 3

range, a measure of dispersion or spread designed for interval level variables. The difference between the highest and lowest values.

Inspection shows, however, that Sample A's values (varying between 1 and 5) are more spread out or dispersed than are those of Sample B (*all* of which are 3). To alleviate this problem, researchers sometimes report **measures of dispersion** for individual variables. The simplest of these is the **range**: the difference between the highest value and the lowest. The range of Sample A would be 5 – 1 = *4*, while that of Sample B is 3 – 3 = *0*. The two ranges tell us that the spread of Sample A is larger than the spread of Sample B.

STOP AND THINK *What is the range of "years of education" for Filinson's whole (20 case) subsample?*

STOP AND THINK AGAIN *Calculate the range of Samples C and D:*

Sample C: 1, 1, 5, 5; Sample D: 1, 3, 3, 5

Having calculated the range, can you think of any disadvantage of the range as a measure of spread or dispersion?

standard deviation, a measure of dispersion designed for interval level variables that accounts for every value's distance from the sample mean.

There are several measures of spread that, like the range, require interval scale variables. The most commonly used is the **standard deviation**. The major disadvantage of the range is that it is sensitive only to extreme values (the highest and lowest). Samples C and D in the last Stop and Think exercise have ranges of 4, but the spreads of these two samples are obviously not the same. In Sample C, each value is two "units" away from the sample mean of 3. In Sample D, two values (the 1 and the 5) are two "units" away, but two values (the 3's) are zero units away. In other words, the average "distance" or "variation" from the mean is greater in Sample C than in Sample D. The average variation is 2 in Sample C, but less than 2 in Sample D. The standard deviation is meant to capture this difference (one that isn't caught by the range) and to assign higher measures of spread to samples like Sample C than to those like Sample D. And, in fact, it does.

It does so by employing a computational formula that, in essence, adds up the "distances" of all individual values from the mean and divides by the number of values—a little like the computation of the mean in the first place. That's the essence. In fact, the computational formula is

$$s = \sqrt{\frac{\Sigma(X - \bar{X})^2}{N}}$$

where s stands for standard deviation
$\bar{X}$ stands for the sample mean
X stands for each value
N the number of sample cases

Although this formula might look a bit formidable, it's not very difficult to use. We'll show you how it works for Sample D.

First notice that the computational formula requires that you compute the sample mean ($\bar{X}$). For Sample D, the mean is 3. Then subtract this mean from each of the individual values, in turn ($X - \bar{X}$). For Sample D (whose values are 1, 3, 3, 5), these differences are –2, 0, 0, and 2. Then square each of those differences [$(X - \bar{X})^2$]. For Sample D, this results in four terms: 4, 0, 0, 4 (–2 squared = 4). Then sum these terms [$\Sigma (X - \bar{X})^2$]. For Sample D, this sum is 8 (4 + 0 + 0 + 4 = 8). Then the formula asks you to divide the sum by the number of cases [$\Sigma (X - X)^2 / N$]. For Sample D, this quotient is 2 (8 / 4 = 2). Then the formula asks you to take the square root of this quotient. For Sample D, this is the square root of 2, or about 1.4. So the standard deviation of Sample D is about 1.4.[4]

STOP AND THINK *Now try to calculate the standard deviation of Sample C. Is the standard deviation of Sample C greater than (as we hoped it would be) or less than the standard deviation of Sample D?*

[4]Another measure of spread, the variance, is simply the standard deviance squared. For Sample D, then, the variance would be 2.

We hope you've found that the standard deviation of Sample C (we calculate it to be 2) is greater than the standard deviation of Sample D. In any case, we're pretty sure you'll see by now that the standard deviation does require variables of at least interval scale. (Otherwise, for instance, how could you calculate a mean?) You might also see why a computer is helpful when you are computing many statistics.

There are few measures of spread for nominal and ordinal level variables. For these, a good but sometimes uneconomical way of showing spread is to speak of the percentage of cases falling into each category. For instance, you might report, referring back to Table 15.1, that 85 percent of the cases in Filinson's subsample were female and 15 percent were male. Compared with the range, which, after all, tells a fair amount about the spread of a variable with a single number, such percentage comparisons are somewhat inelegant, especially when a variable has many categories. But they do make the point, in this case, that most of Filinson's cases were female (and that there is little dispersion in the variable gender).

Bivariate Analyses

Analyzing single variables can raise interesting questions. (Why do you suppose such a high percentage of Filinson's elderly, publicly housed respondents is female?) But the real fun begins when you begin to examine the *relationship* between variables. A relationship exists between two variables when categories of one variable tend to "go together" with categories of another. Filinson, for instance, had expected that residents in elderly housing would report fewer problems with other residents than would their counterparts in mixed-aged housing that accommodated a greater range of age groups. If you think of this hypothesis in the variable language introduced in Chapter 2 and residence (either in elderly or public housing) and the experience of problems with other residents (either problems or no problems) as variables, Filinson is saying she expects the categories of residing in elderly housing and a lack of problems to "go together," and the categories of residing in public housing and problems to go together. One way of depicting this expectation is shown in Figure 15.1.

In particular, Filinson expected that residents of elderly housing would be less likely to have seen other residents being drunk than would residents of mixed-aged housing. One way of showing such a relationship is to **crosstabulate**, or create a bivariate table for, the two variables. Crosstabulation is a technique designed for examining the association between two nominal level variables and is, therefore, in principle applicable to variables of any level of measurement (because all ordinal and interval-ratio variables can also be treated as nominal level variables). It is the tuna-noodle casserole of quantitative data analysis.

crosstabulation, the process of making a bivariate table to examine a relationship between two variables.

Bivariate tables (sometimes called contingency tables) provide answers to questions such as "Is there a difference between older residents of elderly housing and older residents in public housing in their exposure to certain kinds of problems?" Or, more generally to questions such as, "Is there a

FIGURE 15.1

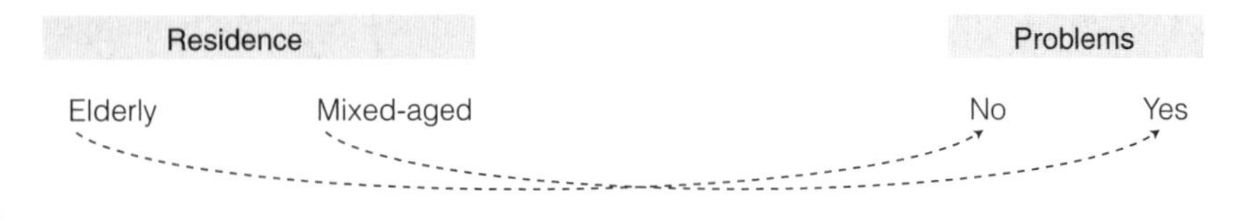

Filinson's Hypothesis about the Relationship Between Residence and the Experience of Problems with Other Residents

difference between sample members that fall into one category of an independent variable and their counterparts in other categories in their experience of another characteristic?" Return to the subsample of Filinson's data and note that the first 10 cases are from mixed-age public housing projects. Of those, only cases 06 and 10 report not having seen residents being drunk. This means that 8 out of 10, or 80 percent, of residents in public housing have seen such behavior (and 2, or 20 percent, haven't). Note that the second 10 cases in the same subsample are from elderly housing projects. Of those, all except cases 14 and 17 report never having seen residents being drunk. This means that 2 out of 10, or 20 percent, of residents in elderly housing have seen drunken behavior (and 8, or 80 percent, haven't). We'll put this information into a bivariate (or contingency) table, Table 15.3.

This table demonstrates that, as Filinson hypothesized, elderly residents of mixed-age housing were more likely to observe other residents engage in drunken behavior than were elderly residents in elderly housing. Type of residence is related to or associated with this type of problem; it makes a difference in the chances that a resident will have been exposed to drunken behavior.

STOP AND THINK *What percentages would you compare from the table to show that residents of mixed-age housing were more likely to observe other residents engage in drunken behavior than were residents in elderly housing?*

Note that the way Filinson formulated her hypothesis forces us to look at the relationship between place of residence and the observation of drunken behavior in a certain way. Using the variable language introduced in Chapter 2, Filinson expected "place of residence" to affect "observation of drunken behavior," rather than "observation of drunken behavior" to affect "place of residence." In other words, "place of residence" was her independent variable and "observation of drunken behavior," her dependent variable. Given Filinson's formulation of her hypothesis and given the analysis of Table 15.3, we're in a good position to say that place of residence might affect the chances of observing drunken behavior.

Let's pursue a few other points about crosstabulation by examining the two computer-produced analyses in Table 15.4. The major difference between this table and Table 15.3 is that Table 15.4 involves two relationships: (1) that between place of residence and sense of privacy (a variable whose values we haven't discussed before) and (2) that between place of residence and observation of drunken behavior. A second difference is that both of the analyses in Table 15.4 are based on all 50 cases in Filinson's sample, rather than the 20 cases shown earlier. Filinson had expected that elderly residents from mixed-age housing would be worse off than residents from elderly housing on both

TABLE 15.3 **Residence by Observation of Drunken Behavior**

Observes Drunken Behavior	*Place of Residence*	
	Mixed-Age Housing	*Elderly Housing*
Yes	8 (80%)	2 (20%)
No	2 (20%)	8 (80%)
Total	10 (100%)	10 (100%)

TABLE 15.4 **Residence by Sense of Privacy and by Observation of Drunken Behavior**

	Place of Residence	
	Mixed-Age Housing	*Elderly Housing*
Sense of Privacy		
Yes	25 (100%)	24 (96%)
No	0 (0%)	1 (4%)
Total	25 (100%)	25 (100%)
	Phi = .14	p > .05
Observes Drunken Behavior		
Yes	20 (80%)	6 (24%)
No	5 (20%)	19 (76%)
Total	25 (100%)	25 (100%
	Phi = .56	p < .05

counts: They would experience less privacy and they would observe more drunken behavior. A third difference is the presence of those curious little symbols (like phi and p) at the bottom of each part of the table.

As Filinson suggests in her article, her hypothesis about residence and privacy was not borne out by her data. All of her respondents from mixed-age housing report having enough privacy; 96 percent of the ones from elderly housing report having enough. To the extent that there's a difference in the experience of privacy, it actually seems to favor residents of mixed-age housing, but the difference is so small (4 percent—based on one case), we are inclined to chalk it up as unimportant (perhaps attributable to an idiosyncrasy of the one person in elderly housing who reported not having enough privacy). There seems, then, to be no real relationship between residence and sense of privacy in Filinson's sample.

But, also as Filinson suggests in her article, and as our analysis of fewer cases (in Table 15.3) had suggested, her hypothesis about residence and the observation of drunken behavior is well borne out by her data. Although 80 percent of her mixed-age housing respondents reported observing drunken behavior, only 24 percent of her elderly housing respondents did so. This difference is so large that it would be foolish to deny its meaningfulness. Public housing residents who live with mixed-age populations seem much more likely to observe drunken behavior than do their elderly housing counterparts. There *does* seem to be a real relationship, then, between residence and the observation of drunken behavior.

Measures of Association

measures of association, measures that give a sense of the strength of a relationship between two variables.

Before we leave the topic of bivariate analyses, let's note a few more important points. Table 15.4 contains some curious new symbols that are worth mentioning, mainly because they illustrate a whole class of others. One such class is called **measures of association**, of which "phi" is a specific example. Measures of association give a sense of the strength of a relationship between two variables, of how strongly two variables "go together" in the sample. Phi can vary between 0 and 1, with 0 indicating that there is absolutely no relationship between the two variables, and 1 indicating that there is a perfect relationship. A perfect relationship exists between two variables when change in one variable is always associated with a predictable change in the other variable. The closer phi is to 0, the weaker the relationship; the farther from 0 (closer to 1), the stronger the relationship.

STOP AND THINK *You'll note, from Table 15.4, that phi for the relationship between place of residence and sense of privacy is .14 and that phi for residence and observation of drunken behavior is .56. Which relationship is stronger?*

Measures of Correlation

Phi is a measure of association that can be used when both variables are nominal level variables (and have two categories each). Statisticians have cooked up literally scores of measures of association, many of which can be distinguished from others by the levels of measurement for which they're designed: some for nominal, some for ordinal, and some for interval level variables. A particular favorite of social science researchers is one called Pearson's r, designed for examining relationships between interval level variables. Pearson's r falls in a special class of measures of associations: It's a **measure of correlation**. As such, it not only provides a sense of the strength of the relationship between two variables, it also provides a sense of the *direction* of the association. When variables are intervally (or ordinally) scaled, it is meaningful to say that they can go "up" and "down," as well as to say they vary from one category to another.

measures of correlation, measures that provide a sense not only of the strength of the relationship between to variables, but also of its direction.

Suppose, for instance, that you had data about the education and annual income of three people. Suppose further that the first of your persons had 12 years of education and made $15,000 in income, the second had 13 years of education and made $20,000, and the third had 14 years of education and made $25,000. In this case, not only could you say that the two variables (education and income) were related to each other (that every time education changed, income changed), but you could also say that they were

FIGURE 15.2

A Graph of the Relationship Between Education and Income for Our Hypothetical Sample

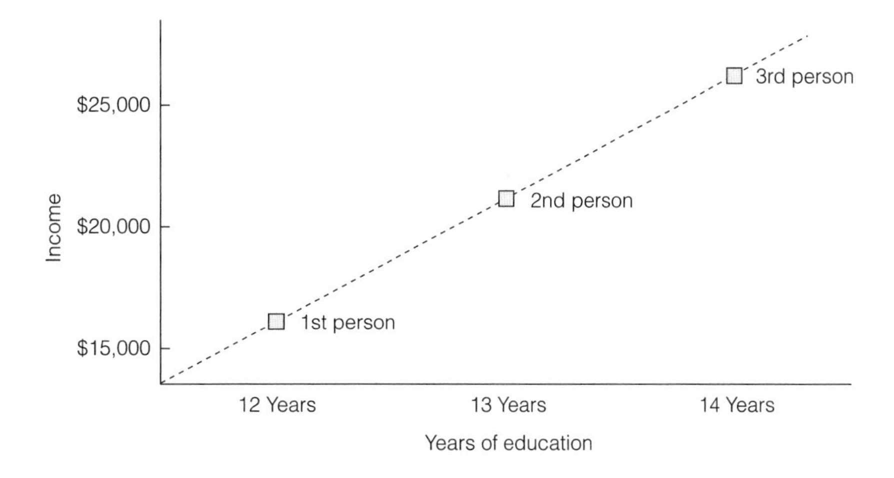

positively or *directly* related to each other—that as education rises so does income. You might even be tempted to graph such a relationship, as we have done in Figure 15.2.

One feature of our graphical representation of the relationship between education and income is particularly striking: All the points fall on a straight line. And one striking feature of this line is that it rises from the bottom left of the graph to the top right. When such a situation exists (that is, when the data points from two variables fall on a straight line that rises from bottom left to top right), the correlation between two variables is said to be perfect and positive, and the Pearson's r associated with their relationship is 1.

STOP AND THINK *Suppose your sample included three individuals, one of whom had 14 years of education and made $15,000, one of whom had 13 years of education and made $20,000, and one of whom had 12 years of education and made $25,000. What would a graphical representation of this data look like? Can you imagine what the Pearson's r associated with this relationship would be?*

If, on the other hand, the data points for two variables fell on a line that went from the "top left" of a graph to the "bottom right" (as they would for the data in the Stop and Think exercise), the relationship would be perfect and negative and the Pearson's r associated with their relationship would be –1. Thus, Pearson's r, unlike phi but like all other measures of correlation, can take on positive *and* negative values, between 1 and –1.

In general, negative values of Pearson's r (less than 0 to –1) indicate that the two variables are *negatively* or *indirectly* related, that is, as one variable goes up in values, the other goes down. Positive values of Pearson's r (greater than 0 to 1) indicate that the two are directly related—as one goes up, the other goes up. In both cases, r's "distance" from 0 indicates the strength of the relationship: The farther away from 0, the stronger the relationship. Thus, a Pearson's r of .70 and a Pearson's r of –.70 indicate relationships of equal strength (quite strong!) but opposite directions. The first (r = .70) indicates that as one

FIGURE 15.3

Idealized Scattergrams of Relationships of R = .70 and -.70

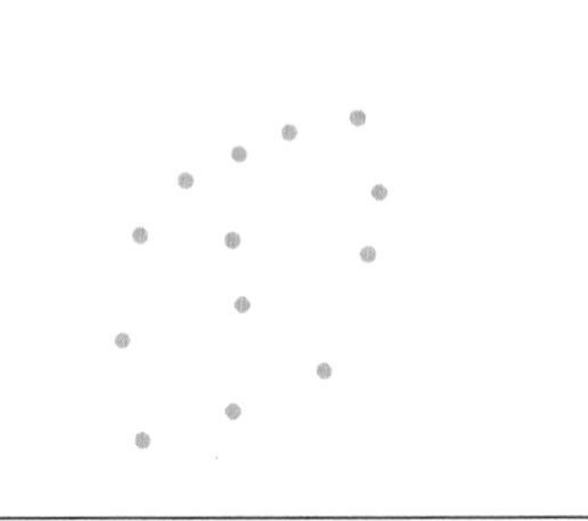

r about equal to .70

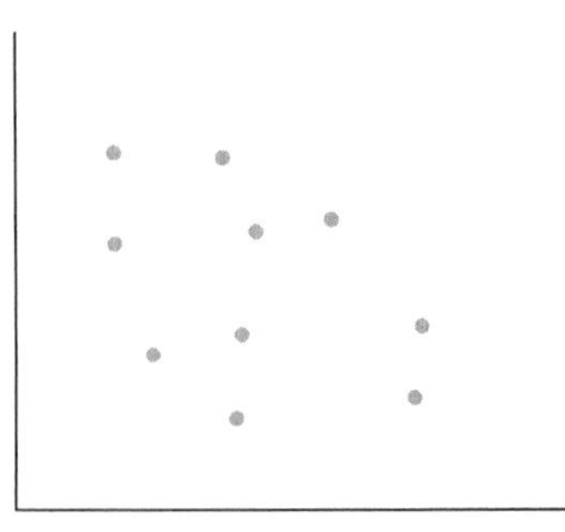

r about equal to – .70

variable's values go up, the other's go up as well (pretty consistently); the second (r = –.70) indicates that as one's values go up, the other's go down (again pretty consistently). Figure 15.3 provides a graphical representation (known as a *scattergram*) of what relationships (with many data points) with r's of .70 and -.70 might look like. Notice that, although in neither case do the points perfectly conform to a straight line, in both cases they all hover around a line. R's of .30 and –.30 would, if you can imagine them, conform even less well to a straight line.

To turn to a real-life example, we've done (using a computer) a correlation analysis of Filinson's variables, "years of education" and "number of confidantes." We've been able to do this because both are interval-level variables. It is possible that as "years of education" goes up, so would "number of confidantes"; but it's also possible that as "years of education" goes up, "number of confidantes" would go down. As it turns out, the Pearson's r relating these two variables is .20, indicating that as respondents' years of education go up, their number of confidantes tends to go up as well (the positive "sign" of .20 indicates a direct relationship), but that the relationship is not a very strong one (.20 is closer to 0 than to 1)—so one would not expect the data points to conform very well to a straight line.

STOP AND THINK *Suppose the Pearson's r for the relationship between years of education and number of confidantes had proven to be –.67, instead of .20. How would you describe the relationship?*

We could tell you lots more about correlation analysis, but any standard statistics book could do so in more detail. We think you'll find correlation analysis a particularly handy form of analysis if you ever have to study the associations among interval level variables.

Inferential Statistics

We draw your attention to a final detail about Table 15.4 before we conclude our discussion of bivariate relationships. On the same row as phi in each part of the table is a funny looking set of symbols, either "p > .05" or "p < .05." These two, standing for "probability is greater than .05" and "probability is less than .05," respectively, enable the reader to make inferences about whether a relationship, like the one shown in the table for the sample (of, in this case, 50 respondents), exists in the larger population from which the sample is drawn. These "probabilities" are estimated using what we've called inferential statistics, in this case chi-square, perhaps the most popular inferential statistic used in crosstabulation analysis. There are almost as many inferential statistics as there are measures of association and correlation, and a discussion of any one of them stands outside the scope of this chapter. Inferential statistics are properly computed when one has a probability sample (see Chapter 5). They are designed to permit inferences about the larger populations from which probability samples are drawn.

You may recall from Chapter 5 however, that, even with probability sampling, it is possible to draw samples whose statistics misrepresent (sometimes even badly misrepresent) population parameters. Thus, for instance, it is possible to draw a sample in which a strong relationship exists between two variables even when no such relationship exists in the larger population. Barring any other information, you'd probably be inclined to infer that a relationship also exists in the larger population. This kind of error—the one where you infer that a relationship exists in a population when it really doesn't—is called a Type I error.[5] Such an error could cause serious problems, practically and scientifically. Suppose, for instance, that you drew a sample of adult Americans in which Republicans proved more likely than Democrats to vote for candidate A. Suppose, further, that you bet a friend $10,000 to her $10 that this relationship would prove to exist in the larger population when election time rolled around. Now, suppose again, that she took the bet and that, when election time rolled around, more Democrats than Republicans voted for A. Practically, and scientifically, you'd be in trouble.[6]

Social scientists, like intelligent bettors, are a conservative lot and want to keep the chances, or probability, of making such an error pretty small—generally lower than 5 times in 100, or "$p < .05$." When the chances are greater than 5 in 100, or "$p > .05$," social scientists generally decline to take the risk of inferring that a relationship exists in the larger population. Thus, for instance, although there is a (very weak) relationship between place of residence and sense of privacy in Filinson's sample (with 100 percent of her public housing residents reporting a sense of privacy and "only" 96 percent of her elderly housing residents doing so), the chances of making an error when inferring such a relationship would exist in the larger population ("$p > .05$") are unacceptably high, and most social scientists wouldn't be tempted to do it.

STOP AND THINK *When "$p < .05$," social scientists generally take the plunge and infer that a relationship, like the one in the sample, exists in the larger population. Would social scientists take such a plunge with the relationship between place of residence and observation of drunken behavior, depicted in Table 15.4?*

[5]One makes a Type II error when one fails to infer that a relationship exists in a larger population when, in fact, it does.

[6]We thank a reviewer, Kevin Thompson of North Dakota State University, for suggesting the betting analogy for showing how terrifying Type I errors should be.

Measures of association, measures of correlation, and inferential statistics are not an exotic branch of statistical cookery. They are the "meat and potatoes" of quantitative data analysis. We hope we've presented enough about them to tempt you into learning more about them. We'd particularly recommend Marija Norusis' *The SPSS Guide to Data Analysis*, which, in addition to offering a pretty good guide to SPSS (the statistical package we used to generate Table 15.4), offers a good guide to the kinds of statistics we've referred to here.

Multivariate Analysis and the Elaboration Model

Examination of bivariate relationships is the core activity of quantitative data analysis. It establishes the plausibility of the hypothesized relationships that inspired analysis in the first place. Filinson believed that elderly residents of mixed-age housing would have certain negative experiences that elderly residents of elderly housing would not. To use the variable language introduced in earlier chapters, she expected that an independent variable, residence of respondent, would be associated with various dependent variables such as the observation of drunken behavior. Her bivariate analyses suggested she was right. But the examination of bivariate relationships is frequently just the beginning of quantitative data analyses. Often the researcher will want to pursue questions like "Is there a third variable, associated with both the independent and dependent variable, that can account for the association between the two?," or "Are there certain conditions under which the association between the independent and dependent variables is particularly strong or weak?," or "Are there any other variables, affected by the independent variable and affecting the dependent variable, that can help me understand how the independent variable affects the dependent variable?" The pursuit of any such question can lead researchers to elaborate on the original bivariate relationship by introducing additional variables. In doing so, they will employ what Paul Lazarsfeld dubbed the **elaboration** model. A complete description of Lazarsfeld's elaboration model is beyond the scope of this book. (We recommend Morris Rosenberg's classic (1968) *The Logic of Survey Analysis* for the fullest available description of it.) But we think you deserve at least a taste of it.

elaboration, the process of examining the relationship between two variables by introducing the control for another variable or variables.

To give you an idea of how elaboration works, recall Filinson's discussion of the relationship between residence of respondent and whether he or she is exposed to drunken behavior. Table 15.4 shows, for instance, that elderly respondents in mixed-age housing were substantially more likely to be exposed to such behavior than were their counterparts in elderly housing. But why? Filinson believes it has to do with characteristics of younger residents in age-integrated public housing, some of whom might have been institutionalized in an earlier period in our nation's history. But one counter-explanation considered by Filinson, undoubtedly in anticipation of what other researchers might claim, is that residence and exposure to drunken behavior are *not* causally related at all. Their association might reflect their common association with a third variable, economic status, that forces them to "go together," as it were. The idea behind this explanation is that because poorer respondents are more likely both to live in mixed-age housing and to see drunken behavior and because wealthier respondents are more likely to live in elderly housing and not see drunken behavior, the original relationship results from the association of both residence and the observation of drunken behavior with economic status. In this case, economic status is hypothesized to be an antecedent **control variable** because it is expected to affect both residence and the observation of drunken behavior. (In the variable language of earlier chapters, it is a possible independent variable for both residence and such observation.) This possibility is depicted in Figure 15.4.

control variable, a variable that is held constant to examine the relationship between two other variables.

Now, let's be clear. Filinson entertained the possibility that economic status might prove to be the reason why residence and the observation of drunken behavior were related, but she did so largely because she anticipated that critics of her research might think it was the reason. She didn't really want economic status to prove to be the reason. (That would have undermined her argument that residence itself has a significant impact for the elderly.) But to show that it wasn't the reason, she needed to understand how one would go about showing that it was (or, at least might be). To do that, she realized, one would have to do three things: (1) show that economic status and residence are related, (2) show that economic status and observation of drunken behavior are related, and (3) show that when economic status is *controlled,* the association between residence and the observations of drunken behavior "disappears." Thanks to Lazarsfeld's elaboration model, Filinson knew that all three of these conditions had to exist to conclude that economic status "explained" the relationship between residence and the observation of drunken behavior. In fact, explanation is the term Lazarsfeld gave to that form of elaboration that shows a spurious relationship: a relationship that is explained away by an antecedent control variable.

When Filinson found that neither of her measures of economic status (either whether a respondent received Medicaid [national health support for the poor] or SSI [or national income support for the poor]) was significantly associated with the observation of drunken behavior, or any other "environmental outcome" variable, she concluded that economic status couldn't explain the original association.

STOP AND THINK *When Filinson claims that, say, the association between the reception of Medicaid support and the observation of drunken behavior is not significant she's referring to the probability of making an appropriate inference to a larger population. Does this lack of significance mean "*$p > .05$*" or "*$p < .05$*?"*

You might be able to imagine how Filinson demonstrated that, say, the reception of Medicaid support is not associated with the observation of drunken behavior (she could have crosstabulated the two variables), but can you imagine what it would mean to show that when economic status (in this case, measured as the reception [or not] of Medicaid support) is *controlled,* the association between residence and the observation of drunken behavior "disappears"? Understanding the introduction of a control variable is really the key to understanding the elaboration model, so we'd like to show you how it's done.

FIGURE 15.4

Economic Status as an Antecedent Control Variable for Residence and Observation of Drunken Behavior

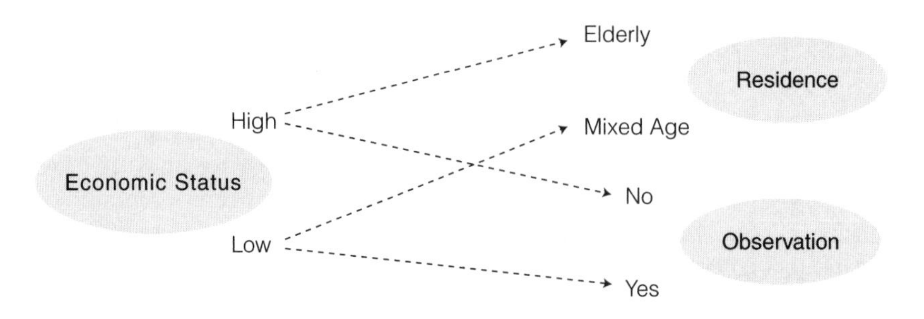

In the context of crosstabulation, the concept of *controlling* a relationship (for example, between residence and observation of drunken behavior) for a third variable (for example, whether a respondent receives Medicaid) is reasonably straightforward. It means looking at the original association separately for cases that fall into each category of the control variable. For the control variable, whether a respondent receives Medicaid, there are two categories: yes, the respondent receives Medicaid, and no, she or he doesn't. Controlling the relationship between residence and the observation of drunken behavior for whether a respondent receives Medicaid means, then, examining that relationship only for those who receive Medicaid, on the one hand, and only for those who don't receive Medicaid, on the other.

But what does it mean to say that a relationship "disappears" when you control for a third variable? Table 15.5 illustrates with a hypothetical example.

The first segment of this table simply replicates the information about the association between residence and the observation of drunken behavior from Table 15.4 and shows that, for the full sample of 50 respondents, a substantial relationship exists. We could emphasize the strength of this relationship by comparing the percentages (for example, by saying that 80 percent of mixed-age housing residents have seen drunken behavior, while only 24 percent of elderly housing residents have) or by looking at the measure of association (a phi of .56 indicates a strong association).

partial relationship, the relationship between an independent and a dependent variable for that part of a sample defined by one category of a control variable.

The second and third segments of this table reflect the introduction of the control variable, whether respondents receive Medicaid. The second segment only presents hypothetical information about respondents who have received Medicaid. Here we examine what is sometimes called a **partial relationship** (or **partial**, for short) because we're only examining the original relationship for part of the sample—for Medicaid recipients. Within this group, the table suggests, all 20 respondents who lived in mixed-age housing had seen drunken behavior, and all 6 respondents who lived in elderly housing had seen such behavior. In other words, there is no relationship between residence and having seen drunken behavior for those who received Medicaid—a fact that is underscored by a phi of .00. Similarly, the third segment of the table presents hypothetical information only about respondents who have not received Medicaid. Within this group too, there is no relationship between residence and the observation of drunken behavior.

STOP AND THINK *How can we tell there's no relationship in the* ***partial*** *involving only those who have not received Medicaid?*

TABLE 15.5

Hypothetical Example
Demonstrating How the Relationship Between Residence and Observation of Drunken Behavior Could "Disappear" With Reception of Medicaid Controlled

Full Sample N = 50	*Mixed-age*	*Elderly*
Observed Behavior?		
Yes	20 (80%)	6 (24%)
No	5 (20%)	19 (76%)
Total	25 (100%)	25 (100%)
	Phi = .56	p < .05
Medicaid Recipients N = 26		
Yes	20 (100%)	6 (100%)
No	0 (0%)	0 (0%)
Total	20 (100%)	6 (100%)
	Phi = .00	p > .05
Not Medicaid Recipients N = 24		
Yes	0 (0%)	0 (0%)
No	5 (100%)	19 (100%)
total	5 (100%)	19 (100%)
	Phi = .00	p > .05

Consequently, because there's no relationship between residence and observation for Medicaid recipients and none for non-Medicaid recipients, the relationship that existed with no control has "disappeared" in each of the controlled situations. Table 15.5, then, makes the kind of demonstration that would lead researchers to doubt that residence "causes" the observation of drunken behavior. It shows a situation where the control for a third variable makes the relationship between an independent and a dependent variable "disappear."

But this table was based on hypothetical data. As it turns out, nothing like what the table shows occurred in reality. Table 15.6 shows what really occurred when we controlled the relationship between residence and the observation of drunken behavior for whether or not respondents received Medicaid, using data Filinson had actually collected. The real data indicate that the partial relationship between residence and the observation of drunken behavior is almost exactly as strong for Medicaid recipients (phi =. 55), on the one hand, and non-Medicaid recipients (phi = .52), on the other, as it had been in the larger sample (phi =.56). Certainly, the relationship doesn't "disappear" for either Medicaid recipients or non-recipients.

Consequently, the introduction of a control for economic status (whether a person receives Medicaid or not) does not do what Filinson feared it might: "explain away" the original relationship. As a result, she's in

TABLE 15.6 **What Really Happened to the Relationship Between Residence and Observation of Drunken Behavior with Reception of Medicaid Controlled Medicaid Recipients**

Medicaid Recipients N = 24	*Mixed Age*	*Elderly*
Observed Behavior?		
Yes	3 (19%)	6 (75%)
No	13 (81%)	2 (25%)
Total	16 (100%)	8 (100%)
	Phi = .55	p < .05
Not Medicaid Recipients N = 26		
Yes	2 (22%)	13 (77%)
No	7 (78%	4 (23%)
Total	9 (100%)	17 (100%)
	Phi = .52	p < .05

a stronger position to assert that the relationship might be "causal" (rather than spurious, or due to the action of a third variable that makes the variables in the relationship appear to vary together).

STOP AND THINK *Of course, you can never entirely eliminate the possibility that a relationship is spurious. In Chapters 2 and 7 we suggested that such an elimination would require showing that there is no antecedent variable whose control makes the original relationship disappear. Can you see why this is not possible?*

To summarize and, well, elaborate, elaboration permits us to better understand the meaning of a relationship (or lack of relationship) between two variables. We've actually discussed two types of elaboration above, but there are several others as well. Let us briefly introduce you to four kinds of elaboration: replication, explanation, specification, and interpretation.

replication, a kind of elaboration in which the original relationship is replicated by all the partial relationships.

1. **Replication**. The original relationship can be replicated in each of the partial relationships. This was essentially true when Filinson controlled the relationship between residence and having seen drunken behavior by reception (or non-reception) of Medicaid (see Table 15.6 above). The partials (with phi's of .55 and .52) were just about the same as the original (phi =.56).

explanation, a kind of elaboration in which the original relationship is explained away as spurious by the control for an antecedent variable.

2. **Explanation**. The original relationship can be *explained away* (or found to be *spurious*) through the introduction of the control variable. Two variables that were related in the bivariate situation might no longer be related when an *antecedent* variable is controlled. This happened in the hypothetical (but unreal) situation, shown in Table 15.5, when the relationship between type of residence and exposure to drunken behavior

"disappeared" for both respondents who had received Medicaid and for those who had not.

specification,
a kind of elaboration which permits the researcher to specify conditions under which the original relationship is particularly strong or weak.

3. **Specification**. Introducing a control variable might permit the researcher to *specify* conditions under which the original relationship is particularly strong or weak. Suppose, for instance, that the control for Medicaid status had led to the finding that for respondents who were on Medicaid, the relationship between residence and exposure to drunken behavior was non-existent (with a phi of .00). Suppose, moreover, that the control had led to the finding that this relationship was very strong (say, with a phi of .80) for non-Medicaid recipients. Then the researcher could specify a condition (that is, when respondents didn't receive Medicaid) that the relationship between residence and exposure to drunken behavior was particularly strong, as well as a condition (that is, when respondents did receive Medicaid), when it was particularly weak.

interpretation,
a kind of elaboration that provides an idea of the reasons why an original relationship exists without challenging the belief that the original relationship is c

4. **Interpretation**. A control variable can actually give the researcher an idea about why the independent variable is related to the dependent variable *without challenging the belief that the one causes the other*, as it does in *explanation*. In this case the control is said to *interpret* rather than *explain* the relationship. Let's change our examples here. Suppose you noticed, after many years of observation, that as the seasons in your area changed from autumn to winter, the leaves fell from (deciduous) trees—at is, that there was a relationship between season and the presence of leaves. Suppose further that you thought a third variable, air tem-re, was the reason why this relationship existed. You might con-relationship between season (your independent variable) and tree leaves (your dependent variable) for air temperature trol variable). You might visualize the relationship among hree variables in the manner we've depicted in Figure 15.5.

otice that in Figure 15.5, the control variable (air temperature) isn't epicted as occurring before the independent variable (season), as it needs to be for explanation. Indeed, it (temperature) is believed to be affected by the independent variable (season) and, in turn, to affect the dependent variable (presence of leaves). It is expected to "come between" the independent and dependent variables and is therefore called an *intervening control* variable.

To show that the model in Figure 15.5 is plausible, a researcher would, as in explanation, have to find that the partial relationships between the independent and dependent variables were much smaller than the original relationship, or that they "disappeared" altogether. Crucially, however, the disappearance in this case would have a very different meaning than it did when the control variable was antecedent to both of the original variables. Now it would suggest that seasonal change affects temperature change, which in turn affects the falling of the leaves. Seasonal change's causal effect on the falling of the leaves would not be questioned; it would simply be interpreted as seasonal change's effect on temperature change and temperature change's effect on leaves.

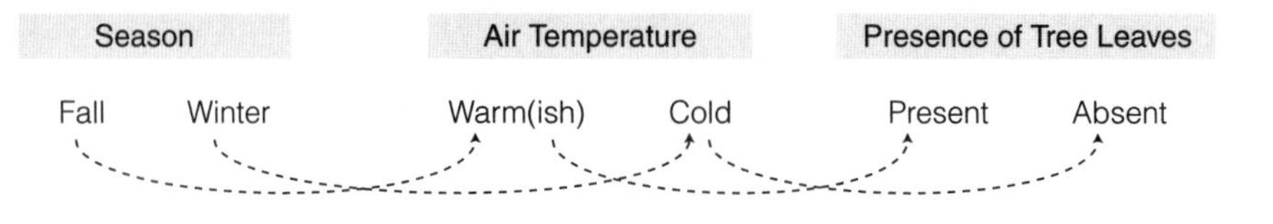

FIGURE 15.5

Air Temperature as an Intervening Control Variable for Season and Presence of Tree Leaves

The purpose of the elaboration model is to give us a way of examining relationships that can be *causal*. One requirement for showing causality is to show that there are no *antecedent* variables that cause the relationship between the independent and the dependent variable to exist. In Chapter 8 we mentioned that experimental designs are especially useful designs for dealing with questions of causation. This is not only because experiments generally ensure the appropriate temporal sequencing of the independent and dependent variables, but also because, through the randomized selection of subjects into experimental and control groups, they minimize the chance that any antecedent variables will end up affecting the relationship between those two variables.

The greatest advantage of the elaboration model is that it permits the examination of causal relationships with data from non-experimental designs. Filinson used the model, for instance, to eliminate one possible reason why the relationship between residence and exposure to drunken behavior might be spurious (or noncausal). She did this with data from a cross-sectional survey design. The greatest problem with the elaboration model, however, is that it might provide the illusion that you have demonstrated causation when no such demonstration is possible. One requirement for demonstrating causation is to show that there is **no** antecedent variable whose control makes the original relationship disappear. The problem is that because there are, in principle, an infinite number of possible antecedent variables for any given relationship, you can never eliminate them all through elaboration. Filinson eliminated one (Medicaid status). But, all that did was make slightly more plausible her belief that residence and exposure to drunken behavior was causal. She could never have explored all possible antecedents to her relationship of concern.

Elaboration with Interval Level Variables[7]

And so it is with multivariate analysis generally. We can establish that relationships are noncausal (through explanation) or make a better case for causality (through replication). We can find conditions (through specification) under which a relationship is particularly strong (or weak) [when one partial is stronger (or weaker) than others], or even find out why one variable affects another (through interpretation). And we can do these things whether our variables are

[7] We thank three reviewers, Lawrence Rosen of Temple University, Michael Kleiman of the University of South Florida, and Cynthia Beall of Case Western Reserve University, for suggesting the value of showing, as we do here, at least one alternative way that social scientists do elaboration.

nominal, ordinal, or interval. Only the statistical procedures, not the logic, change with measurement level.

multiple regression analysis, a method of data analysis that yields, among other things, an estimate of the correlation between each of a series of independent variables and a dependent variable when all other (included) independent variables are controlled.

standardized regression coefficient, (or beta), a product of regression analysis that provides the estimate of the correlation between an independent and dependent variable.

As an example of how multivariate analysis is done with interval level variables, let's look at **multiple regression analysis**. In our discussion of correlation analysis, we suggested that one way of interpreting a Pearson's r, which can vary from –1 to 1, is by how much a series of data points deviate from a straight line. The more they conform to a line, the closer r will be to either –1 or 1; the less their graph approximates a line, the closer r will be to 0.

But what line? One purpose of regression analysis is to calculate (this is usually best done with the aid of a computer) the formula for a line that comes closest to all data points. But that's not the feature of regression analysis that interests us here. One product of regression analysis is something called a **standardized regression coefficient**, and the neat thing about this coefficient is that, when there's only one independent variable, the standardized regression coefficient is exactly the same as Pearson's r. Thus, when we calculated the Pearson's r for the relationship between "years of education" and "number of confidantes" for Filinson's sample (remember, r was .20), we were also calculating the standardized regression coefficient for that relationship. In other words, the standardized regression coefficient and Pearson's r bear exactly the same interpretation when you have one independent variable.

STOP AND THINK *Do you recall what the Pearson's r, and so the standardized regression coefficient, of .20 tells us about this relationship?*

regression analysis, a method of data analysis that yields a formula for the straight line that "best fits" a series of data points, but that also yields estimates of the strength and direction of the relationship between interval level variables.

Regression analysis that involves one independent variable is called simple **regression analysis**. But regression analysis can be used when you have one interval-level dependent variable and more than one interval-level independent variables as well. In this case, it's called multiple regression analysis. One of the most frequently used products of multiple regression analysis is its generation of separate standardized regression coefficients (sometimes called betas) for each of the independent variables. Moreover, and here's the big news, each of these coefficients stands for the relationship of the corresponding independent variable to the dependent variable, *when all the other independent variables are controlled*. Using the language we introduced earlier, each of these standardized regression coefficients represents the controlled association between the relevant independent variable and the dependent variable.

One example that will stand for all (and for those examples we could show you using many other techniques) comes from the article by Steven Messner from Chapter 12. Messner had expected that the more a population (in an SMSA) was exposed to TV violence, the higher its crime rates would be. Instead, he found that the more a population was exposed to such violence, the lower its crime rates were. Messner anticipated that critics might find fault with his results and argue that they were due to the "action" of some other variable, much as Filinson anticipated criticism about her findings about housing and exposure to undesirable behavior among her elderly respondents. So Messner "controlled" for a large number of other independent

TABLE 15.7 **Regression Results for Violent Crime Rates, TV Violence Exposure, and Additional Characteristics of SMSAs (N = 281)[a]**

	Dependent Variables				
Independent Variables	*Criminal Homicide*	*Forcible Rape*	*Robbery*	*Aggravated Assault*	*Violent Crime Index*
Exposure to TV Violence	–.17*	–.17*	–.15*	–.12	–.17*
Percent Black	.43*	.42*	.39*	.32*	.43*
Percent 18–34	–.08	–.18*	–.18*	–.17	–.20*
Males/Females	.14*	.16*	.13*	.19*	.19*
Percent less than HS education	.21*	–.37*	–.06	–.05	–.09
Population (ln)	.29*	.28*	.41*	.09	.30*
Gini Coefficient	.41*	.28*	.36*	.36*	.42*
AFDC Payments	–.18*	–.01	–.03	.03	–.01
Population per sq mi. (ln)	–.09	–.15*	.20*	–.03	.08
Percent Poor	–.12	.05	–.05	.00	–.03
South	–.08	.05	–.23*	.05	–.09
Adj. R^2	.58	.39	.65	.27	.52

Notes: a. Standardized regression coefficients are reported.
* $p < .05$

variables and presented his results in the table which we reproduce as Table 15.7.

This table permits Messner to examine the association of "exposure to TV Violence" with various dependent variables when various independent variables are controlled. Let's look at the last column of the table, labeled "Violent Crime Index," to see how it does this. Here the dependent variable is the "Violent Crime Index" of SMSAs. Reading across the row labeled "exposure to TV Violence" we find a standardized regression coefficient of -.17. This means that the association between "exposure to TV violence" and the "violent crime index" in America's SMSAs remains negative, even when all other independent variables (namely, "Pct. Black" to "South") in the table are controlled. As a result, Messner can conclude that the association between "exposure to TV violence" and "violent crime index" is not "explained away" by any of the other variables.

Note too that, in addition to exposure to TV violence, each of the other independent variables listed (from "Pct. Black" to "South") is associated with a standardized regression coefficient. One benefit of multiple regression is that it permits us to simultaneously examine the controlled association of each independent variable with the dependent variable, when all other variables are controlled. So, for instance, the coefficient of .43 associated with "Pct. Black" indicates that there is a positive relationship between the percentage of an SMSA's population that is black and its violent crime index (that as population goes up, so does crime), even when all other independent variables in the table are controlled. This relationship, Messner suggests, reflects not only group differences between blacks and whites in age, income, occupation, education, family background, and other social characteristics, but also differences in the experience of historical and contemporary patterns of discrimination.

STOP AND THINK *Can you interpret the coefficient of -.20, associated with "Pct. 18–34" (percentage of the population that's between the ages of 18 and 34)?*

Used in this way, multiple regression analysis is similar to a host of other statistical techniques (for example, multiple correlation analysis, analysis of covariance, discriminant analysis) that permit us to examine the association between an individual independent variable and a dependent variable when other independent variables are controlled. Space limitations prevent us from telling you more about the splendors (and provisos) associated with each of them, but we highly recommend Marija Norusis'(1990) *SPSSX Advanced Statistics Guide.*

Summary

The most common goal of quantitative data analysis is to demonstrate the presence or absence of association between variables. To demonstrate such associations, as well as other characteristics of variables, quantitative data analysts typically compute statistics, often with the aid of computers.

Especially when they use computers, quantitative data analysts need to code their data, or prepare them for computer use. Coding frequently involves assigning a number to each piece of data. Codebooks remind analysts of how they've treated data during the coding process.

Statistics can be distinguished by whether they focus on describing a single sample, in which case they are descriptive statistics, or on speaking to the generalizability of sample characteristics to a larger population, in which case they are inferential statistics. Statistics can also be distinguished by the number of variables they deal with at a time: one (as in univariate statistics), two (bivariate statistics), or more than two (multivariate statistics).

The most commonly used univariate statistical procedures in social data analysis are frequency distributions, measures of central tendency, and measures of dispersion. Measures of central tendency, or average, are the mode, the median, and the mean. Measures of dispersion include the range and standard deviation. All the univariate statistical procedures described in this chapter focus on describing a sample and are therefore descriptive, rather than inferential, in nature.

We demonstrated bivariate and multivariate statistical procedures by focusing on crosstabulation, one of many types of procedures that allow you to examine relationships among two or more variables. In passing, we discussed descriptive statistics, like phi and Pearson's r, used to describe sample relationships, and inferential statistics, like chi-square, used to make inferences about population relationships. We also illustrated ways of controlling the relationship between two variables through the introduction of control variables.

SUGGESTED READINGS

Craft, J. 1990. *Statistics and data analysis for social workers*. Itasca, IL: F. E. Peacock.

A thin and readable guide to statistics commonly used in the social sciences.

Norusis, M. 1990. *The SPSS guide to data analysis*. Chicago: SPSS Inc.

An excellent guide to one of the most frequently used computer packages for statistical analysis in the social sciences. Also good for its enlightening discussions of various topics in quantitative data analysis, from coding to all kinds of statistical analysis.

Rosenberg, M. 1968. *The logic of survey analysis*. New York: Basic Books.

A wonderful guide to various forms of the elaboration model.

EXERCISE 15.1

Univariate Analyses

This exercise gives you a chance to employ univariate descriptive statistical analyses on two of Filinson's variables. In your work, please refer to the subsample of 20 cases presented on pages 397 and 398 of this chapter.

Years of education:

1. Using the codebook on page 400, identify, again, the columns in the rectangular data matrix on pages 397 and 398 that are devoted to the years of education residents have. In what columns are these data held?

2. What level of measurement (nominal, ordinal, or interval) is the variable "years of education?"

3. What measures of central tendency could be used to describe "years of education?"

4. Which measure of central tendency would you be most inclined to use? Why?

5. Calculate this measure of central tendency for years of education using the data on all 20 cases depicted on pages 397 and 398. What does this measure show about the distribution of years of education in this sample?

6. What measure of dispersion could you easily use to describe years of education?

7. Calculate a measure of dispersion for years of education. What does this measure show about the distribution of years of education?

Occupation:

1. Identify the column in the matrix on pages 397 and 398 devoted to occupation. In what column is this variable located?

2. Using the codebook, again determine what the numbers in the column stand for.

3. What level of measurement is "occupation," as used here?

4. What measure of central tendency would you use to describe this variable? Why?

5. Calculate the measure of central tendency for occupation.

   (Be sure to translate the code into English. For example, if the code were "1," what would the "1" stand for?)

6. What would be a way of describing the dispersion of occupation for this subsample?

EXERCISE 15.2

A Bivariate Analysis

This exercise will enable you to examine the association between two of Filinson's variables that we might expect to be strongly associated, especially because she argues both are indicators of economic status: whether respondents receive Medicaid and whether they receive Supplemental Security Income.

1. Identify the columns in the rectangular matrix on pages 397 and 398 in which one would expect to find data on Medicaid and Supplemental Security Income, respectively.

2. What do the numbers in these columns indicate?

3. Now fill in the following bivariate table, following steps I through VIII that follow the table.

Receipt of Medicaid by Receipt of Supplemental Security Income

	Does Respondent Receive Medicaid	
	Yes	No
Does Respondent Receive Supplemental Security Income?		
Yes	a)	b)
No	c)	d)
Total	e)	f)

I. Count and place the number of respondents who receive Medicaid and SSI after a) in the table.

II. Count and place the number of respondents who don't receive Medicaid but do receive SSI after b) in the table.

III. Count and place the number of respondents who receive Medicaid but do not receive SSI after c) in the table.

IV. Count and place the number of respondents who don't receive Medicaid and don't receive SSI after d) in the table.

V. Count and place the total number of respondents who receive Medicaid after e) in the table.

VI. Count and place the total number of respondents who don't receive Medicaid after f) in the table.

VII. Calculate the percentage of all respondents who receive Medicaid who also receive SSI. Place this percentage (after the number) after a) in the table.

VIII. Calculate the percentage of all respondents who don't receive Medicaid who do receive SSI. Place that percentage (after the number) after b) in the table.

4. Compare the percentages in a) and b). Interpret your comparison relative to the proposition guiding this exercise that Medicaid recipients should also be SSI recipients.

EXERCISE 15.3

Multivariate Analysis

Now let's see what happens to the relationship between the two measures of economic status when we control for a third variable that might affect both of them: educational attainment.

1. Let's collapse the several categories of educational status into just two: high education and low education. Let's say that everyone in Filinson's sample who had 9 years of education or more qualifies for the high category and that everyone who has 8 years or less qualifies for the low category. Look at the data on pages 397 and 398 and pencil in an **HE** next to each case that falls into the high education category. Pencil in an **LE** next to each case that qualifies for the low education category. How many of Filinson's cases fall into the **HE** category? How many fall into the **LE** category?

2. Now retabulate the data into two tables that look like the one in the previous exercise. In one of the tables, you're only going to put cases that fall into our **HE** category.

   a. Put only **HE**s into this table, following steps I through VIII following the table:

Receipt of Medicaid by Reception of Supplemental Security Income for Highly Educated Respondents

	Does Respondent Receive Medicaid	
	Yes	No
Does Respondent Receive Supplemental Security Income?		
Yes	a)	b)
No	c)	d)
Total	e)	f)

   I. What number of **HE**s receive both Medicaid and SSI? Put this number after a) in the table.

   II. What number of **HE**s don't receive Medicaid but do receive SSI? Place this number after b) in the table.

   III. What number of **HE**s receive Medicaid but don't receive SSI? Place this number after c) in the table.

   IV. What number of **HE**s don't receive Medicaid and don't receive SSI? Place this number after d) in the table.

   V. What is the total number of **HE**s who receive Medicaid? Place this number after e) in the table.

   VI. What is the total number of **HE**s who don't receive Medicaid? Place this number after f) in the table.

VII. What percentage of **HE**s who receive Medicaid also receive SSI? Place this percentage (after the number) after a) in the table.

VIII. What percentage of **HE**s who don't receive Medicaid receive SSI? Place this percentage (after the number) after b) in the table.

b. Use only **LE**s for this table:

Receipt of Medicaid by Receipt of Supplemental Security Income for Less Educated Respondents

	Does Respondent Receive Medicaid	
	Yes	No
Does Respondent Receive Supplemental Security Income?		
Yes	a)	b)
No	c)	d)
Total	e)	f)

Follow the previous steps I through VIII for this table, substituting LE for HE in each step.

c. Do either of the partials "disappear" ?

d. Is the partial relationship between the reception of Medicaid and the reception of SSI stronger for **HE**s or **LE**s? How can you tell?

e. Specification is that kind of elaboration that permits the researcher to specify conditions when a relationship is particularly strong (or weak). Specify an educational condition under which the relationship between receiving Medicaid and receiving SSI seems particularly strong for Filinson's subsample.

EXERCISE 15.4

Interpreting Quantitative Analysis

This exercise gives you a chance to use the skills you've developed in this chapter to interpret quantitative analyses used by other researchers. Return to either of two articles in this book: Adler and Foster's "A Literature-Based Approach to Teaching Values to Adolescents" in Chapter 8 or Clark, Lennon and Morris' "Engendering Junior" in Chapter 13. In either case, focus on the analyses presented in the first table of the article. Review the article well enough so that you can answer the following questions about the table.

1. What research question is addressed in the table?

2. What is the "independent" variable in the table?

3. What are the "dependent" variables in the table?

4. Discuss, in your owns words, how the authors demonstrate that a relationship exists or doesn't exist between the independent variable and one of the dependent variables.

5. What answer does this demonstration suggest to the research question that was addressed in the table?

EXERCISE 15.5

Analyzing Data on the Web

In this exercise we'll introduce you to a Web site that a colleague, Sandra Enos, showed us at the University of California, Berkeley (http://socrates.berkeley.edu:7502/archive.htm) that not only provides codebooks for data collected from several national surveys but also permits you to analyze the data while you visit the site! You might want to do some analyses of your own, but first let us guide you through one involving General Social Survey (GSS) data on two variables: marital status (called "marital" in the codebook) and labor force status ("wrkstat").

1. Before you gain access to the site, consider the U.S. adult population, which the GSS samples randomly, and try to guess which marital status (married, widowed, divorced, separated, or never married) would be the modal, or most common one. State your guess. Gain access to the "SDA Archive" site (http://socrates.berkeley.edu:7502/archive.htm). Choose the GSS option, make sure the setting is on "Browse codebook," and click start. Now, using the "alphabetical variable list," find the variable named "marital." Which, in fact, is the modal marital status among adults in the United States? Were you right?

2. Now consider the variable "labor force status" ("wrkstat" in the codebook). Which labor force status would you expect to be most common among adults in the United States: working full time, working part time, temporarily not working, retired, in school, keeping house, or something else? State your guess. Now return to the "alphabetical variable list" and find the variable named "wrkstat". Which, in fact, is the modal labor force status among adults in the United States? Were you right?

3. Now consider both marital status and labor force status: Which marital status would you expect to have the greatest representation in the full-time labor force? State your guess in the form of a hypothesis. Now return to the original GSS menu (where you found the "browse codebook" option before). This time, click "run crosstabulation" and click "start". Where the menu asks for "vertical" variable, type "marital".

Where it asks for "horizontal" variable, type "wrkstat". Then ask for the "horizontal" percentage and have the site "run the table." Which marital status does, in fact, have the greatest percentage of its members employed "full-time"? Were you right?

16 Qualitative Data Analysis

Qualitative Data Analysts on Qualitative Data Analysis

Qualitative data are sexy. They are a source of well-grounded, rich descriptions and explanations of processes in identifiable local contexts . . . Then, too, good qualitative data are more likely to lead to serendipitous findings and to new integrations . . . Finally, the findings from qualitative studies have a quality of "undeniability." Words, especially organized into incidents or stories, have a concrete, vivid, meaningful flavor that often proves far more convincing to a reader . . . than pages of summarized numbers (Miles and Huberman, 1994: 1).

In qualitative field studies, analysis is conceived as an *emergent* product of a process of gradual induction. Guided by the data being gathered and the topics, questions, and evaluative criteria that provide focus, analysis is the fieldworker's *derivative ordering* of the data (Lofland and Lofland, 1995: 181).

In ethnography the analysis of data is not a distinct stage of the research. It begins in the pre-fieldwork phase, in the formulation and clarification of research problems, and continues into the process of writing up (Hammersley and Atkinson, 1983: 174).

Introduction

As we indicated in the last chapter and as the quotation from Miles and Huberman (1994) suggests, a major distinction between qualitative and quantitative data analyses lies in their products: words, "especially organized into incidents or stories," on the one hand, and "pages of summarized numbers," on the other. Even though, as Miles and Huberman argue, the former are frequently more tasty to readers than the latter, the canons, or recipes, guiding qualitative data analysis are less well defined than those for its quantitative counterpart. As Miles and Huberman elsewhere observe, "qualitative researchers . . . are in a more fluid—and a more pioneering—position" (1994: 12). There is, however, a growing body of widely accepted principles used in qualitative analysis. We turn to these in this chapter.

If the outputs of qualitative data analyses are usually words, the inputs (the data themselves) are also usually words—typically in the form of extended texts. These data (the words) are almost always derived from what the researcher has observed (for example, through the observations techniques discussed in Chapter 11), heard in interviews (discussed in Chapter 10), or found in documents (discussed in Chapter 13). Thus, the data analyzed by Adler and Adler were their observations and interviews surrounding children's afterschool activities, and the data analyzed by Enos were the words of incarcerated women. The outputs of both data analyses were the essays themselves (see Chapters 10 and 11).

But, although there might be general agreement about the nature of the inputs and outputs of qualitative analyses (words), there are nonetheless some fairly fundamental disagreements among qualitative data analysts. One basic division involves theoretical perspectives about the nature of social life. To illustrate, and following Miles and Huberman (1994: 8ff.), look at two general approaches: a social anthropological and an interpretative approach. Social anthropologists (and others, like grounded theorists and life historians) believe that there exist behavioral regularities (for example, rules, rituals, relationships, and so on) that affect everyday life and that it should be the goal of researchers to uncover and explain those regularities. Interpretivists (including phenomenologists and symbolic interactionists) believe that actors, including researchers themselves, are forever interpreting situations, and that these, often quite unpredictable, interpretations largely affect what goes on. As a result, interpretivists see the goal of research to be their own (self-conscious) "accounts" of how others (and they themselves) have come to these interpretations (or "accounts"). The resulting concerns with lawlike relationships, by social anthropologists, and with the creation of meaning, by interpretivists, can lead to distinct data-analytic approaches. Given this difference, our discussion in this chapter will generally be more in line with the beliefs and approaches of the social anthropologists (and those with similar perspectives) than it is with those of the interpretivists. Nevertheless, we believe that much of what we say is applicable to either approach and that their differences can be overstated. Thus, although the Adlers clearly sought to discover lawlike patterns in the experience of afterschool activities (remember their concept of "funneling"?), they were nonetheless sensitive to variation in the interpretations or accounts given by children, and adults, of those activities.

STOP AND THINK *When the Adlers pursued lawlike patterns in the experience of afterschool activities, were they doing what the social anthropologists or the interpretivists advocate? What about when they looked to the accounts given by children and adults of those activities?*

The quotations at the beginning of the chapter from Lofland and Lofland (1995) and Hammersley and Atkinson (1983) seem to point to another debate concerning qualitative analysis: whether qualitative analysis *emerges from* or *is generative of* the data collected. At issue seems to be, to paraphrase the more familiar chicken-or-egg dilemma, the question of which comes first: data or ideas about data (for example, theory). Lofland and Lofland stress the creative, after-the-fact nature of the endeavor by describing it as a "process of gradual induction"—of building theory from data. Hammersley and Atkinson emphasize how data can be affected by conceptions of the research topic, even before one enters the field. These positions, however different in appearance, are, in our opinion, both true. Rather than being in conflict, they are two sides of the same coin. The Adlers, in discussing their own practice in studying afterschool activities (see Chapter 11), have told us about their efforts to immerse themselves in their setting first, *before* examining whatever literature is available on their topic, to better "grasp the field as members do" without bias. But *something* compelled their interest in afterschool activities, and whatever that was surely entailed certain preconceptions. Enos (the author of the focal research report in Chapter 10) is much more explicit about her preconceptions and, in conversation, has confided that one of her expectations before observing mother-child visits was that she'd see mothers attempting to create "quality time" with their children during the visits. Still, she remained open to the "serendipitous findings" and "new integrations" mentioned by Miles and Huberman. When she found mothers casually greeting their kids with "high-fives," rather than warmly embracing them, as she'd expected, she dutifully took notes. Because Enos has been explicit about various stages of her data analysis, and because she's been willing to share her field notes with us, we use her data to illustrate our discussion of qualitative data analysis in this chapter, much as we used Filinson's to focus our discussion in the previous chapter.

Are There Predictable Steps in Qualitative Data Analysis?

Quantitative data analysis often follows a fairly predictable pattern: Researchers almost always begin by coding the data (either for computer or hand computation), move on to some univariate analyses, perhaps followed by bivariate and multivariate analyses. The "steps" involved in executing good qualitative analyses, however, are much less predictable, and a lot "more fluid," as Miles and Huberman suggest. Miles and Huberman, in fact,

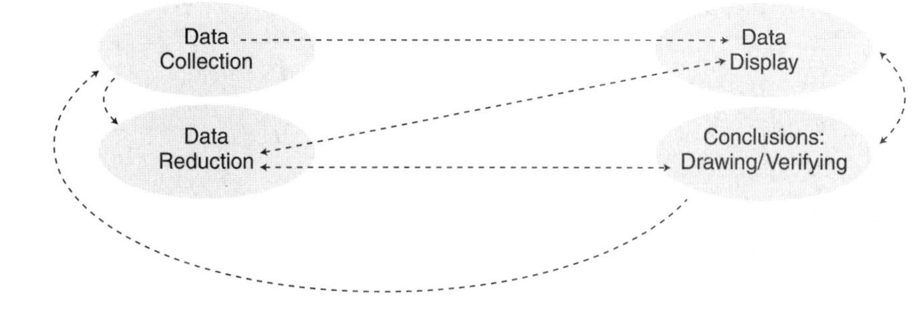

FIGURE 16.1

Components of Data Analysis: Interactive Model From Miles and Huberman: 12. 1994.

prefer to see qualitative analysis as consisting of "concurrent flows of activity," rather than as "steps." They dub these flows "data reduction," "data display," and "conclusion drawing/verification," and depict the interactions among these flows as we do in Figure 16.1.

This depiction of qualitative data analysis involves several notable assumptions. First, even data collection is not immune to the "flows" of data analysis, being itself subject to the (perhaps tentative) conclusions a researcher entertains at various stages of a project (note the arrow from "conclusions" to "data collection"). Enos had expected prison mothers to greet their children with warm embraces (even before she entered the field), and so focused attention on the gestures of greeting and, as you'll see, collected data about them. Second, Figure 16.1 suggests that data reduction and data display (both discussed later) are not only products of data collection but of one another and of conclusions, again even if such conclusions are not yet fully developed. Thus Enos' coding of her data (a data reduction process described later) not only contributed to her final conclusions about mothering in prison, but might have been itself a product of earlier, perhaps more tentative, conclusions she'd reached. Finally, the figure emphasizes that the conclusions a researcher draws are not just products of earlier flows (data reduction and data display) but continually inform both of the other "flows," as well as the data collected in the first place. We'll refer to this figure throughout this chapter.

Data Collection and Transcription

We've discussed at some length the process of collecting qualitative data in earlier chapters: through qualitative interviewing (Chapter 10), observation techniques (Chapter 11), and studying the content of texts (Chapter 13). So far, however, we haven't shown you any actual data. We'd like to present a small portion of Sandra Enos' fieldnotes, as she composed them on the evening after an interaction with one mother in the prison parenting program. These notes were based on observation and conversation in the off-site parenting program, where women got together with their children. The excerpt in Figure 16.2 is from nearly 150 pages of field notes Enos took before she engaged in the qualitative interviewing she reports on in the essay within Chapter 10.

FIGURE 16.2

A Segment from Enos' Field Notes

The main part of the Parenting
Program takes place in a large room
that served as a cafeteria and large
meeting area previously. The
kitchen remains in service although
it is not used for the program.
Food for lunch and breakfast is
brought over from institutional
kitchens.
The large room has a linoleum floor,
tables and chairs and a small
playing area that is carpeted.
Because the room is so large, it is
hard to get a feeling of intimacy.
With so few participants, women and
their children can remain virtually
alone during the day. Smoking is
allowed in one area of the room,
away from the children's play area.
While the women are waiting for
their children, they may sweep up
the area, straighten out toys, smoke
a cigarette or have some breakfast.
I spoke to a woman who had remained
silent in the group. She was
sitting alone at a table. I asked
if I could talk with her. She has
six children, two of her children
will be visiting today. Her mother
has always had her kids. Last year
her mother passed away and now her
sister is taking care of the kids.
When she's released, she'll have
responsibility for the whole bunch.
She says it was hard to see your
kids while you were in jail but that
if you were DCYF[1] involved sometimes
it was easier. (This has to do with
termination of rights.) She
remarked that it was important not
to be a "stranger" to your kids
while you were in jail. It made
everything worse. I asked about
telling your kids where you are.
She said she just straight out told
them and they accepted. "Mommy's in
jail. That's it." She thought
having such direct experience with
jail that maybe her kids would
"avoid" it when they got older.
I asked her when she'd be released.
In a few months, but she'd be
going to a drug treatment program.

[1]Department of Children, Youth & Families.

FIGURE 16.2

A Segment from Enos' Field Notes *(continued)*

She's been in prison many times before and this time she's going to do things right. She's made a decision to change, to do something new.
She is very happy with this program. She likes the fact that there are no officers around, that kids can run and scream, that there is an extended time out of the joint and that she can spend more time with her kids.
Her children arrived, a boy five years old or so and his older sister. They walked over to us.
The mother greeted her son, "Hey, Dude! How you doing?" He didn't say anything. The woman asked for the children's coats and hung them on the chair backs. She introduced me to the kids, who were quiet and beautiful, very well cared for. I complimented her on these kids and excused myself so she could be alone with them.

The physical appearance of these fieldnotes brings up the issue of computer usage in qualitative data analysis. Enos did all her written work, note collection and analysis with an elementary word-processing program—a practice that is extremely common these days. Enos did not use one of the increasingly available software packages (for example, Kwalitan, MAX, QUALPRO) designed to facilitate the processing of qualitative data. To those students who think they might be interested in software for analyzing qualitative data, we recommend Weitzman and Miles (1995) and Richards and Richards (1994) for discussions of what's available. We believe that much good qualitative analysis can be done without the use of anything more sophisticated than a word processor—and are inclined to point out that much has been achieved with nothing more sophisticated than pen and paper. But on to Enos' work . . .

STOP AND THINK *One of Enos' expectations had been that women would greet their children with hugs. How did this woman greet her son? How did he greet her?*

The physical appearance of these fieldnotes might be the first thing that strikes you. Enos uses only half a page, the left column, for her field notes in anticipation of a need to make notes about the notes later on. (See the data reduction section later.) Leaving room for such notes about notes is excellent practice. Enos didn't have an interview guide at this stage of her investigation, but she evidently had certain questions in mind as she approached this woman. Her notes indicate that she asked about how the woman explained her incarceration to her children, as well as about her expectations for release and her current childcare arrangements.

Enos' interest in the aforementioned questions indicates that she's probably engaged in a fair amount of what Miles and Huberman (1994: 16) call **anticipatory data reduction**—that is, decision making about what kinds of data she'd like to collect. Admittedly her research strategy (more or less "hanging out" at an off-site parenting program for female prisoners) is pretty loose at this stage.

anticipatory data reduction, decision making about what kinds of data are desirable that is done in advance of collecting data.

There is quite a bit a debate about the relative merits of "loose" and "tight" (those that entail explicit conceptual frameworks and standard instruments) research strategies in field research. Those advocating relatively loose strategies or "designs" (for example, Wolcott, 1982) point to the numbers of questions, as well as answers, that can be generated through them. Susan Chase, for instance, in the study mentioned in Chapter 10, was primarily interested in getting female school superintendents to tell their life stories and was willing to use any question that stimulated such stories. She was willing to "hang out" during these interviews, though she does confess that she eventually found asking female superintendents to describe, generally, their work histories, was a pretty good way to induce "lively, lengthy, and engrossing stor[ies]" (Chase, 1995a : 8).

But the advantages of relatively tight strategies, especially for saving time spent in the field, are worth noting. Enos developed an appreciation of tight strategies during the course of her fieldwork. In personal communication, she's indicated how tempting it is to pursue every idea that occurs in a setting:

> So many questions occurred to me after I left the site. I would consider what women had said to me and then try to figure out if that was related to race/ethnicity, to class, to previous commitments to prison, to number of children, to street smarts, to sophistication about the system, and so on and so on. (Enos, personal communication, 1996)

In the end, she realized that she couldn't pursue all those questions, that she needed, in her words, "to keep on track." She claims that a most important task was to continually sharpen her focus as the fieldwork went on and that this meant "finally coming to the conclusion that you can't look at everything" (personal communication, 1996).

By the time Enos entered the interview stage of her project, she'd developed a very tight design indeed, having, among other things, composed the interview guide shown at the end of her report in Chapter 10. This guide led Enos to ask very specific questions of her respondents, about their age of first incarceration, and so on.

Data Reduction

data reduction, the various ways in which a researcher orders collected and transcribed data.

Qualitative **data reduction** refers to the various ways in which a researcher orders collected data. Just as questioning 1,000 respondents in a closed-ended questionnaire can generate more data on most variables than the researcher can summarize without statistics, fieldnotes or transcripts can generate volumes of information that need reduction as well. Some researchers can do a reasonably good job of culling out essential materials, or at least tell compelling stories, without resorting to systematic data reduction procedures in advance of report writing. But most can't, or don't. Especially in long-term projects with ongoing data collection, most researchers engage in frequent data reduction. In the following section, we'll focus on two common data reduction procedures: coding and memoing. In doing so, we'll be emphasizing one thing, implicit in Figure 16.1: Data reduction is affected by the kinds of data you collect but is also affected by, and affects, the conclusions you draw.

Coding

coding,
the process of associating words or labels with passages and then collecting similarly labeled passages into files.

Coding, in both quantitative and qualitative data analysis, refers to assigning observations, or data, to categories. In quantitative analysis, the researcher usually assigns the data to categories of variables she has already created, categories that she frequently gives numbers to in preparation for computer processing (see Chapter 15). In qualitative analysis, coding is somewhat more open-ended because both the relevant variables and their significant categories are apt to remain in question longer. Here coding usually refers to two distinct, but related, processes: The first is associating words or labels with passages in one's fieldnotes or transcripts; the second is collecting similarly labeled passages into files. The goal of coding is to create categories (words or labels) that can be used to organize information about different cases (whether situations, individuals, or so on). When you don't have much of a framework for analysis, as Enos didn't initially, the codes (words or labels), at least initially, can feel a little haphazard and arbitrary. They can nonetheless be a useful guide to future analysis. Thus, Enos' own coding of the middle paragraph of her notes (from Figure 16.2) looked something like what appears in Figure 16.3.

Here we see Enos, in her first informal interview, developing codes that will appear again and again in the margins of her fieldnotes: codes that permit her to characterize how social mothering is managed while a biological mother is in prison. The codes "permanent substitute" and "temporary substitute," referring to this particular woman's mother and sister, respectively, become categories for classifying the kinds of mothering that all other children receive.

STOP AND THINK *Suppose you wanted to compare all the data that Enos coded with the words "temporary substitute." What might you do to facilitate such a comparison?*

Assigning a code to a piece of data (that is, assigning "temporary substitute" to this woman's sister or "permanent substitute" to her mother) is just the first step in coding. The second step is physically putting the coded data together with other data you've coded in the same way. There are basically two ways of doing this: the old-fashioned way, which is to put similar data into a file folder, and the new-fashioned way, which is to do essentially the same thing using a computer program that stores the "files" in the computer itself. Old-fashioned manual filing is simple enough: you create file folders with code names on their tabs, then place a copy of every note you have

FIGURE 16.3

Enos' Coding of Fieldnotes

Fieldnotes	***Codes***
I spoke to a woman who had remained	
silent in the group. She was	
sitting alone at a table. I asked	
if I could talk with her. She has	
six children, two of her children	
will be visiting today. Her mother	women as birthgivers,
has always had her kids. Last year	not caretakers
her mother passed away and now her	
sister is taking care of the kids.	motherhood management
When she's released, she'll have	permanent substitute
responsibility for the whole bunch.	temporary substitute
She says it was hard to see your	
kids while you were in jail but that	
if you were DCYF involved sometimes	
it was easier. (This has to do with	
termination of rights.) She	
remarked that it was important not	
to be a "stranger" to your kids	
while you were in jail. It made	
everything worse. I asked about	
telling your kids where you are.	
She said she just straight out told	
them and they accepted. "Mommy's in	
jail. That's it." She thought	
having such direct experience with	
jail that maybe her kids would	
"avoid" it when they got older.	

coded in a certain way into the folder with the appropriate label. In these days of relatively cheap photocopying and the capacity of computers to make multiple copies, you can make as many copies of each page as there are codes on it, and slip a copy of the page into each file with the corresponding code. (In the bad old days, this process required relatively cumbersome manipulations of carbon paper and scissors.) Thus, Enos might have placed all pages on which she placed the codes "temporary substitute" into one folder and all pages on which she wrote the code "permanent substitute" into another. Later in her analysis, when she wanted to see all the data she'd coded in same way, she'd just have to pull out the file with the appropriate label.

The logic of the new-fashioned, computer-based filing is the same as that of the old-fashioned one. The newer way requires less paper and no actual file folders (and can therefore result in a great savings of money and trees), but it does generally require special computer software (so the initial outlay can be substantial). "Code-and retrieve programs," for instance, like Kwalitan, MAX and NUDIST, help you divide notes into smaller parts, assign codes to the parts, and find and display all similarly coded parts (Miles and Huberman, 1995: 312). Computer-based filing also makes revising codes (see later) simpler. For those who are interested, we again recommend Weitzman and Miles (1995) and Richard and Richards (1994) for discussions of what's available.

Types of Coding

One purpose of coding is to keep facts straight. When you code for this purpose alone, as you might be doing if you were to assign a person's name to all paragraphs about that person, you'd be engaged in purely *descriptive coding*. But because the basic goal of coding is to advance your analysis, you'd eventually want to start doing a more *analytic* type of coding. Enos's simultaneous development of the two codes "permanent substitute" and "temporary substitute," is an example of analytic coding. Charmaz calls this preliminary phase of analytic coding *initial coding* and defines it as the kind through which "researchers look for what they can define and discover in the data" (Charmaz, 1983: 113). This is also a point when one's early, provisional conclusions (see Figure 16.1) can inform one's choice about codes.

Eventually, according to Charmaz, initial coding gives way to *focused coding,* during which the initial codes themselves become the subject of one's coding exercise. At this stage, the researcher engages in a selective process, discarding, perhaps, those codes that haven't been terribly productive (for example, Enos never used the "women as birthgivers" code after its first appearance) and concentrating or elaborating on the more enlightening ones. After Enos saw all the ways in which women found temporary substitutes, for instance, she began to use variations in those ways—for example, relatives, foster care, friends—as codes, or categories, of their own.

Memoing

memos,
more or less extended notes that the researcher writes to help herself or himself understand the meaning of codes.

data displays,
visual images that summarize information.

All data analyses move beyond coding once the researcher tries to make sense of the coded data. In quantitative analyses, the researcher typically uses statistical analyses to make such sense (see Chapter 15). In qualitative analyses, other techniques, such as **memoing** and **data displaying** are used. Memos are more or less extended notes that the researcher writes to help understand the meaning of coding categories. Memos can vary greatly in length (from a sentence to a few pages) and their purpose, again, is to advance the analysis of one's data (notes). An example (labeled distinctly as a theoretical note) of a memo appears about 100 pages after Enos' write-up of the encounter with the woman whose children had been (permanently) cared for by her mother until the mother died, but were now being (temporarily) cared for by her sister. It appears to be a reflection on all the material that's been coded "temporary" and "permanent substitute" throughout the fieldnotes.

This memo is shown in Figure 16.4.

STOP AND THINK *Enos tells herself to think about the literature on family breakdown, something that the mother's imprisonment can also lead to. What do you think are some of the more familiar causes of family breakdown?*

Notice that Enos makes meaning by drawing an analogy between family separations brought about by imprisonment and those brought about by divorce, separation, death, and so on. She's focusing her attention on what happens to the care of children when they lose daily access to a parent (in

FIGURE 16.4

Illustration of Memo from Enos

Theoretical note
Consider the work on the breakdown of the family and compare that to the efforts of these families to keep their children in their midsts. These children are being raised by persons other than their birth mothers but they are well dressed, taken care of, appear healthy, and so forth. If they are on welfare, there appears to be another source of support.

this case, the mother). This has clearly become a central concern for future consideration. With this memo as a springboard, it's not surprising that childcare and relationships with family members are a key issue in the research she reports on in Chapter 10.

Because fieldwork and coding can be very absorbing, it's easy to lose track of the forest for the trees. The practice of memoing, of regularly drawing attention to the task of analysis through writing, is one way that many qualitative researchers force themselves to "look for the big picture." Memos can be thought of as the building blocks of the larger story one eventually tells. Some memos, such as Enos', still need some refinement (for example, through the reading of the literature on family breakdown), but others qualify for inclusion in the final report without much alteration. Some, especially those that are memos about linkages among memos, can provide the organizing ideas for the final report. Memos are a form of writing and have the potential for creative insight and discovery that writing always brings.

Memos can be potent theoretical building blocks, but they can be vital methodological building blocks as well. Enos tells us, for instance, that memos were a particularly useful way for her to figure out what she needed to "learn next" during her study, as well as how to actually learn it (personal communication, 1996). One memo (presented in Figure 16.5) shows Enos deciding to adopt a more formal interview strategy than she'd used before. This memo constituted a transition point between the fieldwork and the interview stages of Enos' study.

Data Displays

If memos, in their relative simplicity, are analogous to the relatively simple univariate examinations of quantitative analyses, data displays can be seen as a rough analogue to the more complex bivariate and multivariate analyses used there (see Chapter 15). Miles and Huberman (1994) are particularly adamant about the value of data displays for enhancing both the processes

FIGURE 16.5

Methodological Memo From Enos

Methodological note While it has been easy to use my role as researcher and "playmate" to gain some information about mothers and children in prison, it has also been frustrating from a data collection perspective. I am involved in some long and wide-ranging conversations with the women and trying to remember the tone, the pacing and the content of discussion while engaging in these as an active member. I can disappear for a few minutes to an adjoining room to scribble notes but this is difficult. I focus on the points that are most salient to my interests but can get sidetracked because so much is of interest to me. A more regimented conversation in an atmosphere that is more controlled than this one would probably yield "better" notes.

of data reduction, admittedly a means to an end, and conclusion making and verifying, the ultimate goals of data analysis.

A data display is a visual image that summarizes a good deal of information. Its aim is to present the researcher herself with a visual image of what's going on and help her to discern patterns. Miles and Huberman's figure (our Figure 16.1) itself can be seen as an example of such a display, summarizing the interaction among various processes in qualitative analysis.

The crucial feature of data displays is that they are tentative summaries of what the researcher thinks he knows about his subject. Therefore data displays permit him to examine what he thinks by engaging in more data examination (and reduction) and, eventually, to posit conclusions. Enos, for instance, tends to use the space on the right half of her fieldnote pages not only for coding but also for data displays, often in the form of exploratory typologies about the kinds of things she's seeing. Thus, next to notes she's taken about one mother, "Pam," and her interaction with her children, "Tom" (a baby) and "Joy" (a three-year-old), we find a display (in Figure 16.6) about how mothers seem to manage their weekly interactions with their children in the parenting program.

It's pretty clear that Enos' coding of the first paragraph (referring to Pam's attempt to normalize her interaction with Tom and Joy) with a label ("attempt at normalization") she's used before has pushed her, by the time she rereads the second paragraph, to play with a list of ways she's seen mothers interact with their children at the parenting program. (Nothing in the second paragraph, in fact, is likely to have sparked this reflection.) This

FIGURE 16.6

Enos' Emergent Typology of Mothering Behavior in the Parenting Program

There were a number of infants visiting today. One of them, Tom,[2] is the son of Pam who also has a three year old girl. Pam plays actively with her children, is with them on the carpeted area, holding the baby and roughhousing with her daughter. Joy is an active and very outgoing kid who asks to have her hair fixed by her mom who is struggling to feed the baby and pay attention to her daughter.	attempt at normalization
The baby is tiny for his age and pale. These children are being taken care of by Pam's boyfriend's mother. Pam has a problem with cocaine and got high the day before Tom was born. She says that before he was born all she was interested in was getting high and the "chase." During her pregnancy with Joy, she quit smoking and was drug free.	Setting management Mother Behavior 1) no playing with child 2) active playing 3) sporadic conversation 4) steady conversation 5) steady interaction 6) kid in other's care

typology is the kind of data display that Miles and Huberman advocate. It pictorially summarizes observations Enos has made about the varieties of mothering behavior she's seen and its presence facilitates her evaluation of its adequacy for subsequent presentation. Something like this display guided Adler and Adler (see their report in Chapter 10) when they laid out the various stages (recreational, competitive, and elite) of commitment to afterschool activities shown by the children they studied.

The forms in which data displays occur are limited only by your imagination. Figure 16.1 is a flow chart; Enos's display is a more conventional-looking typology. Miles and Huberman (1994) also advocate using various kinds of matrices, something that Roger and a student co-author, Heidi Kulkin, used in preparing their findings about 16 recent young-adult novels about non-White, non-American, or non-heterosexual main characters. Heidi and Roger found themselves most interested in themes of oppression and resistance in such novels and needed a quick way of summarizing the variation they'd discovered. The matrix shown in Table 16.1 worked. Preparing this matrix was helpful in two ways. First, it revealed some things they definitely wanted to say in their final analysis. It gave them the idea, for instance, of organizing their written presentation into four parts: stories about contemporary American charac-

[2] All names are pseudonyms.

TABLE 16.1 **Content Analytic Summary Matrix Used in Preparation of Clark and Kulkin's (1996) "Toward Multicultural Feminist Perspective on Fiction for Young Adults"**

Place	*Contemporary*	*Historical*
	No. of books: 4	No. of books: 3
United States	Varieties of oppression: racism, classism, cultural imperialism, heterosexism	Varieties of oppression: sexism, racism
	Strategies of resistance: escape, resort to traditional custom, communion with similarly oppressed others	Strategies of resistance: escape, cooperation, extraordinary self-sacrifice, learning, writing
	No. of books: 4	No. of books: 5
Outside United States	Varieties of oppression: sexism, classism, racism, imperialism	Varieties of oppression: all the usual suspects
	Strategies of resistance: education, communion with similarly oppressed persons, creative expression, self-reliance, self-sacrifice	Strategies of resistance: escape, communion with similarly oppressed others, writing, self-definition, education, self-sacrifice

ters, historical American characters, non-American characters, and historical non-American characters. Reviewing the table also led them to valid observations of between-group differences for their sample. Thus, for instance, they observed that authors writing about contemporary American characters were less likely to emphasize the importance of formal education and extraordinary self-sacrifice than those writing about other subjects.

Preparing the matrix, then, led Heidi and Roger to some observations they directly translated into their final report (Clark and Kulkin, 1996). But it also helped them in another, perhaps more important, way: It warned them against saying certain things in that report. Thus, for instance, although each mode of oppression and resistance listed in each "cell" of the matrix was prominent in some novel that fell within that cell, each was not equally evident in all the novels of the cell. Heidi and Roger knew, then, that their final written presentation would have to be much more finely attuned to variation within each of their "cells" than the matrix permitted.

Data displays of various kinds, then, can be useful tools when you analyze qualitative data. Miles and Huberman would take this as too weak an assertion. They claim

> You know what you display. Valid analysis requires, and is driven by, displays that are focused enough to permit a viewing of a full data set in the same location, and are arranged systematically to answer the research questions at hand (Miles and Huberman, 1994: 91–92).

Although we think this position is overstated, if only because we know of good qualitative analyses that make no explicit reference to data displays, we think you could do worse than try to create data displays of your own while you're engaged in qualitative data analysis. When you do, you might well look to Miles and Huberman (1994) for advice.

Conclusion Drawing and Verification

The most important phase of qualitative data analysis is simultaneously its most exciting and potentially frightening: the phase during which you draw and verify conclusions. Data displays are one excellent means to this end because they force you to engage in two essential activities: pattern recognition and synthesis. Coding, of course, is a small step in the ongoing effort to do these things, and memoing is a larger one. But you can hardly avoid these activities when you're engaged in creating a data display.

Enos emphasizes that one very important resource for verifying conclusions, especially for researchers who are doing announced field observations or qualitative interviews, is the people being observed or interviewed. These people, she reminds us, can quickly tell you "how off-base you are":

> This was very helpful to me, as was asking questions they could theorize about, like "It seems that white women use foster care a lot more than black women do. Is that right?" [And, if they agree, "Why is that the case?"] (personal communication, 1996)

Elliot Liebow (1993) advocates, however, an "Ask me no questions, I'll tell you no lies" principle, especially relative to participant observation. In his study of homeless women, he consciously decided not to bring the kind of theoretical questions to the women that Enos recommends. In doing so, he claimed he was less concerned with the "lies" he might receive in return for his questions than with "contaminating the situation with questions dragged in from the outside" (Liebow, 1993: 321).

Whatever you decide to do about the issue of "bouncing ideas" off of those you're studying, we think you'll find that a vital phase of conclusion drawing and verification comes when you compose your final report. The very act of writing, we (and others [for example, Lofland and Lofland, 1995]) find, is itself inseparable from the analysis process. As a result, we divide the remainder of the section on drawing and verifying conclusions into a discussion of two phases: the pre-report writing and the report writing phase.

Pre-Report Writing

Lofland and Lofland (1995) point out that the growing use of computers for qualitative data analysis has helped underscore two points. The first is that qualitative analysis can be facilitated by the kinds of things computers do, like storing and retrieving information quickly. The new programs that facilitate data coding testify to these strengths. But the second is that good qualitative analysis requires what computers aren't really much good at. One of these tasks, and something that really distinguishes qualitative research from other brands of social research, is deciding what problems are important. Enos began her study of women in prison with an open mind about what the real problems were. The same was true of Adler and Adler (see Chapter 11) as they approached children's afterschool activities and Clark and Kulkin (1996) as they began to read nontraditional novels for young adults. These authors all came to their settings with open eyes, open ears, and, particularly, open minds about what they'd find to be the most important problems.

STOP AND THINK *Contrast this openness with the hypothesis-driven approaches of Adler and Foster (see Chapter 8), Messner (see Chapter 12), and Clark, Lennon, and Morris (see Chapter 13).*

Finding out just what problem you should address requires something more than what computers are good at: The imagination, creativity, and moral insight of the human brain.

Imagination and creativity are necessary even after you've chosen your topic and are well into the data analysis. One way to stimulate such imagination and creativity is to use the kinds of data displays advocated by Miles and Huberman. But, as Lofland and Lofland (1995: 202–203) point out, other practices can create fertile conditions for qualitative analysis. One practice is to *alter data displays*. If you've created a data display that, somehow, isn't useful, try a different form of display. We can imagine, for instance, Miles and Huberman building the flow chart that we've reproduced in Figure 16.1, indicating the various interrelationships that might exist among data collection, data reduction, data display, and conclusions, out of a simple typology of (with no indication of relationships among) activities involved in qualitative analysis: data collection, data reduction, data display, and conclusions. Note that altering data displays involves conclusion verification and dismissal, both important goals of data analysis.

Another pre-report writing activity, advocated by Lofland and Lofland, and before them by Glaser and Strauss (1967), is that of *constant comparison*. The key question here is this: how is this item (for example, person, place, or thing) similar to or different from itself at a different time or comparable items in other settings? One can, again, easily imagine how such comparisons led Enos to her eventual interest (see Chapter 10) in how substitute mothering differs for women from different racial backgrounds, or the Adlers (Chapter 11) to their focus on how the meaning of afterschool activities changed with the age of children, or to Clark and Kulkin's interest in how themes of oppression and resistance varies by geographical or historical context. Constant comparison is also implicit in Janet Mancini Billson's (1991) progressive verification method for analyzing individual and joint interviews (see Chapter 10).

Lofland and Lofland (1995: 202–203) actually recommend several other pre-report writing techniques for developing creative approaches to your data, but we'd like to highlight just one more of these: *speaking and listening to fellow analysts*. Conversations, particularly with friendly fellow analysts (and these needn't, of course, be limited to people who are actually doing qualitative data analysis but can include other intelligent friends) can not only spark new ideas (remember the adage "two heads are better than one?") but also provide the kind of moral support we all need when ideas aren't coming easily. Friendly conversation is famous not only for showing people what's "right in front of their noses," but also for providing totally new insights.

Report Writing

A qualitative report writer on report writing had this to say:

> When I began to write, I still had no order, outline, or structure in mind . . . I plunged ahead for many months without knowing where I was going, hoping and believing that some kind of structure and story line would eventually jump out. My original intention was to write a flat descriptive study and let the women speak for themselves—that is, to let the descriptive data and anecdotes drive the writing. In retrospect, that was a mistake. Ideas drive a study, not observations or unadorned facts . . . I shuffled my remaining cards again and discovered that many cards wanted to be grouped under Problems in Day-to-Day Living, so I undertook to write that chunk. The first paragraph asked the question, How do the women survive the inhuman conditions that confront them? After I wrote that paragraph it occurred to me that most situations, experiences, and processes could be seen as working for or against the women's survival and their humanity— that the question of survival might serve as the opening and organizing theme of the book as a whole and give the enterprise some kind of structure. (Liebow, 1993: 325–26)

Most authorities on qualitative data analysis attest to the difficulty of distinguishing between report writing and data analysis (see Hammersley and Atkinson, 1983: 208, and Lofland and Lofland, 1995: 203). On the one hand, data collection, data reduction, data display, and other pre-report writing processes all entail an enormous amount of writing. On the other, whatever else it is, *report writing is a form of data analysis*. Liebow's confession that he came to the report-writing phase of *Tell Them Who I Am*, his touching portrayal of the lives of homeless women, with "no order, outline, or structure in mind" could probably have honestly been made by many authors of qualitative social science—and, frankly, by many authors of quantitative social science as well. The discovery, through writing, of one vital organizing question or another is probably the rule rather than the exception for qualitative analyses. Consequently, we could not, in good conscience, have ended our chapter on qualitative data analysis without saying a few words about the importance of report writing as a stage in data analysis.

The writing involved in pre-report phases and the report-writing phase can be distinguished, notably by their audiences. Early phases involve writing in which the researcher engages in conversations with herself. We're privy to Enos's codes, her memos, and her typologies, for instance, only because she's generously permitted us to peer over her shoulder, as it were. The intended audience was Enos and Enos alone. Report writing, on the other hand, is much more clearly a form of communication between author and some other audience. This is not to say that the author herself won't gain new insights from the process. We've already said that she will. But it does imply that at least a nodding acquaintance with certain conventions of social science writing is useful, even if you only intend to "play" with those conventions for effect. As a result, we'd like to introduce you to a few such conventions.

There are few absolutes here, however. Perhaps the most important rule is to give yourself oodles of time to write. If you think it might take about 25 hours to write a report, give yourself 50. If you think it might take 50, give yourself 100. Writing's hard work, so give yourself plenty of time to do it.

Liebow's confession suggests a few other guidelines. One is that, even if you're not sure just where a particular segment of a report will fall within the whole, as long as you've got something to say, write. Don't insist on seeing the forest, before you write, especially if you've got a few good trees. Liebow wrote and wrote, long before he'd gotten a sense of his main theme. He wrote with the kind of hope and belief that an experienced writer brings: that a story line will emerge. But in writing what he took to be a minor theme, the "Problem of Day-to-Day Living," he first glimpsed what would be the organizing principle of his whole book, "the question of survival." Once he discovered this principle, he found that much of what he'd written before was indeed useful.

Another principle, implicit in Liebow's confession, is that you needn't look for perfection as you write. Especially if you have access to a word-processor and time (see our advice of two paragraphs earlier), you can write relatively freely, in the certain knowledge that whatever you write can be revised. Don't assume you need to say what you want to say in the best (the most grammatical, the most concise, the most stylish) way possible the first time around. Assure yourself that you can always go back, revise, and rethink as necessary. Such assurances are terribly liberating. Just remember to keep your implicit promise to yourself and revise whenever you can.

We'd like to tell you about some organizational conventions of report writing, even though we feel a bit presumptuous in doing so. After all, what will work for book-length manuscripts (such as Liebow's) probably won't work equally well for article-length reports (like Enos' in Chapter 10 and Adler and Adler's in Chapter 11). Moreover, generally speaking, research reports involving qualitative data are less formulaic than are those involving quantitative data. Nonetheless, because report writing does involve an effort to reach out to an audience, and because social-science report writing frequently involves an effort to reach out to a social science audience, it's not a bad idea to have some idea of the forms expected by such audiences.

In the spirit of Lofland and Lofland (1995), we'd like to suggest a checklist of features that often appear in article-length research reports about qualitative data analyses.[3] (Books tend to be much less predictable in their formats.) Many readers will be looking for them, so we think it's a good idea to include most of them most of the time. These features are the following:

1. *A title* that tells what your research is about. We think the title should give the reader some idea of what you have to say about your subject. Two good examples are from reports presented in Chapters 10 and 11:

- "Managing Motherhood in Prison" (Enos, Chapter 10)
- "Social Reproduction and the Corporate Other: The Institutionalization of Afterschool Activities" (Adler and Adler, Chapter 11)

2. *An abstract*, or a short (usually more than 100 but less than 150 words) account of what the paper does. Such accounts should state the paper's major argument and describe its methods. We haven't included abstracts for any of our focal research articles because we wanted to keep the overall length of this book manageable. But we've been very self-conscious about not following this convention and think we should give you at least one good example of an abstract, this one from Adler and Adler's original version of "Social Reproduction and the Corporate Other":

> This article identifies and describes a phenomenon that has arisen over the course of the last generation: an institutionalized "afterschool" period marked by children's involvement in adult-organized and -supervised activities. We trace the historical development of this period and examine its socializing influences on children. Children experience passage through an "extracurricular career" that begins with a recreational ambiance but progresses into competitive and finally elite activities as they grow older and become more skilled. Along this route, their leisure activities become less spontaneous and more rationalized, focused and professionalized. Adults' incursions into children's play thus represent a means for them to reproduce the existing social structure and to socialize young people to the corporate work values of American culture (Adler and Adler, 1994: 309).

Although this abstract is a little longer than many and not focused on method, we think it does a wonderful job of encapsulating key points of the article. Abstracts frequently come before the main body of a report—just after the title—but are generally (and ironically) written *after* the rest of the report's been put together, so the author knows what points to summarize.

[3]These features frequently appear in reports on quantitative data analyses too. Check the reports in Chapters 5, 7, 8, 12, and 13 for examples. Thus, although we emphasize the significance of report writing for qualitative analyses in this chapter, we do not mean to de-emphasize its importance for quantitative data as well.

Here again, however, there are no rules: Drafting an abstract can also be a useful heuristic earlier in the process—to help the writer focus (in a process sometimes called "nutshelling"). Notice, however, that in this abstract, Adler and Adler have actually engaged in a final act of conclusion drawing—the ultimate goal, after all, of data analysis itself.

3. *Overview of the problem* and *literature review*. These two features are sometimes treated distinctly but are more often combined, as they are in Enos' piece in Chapter 10 and Adler and Adler's paper in Chapter 11. The basic function of these features is to justify your research. The literature review, in particular, should justify the current research through what's already been written or not written about the topic. Enos does this with three kinds of "literatures": a literature that implies family forms should have certain "normative" aspects, such as mothers who engage in full-time mothering; a literature that suggests we look seriously at alternate family forms, especially as these have emerged in various racial/ethnic groups; and literatures that have discussed the "imprisonment boom" in the United States and have pointed to the special problems of women in prison. If you've never written a research report before, we think you could do worse than study the ways in which the authors of articles in this book have used literature reviews to justify their projects. Literature reviews are an art form, and social science literature reviews are a specialized genre of the larger one. As an art form, there are few hard-and-fast rules, but there are some. One is this: Never simply list the articles and books you've read in preparation for your report, even if you include some of what they say while you're listing them. Always use your literature review to build the case that your research has been worth doing.

Literature reviews themselves offer an excellent chance to engage in data analysis. Whether your review occurs before, after, or throughout your field research, it will compel you to make decisions about what your study's really about. Thus, for instance, to use whatever tools libraries supply for gaining access to lists of articles and books (and these can include such hard-copy reference tools as the *Reader's Guide to Periodical Literature*, the *Social Science and Humanities Index*, *Psychological Abstracts*, or *Books in Print* and card catalogues or such computer-based sources as *InfoTrac*, *Sociofile*, and *ERIC*, and computer-based card catalogues) about your topics of interest, you need to develop a list of possible topics of interest. A reference librarian can introduce you to the tools, but only you can come up with a list of appropriate topics. Thus Enos, in preparing her report had to decide that her report was going to be related to what others had to say about *mothering*, *women in prisons*, and perhaps even *race and mothering* before she could seek out the appropriate kinds of articles and books. And Adler and Adler had to recognize that their research interests touched on *afterschool activities* and *adult supervision* and perhaps even *corporate work values*. Choosing such topics is every bit as important a data reduction exercise as, say, coding (described earlier)—though, of course, the former activity might well be informed by the latter, and vice versa.

Perhaps even more important, reviewing articles and books that somehow "touch" on your topic can very well give you insights into just what you're seeing, or might see, in the data. Thus, as Clark and Kulkin were reviewing the literature on *multicultural feminism* during their analysis of novels for young adults, they realized that one focal theme of that literature, "efforts to resist oppression," was a common theme in all the novels they'd been reading as well. This theme became one of the organizing principles of their final report.

4. *Data and Methods*. Most reports of qualitative data analysis, like those of quantitative data analysis, include a section on data and methods. Both Enos (in Chapter 10) and Adler and Adler (in Chapter 11) do so. The kind of introspection and retrospection required of such sections can lead, again, to a kind of analysis that provides new insights about the entire project.

5. *Findings*. Being able to present findings is usually the point of all your previous analysis. You might want to subdivide your findings section, as Adler and Adler do. Still, this is the section in which you'll want to present all those conclusions (perhaps in the form of theories, or stories about what's going on) you've been coming to through earlier analyses. In writing it you'll undoubtedly find others. Our advice: Keep an open mind and listen to yourself as you write.

6. *Summary and Conclusions*. This section should permit you to distill, yet again, your major findings, as Enos and the Adlers do in theirs, and to say something about how those findings are related to the literature or literatures you've reviewed earlier. It's very easy to forget to draw such relationships (that is, between your work and what's been written before), but once you get started, you might find you have a lot more to say that could interest other social scientists than you anticipated.

7. *References*. Don't forget to list the references you've referred to in the paper! Here you'll want to be guided by certain stylistic conventions of, perhaps, a formal guide to stylistic conventions (for example, ASA, 1996). If you're preparing a paper for a course or other academic setting, you might even ask the relevant faculty member what style she prefers.

Again, report writing is an art form. Although guidance, such as what we've presented in this section, can be useful, this is one of those human activities where experience is your best guide. Good luck as you gather such experience!

Summary

The canons for analyzing qualitative data are much less widely accepted than are those for analyzing quantitative data. Nonetheless, the various processes involved in such analysis are broadly accepted in general terms: data collection itself, data reduction, data display, and conclusion drawing and verification. These processes are interactive in the sense that they do not necessarily constitute well-defined and sequential steps and insofar as all the individual processes can affect or be affected by, directly or indirectly, any of the others.

Word-processors are very useful tools in data collection. Soon after each field or interview session, it is useful to word-process fieldnotes or transcripts. It's often a good idea to leave plenty of room in your notes for coding, comments, memos, and displays.

Data reduction refers to various ways a researcher orders collected and transcribed data. We've talked about two subprocesses of this larger process: coding and memoing. Coding is the process of associating word or labels with passages in your fieldnotes and then collecting all passages that are similarly coded into appropriate "files." These files can be traditional file folders or computer files. Codes can be created using predefined criteria, based on a conceptual framework the researcher brings to his data, or using postdefined criteria, that "emerge," as it were, as the researcher reviews his fieldnotes. Codes can be merely descriptive of what the researcher's seen or analytic, in which case the codes actually advance the analysis in some way. Memos, unlike codes, are more or less extended notes that the researcher writes to help herself create meaning out of coding categories. Memos are the building blocks of the larger story the researcher eventually tells about her data.

Data displays are visual images that present information systematically and in a way that allows the researcher to see what's going on. The forms in which data can be displayed are limitless, but we've discussed three types in this chapter: charts, typologies, and matrices.

Data displays are particularly useful for drawing conclusions that are then subject to verification or presentation. We discussed a few other prereport techniques for developing such conclusions: altering data displays, making comparisons, and talking with and listening to fellow analysts. We also described report writing itself as an important stage in the analysis process. Even in this stage, when the researcher is, at least in theory, addressing a more general audience, she is likely to bombard herself with new ways of looking at the data. Preparing titles, abstracts, overviews and literature reviews, methods sections, findings, and conclusions can all, if done thoughtfully, shed much light on your work.

SUGGESTED READINGS

Lofland, J., and L. H. 1995. *Analyzing social settings: A guide to qualitative observation and analysis*. 3rd ed.. Belmont, CA.: Wadsworth.

The third edition of this classic provides an especially strong guide to qualitative data analysis, now that its ninth and tenth chapters, dedicated to this topic, have been revised and expanded.

Miles, M. B., and Huberman, M. A. 1994. *Qualitative data analysis*. 2nd ed. Thousand Oaks, CA.: Sage.

An extraordinarily comprehensive guide to qualitative data analysis. An especially useful resource for understanding data displays.

EXERCISE 16.1

Data Collection and Coding

For this exercise, you'll need to collect data either through observation or qualitative interview. If you prefer to do observations, you'll need to locate a social setting to which you can gain access twice for at least 20 minutes each time. You can return to one of the settings you observed for Exercises 11.1 or 11.2 (a children's afterschool setting or a group you're already a member of), or you can pick some other easily observable activity (for example, a condiment table at a cafeteria, an elevator on campus, a table at a library where several people sit together). You can choose to tell others of your intention to do research on them or not. You can work out some research questions before you go to the setting the first time or not. You can take notes at the setting, tape-record what's going on, or you can choose to do all your note-taking later.

If you prefer to do two interviews, we recommend that you do something like the work you did in Exercise 10.3, pick an occupation you know something, but not a great deal, about (such as police officer, waiter, veterinarian, letter carrier, high school teacher, and so on) and find two people who are currently employed in that occupation and willing to be interviewed about their work. Conduct 20 minute semi-structured interviews with each of these people, finding out how they trained or prepared for the work, the kinds of activities they do on the job, about how much time is spent on each, which activities are enjoyed and which are not, what their satisfactions and dissatisfactions with the job are, and whether or not they would like to continue doing their work for the next 10 years.

Whatever the setting or whoever the interviewee, write up your notes afterwards, preferably using a word-processor and preferably leaving room for coding and memoing (that is, use wide right margins).

After writing up your notes, spend some time coding your data by hand. Remember: This means not only labeling your data but then filing similarly labeled data away (in this case, in file folders you make or purchase).

EXERCISE 16.2

Memoing and Data Displaying

Some time shortly after you've recorded your notes and finished coding them (at least for the first time), reread both the notes and your coding files.

Spend about 20 minutes or so writing memos (either word processing them or hand-writing them) at the end of your notes. You can, of course, spend more than 20 minutes at this activity. The important thing is to develop some facility with the process of memoing.

Once you've written your memos, spend some time trying to create a data display that helps you understand patterns in the observations or interviews you've done. This display can be in the form of a chart, a typology, a graph, or a matrix. The purpose is to help you "see" patterns in the data that you might not have already seen.

EXERCISE 16.3

Report Writing

Having completed exercises 16.1 and 16.2, write a brief report (no more that two or so typed pages) about your project. You can omit a literature review (though, of course, if you found a couple of relevant articles, they might help), but you might otherwise be guided by the list [of items to include in a report] provided near the end of the chapter. (In particular, consider giving your report a title, providing a brief abstract, an overview, and a sense of your methods, highlighting your major findings and your conclusions.) Good luck!

17

Summing Up

Felton Earls, physician and professor at the Harvard School of Public Health, and Albert Reiss, Jr., professor of sociology and lecturer in law at Yale University, were interested in finding ways to predict and prevent crime. Planning a research project to make contributions to solving social problems, one of their research questions was "How do the individual characteristics of mental and physical health, family relationships, school environment, and type of community interact to contribute to antisocial behavior?" Earls and Reiss co-direct the Project on Human Development in Chicago Neighborhoods, which is designed to answer this and other questions. The data collection for their study, with a sample of 11,000 individuals, their families, and communities, began in 1993 and is scheduled to end in 2001 (Earls and Reiss, Jr., 1993).

Christine Angell, a sociology major at Rhode Island College, was interested in the adjustment of recent immigrants to life in the United States. Her research question was, "What factors contribute to the psychological, economic, and social adaptation of immigrants?" To answer this question, Angell designed a project that she completed over two semesters. Her data included a small number of semi-structured interviews and available data from community and state agencies (Angell, 1997).

In this concluding chapter, we'll summarize the research process. When we finish, we think you'll see that although projects like the one by Earls and Reiss differs dramatically from Angell's in purpose, scope, and the practical matters of time, money, and access, they are similar in many ways. The researchers in both projects made decisions about study design, sampling, ways to measure concepts and variables, and methods of data collection that were ethical, practical, and appropriate for the study's purposes.

These two studies are useful examples of the many studies, large and small, that make important contributions to our understanding of the world and to improving our lives. Earls, Reiss, and Angell are scholars whose work is making a difference. Perhaps you'll join them and make your own research contributions in the near future.

Introduction

At the beginning of this book, we invited you to join us in exploring the world of social research. We hope you've come to share our view that research is an adventure into learning about the social world—an adventure that begins in the charted territory of existing research and theory and heads off into the unknown. We hope you'll use this text as a map as you continue the process of exploration and proceed to learn more about and contribute to social science knowledge.

Our primary goals have been to describe the research process as it occurs in the real world and to share useful ways to approach research questions, data collection, and data analyses. By now you have many of the skills you'll need to complete your own research projects and information that can help you evaluate the work of others.

The focal research articles and our discussion of these projects have been central to our approach. We hope you've found each "story behind the story" interesting. All of the authors of the focal research articles have shared their insights and conclusions and disclosed problems and unanticipated circumstances. Their work demonstrates that despite occasional false starts and circuitous routes, research adventures can have successful conclusions.

In this chapter, we will continue our emphasis on planning and carrying out a research agenda. We'll first concentrate on the planning stages, then move to the "doing" part of a project by comparing the methods of data collection and analysis, and end with a discussion of writing research reports.

Planning Research

The Research Question

Beginning at the beginning means starting with a topic. Because the overall goal of all social research is to increase our understanding of human behavior, through description, explanation, and prediction, any aspect of social reality is fair game as a topic of research. Research questions can vary in scope and purpose. Some projects are designed to address very broad and basic questions whose answers will help us describe and understand our social world more completely. Other projects work toward solving social problems or the creation of information that can be used directly by members of communities and participants in programs or organizations.

Research questions and hypotheses emerge from a variety of sources, including personal experiences and interests; scientific curiosity; social, political, economic, and intellectual climates; previous research; chance factors; or the ability to get funding. Although each project has a unique history, some factors are common to most research. In Chapter 11, for example, we highlighted Patricia and Peter Adler's work on children's afterschool activities. Part of the impetus for their study was that they each had been "parent, friend, counselor, coach, volunteer, and carpooler" and wanted to pursue in more detail some observations they had made while in those roles. In Chapter 4, we focused on Rose Weitz's research on people living with HIV, a study that developed after Weitz attended a meeting on AIDS and received encouragement from a civil rights attorney to work on the project. Sandra Enos's work on mothers in prison (Chapter 10) resulted when Enos was granted access by prison administrators to an inmate parenting program, a setting in which she found she could explore two interests—crime and the family.

The Adlers, Weitz, and Enos did not begin with hypotheses; instead they focused on concepts. Enos's work was on mothering in prison; the Adlers observed children's afterschool activities; Weitz described the process of being diagnosed and living with HIV. Each of these projects resulted in detailed descriptions of social phenomena.

Working Inductively or Deductively?

The Adlers, Enos, and Weitz used the inductive method, moving from the specific to the more general. In the inductive approach, observations of the social world are used to create empirical generalizations and are sometimes used to develop theoretical understandings.

In contrast, the deductive method starts with the derivation of specific statements from more general ones; theories or more general knowledge of behavior are used to generate one or more testable hypotheses. This approach is exemplified by the focal research by Gray, Palileo, and Johnson in Chapter 9 and Messner in Chapter 12. Studying the assignment of blame for sexual assault, Gray and associates were interested in predicting the extent to which victims would be blamed for being raped. Working deductively, the researchers took the broad theory of assigning blame (called attribution theory) and derived several testable hypotheses for sexual assault, risk taking, and blaming before collecting data. Messner similarly derived hypotheses about the connection between the exposure to violent TV and violent crime rates.

Although some individual researchers prefer a theory-discovery approach and others use theory-verification, the social science community as a whole does both. Following Wallace (1971), we support the view that the interaction between theory and research can start at any point. We think the link between theory and research is best visualized not as a straight line but as circular with connections between existing theory, logical deduction of hypotheses, observations, empirical generalizations, and theory creation.

Concepts, Variables, and Hypotheses

In the more inductive studies like those by Enos and the Adlers, the focus is on concepts (such as "motherhood in prison" and "competitive afterschool activities"). Some studies are interested in providing an in-depth exploration of one or more concepts (or the ideas about which words or signs referring to significant aspects of a phenomenon have been developed). Other studies, like the one conducted by Gray and her associates, concentrate primarily on *variable* concepts (more commonly called variables), which are characteristics that can vary from one subject to another or for one subject over time. In the study on rape blame, for example, Gray, Palileo, and Johnson classified each member of the sample on each variable. By locating each subject on the variable "risk taking," for example, they could determine if subjects differed in the number of risks they took in putting themselves in situations with some danger of assault.

The use of variables implies the notion that some factors can affect or influence change in others. If we predict that change in one variable will affect change in another variable, then we are working with a causal hypothesis. In a two-variable causal hypothesis, the "affecting" variable is the independent variable, and the "affected variable" is the dependent variable. In the rape blaming study, for example, increases in the independent variable of risk taking were hypothesized to be connected to decreases in the dependent variable, the extent of victim blaming. Another example of a causal hypothesis is the one used by Adler and Foster in Chapter 8. Their field experiment was designed to examine the effect of an independent variable (a literature-based curriculum) on a dependent variable (support for the value of caring for others) in one classroom.

STOP AND THINK *Do you recall the steps in the research process? What decisions, for example, would a researcher like Christine Angell need to make once she selected her concepts and research questions?*

Practical Matters and Ethical Considerations

Once a research question or hypothesis is developed, a researcher must make several choices before collecting data. The planning phase includes selecting a population, sample, and sampling strategy; choosing a study design; picking method(s) of data collection; and, if appropriate, deciding how to measure variables and determine ways to classify units into categories of each variable concept. Each of these choices has many options; some are more likely than others to produce valid, reliable, and generalizable data. Making appropriate choices means selecting feasible and ethical strategies. Making appropriate choices also allows researchers to have confidence in their research conclusions. Theory, validity, practicality, and ethical considerations are all factors when you are planning a research project.

All the studies we've highlighted as focal research have been feasible. That is, the researchers completed their work: They had sufficient time and funding and were able to get access to the information they needed. With additional funding, more time or greater access, however, some of the researchers might have made different choices about sample size, study design, or methods of data collection.

All the researchers whose work we've used in this text have endeavored to conduct their work ethically and to protect the physical, emotional, and psychological well-being of study participants. In each study, the researchers sought to have the balance of benefits versus harm to subjects weighted toward benefits. The researchers acted ethically toward colleagues and the public, even though in some projects like the evaluation of the DARE program by Wysong and associates discussed in Chapter 14, the researchers were in a difficult position relative to sponsors of the project.

Selecting a Sample

Keeping both practical and ethical considerations in mind, one of the first parts of the planning process is the decision about who or what to study. Selecting an element or sampling unit means determining which individuals, organizations, institutions, or collectivities will be the focus of the project. Examples of *sampling units* from the focal research selections are Enos's (Chapter 10) incarcerated mothers, Clark, Lennon, and Morris's (Chapter 13) children's books, and Messner's (Chapter 12) "Designated Market Areas," which were the geographical units comprising the counties served by television stations.

Selecting a *specific* population and sample means determining *which* particular sampling units are of interest. For example, in Hoffnung's focal research on women's lives (Chapter 3), she did not think it feasible to study *all* college seniors in North America or *all* women graduating from schools in the Northeastern United States. Because of considerations of time, money, and access, Hoffnung selected the population of all female seniors attending five colleges in the Northeast in 1993.

Related issues are whether to study every member of the population and, if the sample is to be smaller than the population, selecting a method for picking a sample. The decisions depend partly on whether it is theoretically important to be able to generalize from the sample to a larger group. If it is, then it is more appropriate to select a probability instead of a nonprobability sample. In her work, Hoffnung selected a disproportionate random sample, and, using weighting procedures, was able to make population estimates from her sample data. If selecting a probability sample is not feasible, however, or if the focus of the study is to explore and describe concepts, as was the case for the Adlers and for Enos, then a nonprobability sample may be sufficient.

Purposes of Research and Study Designs

A research question or hypothesis can tell us if a researcher's purpose is to explore, describe, explain, and/or critique an aspect of the social world. In the focal research in Chapter 7, for example, gerontologist Rachel Filinson asked a question with both descriptive and explanatory purposes: "Does the kind of housing for the elderly affect their assessment of residential environmental quality?" Filinson's goal was to describe the elderly residents' assessments of the residential environmental quality *and* to explain what contributed to the assessments.

With the study's purpose(s) in mind, a researcher can select a study design that will allow adequate exploration, description, explanation, or critique of the social phenomena and will provide either basic or applied, more immediately useful knowledge. Crafting a study design means determining the number of times to collect data, the number of samples to be selected, and, when testing a causal hypothesis, whether to control or manipulate the independent variable in some way. Filinson selected a cross-sectional design and, not having the ability to control the independent variable (place of residence), collected data for all variables. Collecting data once with one sample of elderly residents of public and publicly subsidized housing, her cross-sectional design was ideal for description. Because her data analysis included control variables, she was also able to consider possible spuriousness and provide some support for her hypothesis as causal. The design was practical because the cross-sectional is among the least time consuming and costly of designs.

Researchers who want to study people over time prefer to use longitudinal designs and collect data more than once. We've seen the need for over-time data collection in some of our studies, such as the evaluation of DARE conducted by Wysong and his associates and the research on women's lives by Hoffnung.

Often it is unnecessary or impossible to control the independent variable, as in Filinson's research on the impact of housing, or in Messner's study of the effect of television viewing on crime. However, researchers who have access to situations where the independent variable can be controlled in some way should consider using an experimental design, as Adler and Foster did in their evaluation of a reading curriculum. Selecting panel, trend, cohort, or experimental designs can be worth the extra costs of time, money, and effort when greater validity results from using more than one sample, more than one data collection, or attempting to control the independent variable.

STOP AND THINK *Assume your question is "What factors contribute to alcohol consumption among college students?" What study design would you select?*

STOP AND THINK AGAIN *Assume you have decided to do research on the question about alcohol consumption using a cross-sectional study with a random sample of students attending your institution. What other aspects of the study would you need to plan?*

Measurement and Methods of Data Collection

The last aspect of research planning is taking the concepts to be studied and deciding how to measure them. Measurement is a way of making observations about each concept, or, when direct perception is not possible, of drawing inferences. Implementing the process of measurement means making observations and drawing inferences by using one or more methods of data collection. Possible methods include direct observation of behavior, asking questions, and obtaining information from available data, including existing documents and records.

Measurement encompasses the processes of conceptualization and operationalization. First, each concept must be clarified by developing a conceptual definition to narrow the concept's range of meaning. Then, an operational definition can be constructed to identify specific ways to determine the existence or the degree of existence of the phenomena described by the concept.

Each study we've included has focused on at least one concept and has presented measurement schemes of varying clarity and precision. For example, in the study on suicide that we described in Chapter 6, Ian Rockett and Gordon Smith have a clear conceptualization of suicide as "the act of killing oneself intentionally," but their work points out the wide variation in operational definitions used by different nations. Because of differences in medicolegal procedures and decision making, medical examiners have been more likely than coroners to classify similar deaths as suicides, which raises concerns about using official cause of death classifications for research purposes.

In the focal research project described in Chapter 14, Wysong and his associates attempt to assess the effectiveness of the DARE program. They conceptually define one of their concepts, lifetime prevalence of drug use, as all current and past use of tobacco, alcohol, marijuana, hashish, psychedelics, cocaine, crack, tranquilizers, barbiturates, amphetamines, and narcotics (heroin, morphine, and the like). Their operational definition of the concept is the Drug Use Scale, a set of 12 multipart questions with closed-ended answers they included on an anonymous questionnaire. Wysong and his colleagues did not use the DARE scale (the measurement technique promoted by the national DARE organization) because it focuses more on attitudes and knowledge than on behavior.

Studies with very different topics can use the same methods of data collection. Rockett and Smith studied suicide whereas Messner focused on the connection between TV viewing and violent crime; both studies used available data (census reports and Nielsen estimates of audience size, for example). Filinson was interested in residential environmental quality, Enos focused on managing motherhood, Hoffnung concentrated on work and family issues, and Weitz studied living with HIV—and all four researchers used in-person interviews with varying degrees of structure. Gray, Palileo, and Johnson studied assigning rape blame, and Wysong and colleagues evaluated DARE, yet both used group administered questionnaires. The wide range of methods included in the focal research also includes observation, content analysis, and introducing the independent variable: The Adlers collected data as participant observers in afterschool settings, Clark and his colleagues did a content analysis of children's books, and Adler and Foster conducted an experiment on the effects of a seventh-grade reading curriculum. In each focal research selection, the researchers selected methods they thought would provide useful data and made choices that had advantages, disadvantages, and implications for the validity and generalizability of their data.

The text covered the major methods of data collection in the social sciences: questionnaires, interviews, observations, available data, and content analysis. One point that we stressed was that no method is inherently better or worse than the others. Instead, we believe that each method is more or less useful for studying specific topics and situations. We turn now to a summary of the major methods of data collection and the process of selecting appropriate methods for specific studies.

Comparison of Data Collection Methods

STOP AND THINK *If we had a research question that focused on an aspect of social behavior, what are some of the data collection options we would have? How should we decide which method(s) to use?*

The two research projects that we described at the beginning of this chapter focused on both some aspect of social behavior. Felton Earls and Albert Reiss, Jr., are studying criminal behavior; Christine Angell focused on immigrant adaptation to life in the United States. In both projects, the researchers considered various methods of data collection before making their choices.

Our guess is that, like most researchers, in the planning stages of their projects, Angell, Earls, and Reiss asked at least some of the following questions: Can I observe this behavior? Can I observe without disclosing my identity as a researcher or take the role of total participant by pretending to be a member of the community I'm interested in? Or should I become participant observer and ask permission of those I want to observe? Can I ask people about their behavior? Will they give me honest and accurate responses? If I do ask for self-reports, should I conduct semi-structured interviews, or would I be better off using a structured phone interview? How about finding a way to distribute a group administered questionnaire or mailing a survey to people's homes? Is there available data that I could use? Would it be useful to analyze the content of some form of communication? Asking these kinds of questions, researchers can evaluate the methodological appropriateness of each data collection technique.

Taking them one at a time, researchers should consider these data collection issues:

- Which method(s) will allow for the collection of valid information about the social world?
- Which method(s) will allow for the collection of reliable information about the social world?
- To what extent will the results be generalizable?
- To what extent are these data collection methods ethical?
- To what extent are these data collection methods practical?

If we make appropriate choices in selecting our methods of data collection, then we have a good chance of drawing accurate conclusions about empirical reality. Such research will allow us to make worthwhile contributions in describing, explaining, or predicting social phenomena and help us to develop a basic understanding of the social world or provide information to help solve social problems.

If we choose to violate ethical principles to collect valid or generalizable data, we must know what benefits we obtain for this cost. On the other hand, we might have to make compromises in validity or generalizability to conduct research ethically. Finally, if we can't get access to data, or if time and funding are insufficient, then we might end up with incomplete data, even if the data we have are valid and ethical.

To recapitulate some points we made in earlier chapters, we will briefly focus on the validity, generalizability, ethical standards, and feasibility of each data collection method. We offer some general guidelines for selecting data collection methods for specific purposes and for studying specific topics.

Making Observations

Direct observational methods include being a complete participant, participant-as-observer, observer-as-participant, and complete observer. Observing can be a useful and valid way to study behavior because the behavior can be recorded as it occurs. With observation, the researcher does not depend on people's reports of behavior that can be colored by the willingness or ability to share information. Observational techniques usually require little more than the observer's time. Observation methods can be extremely useful for exploratory, descriptive, and theory generating purposes. The researcher selecting observational methods can use the observation period to figure out which concepts and meanings are most relevant to the question under study.

There are, however, some concerns about validity and reliability when you are using this method. People sometimes modify their behavior significantly when being observed; different observers can focus on different aspects of the behavior and setting and can see different things. The method might be impractical in some situations, as when the behavior can't be observed (like voting), when the behavior occurs in private rather than public settings (such as doing housework), and when it is difficult to predict when and where the behavior will occur (such as a theft or assault). Researchers usually find observation to be time-consuming and labor-intensive. Often those using this method will have small and nonrandom samples, which means results with limited generalizability.

The specific choice of observational method should receive careful consideration. Being a complete participant involves serious ethical concerns, including the lack of informed consent. Planning to be participant-as-observer can mean denial of access to a setting. In some situations, the knowledge that there is an observer can change the behavior of the observed. The observer-as-participant has similar problems, plus the validity concern that the researcher is less involved in the setting and more likely to misinterpret comments and actions. Complete observers can't interact with participants and can't ask questions, which makes possible misunderstanding the context and meaning of the observed behavior. In addition, complete observers who don't get permission when observing in nonpublic settings can be violating ethical principles.

Asking Questions

Collecting data by asking questions—the self-report methods—include mailed, group-administered, and individually administered questionnaires, and structured, semi-structured phone, and in-person interviews. These methods can be useful for studying attitudes, beliefs, values, goals, knowledge, life stories, expectations, and social characteristics. They can also be used to find out about an individual's previous experiences or for information about organizations and institutions. When it's either impractical or unethical to observe behavior, questions can be asked about behavior.

Asking questions, especially in a survey with structured interviews and questionnaires, allows you to collect comparable data from large numbers of people even if they are geographically far apart. If a probability sample is selected and the response rate is high, doing a survey permits you to generalize the sample results to the larger population.

On the other hand, there are important concerns about the validity of self-report information. People might misunderstand questions, might not have the information to answer some questions, might not take the survey seriously, or might be reluctant to answer honestly.

Selecting a specific method of asking questions involves considering several issues. Structured interviews and questionnaires can be useful when you are covering many topics or surveying large, representative samples. These forms of research are less expensive and time consuming than are in-person interviews, but the specific order of the questions can affect responses and the use of closed-ended questions might force respondents into categories that misrepresent their opinions.

On the other hand, in more qualitative work, when the goal is not to measure variables but, rather, to see how individuals subjectively see the world or make sense of their lives, it might be more useful to conduct semi- or unstructured interviews. The less structured techniques are usually more expensive because of interviewer time and transcription costs and are far more time consuming to code and analyze. However, less structured interviews can produce more valid data as interviewers can modify their presentations of self, the specific questions, and the order of questions.

Available Data and Content Analysis

Available data are materials that already exist. Most often they are data that have been collected by entities other than the researcher, including governments, businesses, schools, and other researchers. These data can be available at the individual level or available only as statistics about aggregates. Content analysis is typically the systematic study of one kind of available data—material that is some form of communication, including things like speeches, advertisements, newspaper articles, films, and books.

Available data are useful for providing information about individuals, organizations, or institutions and can be used to study both the present and the past. Another advantage is that using these kinds of data does not affect them in any way. In addition, much available data (that is, census reports, crime statistics, children's books) are public, widely available at minimal cost, and can be used with little expenditure of time.

Some documents (that is, tax records, personal diaries, confidential business reports), however, are private and can be quite difficult to obtain or obtainable only by violating ethical standards. Other disadvantages in using available data are that the meanings of some data are difficult to interpret, significant errors or biases may exist in the original data, and analyses are limited by what data exist.

STOP AND THINK *Now that you've reviewed the various techniques for data collection, see if you can apply what you've learned. Assume you are given a semester-long grant by the Alpha Beta Foundation to study alcohol consumption among college students. You decide to do your research on your own campus, and, from the grant, you have the funds to hire others to help with the data collection. You carefully consider your choices and narrow them down to three: Designing a closed-ended anonymous questionnaire on drinking and selecting and attending a random selection of classes to ask students to complete it anonymously; hiring several students who live in the residence halls on campus to work as complete participant observers for a month and having them privately record the amount of alcohol use they see as they go about their daily lives; counting the beer, wine, and liquor bottles in several dozen trash cans on campus for one month. Think about validity, generalizability, being ethical, and practical matters: How do these choices stack up against each other? Which would you use and why?*

Each of these methods has advantages and disadvantages. Although a voluntary, anonymous survey is likely to be ethical, fairly practical, and, if there is a good response rate, produce results that are generalizable, an important issue is validity—will students tell the truth? Do they remember accurately? Will they over- or under-estimate usage? On the other hand, complete participant observers might be able to give accurate results for the behavior they see (assuming they record periodically and do not consume much alcohol themselves), but we don't know how representative their observations are, and more important, the students being observed have not been informed and have not given consent. Analyzing trash is fairly practical and ethical, but collecting data for only one month raises questions of reliability. Counting bottles in the trash also raises concerns about validity as the bottles for some alcoholic beverages consumed by students might have been returned for the deposit, others might be in dorm rooms or in recycling bins, and some might have been left by nonstudents. At best, the data tell us about the campus as a whole rather than about individual student use.

Multiple Methods

STOP AND THINK *When you "stopped and thought" about the proposed study on alcohol use among college students, did you consider using more than one method? Why might you use multiple methods?*

BOX 17.1

Examples of multiple methods

Research Question:

Do nurses treat terminally ill patients differently from those that are expected to recover?

Possible methods:

Being a participant observer in a hospital for at least several weeks coupled with semi-structured, in-person interviews with a sample of nurses and a sample of patients, and an analysis of a sample of patient records.

Research Question:

How do two communities in the same urban area differ in community service, social isolation, and fear of crime?

Possible methods:

Available data (such as the annual reports from local agencies and organizations, editorials, and news stories in local newspapers, National Crime Victim Survey data, and the FBI Uniform Crime Reports), along with observations in public places as a complete observer, and structured phone interviews conducted with a random sample of residents.

Each individual method has advantages and disadvantages, so it is useful to employ more than one method of data collection whenever it is practical. In studying alcohol use on campus, if the resources were available why not use more than one technique? A questionnaire, for example, could be coupled with the analysis of available data or a focus group or both. Together they would give a better overview of the topic than any one of the methods. Box 17.1 includes some other examples of using more than one method of data collection.

Using multiple methods allows the strengths of one method to compensate for the weaknesses of another and increases the validity of the study's conclusions. Of course, additional methods mean increases in time and cost and additional settings mean obtaining access more than once, so practical considerations must be weighed against the increases in validity or generalizability that the use of multiple methods can bring.

Writing the Research Report

After the literature has been reviewed, the decisions concerning strategy have been made, and the data have been collected and analyzed, one last important step remains: reporting the results of a project. The research report can be communicated in a book, a written paper, or an oral presentation. The intended audience(s) can be the teacher of a class, a supervisor at work, an agency that funded the research, an organization that gave access to the data, or the general public. Although the report might differ for each audience, each version should describe the research process and research conclusions in an orderly way.

Students wanting a detailed discussion of the "nuts and bolts" of technical writing, including coverage of footnoting, referencing, and writing style, should consult one of the many excellent guides, such Turabian's (1996) *Manual for Writers of Term Papers, Theses, and Dissertations*, Pyrczak and Bruce's (1992) *Writing Empirical Research Reports: a Basic Guide for Students of the Behavioral Sciences*, or Lester's (1990) *Writing Research Papers: A Complete Guide*. We will briefly cover the general model used by scholarly journals that was exemplified by the focal research selections in this text. Most research reports include the following:

1. *A Title*. This tells the reader the topic of the research.
2. *An Abstract*. This short summary of the project covers its essential aspects, including what's been studied, major questions or hypotheses, and findings.
3. *Introduction*. This section introduces the topic and gives reasons why it is important to study.
4. *Review of the Literature*. This section presents an overview of previous work in the field related to the topic. Concepts, variables, research questions, theories, or hypotheses culled from the existing literature are discussed.
5. *Topics, Research Questions, or Hypotheses*. Qualitative studies will discuss the topic and concepts under consideration and their importance for understanding the social world. In more quantitative studies, the research questions or hypotheses that were developed before data collection are described and variables are identified.
6. *Research Methods*. Research strategies are described and explained. The study design, population, sample, sampling technique, method of data collection, and any measurement techniques are described. Practical and ethical considerations that influenced the choice of methods are discussed.
7. *Research Findings and Discussion*. Data from the research are presented. If quantitative analyses were done, statistical results, tables, or graphs should be included. If qualitative analyses were done, then summary statements or generalizations with evidence from the data should be included. In quantitative analyses, data that support and do not support hypotheses should be presented. In both kinds of studies, interpretations of data should be included.
8. *Conclusions*. The general conclusions are discussed. Answers to research questions and, if hypotheses were tested, thoughts about why they were supported or not supported should be included. Implications of the study for theory construction or modification can be presented. Limitations of the study's conclusions are specified and any "next steps" are identified.
9. *References and Appendices*. All citations are included. It's quite useful to include a copy of a questionnaire, an interview schedule, or a coding scheme if one was used.

Summary: Endings and Beginnings

We've concluded our discussion of research methods with a summary of the process of planning, conducting, and writing about a study. We've talked about selecting research questions, working either inductively or deductively to generate theories or hypotheses, selecting sampling strategies and study designs, developing measurement schemes, selecting methods of data collection, and writing up the results. We've suggested that the multiple criteria of validity, generalizability, feasibility, and ethical considerations be used when deciding about each part of the research design.

Now you can evaluate the methods used in the published research that you'll continue to encounter throughout your life—in other classes, in professional settings, when you read a daily newspaper, or when you watch a news broadcast. In addition, if you've tried out some research techniques and strategies, we hope you've enjoyed the experience and will be able to apply these tools if you are called on to do research in the future.

You've just begun the research adventure. We invite you to continue because we believe that research can provide you with fascinating opportunities to learn about human behavior and the social world.

EXERCISE 17.1

Summing Up

Read the following hypotheses and answer the following questions:

A. It is hypothesized that soccer coaches of younger, coed teams, treat male and female players on their teams differently. More specifically, it is hypothesized that on coed teams, the boys are kept in the game for longer periods of time and are more likely to play offensive positions while girls are more likely to be on the sidelines and to play defensive positions.

B. It is hypothesized that age is related to attitude toward violence on television. More specifically, it is hypothesized that people in their 40s and 50s are more critical of TV violence than are people in their 20s and 30s.

C. It is hypothesized that residential mobility affects academic achievement. Specifically, it is hypothesized that children from families that have lived in no more than two different communities during their elementary school years tend to have higher academic achievement than do children from families that have lived in three or more communities.

1. For each hypothesis, identify the independent and dependent variables.

2. For each variable in each hypothesis, what method(s) of data collection would you use? Suggest a way to measure each of the variables.

3. Justify why you selected the method(s) you described in #2. Focus on the issues of validity, ethical concerns, and practical matters in this and alternative methods.
4. Do you need more than one data collection period? What study design would you use?

EXERCISE 17.2

Alternative Study Designs and Methods of Data Collection

Each of the research projects highlighted in the focal research sections had a specific study design and most used only one method of data collection. In this exercise, we invite you to ask "would've, should've, could've" about one of the focal research articles.

Select an article and discuss the advantages and disadvantages of the study design and method(s) used. Consider a possible alternative design and method of data collection.

1. Write the name of the focal research article selected.
2. What was the actual study design used in this study?
3. What were some advantages of this study design?
4. What were some disadvantages of this study design?
5. Was an alternative design possible? What would some advantages and disadvantages be if it were used?
6. What data collection method or methods were actually used by the researcher(s) in this study?
7. What were some advantages of the method(s) used?
8. What were some disadvantages of the method(s) used?
9. What is an alternative or additional method for data collection? What are some advantages and disadvantages of using this alternative method?

EXERCISE 17.3

Planning a Study

Based on your own interests, select a topic appropriate for social research. Assume that you have the time to do research on this topic and plan a small research project. List your concepts and a research question or hypothesis. Describe your study design, population, and sampling technique, method of data collection, and how you would measure one or two of the concepts. Be sure to justify your choices using the criteria of validity, generalizability, feasibility, and ethical considerations.

Random Number Table

3250	2881	2326	9108	3702	6844	9827	3210	4895
7961	7952	3660	1503	4644	7656	3856	1672	0695
7739	5850	5998	9821	9476	2165	7662	5742	5048
0417	0262	7442	0873	6101	8592	2862	5346	7284
5816	4578	4061	7262	6328	1355	9386	5446	0666
7843	1922	3038	8442	5367	5952	6782	4206	4630
4912	7287	8526	9602	4545	3755	4159	3288	2900
2105	4791	5097	7386	6040	3575	4143	4820	5547
3263	3102	3555	4204	9411	0088	4873	1407	0858
4619	0105	9946	3234	1265	9509	0227	6671	3992
4460	7800	4476	3613	1252	8579	0611	6919	4208
6041	6103	3043	4634	1191	7886	4579	6301	4488
6718	8181	1425	6396	5312	4700	8077	1604	0870
8253	0766	2292	0146	6302	4413	6170	9764	3377
2061	5261	7862	8368	6533	9481	4098	1313	8527
4712	2096	6000	4173	8920	6913	3092	2028	2678
3344	6985	0656	2416	8367	5673	3272	8878	8714
5965	7229	0064	8356	7767	0960	5365	4980	9568
1203	8216	0354	1005	8063	3853	6732	0012	5391
6619	7927	0067	5559	2929	8706	6366	1111	9997
5545	8811	0833	1259	1344	0579	2735	9419	8533
8446	2198	8304	0803	5947	6322	9623	4419	8007
6017	2981	0880	6879	0193	6721	9130	2341	4528
7582	0139	5100	9315	5571	4508	1341	3450	1307
4118	7411	9556	9907	0612	0601	9545	8084	7117
8443	2486	1121	6714	8133	9113	4613	2633	6828
0231	2253	0931	9757	3471	3080	6369	6270	6296
8393	2801	8322	2871	6504	4972	1608	4997	2211
6092	5922	5587	8472	4256	2888	1344	4681	0332
8728	4779	8851	7466	9308	3590	3115	0478	6956
8585	4266	8533	9121	1988	4965	1951	4395	3267
9593	6571	7914	1242	8474	1856	0098	2973	2863
9356	6955	4768	1561	6065	3683	6459	4200	2177
0542	9565	7330	0903	8971	5815	2763	3714	0247
6468	1257	2423	0973	3172	8380	2826	0880	4270
1201	1959	9569	0073	8362	6587	7942	8557	4586
3616	5075	8836	1704	9143	6974	5971	0154	5509
4241	9365	6648	1050	2605	0031	3136	5878	9501
1679	5768	2801	5266	9165	5101	7455	9522	4199
5731	9611	0523	7579	0136	5655	6933	7668	8344
1554	7718	9450	9736	4703	9408	4062	7203	8950
1471	3213	7164	3788	4167	4063	7881	5348	4486
9563	8925	5479	0340	6959	3054	9435	6400	7989
3722	8984	2755	2560	1753	1209	2805	2763	8189
4237	1818	5266	2276	1877	4423	7318	1483	3153
2859	9216	5303	2154	3243	7359	1393	9909	2482
8303	7527	4566	8713	0538	7340	6224	3135	8367

4150	4056	1579	9286	4353	3413	6618	9524	9284
1507	1745	7221	9009	4769	9333	6803	0198	7879
1365	3544	3753	1530	7384	9357	2231	6166	3217
7414	5143	8912	6611	3396	7969	6559	9463	9114
9658	7935	7850	4297	4908	9653	9540	8753	3920
3169	0990	6016	0631	8451	2488	8724	3625	1805
7294	6007	0887	9667	5211	0082	2522	7055	4431
0514	8862	2713	9944	7880	5101	7535	1200	2583
7016	0451	2039	5827	2820	4979	1997	0920	0988
0389	0734	3322	0822	7189	7229	1031	2444	5827
4903	9574	0313	6611	6589	4437	8070	4480	6481
3746	6799	3082	6987	3766	8230	4376	7150	6963
7695	8918	3141	8417	7200	6066	4097	8883	5254
8423	2024	8911	9430	1711	6618	6768	0251	8926
5009	0421	9412	0227	4881	2652	8863	2900	1878
3265	7124	8628	0386	7584	3412	9002	7013	2898
2263	6500	7451	5922	8289	3552	5546	7389	1959
4629	5201	3214	6479	0313	4100	0596	9274	0415
6833	0952	8529	2108	2219	8249	2184	6631	7768
1991	4762	7349	3282	6394	3786	5792	0337	7781
9628	1897	4009	1946	2765	3552	9865	0398	4612
5860	8540	7938	2799	6545	0922	9124	4352	2057
3394	0297	8151	6445	6994	4566	0323	6185	3117
6976	4713	7478	9092	7028	8140	2490	9508	9481
5359	7743	8669	8469	5126	1434	7695	8129	3184
7127	6156	8500	7593	0670	9534	3945	9718	0834
8690	6983	6359	3205	6167	4362	5340	4104	3004
5705	2941	2505	3360	5976	2070	8450	6761	7404
7210	4415	4744	2061	5102	7796	2714	1876	3398
0777	2055	8932	0542	1427	8487	3761	3793	0954
0270	9605	7099	9063	2584	5289	0040	2135	6934
7194	7521	3770	6017	4393	3340	8210	6468	6144
6029	0732	9672	6507	1422	3483	2387	2472	6951
6631	3518	5642	2250	3187	8553	6747	6029	1871
6958	4528	6053	9004	6039	5841	8685	9100	4097
1589	6697	9722	9343	3227	3063	6808	2900	6970
2004	8823	1038	1638	9430	7533	8538	6167	1593
2615	5006	1524	4747	4902	7975	9782	0937	5157
6487	6682	2964	0882	2028	2948	6354	3894	2360
7238	7432	4922	9697	7764	7714	4511	3252	6864
0382	2511	7244	7271	4804	9564	8572	6952	6155
8918	3452	3044	4414	1695	2297	6360	2520	5814
9331	0179	6815	5701	0711	5408	0322	8085	0080
1065	9702	9944	9575	6714	9382	8324	3008	9373
6663	5272	3139	4020	5158	7146	4100	3578	0683
7045	5075	6195	5021	1443	2881	9995	8579	7923
6483	7316	5893	9422	0550	3273	5488	2061	1532
4396	9028	4536	8606	4508	3752	9183	5669	5927
5766	7632	9132	7778	6887	8203	6678	8792	3159
2743	8858	5094	2285	2824	9996	7384	4069	5442
3330	8618	1974	2441	0457	6561	7390	4102	2311
4791	2725	4085	3190	9098	2300	6241	7903	9150
6833	9145	2459	2393	7831	2279	9658	5115	6906
9474	9070	4964	2035	7954	7001	8688	7040	6924
8576	0380	5722	6149	4086	4437	9871	0850	3248

Questionnaires

Questionnaires used by Norma B. Gray, Gloria J. Palileo and G. David Johnson for their research "Explaining Rape Victim Blame: A Test of Attribution Theory" included in Chapter 8.

University of South Alabama
SURVEY ON CAMPUS ISSUES

This is a survey of the opinions and experiences of students at the University of South Alabama. The survey is being conducted by the Department of Sociology and Anthropology. The questions are of a personal nature and your participation is voluntary. We would like for you to answer the following questions as truthfully as you can. To insure your anonymity, please do not write your name anywhere on the survey. Your cooperation is very much appreciated.

Please drop the completed form in a box provided by the monitor.

If you do not wish to participate, please return the questionnaire to the monitor and remain in your seat.

FOR MALE RESPONDENTS

Have you completed this questionnaire in another class?

☐ Yes (Please return this questionnaire to the monitor).
☐ No (Please answer the questions).

Card No. 1 Form Code: M-1:1

1. What is your class standing? (Check one).
   ☐ 1 Freshman ☐ 4 Senior
   ☐ 2 Sophomore ☐ 5 Graduate student
   ☐ 3 Junior

2. How old are you today? ____________

3. Where are you living now? (Check one)
   ☐ 1 Dorm ☐ 4 Off-Campus Housing
   ☐ 2 Sorority or fraternity ☐ 5 Home of parent or relative
   ☐ 3 Campus Housing at Hillsdale

4. What is your marital status? (Check one)
   ☐ 1 Single ☐ 4 Widowed
   ☐ 2 Married ☐ 5 Cohabitating (Living with someone)
   ☐ 3 Divorced/Separated

5. What is your race or ethnic background? (Check one)
   ☐ 1 White, Non-Hispanic
   ☐ 2 Black, Non-Hispanic
   ☐ 3 Hispanic
   ☐ 4 Asian or Pacific Islander
   ☐ 5 American Indian
   ☐ 6 Other (Please specify) ____________

6. Which applies to you? (Check one)
   ☐ 1 Born in the U.S.
   ☐ 2 Foreign-born but U.S. citizen or permanent resident
   ☐ 3 Foreign-born and neither a U.S. citizen or permanent resident

7. What is your best guess of your family's Income last year? (If married, still estimate the income in the family in which you grew up) (Check one)
   ☐ 1 $15,000 or less ☐ 4 $35,001 - $50,000
   ☐ 2 $15,001 - $25,000 ☐ 5 $50,001 - $100,000
   ☐ 3 $25,001 - $35,000 ☐ 6 Over $100,000

8. How often do you drink alcohol? (Check one)
   ☐ 0 None in the past year
   ☐ 1 Less than once a month but at least once in the past year
   ☐ 2 One to three times a month
   ☐ 3 One to two times a week
   ☐ 4 More than twice a week

9. On a typical drinking occasion, how much do you usually drink? (Check one)
   ☐ 0 Never
   ☐ 1 No more than 3 cans of beer (or 2 glasses of wine or 2 distilled spirits)
   ☐ 2 No more than 4 cans of beer (or 3 glasses of wine or 3 drinks of distilled spirits)
   ☐ 3 No more than 5 or 6 cans of beer (or 4 glasses of wine or 4 drinks of distilled spirits)
   ☐ 4 More than 6 cans of beer (Or 5 or more glasses of wine or distilled spirits)

10. What is your sexual orientation? (Check one)
☐ 1 Heterosexual
☐ 2 Gay
☐ 3 Bisexual

The following questlons are about your sexual experlence from age 14 on.

11. Have you engaged in sex play (fondling, kissing, or petting, but not intercourse) with a woman when she didn't want to by overwhelming her with continual arguments and pressure? (From age 14 on)
☐ 0 No ☐ 1 Yes

12. Have you engaged in sex play (fondling, kissing, or petting, but not intercourse) with a woman when she didn't want to by using your position of authority (boss, teacher, camp counselor, supervisor)? (From age 14 on)
☐ 0 No ☐ 1 Yes

13. Have you engaged in sex play (fondling, kissing, or petting, but not intercourse) with a woman when she didn't want to by threatening or using some degree of physical force (twisting her arm, holding her down, etc.)? (From age 14 on)
☐ 0 No ☐ 1 Yes

The followlng are questions about sexual Intercourse. By sexual Intercourse we mean penetration of a woman's vaglna, no matter how sllght, by a man's penis. Ejaculation is not required. Whenever you see the words sexual intercourse, please use this definltion.

14. Have you attempted sexual intercourse (attempted to insert penis) with a woman when she didn't want it by threatening or using some degree of physical force (twisting her arm, holding her down, etc.) but intercourse did not occur? (From age 14 on)
☐ 0 No ☐ 1 Yes

15. Have you attempted sexual intercourse (attempted to insert penis) with a woman when she didn't want it by giving her alcohol or drugs, but intercourse did not occur? (From age 14 on)
☐ 0 No ☐ 1 Yes

16. Have you engaged in sexual intercourse with a woman when she didn't want to by overwhelming her with continual arguments and pressure? (From age 14 on)
☐ 0 No ☐ 1 Yes

17. Have you engaged in sexual intercourse with a woman when she didn't want to by using your position of authority (boss, teacher, camp counselor, supervisor)? (From age 14 on)
☐ 0 No ☐ 1 Yes

18. Have you engaged in sexual intercourse with a woman when she didn't want to by giving her alcohol or drugs? (From age 14 on)
☐ 0 No ☐ 1 Yes

19. Have you engaged in sexual intercourse with a woman when she didn't want to by threatening or using some degree of physical force (twisting her arm, holding her down, etc.)? (From age 14 on)
☐ 0 No ☐ 1 Yes

20. Have you engaged in sex acts (anal or oral intercourse by objects other than the penis) with a woman when she didn't want to by threatening or using some degree of physical force (twisting her arm, holding her down, etc.)? (From age 14 on)
☐ 0 No ☐ 1 Yes

22. Did you answer Yes to any of the questions 11 to 20?
☐ 0 No If no, skip to question #33. ☐ 1 Yes

22. Look back to questions 11 to 20. What is the highest question number to which you marked yes?
☐ 11 ☐ 13 ☐ 15 ☐ 17 ☐ 19
☐ 12 ☐ 14 ☐ 16 ☐ 18 ☐ 20

23. We'd like to ask you about this experience. What was your relationship to the woman at the time of this sexual experience? (Check one)
☐ 1 Stranger
☐ 2 Non romantic acquaintance (friend, neighbor, ex-husband, etc.)
☐ 3 Casual or first date
☐ 4 Romantic acquaintance (steady date, boyfriend, lover)
☐ 5 Husband
☐ 6 Relative (father, stepfather, uncle, brother)

24. How old were you when this experience occurred?
__________ years

25. Did this sexual experience happen before you enrolled at the University of South Alabama or since you have been a student here?
☐ 1 Before enrolling at South Alabama
☐ 2 Since enrolling at South Alabama

26. Did it happen on or off campus?
☐ 1 On USA campus
☐ 2 On another college campus
☐ 3 Off campus

27. Were you using any intoxicants on this occasion? (Check One)
☐ 1 Alcohol
☐ 2 Drugs
☐ 3 Both
☐ 4 None

28. Was the woman using any intoxicants on this occasion? (Check one)
☐ 1 Alcohol
☐ 2 Drugs
☐ 3 Both
☐ 4 None
☐ 5 Don't know

29. What did the woman do in response to your behavior? (Check one for each response)

No	Yes	
☐ 0	☐ 1	Turn cold
☐ 0	☐ 1	Reason, plead, quarrel, tell you to stop
☐ 0	☐ 1	Cry or sob
☐ 0	☐ 1	Scream for help
☐ 0	☐ 1	Run away
☐ 0	☐ 1	Physically struggle, push you away, hit or scratch

30. What effect did her response have on you? (Check one)
☐ 1 I stopped
☐ 2 I became less aggressive
☐ 3 No effect on me
☐ 4 I became more aggressive

31. Could you describe these aspects of the incident? (Check one)

	Not at all	A little	Some.-what	Quite a bit	Very much
a. How aggressive were you?	1 ☐	2 ☐	3 C	4 ☐	5 ☐
b. How clear was the woman that she did not want sex?	1 ☐	2 ☐	3 C	4 ☐	5 ☐
c. How much do you feel responsible for what happened?	1 ☐	2 ☐	3 C	4 ☐	5 ☐
d. How much did the woman resist?	1 ☐	2 ☐	3 C	4 ☐	5 ☐
e. How responsible do you feel the woman is for what happened?	1 ☐	2 ☐	3 C	4 ☐	5 ☐

32. Which of the following statements describes how you view this experience? (Check one).
☐ 1 It definitely was not rape.
☐ 2 Some people would think it was something close to rape.
☐ 3 Many people would think it was rape.
☐ 4 It definitely was rape.

33. Since you have been at South Alabama, have you ever experienced any of the following?

	Yes	No
a. Verbal or non-verbal conduct with an inappropriate focus on gender or sexuality that is intimidating, demanding, hostile, or offensive	☐ 1	☐ 0
b. Unwelcome verbal or physical advances	☐ 1	☐ 0
c. Attempts to subject you to unwanted sexual attention or to coerce you into a sexual relationship	☐ 1	☐ 0
d. Retaliation for a refusal to comply with sexual demands.	☐ 1	☐ 0

If you answered "No" to all 4 questions, skip to #34.

a. Who did this to you? (Check all that apply).
☐ 1 College professor/s
☐ 2 Other University personnel
☐ 3 Fellow students
☐ 4 Person/s not connected with the University

34. Have you ever been a victim of a (non-sexual) crime? If yes, what kind of crime were you a victim of? (Check all that happened to you).
☐ 0 No (Skip to Question 37)
☐ 1 Physical assault
☐ 2 Mugging
☐ 3 Theft/purse snatching
☐ 4 Car theft
☐ 5 Theft of valuables in car
☐ 6 Burglary of residence
☐ 7 Harassing phone cells
☐ 8 Other (Please specify) ______________________

35. Did this/these experience/s happen before you enrolled at South Alabama or since you have been a student here?
☐ 1 Before enrolling at South Alabama
☐ 2 Since enrolling at South Alabama
☐ 3 Both

36. Did it happen on or off campus? (Check all that apply).
☐ 1 On USA campus
☐ 2 On another college campus
☐ 3 Off campus

37. How frequently would you say you do the following?

	Never	Seldom	Sometimes	Often
a. Walk alone on campus at night.	☐ l	☐ 2	☐ 3	☐ 4
b. Walk alone off campus at night.	☐ l	☐ 2	☐ 3	☐ 4
c. Leave dorm/apartment door open.	☐ l	☐ 2	☐ 3	☐ 4
d. Open door/let in strangers.	☐ l	☐ 2	☐ 3	☐ 4
e. Leave car unlocked.	☐ l	☐ 2	☐ 3	☐ 4
f. Go out on blind dates.	☐ l	☐ 2	☐ 3	☐ 4
g. Enter or work alone in an isolated building late at night or week-ends.	☐ l	☐ 2	☐ 3	☐ 4

38. How much would you say you know about rape prevention?
☐ 0 Nothing
☐ 1 A little
☐ 2 Some
☐ 3 A lot/much

39. Where did you get this information about rape prevention?
☐ 1 Rape Prevention Workshop
☐ 2 Television
☐ 3 Printed Materials
☐ 4 Classroom Discussions
☐ 5 Sororlty
☐ 6 Self-defense class
☐ 7 Parent/s
☐ 8 Other (Please specify) ______________________

☐ 9 Does not apply (No knowledge)

40. How much would you say you know about other crime prevention?
☐ 0 Nothing ☐ 2 Some
☐ 1 A little ☐ 3 A lot/much

41. Where did you get this information about crime prevention?
☐ 1 Crime Prevention Workshop
☐ 2 Television
☐ 3 Printed materials
☐ 4 Classroom discussions
☐ 5 Sorority
☐ 6 Self-defense class
☐ 7 Parent/s
☐ 8 Other (Please specify) ____________________
____
____
☐ 9 Does not apply (No knowledge)

Now we want to ask you about some of your opinions. Please tell us If you strongly agree, agree, are undecided, disagree, or strongly disagree wIth each of the following statements.

42. A female who goes to the home or apartment of a man on their first date implies that she is willing to have sex.
☐ 1 Strongly Agree
☐ 2 Agree
☐ 3 Undecided
☐ 4 Disagree
☐ 5 Strongly Disagree

43. A man has a right to have sex with his wife even if it is sometimes against her will.
☐ 1 Strongly Agree
☐ 2 Agree
☐ 3 Undecided
☐ 4 Disagree
☐ 5 Strongly Disagree

44. Any healthy female can successfully resist a rapist if she really wants to.
☐ 1 Strongly Agree
☐ 2 Agree
☐ 3 Undecided
☐ 4 Disagree
☐ 5 Strongly Disagree

45. When females go around braless or wearing short skirts and tight tops, they are asking for trouble.
☐ 1 Strongly Agree
☐ 2 Agree
☐ 3 Undecided
☐ 4 Disagree
☐ 5 Strongly Disagree

46. If a female engages in necking or petting and she lets things get out of hand, it is her own fault if her partner uses some force to get her to have sex with him.
☐ 1 Strongly Agree
☐ 2 Agree
☐ 3 Undecided
☐ 4 Disagree
☐ 5 Strongly Disagree

47. Many females who get raped have done something to provoke it.
☐ 1 Strongly Agree
☐ 2 Agree
☐ 3 Undecided
☐ 4 Disagree
☐ 5 Strongly Disagree

48. Most women, although they might not at first admit it, would enjoy being forced to have intercourse with a man.
☐ 1 Strongly Agree
☐ 2 Agree
☐ 3 Undecided
☐ 4 Disagree
☐ 5 Strongly Disagree

49. If a female gets drunk at a party and has intercourse with a man she's just met there, she can be considered available to other males at the party who want to have sex with her too, whether she wants to or not.
☐ 1 Strongly Agree
☐ 2 Agree
☐ 3 Undecided
☐ 4 Disagree
☐ 5 Strongly Disagree

Thank you for your cooperation.

University of South Alabama
SURVEY ON CAMPUS ISSUES

This is a survey of the opinions and experiences of students at the University of South Alabama. The survey is being conducted by the Department of Sociology and Anthropology. The questions are of a personal nature and your participation is voluntary. We would like for you to answer the following questions as truthfully as you can. To insure your anonymity, please do not write your name anywhere on the survey. Your cooperation is very much appreciated.

Please drop the completed form in a box provided by the monitor.

If you do not wish to participate, please return the questionnaire to the monitor and remain in your seat.

FOR FEMALE RESPONDENTS

Have you completed this questionnaire in another class?

☐ Yes (Please return this questionnaire to the monitor).
☐ No (Please answer the questions).

Card No. 1 Form Code: F-1:3

1. What is your class standing? (Check one).
   ☐ 1 Freshman
   ☐ 2 Sophomore
   ☐ 3 Junior
   ☐ 4 Senior
   ☐ 5 Graduate student

2. How old are you today? ____________________

3. Where are you living now? (Check one)
   ☐ 1 Dorm ☐ 4 Off-Campus Housing
   ☐ 2 Sorority or fraternity ☐ 5 Home of parent or relative
   ☐ 3 Campus Housing at Hillsdale

4. What is your marital status? (Check one)
   ☐ 1 Single ☐ 4 Widowed
   ☐ 2 Married ☐ 5 Cohabitating (Living with someone)
   ☐ 3 Divorced/Separated

5. What is your race or ethnic background? (Check one)
   ☐ 1 White, Non-Hispanic
   ☐ 2 Black, Non-Hispanic
   ☐ 3 Hispanic
   ☐ 4 Asian or Pacific Islander
   ☐ 5 American Indian
   ☐ 6 Other (Please specify) ____________________
   ____

6. Which applies to you? (Check one)
   ☐ 1 Born in the U.S.
   ☐ 2 Foreign-born but U.S. citizen or permanent resident
   ☐ 3 Foreign-born and neither a U.S. citizen or permanent resident

7. What is your best guess of your family's Income last year? (If married, still estimate the income in the family in which you grew up) (Check one)
   ☐ 1 $15,000 or less ☐ 4 $35,001 - $50,000
   ☐ 2 $15,001 - $25,000 ☐ 5 $50,001 - $100.000
   ☐ 3 $25,001 - $35,000 ☐ 6 Over $100,000

8. How often do you drink alcohol? (Check one)
   ☐ 0 None in the past year
   ☐ 1 Less than once a month but at least once in the past year
   ☐ 2 One to three times a month
   ☐ 3 One to two times a week
   ☐ 4 More than twice a week

9. On a typical drinking occasion, how much do you usually drink? (Check one)
   ☐ 0 Never
   ☐ 1 No more than 3 cans of beer (or 2 glasses of wine or 2 distilled spirits)
   ☐ 2 No more than 4 cans of beer (or 3 glasses of wine or 3 drinks of distilled spirits)
   ☐ 3 No more than 5 or 6 cans of beer (or 4 glasses of wine or 4 drinks of distilled spirits)
   ☐ 4 More than 6 cans of beer (Or 5 or more glasses of wine or distilled spirits)

10. What is your sexual orientation? (Check one)
   ☐ 1 Heterosexual
   ☐ 2 Lesbian
   ☐ 3 Bisexual

The following questions are about your sexual experience from age 14 on.

11. How often do you say "No" to a man's sexual advances when you actually don't want him to stop?
   ☐ 0 Never ☐ 3 Often
   ☐ 1 Seldom ☐ 4 Always
   ☐ 2 Sometimes

12. Have you given in to sex play (fondling, kissing, or petting, but not intercourse) when you didn't want to because you were overwhelmed by a man's continual arguments and pressure? (From age 14 on)
   ☐ 0 No ☐ 1 Yes

13. Have you had sex play (fondling, kissing, or petting, but not intercourse) when you didn't want to because a man used his position of authority (boss, teacher, camp counselor, supervisor) to make you? (From age 14 on)
   ☐ 0 No ☐ 1 Yes

14. Have you had sex play (fondling, kissing, or petting) when you didn't want to because a man threatened or used some degree of physical force (twisting your arm, holding you down, etc.)? (From age 14 on)
   ☐ 0 No ☐ 1 Yes

The following are questions about sexual Intercourse. By sexual intercourse we mean penetration of a woman's vagina, no matter how slight, by a man's penis. Ejaculation is not required. Whenever you see the words sexual intercourse, please use this definition.

15. Have you had a man attempt sexual intercourse (attempt to insert his penis) when you didn't want to by threatening or using some degree of force (twisting your arm, holding you down. etc.) but intercourse did not occur? (From age 14 on)
☐ 0 No ☐ 1 Yes

16. Have you had a man attempt sexual intercourse (attempt to insert penis) with you by giving you alcohol or drugs, but intercourse did not occur? (From age 14 on)
☐ 0 No ☐ 1 Yes

17. Have you given in to sexual intercourse when you didn't want to because you were overwhelmed by a man's continual arguments and pressure? (From age 14 On)
☐ 0 No ☐ 1 Yes

18. Have you had sexual intercourse when you didn't want to because a man used his position of authority (boss, teacher, camp counselor, supervisor)? (From age 14 on)
☐ 0 No ☐ 1 Yes

19. Have you had sexual intercourse when you didn't want to because a man gave you alcohol or drugs? (From age 14 on)
☐ 0 No ☐ 1 Yes

20. Have you had sexual intercourse when you didn't want to because a man threatened or used some degree of physical force (twisting your arm, holding you down, etc.) to make you? (From age 14 on)
☐ 0 No ☐ 1 Yes

21. Have you had sex acts (anal or oral intercourse or penetration by objects other than the penis) when you didn't want to because a man threatened or used some degree of physical force (twisting your arm, holding you down, etc.) to make you? (From age 14 on)
☐ 0 No ☐ 1 Yes

22. Did you answer Yes to any of the questions 12 to 21?
☐ 0 No If no, skip to question #36. ☐ 1 Yes

23. Look back to questions 12 to 21. What is the highest question number to which you marked yes?
☐ 12 ☐ 14 ☐ 16 ☐ 18 ☐ 20
☐ 13 ☐ 15 ☐ 17 ☐ 19 ☐ 21

24. We'd like to ask you about this experience. What was your relationship to the man at the time of this sexual experience? (Check one)
☐ 1 Stranger
☐ 2 Non romantic acquaintance (friend, neighbor, ex-husband, etc.)
☐ 3 Casual or first date
☐ 4 Romantic acquaintance (steady date, boyfriend, lover)
☐ 5 Husband
☐ 6 Relative (father, stepfather, uncle, brother)

25. How old were you when this experience occurred?
__________ years

26. Did this sexual experience happen before you enrolled at the University of South Alabama or since you have been a student here?
☐ 1 Before enrolling at South Alabama
☐ 2 Since enrolling at South Alabama

27. Did it happen on or off campus?
☐ 1 On USA campus
☐ 2 On another college campus
☐ 3 Off campus

28. Did you do any of the following to resist his advances? (Check one for each response)

No	Yes
☐ 0	☐ 1 Turn cold
☐ 0	☐ 1 Reason, plead, quarrel, tell him to stop
☐ 0	☐ 1 Cry or sob
☐ 0	☐ 1 Scream for help
☐ 0	☐ 1 Run away
☐ 0	☐ 1 Physically struggle, push him away, hit or scratch

29. What effect did your resistance have? (Check one)
☐ 1 He stopped ☐ 3 No effect on him
☐ 2 He became less aggressive ☐ 4 He became more aggressive

30. Could you describe these aspects of the incident? (Check one)

	Not at all	A little	Some.-what	Quite a bit	Very much
a. How aggressive was the man?	1 ☐	2 ☐	3 C	4 ☐	5 ☐
b. How clear did you make it to the man that you did not want sex?	1 ☐	2 ☐	3 C	4 ☐	5 ☐
c. How much do you feel responsible for what happened?	1 ☐	2 ☐	3 C	4 ☐	5 ☐
d. How much did you resist?	1 ☐	2 ☐	3 C	4 ☐	5 ☐
e. How responsible is he for what happened?	1 ☐	2 ☐	3 C	4 ☐	5 ☐

31. Did you tell your family, friends, boyfriend, or husband about this experience?
☐ 0 No (If no, go to question 32 below). ☐ 1 Yes

a. How did they react? (Check one)
☐ 1 Not at all supportive (i.e., blamed me, angry at me, or discouraged me from talking about it)
☐ 2 A little supportive
☐ 3 Somewhat supportive
☐ 4 Quite a bit supportive
☐ 5 Very much supportive (i.e., responded helpfully, encouraged me to talk)

32. Did you tell a crisis center or emergency staff?
☐ 0 No (If no, go to question 33 below). ☐ 1 Yes

a. How did they react? (Check one)
- ☐ 1 Not at all supportive (i.e., blamed me, angry at me, or discouraged me from talking about it)
- ☐ 2 A little supportive
- ☐ 3 Somewhat supportive
- ☐ 4 Quite a bit supportive
- ☐ 5 Very much supportive (i.e., responded helpfully, encouraged me to talk)

33. Did you tell the police?
☐ 0 No ☐ 1 Yes

a. If no, why not? (Check all that apply)
- ☐ 1 I did not think they would do anything.
- ☐ 2 I was embarrassed.
- ☐ 3 I was afraid he would return and hurt me more.
- ☐ 4 Other reasons (Specify) ____________________

____

____

b. If yes, how did they react? (Check one)
- ☐ 1 Not at all supportive (i.e.. blamed me, angry at me, or discouraged me from talking about it)
- ☐ 2 A little supportive
- ☐ 3 Somewhat supportive
- ☐ 4 Quite a bit supportive
- ☐ 5 Very much supportive (i.e., responded helpfully, encouraged me to talk)

34. Did you tell other people at a campus agency? (Check one)
☐ 0 No (If no, go to question 35). ☐ 1 Yes

a. How did they react? (Check one)
- ☐ 1 Not at all supportive (i.e.. blamed me, angry at me, or discouraged me from talking about it)
- ☐ 2 A little supportive
- ☐ 3 Somewhat supportive
- ☐ 4 Quite a bit supportive
- ☐ 5 Very much supportive (i.e., responded helpfully, encouraged me to talk)

35. Which of the following statements describes how you view this experience? (Check one).
- ☐ 1 It definitely was not rape.
- ☐ 2 Some people would think it was something close to rape.
- ☐ 3 Many people would think it was rape.
- ☐ 4 it definitely was rape.

36. Have you ever studied self defense or taken assertiveness training?
☐ 0 No ☐ 1 Yes

37. Since you have been at South Alabama, have you ever experienced any of the following?

	Yes	No
a. Verbal or non-verbal conduct with an inappropriate focus on gender or sexuality that is intimidating, demanding, hostile, or offensive	☐ 1	☐ 0
b. Unwelcome verbal or physical advances	☐ 1	☐ 0
c. Attempts to subject you to unwanted sexual attention or to coerce you into a sexual relationship.	☐ 1	☐ 0
d. Retaliation for a refusal to comply with sexual demands.	☐ 1	☐ 0

If you answered "No" to all 4 questions, skip to #38.

a. Who did this to you? (Check all that apply).
- ☐ 1 College professor/s
- ☐ 2 Other University personnel
- ☐ 3 Fellow students
- ☐ 4 Person/s not connected with the University

38. Have you ever been a victim of a (non-sexual) crime? If yes, what kind of crime were you a victim of? (Check all that happened to you).
- ☐ 0 No (Skip to Question 41)
- ☐ 1 Physical assault
- ☐ 2 Mugging
- ☐ 3 Theft/purse snatching
- ☐ 4 Car theft
- ☐ 5 Theft of valuables in car
- ☐ 6 Burglary of residence
- ☐ 7 Harassing phone cells
- ☐ 8 Other (Please specify) ____________________

____

____

39. Did this/these experience/s happen before you enrolled at South Alabama or since you have been a student here?
- ☐ 1 Before enrolling at South Alabama
- ☐ 2 Since enrolling at South Alabama
- ☐ 3 Both

40. Did it happen on or off campus? (Check all that apply).
- ☐ 1 On USA campus
- ☐ 2 On another college campus
- ☐ 3 Off campus

41. How frequently would you say you do the following?

	Never	Seldom	Sometimes	Often
a. Walk alone on campus at night.	☐ 1	☐ 2	☐ 3	☐ 4
b. Walk alone off campus at night.	☐ 1	☐ 2	☐ 3	☐ 4
c. Leave dorm/apartment door open.	☐ 1	☐ 2	☐ 3	☐ 4
d. Open door/let in strangers.	☐ 1	☐ 2	☐ 3	☐ 4
e. Leave car unlocked.	☐ 1	☐ 2	☐ 3	☐ 4
f. Go out on blind dates.	☐ 1	☐ 2	☐ 3	☐ 4
g. Enter or work alone in an isolated building late at night or week-ends.	☐ 1	☐ 2	☐ 3	☐ 4

42. How much would you say you know about rape prevention?
☐ 0 Nothing ☐ 2 Some
☐ 1 A little ☐ 3 A lot/much

43. Where did you get this information about rape prevention?
☐ 1 Rape Prevention Workshop
☐ 2 Television
☐ 3 Printed materials
☐ 4 Classroom discussions
☐ 5 Sorority
☐ 6 Self-defense class
☐ 7 Parent/s
☐ 8 Other (Please specify) ____________________
____
____
☐ 9 Does not apply (No knowledge)

44. How much would you say you know about other crime prevention?
☐ 0 Nothing ☐ 2 Some
☐ 1 A little ☐ 3 A lot/much

45. Where did you get this information about crime prevention?
☐ 1 Crime Prevention Workshop
☐ 2 Television
☐ 3 Printed Materials
☐ 4 Classroom Discussions
☐ 5 Sorority
☐ 6 Self-defense class
☐ 7 Parent/s
☐ 8 Other (Please specify) ____________________
____
____
☐ 9 Does not apply (No knowledge)

Now we went to ask you about some of your opinions. Please tell us If you strongly agree, agree, are undecided, disagree, or strongly disagree with each of the following statements.

46. A female who goes to the home or apartment of a man on their first date implies that she is willing to have sex.
☐ 1 Strongly Agree
☐ 2 Agree
☐ 3 Undecided
☐ 4 Disagree
☐ 5 Strongly Disagree

47. A man has a right to have sex with his wife even if it is sometimes against her will.
☐ 1 Strongly Agree
☐ 2 Agree
☐ 3 Undecided
☐ 4 Disagree
☐ 5 Strongly Disagree

48. Any healthy female can successfully resist a rapist if she really wants to.
☐ 1 Strongly Agree
☐ 2 Agree
☐ 3 Undecided
☐ 4 Disagree
☐ 5 Strongly Disagree

49. When females go around braless or wearing short skirts and tight tops, they are asking for trouble.
☐ 1 Strongly Agree
☐ 2 Agree
☐ 3 Undecided
☐ 4 Disagree
☐ 5 Strongly Disagree

50. If a female engages in necking or petting and she lets things get out of hand, it is her own fault if her partner uses some force to get her to have sex with him.
☐ 1 Strongly Agree
☐ 2 Agree
☐ 3 Undecided
☐ 4 Disagree
☐ 5 Strongly Disagree

51. Many females who get raped have done something to provoke it.
☐ 1 Strongly Agree
☐ 2 Agree
☐ 3 Undecided
☐ 4 Disagree
☐ 5 Strongly Disagree

52. Most women, although they might not at first admit it, would enjoy being forced to have intercourse with a man.
☐ 1 Strongly Agree
☐ 2 Agree
☐ 3 Undecided
☐ 4 Disagree
☐ 5 Strongly Disagree

53. If a female gets drunk at a party and has intercourse with a man she's just met there, she can be considered available to other males at the party who want to have sex with her too, whether she wants to or not.
☐ 1 Strongly Agree
☐ 2 Agree
☐ 3 Undecided
☐ 4 Disagree
☐ 5 Strongly Disagree

Thank you for your cooperation.

Hoffnung's Data Collection Questionnaire

Questionnaire used by Michele Hoffnung for the 1997 data collection for her panel on women's lives after college graduation described in the focal research in Chapter 2.

PLEASE PROVIDE **TWO** *ADDRESSES AND INDICATE WHICH ONE IS BEST TO USE:*

MAILING ADDRESS	BACK-UP (PARENTS') ADDRESS
______________________	______________________
______________________	______________________
Phone: ______________	Phone: ______________

E-mail Address (if applicable): ______________________

Your primary occupation this year? ______________________

If you are working for pay:

What is your job? ______________________

Is it related to your career plans? ______________________

If you are in school:

What program?______________________

Where? ______________ Degree expected? ______________

Highest degree you have eamed? ______________ Date ________

Are you? (check as many as apply)

Single _______ Married _______ If so, date of marriage _______

Separated _______ Divorced _______

Engaged _______

In a committed relationship _______ Living with partner _______

Dating _______ Not dating _______

Is your partner / person(s) you are dating? Male _______ Female _______

Are you pregnant? _______

Have you had a child? _______ If so, date(s)? ___________

Are you a step-parent? _______

PLEASE FEEL FREE TO USE ADDITIONAL PAGES.

What area(s) of your life gives you the most pleasure or satisfaction. Why?

Compared to when you graduated from college, would you say your commitment to having a career is: ______ greater, ______ the same,______ less. Please explain.

Compared to when you graduated from college, would you say your relationship with your parents has: _______ improved, _______ stayed the same, _______ become worse. Please explain.

Compared to when you graduated from college, would you say your commitment to being a mother is: _______ greater, _______ the same, _______ less. Please explain.

What is your dream for yourself for the next five years?

Other comments?

Mail to: Michele Hoffnung, Box 119, Quinnipiac College, Hamden, CT 06518 (6/97)

D

Code of Ethics

The American Sociological Association's (ASA's) Code of Ethics sets forth the principles and ethical standards that underlie sociologists' professional responsibilities and conduct. These principles and standards should be used as guidelines when examining everyday professional activities. They constitute normative statements for sociologists and provide guidance on issues that sociologists may encounter in their professional work.

Sections 11–13 of the Code of Ethics of the American Sociological Association, approved by ASA Membership in spring of 1997 are reproduced with the permission of the ASA. The complete Code of Ethics can be found on the ASA's homepage at www.asanet.org.

11. Confidentiality

Sociologists have an obligation to ensure that confidential information is protected. They do so to ensure the integrity of research and the open communication with research participants and to protect sensitive information obtained in research, teaching, practice, and service. When gathering confidential information, sociologists should take into account the long-term uses of the information, including its potential placement in public archives or the examination of the information by other researchers or practitioners.

11.01 Maintaining Confidentiality

(a) Sociologists take reasonable precautions to protect the confidentiality rights of research participants, students, employees, clients, or others.

(b) Confidential information provided by research participants, students, employees, clients, or others is treated as such by sociologists even if there is no legal protection or privilege to do so. Sociologists have an obligation to protect confidential information, and not allow information gained in confidence from being used in ways that would unfairly compromise research participants, students, employees, clients, or others.

(c) Information provided under an understanding of confidentiality is treated as such even after the death of those providing that information.

(d) Sociologists maintain the integrity of confidential deliberations, activities, or roles, including, where applicable, that of professional committees, review panels, or advisory groups (e.g., the ASA Committee on Professional Ethics).

(e) Sociologists, to the extent possible, protect the confidentiality of student records, performance data, and personal information, whether verbal or written, given in the context of academic consultation, supervision, or advising.

(f) The obligation to maintain confidentiality extends to members of research or training teams and collaborating organizations who have access to the information. To ensure that access to confidential information is restricted, it is the responsibility of researchers, administrators, and principal investigators to instruct staff to take the steps necessary to protect confidentiality.

(g) When using private information about individuals collected by other persons or institutions, sociologists protect the confidentiality of individually identifiable information. Information is private when an individual can reasonably expect that the information will not be made public with personal identifiers (e.g., medical or employment records).

11.02 Limits of Confidentiality

(a) Sociologists inform themselves fully about all laws and rules which may limit or alter guarantees of confidentiality. They determine their ability to guarantee absolute confidentiality and, as appropriate, inform research participants, students, employees, clients, or others of any limitations to this guarantee at the outset consistent with ethical standards set forth in 11.02(b).

(b) Sociologists may confront unanticipated circumstances where they become aware of information that is clearly health- or life-threatening to research participants, students, employees, clients, or others. In these cases, sociologists balance the importance of guarantees of confidentiality with other principles in this Code of Ethics, standards of conduct, and applicable law.

(c) Confidentiality is not required with respect to observations in public places, activities conducted in public, or other settings where no rules of privacy are provided by law or custom. Similarly, confidentiality is not required in the case of information available from public records.

11.03 Discussing Confidentiality and Its Limits

(a) When sociologists establish a scientific or professional relationship with persons, they discuss (1) the relevant limitations on confidentiality, and (2) the foreseeable uses of the information generated through their professional work.

(b) Unless it is not feasible or is counter-productive, the discussion of confidentiality occurs at the outset of the relationship and thereafter as new circumstances may warrant.

11.04 Anticipation of Possible Uses of Information

(a) When research requires maintaining personal identifiers in data bases or systems of records, sociologists delete such identifiers before the information is made publicly available.

(b) When confidential information concerning research participants, clients, or other recipients of service is entered into databases or systems of records available to persons without the prior consent of the relevant parties, sociologists protect anonymity by not including personal identifiers or by employing other techniques that mask or control disclosure of individual identities.

(c) When deletion of personal identifiers is not feasible, sociologists take reasonable steps to determine that appropriate consent of personally-identifiable individuals has been obtained before they transfer such data to others or review such data collected by others.

11.05 Electronic Transmission of Confidential Information

Sociologists use extreme care in delivering or transferring any confidential data, information, or communication over public computer networks. Sociologists are attentive to the problems of maintaining confidentiality and control over sensitive material and data when use of technological innovations, such as public computer networks, may open their professional and scientific communication to unauthorized persons.

11.06 Anonymity of Sources

(a) Sociologists do not disclose in their writings, lectures, or other public media confidential, personally identifiable information concerning their research participants, students, individual or organizational clients, or other recipients of their service which is obtained during the course of their work, unless consent from individuals or their legal representatives has been obtained.

(b) When confidential information is used in scientific and professional presentations, sociologists disguise the identity of research participants, students, individual or organizational clients, or other recipients of their service.

11.07 Minimizing Intrusions on Privacy

(a) To minimize intrusions on privacy, sociologists include in written and oral reports, consultations, and public communications only information germane to the purpose for which the communication is made.

(b) Sociologists discuss confidential information or evaluative data concerning research participants, students, supervisees, employees, and individual or organizational clients only for appropriate scientific or professional purposes and only with persons clearly concerned with such matters.

11.08 Preservation of Confidential Information

(a) Sociologists take reasonable steps to ensure that records, data, or information are preserved in a confidential manner consistent with the requirements of this Code of Ethics, recognizing that ownership of records, data, or information may also be governed by law or institutional principles.

(b) Sociologists plan so that confidentiality of records, data, or information is protected in the event of the sociologist's death, incapacity, or withdrawal from the position or practice.

(c) When sociologists transfer confidential records, data, or information to other persons or organizations, they obtain assurances that the recipients of the records, data, or information will employ measures to protect confidentiality at least equal to those originally pledged.

12. Informed Consent

Informed consent is a basic ethical tenet of scientific research on human populations. Sociologists do not involve a human being as a subject in research without the informed consent of the subject or the subject's legally authorized representative, except as otherwise specified in this Code. Sociologists recognize the possibility of undue influence or subtle pressures on subjects that may derive from researchers' expertise or authority, and they take this into account in designing informed consent procedures.

12.01 Scope of Informed Consent

(a) Sociologists conducting research obtain consent from research participants or their legally authorized representatives (1) when data are collected from research participants through any form of communication, interaction, or intervention; or (2) when behavior of research participants occurs in a private context where an individual can reasonably expect that no observation or reporting is taking place.

(b) Despite the paramount importance of consent, sociologists may seek waivers of this standard when (1) the research involves no more than minimal risk for research participants, and (2) the research could not practicably be carried out were informed consent to be required. Sociologists recognize that waivers of consent require approval from institutional review boards or, in the absence of such boards, from another authoritative body with expertise on the ethics of research. Under such circumstances, the confidentiality of any personally identifiable information must be maintained unless otherwise set forth in 11.02(b).

(c) Sociologists may conduct research in public places or use publicly available information about individuals (e.g., naturalistic observations in public places, analysis of public records, or archival research) without obtaining consent. If, under such circumstances, sociologists have any doubt whatsoever about the need for informed consent, they consult with institutional review boards or, in the absence of such boards, with another authoritative body with expertise on the ethics of research before proceeding with such research.

(d) In undertaking research with vulnerable populations (e.g., youth, recent immigrant populations, the mentally ill), sociologists take special care to ensure that the voluntary nature of the research is understood and that consent is not coerced. In all other respects, sociologists adhere to the principles set forth in 12.01(a)-(c).

(e) Sociologists are familiar with and conform to applicable state and federal regulations and, where applicable, institutional review board requirements for obtaining informed consent for research.

12.02 Informed Consent Process

(a) When informed consent is required, sociologists enter into an agreement with research participants or their legal representatives that clarifies the nature of the research and the responsibilities of the investigator prior to conducting the research.

(b) When informed consent is required, sociologists use language that is understandable to and respectful of research participants or their legal representatives.

(c) When informed consent is required, sociologists provide research participants or their legal representatives with the opportunity to ask questions about any aspect of the research, at any time during or after their participation in the research.

(d) When informed consent is required, sociologists inform research participants or their legal representatives of the nature of the research; they indicate to participants that their participation or continued participation is voluntary; they inform participants of significant factors that may be expected to influence their willingness to participate (e.g., possible risks and benefits of their participation); and they explain other aspects of the research and respond to questions from prospective participants. Also, if relevant, sociologists explain that refusal to participate or withdrawal from participation in the research involves no penalty, and they explain any foreseeable consequences of declining or withdrawing. Sociologists explicitly discuss confidentiality and, if applicable, the extent to which confidentiality may be limited as set forth in 11.02(b).

(e) When informed consent is required, sociologists keep records regarding said consent. They recognize that consent is a process that involves oral and/or written consent.

(f) Sociologists honor all commitments they have made to research participants as part of the informed consent process except where unanticipated circumstances demand otherwise as set forth in 11.02(b).

12.03 Informed Consent of Students and Subordinates

When undertaking research at their own institutions or organizations with research participants who are students or subordinates, sociologists take special care to protect the prospective subjects from adverse consequences of declining or withdrawing from participation.

12.04 Informed Consent with Children

(a) In undertaking research with children, sociologists obtain the consent of children to participate, to the extent that they are capable of providing such consent, except under circumstances where consent may not be required as set forth in 12.01(b).

(b) In undertaking research with children, sociologists obtain the consent of a parent or a legally authorized guardian. Sociologists may seek waivers of parental or guardian consent when (1) the research involves no more than minimal risk for the research participants, and (2) the research could not practicably be carried out were consent to be required, or (3) the consent of a parent or guardian is not a reasonable requirement to protect the child (e.g., neglected or abused children).

(c) Sociologists recognize that waivers of consent from a child and a parent or guardian require approval from institutional review boards or, in the absence of such boards, from another authoritative body with expertise on the ethics of research. Under such circumstances, the confidentiality of any personally identifiable information must be maintained unless otherwise set forth in 11.02(b).

12.05 Use of Deception in Research

(a) Sociologists do not use deceptive techniques (1) unless they have determined that their use will not be harmful to research participants; is justified by the study's prospective scientific, educational, or applied value; and that equally effective alternative procedures that do not use deception are not feasible, and (2) unless they have obtained the approval of institutional review boards or, in the absence of such boards, with another authoritative body with expertise on the ethics of research.

(b) Sociologists never deceive research participants about significant aspects of the research that would affect their willingness to participate, such as physical risks, discomfort, or unpleasant emotional experiences.

(c) When deception is an integral feature of the design and conduct of research, sociologists attempt to correct any misconception that research participants may have no later than at the conclusion of the research.

(d) On rare occasions, sociologists may need to conceal their identity in order to undertake research that could not practicably be carried out were they to be known as researchers. Under such circumstances, sociologists undertake the research if it involves no more than minimal risk for the research participants and if they have obtained approval to proceed in this manner from an institutional review board or, in the absence of such boards, from another authoritative body with expertise on the ethics of research. Under such circumstances, confidentiality must be maintained unless otherwise set forth in 11.02(b).

12.06 Use of Recording Technology

Sociologists obtain informed consent from research participants, students, employees, clients, or others prior to videotaping, filming, or recording them in any form, unless these activities involve simply naturalistic observations in public places and it is not anticipated that the recording will be used in a manner that could cause personal identification or harm.

13. Research Planning, Implementation, and Dissemination

Sociologists have an obligation to promote the integrity of research and to ensure that they comply with the ethical tenets of science in the planning, implementation, and dissemination of research. They do so in order to advance knowledge, to minimize the possibility that results will be misleading, and to protect the rights of research participants.

13.01 Planning and Implementation

(a) In planning and implementing research, sociologists minimize the possibility that results will be misleading.

(b) Sociologists take steps to implement protections for the rights and welfare of research participants and other persons affected by the research.

(c) In their research, sociologists do not encourage activities or themselves behave in ways that are health- or life-threatening to research participants or others.

(d) In planning and implementing research, sociologists consult those with expertise concerning any special population under investigation or likely to be affected.

(e) In planning and implementing research, sociologists consider its ethical acceptability as set forth in the Code of Ethics. If the best ethical practice is unclear, sociologists consult with institutional review boards or, in the absence of such review processes, with another authoritative body with expertise on the ethics of research.

(f) Sociologists are responsible for the ethical conduct of research conducted by them or by others under their supervision or authority.

13.02 Unanticipated Research Opportunities

If during the course of teaching, practice, service, or non-professional activities, sociologists determine that they wish to undertake research that was not previously anticipated, they make known their intentions and take steps to ensure that the research can be undertaken consonant with ethical principles, especially those relating to confidentiality and informed consent. Under such circumstances, sociologists seek the approval of institutional review boards or, in the absence of such review processes, another authoritative body with expertise on the ethics of research.

13.03 Offering Inducements for Research Participants

Sociologists do not offer excessive or inappropriate financial or other inducements to obtain the participation of research participants, particularly when it might coerce participation. Sociologists may provide incentives to the extent that resources are available and appropriate.

13.04 Reporting on Research

(a) Sociologists disseminate their research findings except where unanticipated circumstances (e.g., the health of the researcher) or proprietary agreements with employers, contractors, or clients preclude such dissemination.

(b) Sociologists do not fabricate data or falsify results in their publications or presentations.

(c) In presenting their work, sociologists report their findings fully and do not omit relevant data. They report results whether they support or contradict the expected outcomes.

(d) Sociologists take particular care to state all relevant qualifications on the findings and interpretation of their research. Sociologists also disclose underlying assumptions, theories, methods, measures, and research designs that might bear upon findings and interpretations of their work.

(e) Consistent with the spirit of full disclosure of methods and analyses, once findings are publicly disseminated, sociologists permit their open assessment and verification by other responsible researchers with appropriate safeguards, where applicable, to protect the anonymity of research participants.

(f) If sociologists discover significant errors in their publication or presentation of data, they take reasonable steps to correct such errors in a correction, a retraction, published errata, or other public fora as appropriate.

(g) Sociologists report sources of financial support in their written papers and note any special relations to any sponsor. In special circumstances, sociologists may withhold the names of specific sponsors if they provide an adequate and full description of the nature and interest of the sponsor.

(h) Sociologists take special care to report accurately the results of others' scholarship by using correct information and citations when presenting the work of others in publications, teaching, practice, and service settings.

Data Sharing

(a) Sociologists share data and pertinent documentation as a regular practice. Sociologists make their data available after completion of the project or its major publications, except where proprietary agreements with employers, contractors, or clients preclude such accessibility or when it is impossible to share data and protect the confidentiality of the data or the anonymity of research participants (e.g., raw field notes or detailed information from ethnographic interviews).

(b) Sociologists anticipate data sharing as an integral part of a research plan whenever data sharing is feasible.

(c) Sociologists share data in a form that is consonant with research participants' interests and protect the confidentiality of the information they have been given. They maintain the confidentiality of data, whether legally required or not; remove personal identifiers before data are shared; and if necessary use other disclosure avoidance techniques.

(d) Sociologists who do not otherwise place data in public archives keep data available and retain documentation relating to the research for a reasonable period of time after publication or dissemination of results.

(e) Sociologists may ask persons who request their data for further analysis to bear the associated incremental costs, if necessary.

(f) Sociologists who use data from others for further analyses explicitly acknowledge the contribution of the initial researchers.

Glossary

access The ability to obtain the information needed to answer a research question.

accretion measures Indicators of a population's activities created by its deposits of materials.

alternative-form method A method of checking the reliability of a measure that involves comparing its results with those received using an alternate form of the measure.

anonymity When no one, including the researcher, knows the identities of research subjects.

anticipatory data reduction Decision-making done in advance of collecting data about what kinds of data are desirable.

available data Data that are available to the researcher.

basic research Research designed to add to our knowledge and understanding of the social world for the sake of that knowledge and understanding.

biased In sampling, the quality of being systematically unrepresentative.

bivariate analyses Data analyses that focus on the association between two variables.

case study A research strategy that focuses on one case (an individual, a group, an organization, and so on) within its social context at one point in time, even if that one time spans months or years.

causal hypothesis A statement that hypothesizes that the independent variable causes or affects the dependent variable.

causal relationship A nonspurious relationship between an independent and dependent variable with the independent variable occurring before the dependent variable.

codebook A guide for translating computer-ready data back into human-understandable language.

coding The process by which raw data are given a standardized form. In quantitative analyses, this frequently means makings data computer usable. In qualitative analyses, this means associating words or labels with passages and then collecting similarly labeled passages into files.

control variable A variable that is held constant to examine the relationship between two other variables.

controlled experiment An experimental design with two or more randomly selected groups (an experimental and control group) in which the researcher controls or "introduces" the independent variable and measures the dependent variable at least two times (pre- and post-test measurement).

closed-ended questions Questions that include a list of predetermined answers.

cluster sampling A probability sampling procedure that involves the random selection of clusters of elements from a population and the subsequent selection of every element in each cluster for inclusion in the sample.

cohort People born within a given time frame or experiencing a life event at approximately the same time.

cohort study A specific kind of trend study that studies a cohort over time.

complete observer role Being an observer of a situation without becoming part of it.

complete participant role Being, or pretending to be, a genuine participant in a situation one observes.

composite measure A measure made up of more than one indicator.

concepts Words or signs that refer to phenomena that share common characteristics.

conceptualization The process of clarifying just what we mean by a concept.

conceptual definition A definition of a concept in terms of other concepts. Also a theoretical definition.

confidence interval A range of values within which the population parameter is expected to lie.

confidence level The estimated probability that a population parameter will fall with a given confidence interval.

confidentiality When no third party knows the identities of the research subjects.

construct validity How well a measure of a concept is associated with a measure of another concept that some theory says it should be associated with.

content analysis A method of data collection in which some form of communication is studied systematically.

content validity How well a measure covers the range of meanings associated with a concept.

contingency questions Questions that depend on the answers to previous questions.

controlled observation Observation that involves clear decisions about what is to be observed.

convenience sampling A nonprobability sampling procedure that involves selecting elements that are readily accessible to the researcher.

correlation A statistical technique for assessing the strength and direction of the association between two variables.

cost-benefit analysis Research that compares a program's costs to its benefits.

cost-effectiveness analysis Comparisons of program costs in delivering desired benefits based on the assumption that the outcome is desirable.

criteria for causal relationships Criteria needed to support a causal relationship between two variables. These include the occurrence of the independent variable before the dependent variable, a relationship between the independent and dependent variables, and support for the conclusion that the apparent relationship is not caused by the effect of a third variable.

cross-sectional study A study design in which data are collected for all the variables of interest using one sample at one time.

crosstabulation The process of making a bivariate table for the examination of a relationship between two variables.

criterion validity How well a measure is associated with behaviors you'd expect it to be associated with.

data displays Visual images that summarize a good deal of information.

data reduction The various ways in which a researcher orders collected and transcribed data.

deductive reasoning Reasoning that moves from more general to less general statements.

dependent variable A variable that a researcher sees as being affected or influenced by another variable (contrast with independent variable).

demand characteristics Characteristics that the observed assume simply as a result of being observed.

descriptive statistics Statistics used to describe and interpret sample data.

double-blind experiment An experiment in which neither the subjects nor the research staff who interact with them knows the memberships of the experimental or control groups.

ecological fallacy The fallacy one commits when making inferences about individuals from information about groups or aggregates.

elaboration The process of examining the relationship between two variables by introducing the control for another variables or variables.

element A kind of thing a researcher wants to sample. Also called a sampling unit.

empirical generalization A statement that summarizes a set of individual observations.

erosion measures Indicators of a population's activities created by its selective wear on its physical environment.

ethical principles in research The set of values, standards, and principles used to determine appropriate and acceptable conduct in all stages of the research process.

evaluation research Research specifically designed to assess the impact of a specific program, policy, or legal change.

exhaustive and mutually exclusive answer categories Closed-ended answer categories that allow respondents to select one and only one answer.

exhaustiveness Refers to the capacity of a variable's categories to permit the classification of every unit of analysis.

experimental design A study design that calls for the control or manipulation of the independent variable in some way.

experimenter expectations When expected behaviors or outcomes are communicated to subjects by the researcher.

explanation A kind of elaboration in which the original relationship is explained away as spurious by the control for an antecedent variable.

explanatory study A study that seeks to explain the cause of a phenomenon, and typically asks a "what causes what?" or "why is it this way" research question.

face validity The degree to which a measure seems to be measuring what it's supposed to be measuring.

feasibility Whether it is practical to complete a study in terms of access, time, and money.

field experiment An experiment done in the "real world" of classrooms, offices, factories, homes, playgrounds, and the like.

formative analysis Evaluation research focused on the design or early implementation stages of a program or policy.

frequency distribution A way of showing the number of times each category of a variable occurs in a sample.

generalizability The ability to apply the results of a study to groups or situations beyond those actually studied.

going native A danger of complete participation, the condition of over-identifying with other participants.

grounded theory Theory derived from data in the course of a particular study.

group administered questionnaire Questionnaire administered to respondents in a group setting.

history The effects of general historical events on study participants.

hypothesis A testable statement of how two or more variables are expected to be related to one another.

independent variable A variable that a researcher sees as affecting or influencing another variable (contrast with dependent variable).

index A composite measure that is constructed by the addition of scores from several indicators.

indicators Observations that we think reflect the presence or absence of the phenomenon to which a concept refers.

individually administered questionnaire Questionnaire that is hand delivered to a respondent and picked up after completion.

inductive reasoning Reasoning that moves from less general to more general statements.

inferential statistics Statistics used to make inferences about the population from which the sample was drawn.

informants Participants in a study situation who are interviewed for an in-depth understanding of the situation.

informed consent The principle that potential participants are given adequate and accurate information about a study before they are asked to agree to participate.

in-person interview An interview conducted face-to-face.

internal validity Agreement between a study's conclusions about causal connections and what is actually true.

interpretation A kind of elaboration that provides an idea of the reasons why an original relationship exists without challenging the belief that the original relationship is causal.

intersubjectivity Agreements about reality that result from comparing the observations of more than one observer.

interval measure A level of measurement that describes a variable whose categories have names, whose categories can be rank-ordered in some sensible way, and whose adjacent categories are a standard distance from one another.

interview A data collection method in which respondents answer questions asked by an interviewer.

interview guide The list of topics to cover and the order in which to cover them that can be used to guide less structured interviews.

interview schedule The set of questions read to a respondent in a structured or semi-structured interview.

interviewer effect The change in a respondent's behavior or answers that is the result of being interviewed by a specific interviewer.

laboratory experiment An experiment done in a setting that allows the researcher control over the conditions, such as in a university or medical laboratory.

levels of measurement Types of categorization used in variable creation (that is, nominal, ordinal, interval, and ratio levels).

keywords The terms for the concepts of interest in searching for sources in a literature review.

literature review Reading, summarizing, and synthesizing existing work on a topic.

longitudinal research A research design in which data are collected at least two different times.

mailed questionnaire A questionnaire mailed to the respondent's home or place of business.

matching Assigning members of the sample to groups by matching members of the sample on one or more characteristics and separating the pairs into two groups with one group randomly selected to become the experimental group.

maturation The biological and psychological processes that cause people to change over time.

mean The measure of central tendency designed for interval level variables. The sum of all values divided by the number of values.

measurement Classifying subjects by categories to represent variable concepts.

memos More or less extended notes that the researcher writes to help herself understand the meaning of codes.

median The measure of central tendency designed for ordinal level variables. The middle value when all values are arranged in order.

mode The measure of central tendency designed for nominal level variables. The value that occurs most frequently.

multidimensionality The degree to which a concept has more than one discernible aspect.

multiple regression analysis A method of data analysis that yields, among other things, estimates of the correlation between each of a series of independent variables and a dependent variable when all other (included) independent variables are controlled.

multistage sampling A probability sampling procedure that involves several stages, such as the random selection of clusters from a population and then the random selection of elements from each of the clusters.

multivariate analyses Analyses that permit the examination of the relationship between variables while investigating the role of other variables.

mutual exclusiveness The capacity of a variable's categories to permit the classification of each unit of analysis by one and only one category.

nominal measure A level of measurement that describes a variable whose categories have names.

nonparticipant observation Observation made by an observer who remains as aloof as possible from those observed.

nonprobability sample A sample that has been drawn in a way that doesn't give every member of the population a known chance of being selected.

nonresponse bias The bias that results from differences between those who participate in a survey and those who don't.

objectivity The ability to see the world as it really is.

observation techniques Methods of collecting data by observing people, most typically in their natural settings.

observer-as-participant role Being primarily a self-professed observer, while occasionally participating in the situation.

open-ended question Question that allows respondents to answer in their own words.

operationalization The process of defining specific ways to infer the absence, presence, and/or degree of presence of a phenomenon.

operational definition Declarations of the specific ways in which the absence, presence, and/or the degree of presence of a phenomenon will be determined in a specific instance.

ordinal measure A level of measurement that describes a variable whose categories have names and whose categories can be rank-ordered in some sensible way.

outcome evaluation Research that is designed to "sum up" the effects of a program, policy, or law in accomplishing the goal or intent of the program, policy, or law.

panel attrition The loss of subjects from a study due to disinterest, death, illness, or inability to locate them.

panel study A study design in which data are collected about one sample at least two times where the independent variable is not controlled by the researcher.

parameter A summary of a variable characteristic in a population.

partial relationship The relationship between an independent and a dependent variable for that part of a sample defined by one category of a control variable.

participant observation Observation performed by observers who take part in the activities they observe.

participant-as-observer role Being primarily a participant, while admitting an observer status.

phone interview An interview conducted over the telephone.

physical traces Physical evidence left by humans in the course of their everyday lives.

placebo A possible, simulated treatment of the control group, which is designed to appear authentic.

population The group of elements from which a researcher samples and to which she or he might like to generalize.

post-test The measurement of the dependent variable that occurs after the introduction of the stimulus or the independent variable.

pre-test The measurement of the dependent variable that occurs before the introduction of the stimulus or independent variable.

primary data Data that the same researcher collects and uses.

probability sample A sample drawn in a way to give every member of the population a known (non-zero) chance of inclusion.

process evaluation Research that monitors a program or policy to determine if it is implemented as designed.

protecting study participants from harm The principle that researchers should see to it that subjects are not harmed, physically, psychologically, emotionally, legally, socially, or financially as a result of their participation.

purposive sampling A nonprobability sampling procedure that involves the selection of elements based on the judgment of the researcher about which elements will facilitate his or her investigation.

qualitative data analysis Analysis that tends to result in the interpretation of action or representations of meanings in the researcher's own words.

qualitative interview A data collection method in which an interviewer adapts and modifies the interview for each interviewee.

quantitative data analysis Analysis that tends to be based on the statistical summary of data.

quasi-experiment An experimental design that is missing one or more aspects of the classic controlled experiment.

questionnaire A data collection method in which respondents read and answer questions in a written format.

quota sampling A nonprobability sampling procedure that involves describing the target population in terms of what are thought to be relevant criteria and then selecting sample elements to represent the "relevant" subgroups in proportion to their presence in the target population.

randomization A technique for selecting experimental and control groups from the sample to maximize the chance that the groups be similar at the beginning of the experiment.

range A measure of dispersion or spread designed for interval level variables. The difference between the highest and lowest values.

rapport A sense of interpersonal harmony, connection, or compatibility between an interviewer and a respondent.

ratio measure A level of measurement that describes a variable whose categories have names, whose categories may be rank-ordered, whose adjacent categories are a standard distances from one another, and one of whose categories is an absolute zero point—a point at which there is a complete absence of the phenomenon in question.

regression analysis A method of data analysis that yields a formula for the straight line that "best fits" a series of data points, but that also yields estimates of the strength and direction of the relationship between interval level variables.

reliability The degree to which a measure yields consistent results.

replication A kind of elaboration in which the original relationship is replicated by all of the partial relationships.

research costs All monetary expenditures needed for planning, executing, and reporting research.

research question A question about one or more concepts that is answerable with research.

research topic A concept, subject, or issue that can be studied through research.

respondent The participant in a survey who completes a questionnaire or interview.

response errors Responses that provide inaccurate information.

response rate The percentage of the sample contacted that actually participates in a study.

sample A number of individual cases drawn from a larger population.

sampling distribution The distribution of a sample statistic (like the average) computed from many samples.

sampling error The error that occurs when a sample statistic is not the same as the population parameter it's meant to estimate.

sampling frames A group of sampling units or elements from which a sample is actually selected (see also study population).

sampling variability The variability in sample statistics that occurs when different samples are drawn from the same population.

sampling unit The kind of thing a researcher wants to sample. Also called an element.

scales Complex measures whose construction is more rigorous than those of other indexes, often in the effort to focus on unidimensional concepts, sometimes in the effort to assure reliability and validity.

screening question A question that asks for information before asking the question of interest.

secondary data Data that have been collected by someone else.

selection bias A bias in the way the experimental and control or comparison group are selected which is responsible for preexisting differences between the groups.

self-report method Another name for questionnaires and interviews as respondents are most often asked to report their own characteristics, behaviors, and attitudes.

semi-structured interview Interview with an interview guide containing primarily open-ended questions that can be modified for each interview.

simple random sampling A probability sampling procedure that ensures that every member of a study population has an equal chance of selection.

snowball sampling A nonprobability sampling procedure that involves using members of the group of interest to identify other members of the group.

specification A kind of elaboration that permits the researcher to specify conditions under which the original relationship is particularly strong or weak.

split-half method A method of checking the reliability of several measures by dividing the measures into two sets of measures and determining whether the two sets are associated with each other.

spurious relationship A non-causal relationship between two variables.

stakeholders People or groups that participate in or are affected by a program or its evaluation, such as funding agencies, policymakers, sponsors, program staff, and program participants.

standard deviation A measure of dispersion designed for interval level variables and that takes into account every value's distance from the sample mean.

standardized regression coefficient (or beta) A product of regression analysis that provides the estimate of the correlation between an independent and dependent variable.

statistic A summary of a variable in a sample.

stimulus The experimental condition of the independent variable that is controlled or "introduced" by the researcher in an experiment.

stratified sampling A probability sampling procedure that involves dividing the population in groups or strata defined by the presence of certain characteristics and then random sampling from each of the strata.

structured interview A data collection method in which an interviewer reads a standardized interview schedule to the respondent and records the answers.

study design A research strategy specifying the number of cases to be studied, the number of times data will be collected, the number of samples that will be used, and whether or not the researcher will try to control or manipulate the independent variable in some way.

study population The group of sampling units or elements from which a sample is actually selected (see also sampling frame).

survey A study in which the same data are collected from all members of the sample and analyzed using statistics.

systematic observations Observations that employ an explicit plan for selecting, recording, and coding data.

systematic sampling A probability sampling procedure that involves selecting every kth element from a list of population elements, after the first element has been randomly selected.

testing effect The sensitizing effect on subjects of the pre-test.

test-retest method A method of checking the reliability of a test that involves comparing its results at one time with results, using the same subjects, at a later time.

theory A story about how and why something is as it is.

theoretical saturation The point where new interviewees or settings look a lot like interviewees or settings one had observed before.

thick description Reports about behavior that provide a sense of things like the intentions, motives, and meanings behind the behavior.

time expenditures The time it takes to complete all activities of a research project from the planning stage to the final report.

trend study A study design in which data are collected at least two times with a new sample selected each time.

units of analysis The kind of thing a researcher wants to analyze. The term is used at the analysis stage, whereas the terms element or sampling unit can be used at the sampling stage.

units of observation The units from which information is collected.

univariate analyses Analyses that tell us something about one variable.

unobtrusive measures Indicators of interactions, events, or behaviors whose creation does not affect the actual data collected.

unstructured interview A data collection method in which the interviewer starts with only a general sense of the topics to be discussed and creates questions as the interaction proceeds.

validity The degree to which a measure measures what we think it's measuring.

variable A characteristic that may vary from one subject to another or for one subject over time.

vraisemblance The verisimilitude, or sense of truth, that a reporter can convey through his or her writing.

References

Abelson, R. P., E. F. Loftus, & A. G. Greenwald. 1992. Attempts to improve the accuracy of self-reports of voting. In *Questions about questions,* 138–153, edited by J. M. Tanur. New York: Russell Sage Foundation.

Acker, J., K. Barry, & J. Esseveld. 1991. Objectivity and the truth problems in doing feminist research. In *Beyond methodology: feminist scholarship as lived research,* 133–153, edited by M. M. Fonow & J. A. Cook, Bloomington, Indiana: Indiana University Press.

Adler, E. S. 1981. The underside of married life: Power, influence and violence. In *Women and crime in America,* 300–319, edited by L. H. Bowker. New York: Macmillan.

Adler, E. S., M. Bates, & J. M. Merdinger. 1985. Educational policies and programs for teenage parents and pregnant teenagers. *Family Relations* 34: 183–198.

Adler, E. S., & R. Clark. 1991. Adolescence: A literary passage. *Adolescence,* 26: 757–768.

Adler, E. S., & J. S. Lemons. 1990. *The elect: Rhode Island's women legislators, 1922–1990.* Providence: League of Rhode Island Historical Societies.

Adler, P., & P. Adler. 1995. Personal communication.

Adler, P. A. 1985. *Wheeling and dealing.* New York: Columbia University Press.

Adler, P. A., & P. Adler. 1987. *Membership roles in field research.* Newbury Park, CA: Sage.

Adler, P. A., & P. Adler. 1994. Observational techniques. In *Handbook of qualitative research,* 377–392, edited by N. K. Denzin & Y. S. Lincoln. Thousand Oaks, CA: Sage.

Adler, P. A., & P. Adler, 1996. Parent-as-researcher: The politics of researching in the personal life. *Qualitative Sociology* 19(1): 35–58.

Albig, W. 1938. The content of radio programs—1925–1935. *Social Forces* 16: 338–349.

Allen, K. R., & D. H. Demo. 1995. The Families of lesbians and gay men: A new frontier in family research. *Journal of Marriage and the Family* 57: 111–128.

American Sociological Association. 1996. *American Sociological Association style guide.* Washington, DC: American Sociological Association.

American Sociological Association. 1997. *Data resources for sociologists.* Washington, DC: American Sociological Association.

Angell, C. S. 1997. *Immigrant adaptation: Seven case studies.* (unpublished).

Aniskiewicz, R., & E. Wysong. 1990. Evaluating DARE: drug education and the multiple meanings of success. *Policy Studies Review* 9: 727–747.

Applewhite, S. L. 1997. Homeless veterans: perspectives on social service use. *Social Work* 42: 19–30.

Aquilino, W. S. 1993. Effects of spouse presence during the interview on survey responses concerning marriage. *Public Opinion Quarterly* 57: 358–376.

Armstrong, J. S., & E. J. Lusk. 1987. Return postage in mail survey: A meta-analysis. *Public Opinion Quarterly* 51: 233–248.

Asheim, L. 1950. From book to film. In *Reader in public opinion and communication,* 299–306, edited by B. Berelson & M. Janowitz. New York: Free Press.

Asher, R. 1992. *Polling and the public.* Washington, D. C. Congressional Quarterly, Inc.

Atchley, R. C. 1994. *Social forces & aging.* Belmont, CA: Wadsworth.

Aquilino, W. S. 1994. Interview mode effects in surveys of drug and alcohol use. *Public Opinion Quarterly* 58: 210–240.

Axelrod, R. M. 1984. *The evolution of cooperation.* New York: Basic.

Babbie, E. 1997. *The practice of social research.* Belmont, CA: Wadsworth.

Baize, H. R. J., & J. E. Schroeder. 1995. Personality and mate selection in personal ads: evolutionary preferences in a public mate selection process. *Journal of Social Behavior and Personality,* 10: 517–536.

Bales, R. F. 1950. *Interaction process analysis.* Reading, MA: Addison-Wesley.

Ballmer-Cao, Thank-Huyen, J. Scheiddeger, V. Bornschier, & P. Heintz. 1979. *Compendium of data for world-system analyses.* Zurich: Soziologogisches Institut der Universität Zurich.

Barry, D. 1987. *Bad habits: a 100% fact-free book.* New York: Henry Holt.

Bart, P. B., & Patricia H. O'Brien. 1985. *Stopping rape: Successful survival strategies.* New York: Pergamon.

Batchelor, J. A., & C. M. Briggs, 1994. Subject, project or self? Thoughts on ethical dilemmas for social and medical researchers. *Social Science & Medicine* 39: 949–954.

Baumrind, D. 1964. Some thoughts on ethics of research: After reading "Milgram's behavioral study of obedience." *American Psychologist* 19: 421–423.

Becker, H. S., B. Geer, E. C. Hughes, & A. L. Strauss. 1961. *Boys in white: student culture in medical school.* Chicago: University of Chicago Press.

Bennett, L. A., & K. McAvity. 1994. Family research: A case for interviewing couples. In *Psychosocial Interior of the Family, edited by* G. Handel & G. G. Whitchurch. New York: Aldine de Gruyter.

Benney, M., & E. C. Hughes. 1956. Of sociology and the interview. *American Journal of Sociology* 62: 137–142.

Berg, B. L. 1989. *Qualitative research methods of the social sciences.* Boston: Allyn & Bacon.

Berger, P. L. 1963. *Invitation to sociology: A humanistic perspective.* Garden City, NY: Anchor.

Berk, R. A., G. K., Smyth, & L. W. Sherman. 1988. When random assignment fails: some lessons from the Minneapolis spouse abuse experiment. *Journal of Quantitative Criminology* 4: 209–223.

Billson, J. M. 1991. The progressive verification method toward a feminist methodology for studying women cross-culturally. *Women's Studies International Forum,* 14: 201–215.

Billson, J. M. 1995. *Keepers of the culture: Women in a changing world.* New York: Lexington.

Bishop, G. F. 1987. Experiments with the middle response alternative in survey questions. *Public Opinion Quarterly* 51: 220–232.

Blaisure, K. R., & K. R. Allen. 1995. Feminists and the ideology and practice of marital equality. *Journal of Marriage and the Family* 57: 5–20.

Blalock, H. M. 1972. *Social statistics.* New York: McGraw-Hill.

Blum, L. 1982. Feminism and the mass media: A case study of "The women's room" as novel and television Film. *Berkeley Journal of Sociology* 27: 1–24.

Booth, R. E., S. K. Koester, C. S. Reichardt, & J. T. Brewster. 1993. Quantitative and qualitative methods to assess behavioral change among injection drug users. *Drugs and Society,* 7: 161–183.

Bornschier, V., & C. Chase-Dunn. 1985. *Transnational corporations and underdevelopment.* New York: Praeger.

Bourgoin, N. 1995. Suicide in prison. Some elements of a strategic analysis. *Cahiers Internationaux de Sociologie* 98: 59–105.

Bower, R. T., & P. de Gasparis. 1978. *Ethics in social research: protecting the interests of human subjects.* New York: Praeger.

Bradburn, N. M., & S. Sudman. 1988. *Polls and surveys.* San Francisco: Jossey-Bass.

Bradburn, N. M., & S. Sudman. 1979. *Improving interview methods and questionnaire design.* San Francisco: Jossey-Bass.

Breault, K. D., & K. Barkey. 1982. A comparative analysis of Durkheim's theory of egoistic suicide. *Sociological Quarterly* 23: 321–331.

Brenner, M. 1985. Intensive interviewing. In *The Research Interview Uses and Approaches,* edited by M. Brenner, J. Brown, & D. Canter. London: Academic.

Bresnen, M. 1988. Insights on site: research into construction project organizations. In *Doing Research in Organizations,* edited by Alan Bryman. London: Routledge.

Brief, A. P., R. T. Buttram, J. D. Elliot, R. M. Reizenstein, & R. L. McCline. 1995. Releasing the beast: a study of compliance with orders to use race as a selection criterion. *Journal of Social Issues* 51: 177–193.

Broderick, C. 1988. To arrive where we started: The field of family studies in the 1930s. *Journal of Marriage and the Family* 50: 569–584.

Brouwer, M., C. Clark, G. Gerbner, & K. Krippendorff. 1969. The television world of violence. In *Mass media and violence: a report to the national commission on the causes and prevention of violence,* edited by R. K. Baker & S. J. Ball. Washington, D.C. Government Printing Office.

Brown, J., & B. G. Gilmartin. 1969. Sociology today: lacunae, emphases and surfeits. *American Sociologist,* 4, 283–291.

Brown, S. R., & L. E. Melamed, 1990. *Experimental design and analysis.* Newbury Park, CA: Sage.

Browne, J. 1976. The used car game. In *The Research Experience,* 60–84, edited by M. P. Golden. Itasca, Illinois: F. E. Peacock.

Bryan, J. A., & M. A. Test, 1982. Models and helping: naturalistic studies in aiding behavior. In *Experimenting in Society,* 139–146, edited by J. W. Reich. Glenview, IL: Scott, Foresman.

Brydon-Mill, M. 1993. Breaking down the barriers: Accessibility self-advocacy in the disabled community. In *Voices of change, participatory research in the United States and Canada,* edited by P. Park, M. Brydon-Miller, B. Hall, & T. Jackson. Westport, CT: Bergin & Garvey.

Budiansky, S. 1994. Blinded by the cold-war light. *U S News & World Report,* 116: 6–7.

Burgess, A. W., & L. L. Holmstrom. 1979. *Rape, crisis and recovery.* Bowie, MD: R. J. Brady.

Burgess, R. G. 1984. Autobiographical accounts and research experience. In *The research process in educational settings: ten case studies,* edited by R. G. Burgess. Lewes: Falmer.

Campbell, D. R. 1995. *The student's guide to doing research on the Internet.* Reading, MA: Addison-Wesley.

Campbell, D. T., & J. C. Stanley. 1963. *Experimental and quasi-experimental designs for research.* Chicago: Rand McNally.

Caporaso, J. A. 1973. Quasi-experimental approaches to social science: Perspectives and problems. In *Quasi-experimental approaches testing theory and evaluating policy,* 3–38, edited by J. Caporaso & L. L. Roos, Jr. Evanston, IL: Northwestern University Press.

Carmines, E. G., & R. A. Zeller. 1979. *Reliability and validity assessment.* Beverly Hills, CA: Sage.

Chafetz, J. S., & A. G. Dworkin. 1986. *Female revolt: Women's movements in world and historical perspective.* Totowa, NJ: Rowman & Allanheld.

Chambers, D. E., K. R. Wedel, & M. K. Rodwell. 1992. *Evaluating social programs.* Boston: Allyn and Bacon.

Charmaz, K. 1983. The grounded theory method: An explication and interpretation. In *Contemporary field research: a collection of readings,* 109–126, edited by R. M. Emerson. Boston: Little, Brown.

Chase, S. E. 1995a. *Ambiguous empowerment: the work narratives of women school superintendents.* Amherst: University of Massachusetts Press.

Chase, S. E. 1995b. Taking narrative seriously: Consequences for method and theory in interview studies. In *Interpreting experience: the narrative study of lives,* edited by R. Josselson & A. Lieblich. Thousand Oaks, CA: Sage.

Clandinin, D. J., & F. M. Connelly. 1994. Personal experience methods. In *Handbook of Qualitative Research,* 413–427, edited by N. K. Denzin & Y. S. Lincoln. Thousand Oaks, CA: Sage.

Clark, B. L. Forthcoming. *Kiddie lit.* Ithaca: Cornell University Press.

Clark, H. H., & M. F. Schober. 1992. Asking questions and influencing answers. In *Questions about questions,* 15–48, edited by J. M. Tanur.New York: Russel Sage Foundation.

Clark, R. 1982. Birth order and eminence: A study of elites in science, literature, sports, acting and business. *The International Review of Modern Sociology* 12: 273–289.

Clark, R. 1992. Economic dependence and gender differences in labor force sectoral change in non-core nations. *The Sociological Quarterly* 33: 83–98.

Clark, R. 1999 (forthcoming). Diversity in sociology: Problem or solution? *American Sociologist.* 30.

Clark, R., & J. Carvalho.1995. *Female revolt revisited.* (unpublished).

Clark, R., & T. Clifford, (in press). Towards a resources and stressors model: the psychological adjustment of adult children of divorce. *Journal of Divorce and Remarriage.*

Clark, R., & K. Heidi. 1996. Toward a multicultural feminist perspective on fiction for young adults. *Youth & Society* 27(3): 291–312.

Clark, R., R. Lennon, & L. Morris. 1993. Of Caldecotts and Kings: Gendered images in recent american children's books by black and non-black illustrators. *Gender & Society* 7: 227–245.

Clark, R., & T. Ramsbey. 1986. Social interaction and literary creativity. *The International Review of Modern Sociology,* 16: 105–117.

Clark, R., T. Ramsbey, & E. S. Adler. 1991. Culture, gender and labor force participation. *Gender & Society* 5: 47–66.

Clark, R., & G. Rice. 1982. Family constellations and eminence: the birth orders of Nobel Prize Winners. *The Journal of Psychology* 110: 281–287.

Clifford, T. & R. Clark. 1995. Family climate, family structure and self-esteem in college students: the physical versus psychological-wholeness divorce debate revisited. *Journal of Divorce and Remarriage* 23: 97–111.

Collins, R., J. Chafetz, R Blumberg, S. Coltrane, & J. Turner. 1993. Toward an integrated theory of gender stratification. *Sociological Perspectives* 36: 185–216.

Comstock, G. 1980. *Television in America.* Beverly Hills, CA: Sage.

Conklin, J. E. 1992. *Criminology.* New York: Macmillan.

Converse, J. M. 1987. *Survey research in the United States roots and emergence 1890–1960.* Berkeley: University of California Press.

Converse, J. M., & H. Schuman. 1974. *Conversations at random: Survey research as interviewers see it.* New York: John Wiley & Sons.

Cook, T. D., & D. T. Campbell. 1979. *Quasi-experimentation: Design and analysis issues for field settings.* Chicago: Rand McNally.

Cowan, G., & R. R. Campbell. 1994. Racism and sexism in interracial pornography: A content analysis. *Psychology of Women Quarterly* 1918: 323–338.

Dabbs, J. M. 1982. Makings things visible. In *Varieties of qualitative research,* edited by J. Van Maanen.Beverly Hills, CA: Sage.

Daniels, A. K. (1967). The low-caste stranger in social research. In *Ethics, Politics and Social Research,* edited by G. Sjoberg. Cambridge, MA: Schenkman.

Danigelis, N., & W. Pope. 1979. Durkheim's theory of suicide applied to the family: an empirical test. *Social Forces* 57: 1081–1106.

Davis, A. J. 1984. Sex-differentiated behaviors in non-sexist picture books. *Sex Roles* 11: 1–15.

Davis, D. W. 1997. The direction of race of interviewer effects among African-Americans: Donning the black mask. *American Journal of Political Science* 41: 309–323.

Davis, J. A., & T. W. Smith. 1985. *General social surveys, 1972–1985.* Chicago: National Opinion Research Center.

Davis, J. A., & T. W. Smith. 1992. *The NORC general social survey; a user's guide.* Newbury Park, CA: Sage.

Deegan, M. J. 1983. A feminist frame analysis of 'Star Trek'. *Free Inquiry in Creative Sociology* 11: 182–188.

Denzin, N. 1970. Sociological methods. Chicago: Aldine.

Denzin, N. 1989. *The research act: A theoretical introduction to sociological methods.* Englewood Cliffs, NJ: Prentice-Hall.

Denzin, N. K. 1996. *The cinematic society. the voyeur's gaze.* Urbana: University of Illinois Press.

Denzin, N. K., & Y. S. Lincoln, eds. 1994. *Handbook of qualitative research.* Thousand Oaks, CA: Sage.

Deutscher, I. 1978. Asking questions cross-culturally: Some problems of linguistic comparability. In *Sociological methods: A sourcebook,* edited by N. K. Denzin. New York: McGraw-Hill.

Diener, E. 1978. *Ethics in Social and Behavioral Research.* Chicago: University of Chicago Press.

Dillman, D. A. 1978. *Mail and telephone surveys: The total design method.* New York: Wiley.

Dillman, D. A. 1991. The design and administration of mail surveys. In *Annual Review of Sociology* 1991, pp. 225–249, edited by W. R. Scott & J. Blake. Palo Alto, CA: Annual Reviews.

Dillman, D. A., M. D. Sinclair, & J. R. Clark. 1993. Effects of questionnaire length, respondent-friendly design and a difficult question on response rates for occupant-addressed census mail surveys. *Public Opinion Quarterly* 57: 289–304.

Douglas, J. D. 1976. *Investigating social research: Individual and team field research.* Beverly Hill, CA: Sage.

Dovring, K. 1973. Communication, dissenters and popular culture in eighteenth-century Europe. *Journal of Popular Culture* 7: 559–568.

Durkheim, É. 1938. *The rules of sociological method.* New York: Free Press.

Durkheim, É. 1951. *Suicide.* Glencoe, IL: Free Press.

Durkheim, É. [1897] (1964)]. *Suicide.* Reprint, Glencoe, IL: Free Press.

Earls, F. J., & A. J. Reiss Jr., 1994. Breaking the cycle: Predicting and preventing crime. National Institute of Justice.

Don't let it happen again. 1997. *Economist* 343: 27–28.

Eisenman, R. 1993. Success at getting grants from the National Institutes of Health by Whites, Hispanics, Asians, and Blacks: Implications for scientific creativity. *The Mankind Quarterly 34,* 95–100.

Ellison, C. G., & D. A. Gay. 1989. Black political participation revisited: A test of compensatory, ethnic community, and public arena models. *Social Science Quarterly* 70: 101–119.

Ennett, S. T., N. S. Tobler, C. L. Ringwalt, & R. L. Flewelling. 1994. How effective is drug abuse resistance education? A meta-analysis of Project DARE outcome evaluations. *American Journal of Public Health* 84: 1394–1401.

Erikson, K. T. 1976. *Everything in its path: Destruction of community in the Buffalo Creek flood.* New York: Simon and Schuster.

Eron, L. D. 1982. Parent-child interaction, Television violence, and aggression of children. *American Psychologist,* 37: 197–211.

Eron, L. D., L. R. Huesman, P. Brice, P. Fischer, & R. Mermelstein. 1983. Age trends in the development of aggression, sex typing, and related television habits. *Developmental Psychology* 19: 71–77.

Federal Bureau of Investigation. 1995. *Uniform crime reports for the United States.* Washington, D.C.: U. S. Government Printing Office.

Finch, J. 1984. 'It's great to have someone to talk to': The ethics and politics of interviewing women. In *Social Researching: Politics, Problems and Practice,* edited by C. Bell & H. Roberts. London: Routledge & Kegan Paul.

Finch, J., & L. Wallis. 1993. Death, inheritance and the life course. *Sociological Review Monograph,* 50–68.

Finkel, S. E., T. M. Gutterbok, & M. J. Borg. 1991. Race-of-interviewer effects in a preelection poll: Virginia 1989. *Public Opinion Quarterly* 55: 313–330.

Firestone, W. A. 1993. Alternative arguments for generalizing from data as applied to qualitative research. *Educational Researcher* 22: 437–466.

Fischer, J., & K. Corcoran. 1994. *Measures for clinical practice: A sourcebook.* New York: Free Press.

Fissell, M. 1995. Gender and generation: representing reproduction in early modern England. *Gender and History* 7: 433–456.

Fontana, A., & J. H. Frey. 1994. Interviewing: the art of science. In *Handbook of qualitative research,* 361–376, edited by N. K. Denzin & Y. S. Lincoln.Thousand Oaks, CA: Sage.

Fowler, F. J., Jr. 1988. *Survey research methods.* Newbury Park, CA: Sage.

Fox, D. 1990. Health policy and the politics of research in the United States. *Journal of Health Politics, Policy and Law* 15: 481–499.

Fox, R. J., M. R. Crask, & J. Kim. 1988. Mail survey response rate. *Public Opinion Quarterly* 52: 467–491.

Frankenberg, R. 1993. *White women, race matters: The social construction of whiteness.* Minneapolis: University of Minnesota Press.

Frankfort-Nachmias, C., & D. Nachmias. 1992. *Research methods in the social sciences.* New York: St. Martin's.

French, M. T. 1995. Economic evaluation of drug abuse treatment programs: Methodology and findings. *American Journal of Drug and Alcohol Abuse* 21(Feb.): 111–135.

Frey, J. H. 1983. *Survey research by telephone.* Beverly Hills, CA: Sage.

Freyer, F. J. 1996. AMA's advice: Turn off the TV. *Providence Journal,* 10th ed, (1996): 1

Friedan, B. 1963. *The feminine mystique.* New York: Norton.

Friedman, J., & M. Alicea. 1995. Women and heroin: The path of resistance and its consequences. *Gender & Society* 9: 432–449.

Gano-Phillips, S., & F. D. Fincham. 1992. Assessing marriage via telephone interviews and written questionnaires: A methodological note. *Journal of Marriage and the Family,* 54: 630–635.

Gans, H. 1962. *Urban villagers.* New York: Free Press.

Gans, H. J. 1982. *The Urban villagers: Group and class in the life of Italian Americans.* New York: Free Press.

Geer, J. G. 1988. What do open-ended questions measure? *Public Opinion Quarterly* 52: 365–371.

Geer, J. G. 1991. Do open-ended questions measure "salient" issues? *Public Opinion Quarterly* 55: 360–370.

Geertz, C. 1983. *Local knowledge: further essays in interpretive anthropology.* New York: Basic Books.

Gilmore, S., J. DeLamater, & D. Wagstaff. 1997. Sexual decision making by inner city black adolescent males: a focus group study. *Journal of Sex Research* 33: 363–371.

Giordano, P. C. 1995. The wider circle of friends in adolescence. *American Journal of Sociology* 101: 661–697.

Glantz, L. H. 1996. Conducting research with children: Legal and ethical issues. *Journal of the American Academy of Child and Adolescent Psychiatry* 35: 1283–1292.

Glaser, B. G., & A. L. Strauss. 1967. *The discovery of grounded theory: Strategies for qualitative research.* New York: Aldine.

Glazer, N. 1977. General perspectives. In *Woman in a man-made world,* 1–4, edited by Nona Glazer & Helen Youngelson. Chicago: Rand McNally.

Gold, R. 1958. Roles in sociological field observations. *Social Forces,* 36: 217–223.

Goode, E. 1996. The ethics of deception in social research: A case study. *Qualitative Sociology* 19: 11–33.

Goodsell, C. B. 1984. Welfare waiting rooms. *Urban Life* 12: 467–477.

Gorden, R. L. 1975. *Interviewing, strategy, techniques and tactics.* Homewood, IL: Dorsey.

Gordon, M. T., & S. Riger. 1989. *The female fear.* New York: Free Press.

Gordy, L. L., & A. M. Pritchard. 1995. Redirecting our voyage through history: A content analysis of social studies textbooks. *Urban Education,* 30: 195–218.

Gottfredson, M. R., & T. Hirschi. 1990. *A general theory of crime.* Stanford, CA: Stanford University Press.

Gould, S. J. 1995. *Dinosaur in a haystack: Reflections in natural history.* New York: Harmony Books.

Graham, L. 1990. *A year in the life of Dr. Lillian Moller Gilbreth: Four representations of the struggle of a woman scientist.* (unpublished).

Grant, L., K. B. Ward, & X. L. Rong. 1987. Is there an association between gender and methods in sociological research? *American Sociological Review* 52: 856–862.

Gray, N. B., G. J. Palileo, & G. D. Johnson. 1993. Explaining rape victim blame: A test of attribution theory. *Sociological Spectrum* 13: 377–392.

Groves, R. M. 1989. *Survey errors and survey costs.* New York: John Wiley & Sons.

Hacker, D. 1998. *Research and documentation in the electronic age.* Boston: Bedford Books.

Hall, L. D., & K. P. Marshall. 1992. *Computing for social research: Practical approaches.* Belmont, CA: Wadsworth.

Hammersley, M. 1995. *The politics of social research.* London: Sage.

Hammersley, M., & P. Atkinson. 1983. *Ethnography: Principles in practice.* New York: Tavistock.

Hammersley, M., & P. Atkinson. 1992. *Ethnography: Principles in practice.* London: Routledge.

Harkess, S., & C. A. B. Warren. 1993. The social relations of intensive interviewing. *Sociological Methods & Research* 21: 317–339.

Harmon, C. 1996. *Using the internet, online services and CD-ROMS for writing research and term papers.* New York: Neal-Schuman Publishers.

Harris, L., and Associates. 1975. *The myth and reality of aging in America.* Washington, D. C.: National Council on Aging.

Hatry, H. P., K. E. Newcomer, & J. S. Wholey. 1994. Conclusion: Improving evaluation activities and results. In *Handbook of Practical Program Evaluation,* 590–602, edited by J. P. Wholey, H. P. Hatry & K. E. Newcomer. San Francisco: Jossey-Bass.

Heise, D. R. 1969. Separating reliability and stability in test-retest correlation. *American Sociological Review,* 34: 93–101.

Henry, G. T. 1990. *Practical sampling.* Newbury Park: Sage.

Herrera, C. D. 1997. A historical interpretation of deceptive experiments in American psychology. *History of the Human Sciences* 10: 23–36.

Herzog, A. R., & W. L. Rogers. 1988. Interviewing older adults. *Public Opinion Quarterly*, 52: 84–99.

Hess, I. 1985. *Sampling for social research surveys 1947–1980.* Ann Arbor: Institute for Social Research.

Hessler, R. M. 1992. *Social research methods.* St. Paul, MN: West.

Hewa, S. 1993. Sociology and public policy: the debate on value-free social science. *International Journal of Sociology and Social Policy* 13: 64–82.

Hill, E. 1995. Labor market effects of women's post-school-age training. *Industrial and Labor Relations Review* 49: 138–149.

Hippler, H., & N. Schwarz, 1987. Response effects in surveys. In *Social Information Processing and Survey Method,*102–122, edited by H. Hippler, N. Schwarz, & S. Sudman. New York: Springer.

Hirschi, T. 1969. *Causes of delinquency.* Berkeley: University of California Press.

Ho, M. L. 1984. Patriarchal ideology and agony columns. In *Looking Back: Some Papers from the BSA 'Gender and Society' Conference,* edited by S. Webb & C. Pearson.

Hochschild, A. R. 1983. *The managed heart: commercialization of human feeling.* Berkeley: University of California Press.

Hochschild, A. R, with A. Machung. 1989. *The second shift.* New York: Avon Books.

Hoffnung, M. 1995. *Who are our students? Comparisons of women attending single sex and coeducational colleges.* Oakland, CA: Presented at the meetings of the Association for Women in Psychology.

Hoffnung, M. 1997. Personal communication.

Hoffnung, M. 1998. Personal communication.

Holsti, O. R. 1969. *Content analysis for the social sciences.* Reading, MA: Addison-Wesley.

Horley, J. 1984. Life satisfaction, happiness, and morale: Two problems with the use of subjective well-being indicators. *Gerontologist* 24: 124–127.

Howe, M., J. A. 1977. *Television and children.* London: New University Education.

Hubbard, A. 1996. School board approves random drug testing. *The Kokomo Tribune*, February 6, 1996: 1.

Hudson, W. 1982. *Clinical measurement package: A field manual.* Chicago: Dorsey.

Hughes, E. C. 1960. Introduction: the place of field work in social science. In *Field Work: An Introduction to the Social Sciences,* iii-xiii, edited by B. H. Junker. Chicago: University of Chicago.

Humphreys, L. 1975. *Tearoom trade: Impersonal sex in public places.* Chicago: Aldine.

Hunt, M. 1985. *Profiles of social research: the scientific study of human interactions.* New York: Russell Sage Foundation.

Hyman, H. H. 1975. *Interviewing in social research.* Chicago: The University of Chicago Press.

International Labor Office. 1977. *Labor force estimates and projections, 1950–2000.* Geneva: International Labor Office.

Jacobs, S. 1997. Boom days for welfare studies. *The Boston Globe,* April 27, 1997: A1-A7.

James, J. M., & R. Bolstein.1990. The effect of monetary incentives and follow-up mailings on the response rate and response quality in mail surveys. *Public Opinion Quarterly* 54: 346–361.

Jary, D., & J. Jary.1991. *The HarperCollins dictionary of sociology.* New York: HarperPerennial.

Jayartane, T. E., & A. J. Sewart. 1989. Quantitative and qualitative method in the social sciences. In *Beyond Methodology Feminist Scholarship as Lived Research,* 85–106, edited by M. M. Fonow & J. A. Cook. Bloomington: Indiana University Press.

Jensen, L., & D. K. McLauglin. 1997. The escape from poverty among rural and urban elders. *The Gerontologist* 37: 462–469.

Johnson, T. P., J. Hougland, & R. Clayton. 1989. Obtaining reports of sensitive behavior: a comparison of telephone and face-to-face interviews. *Social Science Quarterly* 70: 174–183.

Jones, J. H. 1981. *Bad blood: the Tuskegee syphilis experiment.* New York: Free Press.

Jowell, R., J. Berry, & J. Goldman. 1989. *British social attitudes: special international report.* Hants, England: Gower.

Judd, C. M. 1987. Combining process and outcome evaluation. In *Multiple Methods in Program Evaluation,* 23–41, edited by M. M. Mark & R. L. Shotland. San Francisco: Jossey-Bass.

Kaldenberg, D. O., H. F. Koenig, & B. W Becker. 1994. Mail survey response rate patterns in a population of the elderly. *Public Opinion Quarterly* 58: 68–76.

Kammeyer, K. C. W., & J. A. Roth. 1971. Coding response to open-ended questions. In *Sociological Methodology,* edited by H. L. Costner. San Francisco: Jossey-Bass.

Kane, E. W., & L. J. Macaulay. 1993. Interviewer gender and gender attitudes. *Public Opinion Quarterly* 57: 1–28.

Kane, E. W., & H. Schuman. 1991. Open survey questions as measures of personal concern with issues: a reanalysis of Stouffer's *Communism, conformity and civil liberties*. In *Sociological Methodology,* edited by P. V. Marsden, *1991*. Washington, D.C.: American Sociological Association.

Karney, B. R., J. Davila, C. L. Cohan, K. T. Sullivan, M. D. Johnson, & T. N. Bradbury. 1995. An empirical investigation of sampling strategies in marital research. *Journal of Marriage and the Family* 57: 909–920.

Kaufman, D. R. 1985. Women who return to orthodox Judaism: A feminist analysis. *Journal of Marriage and the Family*, 47: 543–551.

Kelly, L. 1988. *Surviving sexual violence.* Minneapolis: University of Minnesota Press.

Kenschaft, L. Forthcoming. *Intellect and intimacy: The marriage of Alice Freeman Palmer and George Herbert Palmer.* Boston University.

Kimmel, A. J. 1988. *Ethics and values in applied social research.* Newbury Park, CA: Sage.

King, A. C. 1994. Enhancing the self-report of alcohol consumption in the community: two questionnaire formats. *American Journal of Public Health* 84: 294–296.

Kinsey, A. C., W. B. Pomeroy, C. E. Martin, & P. H. Gebhard. 1953. *Sexual behavior in the human female.* Philadelphia: W. B. Saunders.

Kirp, D. L. 1995. Blood, sweat and tears: The Tuskegee experiment and the era of AIDS. *Tikkun*, 10: 50–55.

Knowlton, S. R. 1997. How students get lost in cyberspace. *New York Times,* November 2, 1997 ed, Section 4 A: 18–19.

Knox, R. 1997. Nurses' health study has worldwide impact. *The Boston Globe,* February 17 ed.: A1-B5.

Kolbert, E. 1994. Television gets closer look as a factor in real violence. *New York Times,* 14th ed, 1-D18.

Koppel, R., & A. Hoffman. 1996. Dislocation policies in the USA: What should we be doing? *Science* 544 (March): 111–125.

Krippendorff, K. 1980. *Content analysis: an introduction to its methodology.* Beverly Hills, CA: Sage.

Krueger, R. A. 1988. *Focus groups.* Newbury Park, CA: Sage.

Krysan, M., & H. Schuman. 1992. The behavior of respondents, nonrespondents, and refusers across mail surveys. *Public Opinion Quarterly* 56: 530–535.

Krysan, M., H. Schuman, L. J. Scott, & P. Beatty. 1994. Response rates and response content in mail versus face-to-face surveys. *Public Opinion Quarterly,* (Fall): 381–400.

Kuhn, M. H. 1962. The interview and the professional relationship. In *Human Behavior and Social Processes,*193–206, edited by A. M. Rose. Boston: Houghton Mifflin.

Kurland, D., & D. John. 1997. *Internet guide for sociology.* Belmont, CA: Wadsworth.

Kuzel, A. J. 1992. Sampling in qualitative inquiry. In *Doing qualitative research.* 31–44, edited by B. F. Crabtree & W. L. Miller, Newbury Park: Sage.

Langer, E. J., & J. Rodin, 1982. The effects of choice and enhanced personal responsibility for the aged: A field experiment in an institutional setting. In *Experimenting in Society,* 77–84, edited by J. W. Reich. Glenview, IL: Scott, Foresman.

Larson, R., & M. H. Richards, 1994. *Divergent Realities.* New York: Basic.

Lasswell, H. D. 1949. Detection: propaganda detection and the courts. In *The language of politics: studies in quantitative semantics,* 173–232, edited by H. D. Lasswell, N. Leites, R. Fadner, J. M. Goldsen, A. Gray, & I. L. Janis, New York: George Stewart.

Lasswell, H. D. 1965. Detection: Propaganda detection and the courts. In *The Language of Politics: Studies in Quantitative Semantics,* edited by Harold Lasswell et al., Cambridge, MA: MIT Press.

Laumann, E. O., J. H., Gagnon, R. T. Michael, & S. Michaels. 1994. *The Social Organization of Sexuality.* Chicago: University of Chicago Press.

Lavrakas, P. J. 1987. *Telephone survey methods sampling, selection and supervision.* Newbury Park, CA: Sage.

Lavrakas, P. J. 1993. *Telephone survey methods: Sampling, selection, and supervision.* Newbury Park: Sage.

Lazarsfeld, P. F. 1944. The controversy over detailed interviews: An offer for negotiation. *Public Opinion Quarterly* 8: 38–60.

Lazarsfeld, P. F., & A. Barton. 1951. Qualitative Measurement in the Social Sciences: Classifications, Typologies, and Indices. In *The Policy Sciences,* edited by D. Lerner & H. S. Lasswell. Stanford, California: Stanford University Press.

Lee, R. M. 1993. *Doing research on sensitive topics.* Newbury Park: Sage.

Lee, V. E., & J. B. Smith. 1995. Effects of high school restructuring and size on early gains in achievement and engagement. *Sociology of Education* 68: 241–270.

Leff, L. S. 1984. *Data processing: the easy way.* New York: Barron's.

Lester, J. D. 1990. *Writing research papers: A complete guide.* Glenview, IL: Scott, Foresman/Little, Brown Higher Education.

Lever, J. 1981. Multiple methods of data collection. *Urban Life* 10: 199–213.

Levin, H. M. 1987. Cost-benefit and cost-effectiveness analysis. In *Evaluation Practice in Review,* edited by D. S. Corday, H. S. Bloom, & R. J. Light. San Francisco: Jossey-Bass.

Levin, W. 1988. Age stereotyping: college student evaluations. *Research in Aging,* 134–148.

Liang, J. 1984. Dimensions of the life satisfaction index A: a structural formulation. *Journal of Gerontology,* 39: 613–622.

Liebow, E. 1967. *Tally's corner: A study of Negro streetcorner men.* Boston: Little, Brown.

Liebow, E. 1993. *Tell them who I am, The lives of homeless women.* New York: Free Press.

Lindsey, E. W. 1997. Feminist issues in qualitative research with formerly homeless mothers. *Affilia Journal of Women and Social Work* 12: 57–76.

Lofland, J. 1984. *Analyzing social situations.* Belmont, CA: Wadsworth.

Lofland, J., & Lofland, L. H. 1995. *Analyzing social settings: A guide to qualitative observation and analysis.* Belmont, CA: Wadsworth.

Lofland, L. H. 1972. Self-management in public settings: Part I. *Urban Life and Culture* 93–108.

Lofland, L. H. 1973. *A world of strangers: Order and action in urban public space.* New York: Basic.

Loftus, E. F., M. R. Klinger, K. D. Smith, & J. Fielder. 1990. A tale of two questions: Benefits of asking more than one question. *Public Opinion Quarterly* 54: 330–345.

Lopata, H. Z. 1980. Interviewing American widows. In *Fieldwork Experience Qualitative Approaches to Social Research,* edited by W. B. Shaffir, R. A. Stebbins, & A. Turowetz. New York: St. Martin's.

Lundberg, O., & M. Thorslund. 1996. Fieldwork and measurement considerations in surveys of the oldest old. *Social Indicators Research* 37, 165–188.

Lynch, J. P. 1996. Clarifying divergent estimates of rape from two national surveys. *Public Opinion Quarterly* 60 (Fall): 410–431.

Lynd, R. S., & H. M. Lynd. 1956. *Middletown; a study in American culture.* New York: Harcourt, Brace, Jovanovich.

Lystra, K. 1989. *Searching the heart: Women, men and romantic love in nineteenth-century America.* New York: Oxford University Press.

Macleod, J. 1987. *Ain't no makin' it: Leveled aspiration in low-income neighborhoods.* Boulder: Westview.

Maher, F. A., & M. K. Tetreault. 1994. *The Feminist classroom.* New York: Basic.

Malinowski, B. 1922. *Argonauts of the Western Pacific.* New York: Dutton.

Manoff, R. K., & M. Schudson. 1987. *Reading the news.* New York: Pantheon.

Marsh, C. 1982. *The Survey method.* London: Allen & Unwin.

Marshall, S. 1986. Development, dependence, and gender inequality in the third world. *International Studies Quarterly* 29: 217–240.

Martin, L. L., T. Abend, C. Sedikides, & J. D. Green. 1997. How would I feel if ... ? Mood as input to a role fulfillment evaluation process. *Journal of Personality and Social Psychology* 73: 242–254.

Martineau, H. 1962. *Society in America*. New York: Anchor.

Maruyama, G., & S. Deno. 1992. *Research in educational settings*. Newbury Park, CA: Sage.

Marx, K. 1961. *Capital*. London: Lawrence and Wishart.

Marx, Karl. [1867] 1967. *Das Capital*. Repr. New York: International.

Matcha, D. A. 1995. Obituary analysis of early 20th century marriage and family patterns in northwest Ohio. *Omega* 30: 121–130.

McCall, G. J. 1984. Systematic field observation. *Annual Review of Sociology*, 10: 263–282.

McKissack, P. 1988. *Mirandy and brother wind*. New York: Alfred A. Knopf.

McLanahan, S., & G. Sandefur. 1994. *Growing up with a single parent*. Cambridge, MA: Harvard University Press.

McMillen, L. 1992. Anthropology professor studies garbage for insights into modern society. *The Chronicle of Higher Education, September 9,* 3.

McNabb, S. 1995. Social research and litigation: Good intentions versus good ethics. *Human Organization* 54: 331–335.

Mead, M. 1963. *Sex and temperament in three primitive societies*. New York: Morrow.

Menard, S. 1991. *Longitudinal research*. Newbury Park, CA: Sage.

Merton, R. K. 1937. Social structure and anomie. *American Sociological Review*, 3: 672–682.

Merton, R. K., M. Fiske, & P. L. Kendall. 1956. *The focused interview*. Glencoe, IL: Free Press.

Messner, S. F. 1995. Personal Communication.

Michael, R. T., J. H. Gagnon, E. O. Laumann, & G. Kolata. 1994. *Sex in America*. Boston: Little, Brown.

Mies, M. 1991. Women's research or feminist research? In *Beyond Methodology Feminist Scholarship as Lived Research,* 60–84, edited by M. M. Fonow & J. A. Cook. Bloomington: Indiana University Press.

Miles, M. B., & A. M. Huberman, 1994. *Qualitative data analysis*. Thousand Oaks: Sage.

Milgram, S. 1974. *Obedience to authority: An experimental view*. New York: Harper & Row.

Milgram, S., & R. L. Shotland, 1973. *Television and antisocial behavior*. New York: Academic.

Miller, D. C. 1991. *Handbook of research design and social measurement*. Newbury Park: CA: Sage.

Miller, E. M. 1986. *Street woman*. Philadelphia: Temple University Press.

Mishler, E. G. 1986. *Research interviewing context and narrative*. Cambridge, MA: Harvard University Press.

Mitchell, A. 1997. Survivors of Tuskegee study get apology from clinton. *The New York Times,* 146: 9–10.

Morgan, D.L. 1996. Focus groups. *Annual Review of Sociology* 22: 129–153.

Morgan, D. L. 1995. *Focus groups as qualitative research*. Newbury Park, CA: Sage.

Morgan, D. L. 1988. *Focus groups as qualitative research*. Newbury Park, CA: Sage.

Morgan, R., ed. 1984. *Sisterhood is global: The international woman's movement anthology*. Garden City, NY: Anchor.

Morrow, R. 1994. *Critical theory and methodology.* Thousand Oaks, CA: Sage.

Mosteller, F. 1955. Use as evidenced by an examination of wear and tear on selected sets of ESS. In *A study of the need for a new encyclopedic treatment of the social sciences,* edited by K. Davis. (unpublished).

Naisbitt, J., & P. Aburdene, 1990. *Megatrends 2000: Ten new directions for the 1990's.* New York: Morrow.

National Academy of Sciences. 1993. Methods and values in science. In *The "Racial" Economy of Science,* 341–343, edited by S. Harding. Bloomington: Indiana University Press.

National Center for Educations Statistics. 1982. *Conferences, critiques and references on the subject of public and private schools: The Coleman report.* Washington, D.C.

Ness, E. 1967. *Sam, bangs, and moonshine.* New York: Holt, Rinehart & Winston.

Neugarten, B., J. Havighurst, & S. Tobin, 1961. The measurement of life satisfaction. *Gerontology* 16: 134–143.

Newman, F. L., & M. J. Tejeda. 1996. The need for research that is designed to support decisions in the delivery of mental health services. *American Psychologist,* 51(Oct.): 1040–1049.

Newman, W. L. 1994. *Social research methods.* Needham Heights, MA: Allyn and Bacon.

Norusis, M. J. 1990. *The SPSS guide to data analysis.* Chicago: SPSS Inc.

Nunnally, J. C. 1964. *Educational measurement and evaluation.* New York: McGraw-Hill.

Nunnally, J. C. 1978. *Psychometric theory.* New York: McGraw-Hill.

Nyden, P. A., A. Figerts, M. Shibley, & D. Burrows, 1997. *Building community.* Thousand Oaks, CA: Forge.

Oakley, A. 1981. Interviewing women: a contradiction in terms. In *Doing Feminist Research,* 30–61, edited by H. Roberts (Ed.), London: Routledge & Kegan Paul.

Oldendick, R. W., & M. W. Link. 1994. The answering machine generation. *Public Opinion Quarterly* 58: 264–273.

Olson, S. 1976. *Ideas and data: the process and practice of social research.* Homewood, IL: Dorsey.

Orlans, H. 1967. Ethical problems in the relations of research sponsors and investigators. In *Ethics, Politics and Social Research,* 3–24, edited by G. Sjoberg. Cambridge, MA: Schenkman.

Orum, A. M., J. R. Feagin, & G. Sjoberg. 1991. Introduction: The nature of the case study. In *A case for the case study,* 1–26, edited by J. R. Feagin, A. M. Orum, & G. Sjoberg. Chapel Hill: University of North Carolina Press.

Osgood, C. E., & E. G. Walker. 1959. Motivation and language behavior: content analysis of suicide notes. *Journal of Abnormal Social Psychology* 59: 58–67.

Ostrander, S. A. 1995. "Surely you're not in this just to be helpful": Access, rapport and interviews in three studies of elites. In *Studying elites using qualitative methods,* edited by Rosanna Hertz & Jonathan B. Imber. Thousand Oaks: Sage.

Pain, R. H. 1997. "Old age" and ageism in urban research: The case of the fear of crime. *International Journal of Urban and Regional Research* 21: 117–128.

Park, P. 1993. What is participatory research? A theoretical and methodological perspective. In *Voices of change, participartory research in the United*

States and Canada, edited by P. Park, M. Brydon-Miller, B. Hall, and T. Jackson. Westport, CT: Bergin & Garvey.

Patton, M. Q. 1990. *Qualitative evaluation and research methods.* Newbury Park, CA: Sage.

Paulsen, G. 1993. *Nightjohn.* New York: Delacorte.

Peirce, K., & E. Edwards. 1988. Children's construction of fantasy stories: gender differences in conflict resolution strategies. *Sex roles* 18: 393–404.

Philliber, S. G., M. R. Schwab, & G. S. Sloss. 1980. *Social research.* Itasca, IL: F. E. Peacock.

Pillemer, K., & D. Finkelhor, 1988. The prevalence of elder abuse: A random sample survey. *The Gerontologist* 28: 51–57.

Poe, G. S., I. Seeman, J. McLaughin, E. Mehl, & M. Dietz. 1988. "Don't know" boxes in factual questions on a mail questionnaire. *Public Opinion Quarterly* 52: 212–222.

Potuchek, J. L. 1992. Employed wives' orientations to breadwinning: a gender theory analysis. *Journal of Marriage and the Family* 54: 548–558.

Powell, E. H. 1958. Occupation, status, and suicide: Toward a redefinition of anomie. *American Sociological Review* 23: 131–139.

Powell, R. A., H. M. Single, & K. R. Lloyd. 1996. Focus groups in mental health research: Enhancing the validity of user and provider questionnaires. *International Journal of Social Psychiatry* 42: 193–206.

Ptacek, J. 1988. Why do men batter their wives? In *Feminist Perspectives on Wife Abuse,* edited by K. Yllo & Michele Bograd. Newbury Park, CA: Sage.

Pyrczak, F., & R. R. Bruce. 1992. *Writing empirical research reports: A basic guide for students of the social and behavioral sciences.* Los Angeles, CA: Pycrczak.

Rasinski, K. A. 1989. The effect of question wording on public support for government funding. *Public Opinion Quarterly* 53: 388–394.

Rasinski, K. A., D. Mingay, & N. M. Bradburn. 1994. Do respondents really "mark all that apply" on self-administered questions? " *Public Opinion Quarterly,* 58: 400–408.

Rathje, W. L., & C. Murphy. 1992. *Rubbish! The archaeology of garbage.* New York: HarperCollins.

Reinharz, S. 1985. Feminist distrust, content and context in sociological work. In *The Self in Social Inquiry,* 153–172, edited by D. Berg & K. Smith. Beverly Hills, CA: Sage.

Reinharz, S. 1988. Controlling women's lives: a cross-cultural interpretation of miscarriage accounts. In *Research in the sociology of health care,* 2–37, edited by D. Wertz, Greenwich, CN: JAI.

Reinharz, S. 1992. *Feminist methods in social research.* New York: Oxford University Press.

Reisman, C. K. 1987. When gender is not enough: Women interviewing women. *Gender & Society,* 1: 172–207.

Reynolds, P. D. 1982. *Ethics and social research.* Englewood Cliffs, N. J. Prentice-Hall.

Ribbens, J. 1989. Interviewing: An "unnatural situation?" *Women's Studies International Forum* 12: 579–592.

Richards, T. J., & L. Richards. 1994. Using computers in qualitative research. In *Handbook of qualitative research,* 445–462, edited by N. K. Denzin & Y. S. Lincoln. Thousand Oaks, CA: Sage.

Richardson, L. 1990. *Writing strategies: reaching diverse audiences.* Newbury Park, CA: Sage.

Riemer, J. W. 1977. Varieties of opportunistic research. *Urban Life* 5: 467–477.

Rockett, I. R. F. 1995. Personal communication.

Rocky Mountain Behavioral Science Institute (RMBSI). 1995. A model for evaluating DARE & other prevention programs. *News & Views* 22: 1–2.

Rodrigues, D. 1997. *The research paper and the world wide web.* Upper Saddle River, NJ: Prentice-Hall.

Rogers, J. K., & K. D. Henson, 1997. "Hey, Why don't you wear a shorter skirt?" Structural vulnerability and the organization of sexual harassment in temporary clerical employment. *Gender & Society* 11: 215–237.

Rollins, J. 1985. *Between women.* Philadelphia: Temple University Press.

Rosenhan, D. L. (1971). On being sane in insane places, *Science* 179: 250–258.

Rosenberg, J. F. 1967. *The impoverished students' book of cookery, drinkery, & housekeepery.* Garden City, N. Y. Doubleday.

Rosenberg, M. 1968. *The logic of survey analysis.* New York: Basic.

Rosengren, K. E. 1981. Advances in Scandinavian content analysis. In *Advances in Content Analysis,* 9–19, edited by K. E. Rosengren. Beverly Hills, CA: Sage.

Ross, O., & N. A. Kreitman, 1975. A further investigation of differences in the suicide rates of England and Wales and Scotland. *British Journal of Psychiatry* 127: 575–582.

Rossi, P. H., & H. E. Freeman. 1993. *Evaluation a systematic approach.* Newbury Park, CA: Sage.

Rossi, P. H., & J. D. Wright, 1984. Evaluation research: an assessment. In *Annual Review of Sociology,* vol. 10, 332–352, edited by R. H. Turner & J. F. Short. Palo Alto: Annual Reviews.

Rubin, L. B. 1976. *Worlds of pain: Life in the working-class family.* New York: Basic.

Rubin, L. B. 1983. *Intimate strangers: Men and women Together.* New York: Harper & Row.

Rubin, L. B. 1994. *Families on the fault line: America's working class speaks about the family, the economy, race, and ethnicity.* New York: HarperCollins.

Runcie, J. F. 1980. *Experiencing social research.* Homewood, IL: Dorsey.

Rutman, L. 1984. Evaluability assessment. In *Evaluation Research Methods,* 9–38, edited by L. Rutman. Beverly Hills: Sage Publications.

Saris, W. E. 1991. *Computer-assisted interviewing.* Newbury Park, CA: Sage.

Schaeffer, N. C., J. A., Seltzer, & M. Klawitter. 1991. Estimating nonresponse and response bias. *Sociological Methods & Research* 20: 30–59.

Scheirer, M. A. 1994. Designing and using process evaluation. In *Handbook of Practical Program Evaluation,* 40–68, edited by J. P. Wholey, H. P. Hatry, & K. E. Newcomer. San Francisco: Jossey-Bass.

Schleef, D. J. 1995. *Lawyers on line: Professional identity and boundary maintenance in cyberspace.* (unpublished)

Schuerman, J. R., T. L. Rzepnicki, & J. H. Littell, 1994. *Putting families first: An experiment in family preservation.* New York: Aldine de Gruyer.

Schuman, H., & S. Presser. 1981. *Questions and answers in attitude surveys.* New York: Academic.

Schwarz, N., & H. Hippler. 1987. What responses may tell your respondents: informative functions of response alternatives. In *Social Information Processing and Survey Method,*163–178, edited by H. Hippler, N. Schwarz, & S. Sudman. New York: Springer.

Sechrest, L. 1965. *Situational sampling and contrived situations in the assessment of behavior.* (unpublished)

Seidman, I. E. 1991. *Interviewing as qualitative research.* New York: Teachers College, Columbia University.

Selltiz, C., L. S. Wrightsman, & S. W. Cook. 1976. *Research methods in social relations.* New York: Holt, Rinehart & Winston.

Seltzer, R. 1993. AIDS, homosexuality, public opinion and changing correlates over time. *Journal of Homosexuality* 26: 85–97.

Shadish, W. R., Jr., T. D. Cook, & L. C. Leviton. 1991. *Foundations of program evaluation.* Newbury Park, CA: Sage.

Sheirer, M. A. 1994. Designing and using process evaluation. In *Handbook of Practical Program Evaluation,* 40–68, edited by J. P. Wholey, H. P. Hatry, & K. E. Newcomer. San Francisco: Jossey-Bass.

Sheskin, A. 1979. *Cryonics: A sociology of death and bereavement.* New York: John Wiley.

Shingles, R. A. 1973. Organizational membership and attitude change. In *Quasi-Experimental Approaches Testing Theory and Evaluating Policy,* 226–270, edited by J. Caporaso & L. L. Roos, Jr. Evanston, IL: Northwestern University Press.

Shryock, H. S., & J. S. Siegel. 1973. *The methods and materials of demography.* Washington, D.C.: U. S. Government Printing Office.

Simon, B. 1998. Two ways to take stock. *PC Magazine* 17: 73.

Sinclair, B., & D. Brady. 1987. Studying members of the United States Congress. In *Research Methods for Elite Studies,* 61–71, edited by G. Moyser & M. Wagstaffe. London: Allen & Unwin.

Singleton, R., Jr., B. C. Straits, M. M. Straits, & R. J. McAllister. 1988. *Approaches to social research.* New York: Oxford University Press.

Small, S. A. 1995. Action-oriented research: Models and methods. *Journal of Marriage and the Family* 57 (November): 941–955.

Smith, D. 1989. Sociological theory: Methods of writing patriarchy. In *Feminism and Sociological Theory,* 34–64, edited by R. A. Wallace, Beverly Hills, CA: Sage.

Smith, E. A., & Zabin, L. S. 1993. Marital and birth expectations of urban adolescents. *Youth and Society,* 25: 62–74.

Smith, P. M., & B. B. Torrey. 1996. The future of the behavioral and social sciences. *Science* 271: 611–612.

Smith, T. W. 1989. That which we call welfare by any other name would smell sweeter: an analysis of the impact of question wording on response patterns. In *Survey Research Methods,* 99–107, edited by E. Singer & S. Presser. Chicago: University of Chicago Press.

Spender, D. 1981. The gatekeepers: a feminist critique of academic publishing. In *Doing feminist research,* edited by Helen Roberts. London: Routledge & Kegan Paul.

Spitze, G. 1988. Women's employment and family relations: a review. *Journal of Marriage and the Family* 50: 595–618.

Srole, L. 1956. Social integration and certain corollaries: An exploratory study. *American Sociological Review,* 21: 709–716.

St. Jean, Y. 1998. Let people speak for themselves, interracial unions and the general social survey. *Journal of Black Studies* 28: 398–414.

Stacey, J. 1990. *Brave new families: stories of domestic upheaval in late twentieth century America.* New York: Basic.

Starr, P. 1987. The sociology of official statistics. In *The politics of numbers,* 7–60, edited by W. Alonso & P. Starr. New York: Russell Sage Foundation.

Steeh, C. G. 1989. Trends in nonresponse rates, 1952–1979. In *Survey Research Methods,* 32–49, edited by E. Singer & S. Presser. Chicago: University of Chicago Press.

Stewart, D. W., & P. N. Shamdasani. 1990. *Focus groups: Theory and practice.* Newbury Park, CA: Sage.

Stouffer, S. A. 1950. Some observations on study design. *American Journal of Sociology* 55: 355–361.

Straus, M. A. 1992. Sociological research and social policy: The case of family violence. *Sociological Forum,* 7: 211–237

Strauss, A. 1987. *Qualitative analysis for social scientists.* Cambridge: Cambridge University Press.

Stricker, F. 1979. Cookbooks and lawbooks: The hidden history of career women in twentieth century America. In *A Heritage of Her Own: Toward a New Social History of American Women,* edited by N. Cott & E. Pleck. New York: Simon & Schuster.

Strube, G. 1987. Answering survey questions: The role of memory. In *Social Information Processing and Survey Method,* 86–101, edited by H. Hippler, N. Schwarz, & S. Sudman. New York: Springer.

Suchman, L., & B. Jordan. 1992. Validity and the collaborative construction of meaning in face-to-face surveys. In *Questions About Questions,* 241–270, edited by J. M. Tanur. New York: Russell Sage Foundation.

Sudman, S. 1967. *Reducing the cost of surveys.* Chicago: Aldine.

Sullivan, T. J. 1992. *Applied sociology, research and critical thinking.* New York: Macmillan.

Tavris, C. 1996. The mismeasure of woman. In *The Meaning of Difference: American Constructions of Race, Sex and Gender,* edited by K. E. Rosenblum & T. C. Travis (Eds.), New York: McGraw-Hill.

Tavris, C., & C. Wade. 1995. *Psychology in perspective.* New York: HarperCollins.

Taylor, L. 1995. *Occasions of faith: An anthropology of Irish Catholics.* Philadelphia: University of Pennsylvania Press.

Thomas, B. M., & I. R. H. Rockett. 1995. *The validity of elderly female suicide statistics: a cross-national study.* (unpublished)

Thompson, H. A. 1995. *Internet resources: A subject guide.* Chicago: Association of College & Research Libraries.

Thurlow, M. B. 1935. An objective analysis of family life. *Family,* 16: 13–19.

Touliatos, J., B. F. Perlmutter, & M. A. Straus. 1990. *Handboook of family measurement techniques.* Newbury Park, CA: Sage.

Turabian, K. L., 6th edition, revised by John Grossman and Alice Bennett 1996. *A manual for writers of term papers, theses, and dissertations.* Chicago: University of Chicago Press.

U. S. Bureau of the Census. 1975. *Historical statistics of the United States: Colonial times to 1970.* Washington, D.C.: U. S. Government Printing Office.

U. S. Bureau of the Census. 1992. *Statistical abstract of the United States.* Washington, D.C.: U. S. Government Printing Office.

U. S. Bureau of the Census. 1995a. *County and city data book.* Washington, D.C.: U. S. Government Printing Office.

U. S. Bureau of the Census. 1995b. *Statistical abstract of the United States, various years.* Washington, D.C.: U. S. Government Printing Office.

U. S. Bureau of the Census. 1995c. *Current population survey 1991/993.* Washington, D.C.: U. S. Bureau of the Census.

Umennachi, N. 1995. Africa: "Tragic consequences" of the development strategies of founding fathers. *21st Century Afro Review* 1: 179–223.

United Nations. 1995a. *Statistical yearbook.* New York: United Nations.

United Nations. 1995b. *Demographic yearbook.* New York: United Nations.

Unnithan, N. P. 1986. Research in a correctional setting: Constraints and biases. *Journal of Criminal Justice*, 14: 301–412.

Vanneman, A. 1995. Editorial: evaluating the evaluators. *Youth Today*, 2: 2.

Verhovek, S. H. 1997. In poll, Americans reject means but not ends of racial diversity. *The New York Times*, December 14, 1997 ed, (1997): 1–32.

Vidich, A. J., & J. Bensman. 1964. The Springdale case: Academic bureaucrats and sensitive townspeople. In *Reflections on community studies*, edited by A. Vidich, J. Bensman, & Maurice R. Stein. New York: John Wiley and Sons.

Vigderhous, G. 1989. Scheduling telephone interviews: A study of seasonal patterns. In *Survey research methods*, 69–78, edited by E. Singer & S. Presser. Chicago: University of Chicago Press.

Wallace, W. 1971. *The logic of science in sociology.* Chicago: Aldine Atherton.

Walton, J. 1992. Making the theoretical case. In *What is a case? Exploring the foundations of social inquiry*, 121–138, edited by C. C. Ragin & H. S. Becker. New York: Cambridge University Press.

Walworth, A. 1938. *Social histories at war: A study of the treatment of our wars in the secondary school history books of the United States and in those of its former enemies.* Cambridge, MA: Harvard University Press.

Ward, K. 1984. *Women in the world-system: Its impact on status and fertility.* New York: Praeger.

Wasburn, P. C. 1995. Top of the hour radio newscasts and the public interest. *Journal of Broadcasting and Electronic Media* 39: 73–91.

Wax, R. H. 1971. *Doing fieldwork: Warnings and advice.* Chicago: University of Chicago.

Webb, E. J., D. T. Campbell, R. D. Schwartz, & L. Sechrest. 1966. *Unobtrusive-measure: Nonreactive research in the social sciences.* Chicago: Rand McNally.

Weber, M. 1947. *The theory of social and economic organization.* New York: Free Press.

Weisberg, H. F., J. A. Krosnick, & B. D. Bowen. 1989. *An introduction to survey research and data analysis.* Glenview, IL: Scott, Foresman.

Weiss, R. S. 1990. *Staying the course: The emotional life and social situation of men who do well at work.* New York: Free Press.

Weiss, R. S. 1994. *Learning from strangers: The art and method of qualitative interview studies.* New York: Free Press.

Weitz, R. 1991. *Life with AIDS.* New Brunswick, NJ: Rutgers University Press.

Weitzman, E., & M. B. Miles. 1995. *Computer programs for qualitative analysis.* Thousand Oaks, CA: Sage.

Weitzman, L., D. Eifler, E. Hokada, & C. Ross. 1972. Sex-role socialization in picture books for preschool children. *American Journal of Sociology* 77: 1125–1150.

Weitzman, L. J. 1985. *The divorce revolution*. New York: Free Press.

Wescott, M. 1977. Feminist criticism of the social sciences. *Harvard Educational Review* 49: 422–430

West, C., & D. H. Zimmerman. 1987. Doing gender. *Gender & Society* 1: 125–151.

Whitlock, F. F. 1995. Migration and suicide. *Medical Journal of Australia* 2: 840–848.

Whyte, W. F. 1943. *Street-corner society: The social structure of an Italian slum*. Chicago: University of Chicago Press.

Whyte, W. F. 1955. *Street corner society*. Chicago: University of Chicago Press.

Whyte, W. F. 1991. Introduction. In *Participatory action research*, edited by W. F. Whyte. Newbury Park, CA: Sage.

Whyte, W. F., D. J. Greenwood, & P. Lazes. 1991. Participatory action research: Through practice to science in social research. In *Participatory action research*, edited by W. F. Whyte. Newbury Park, CA: Sage.

Willcox, W. F. 1930. Census. *Encyclopedia of the Social Sciences*, 295–300. New York: MacMillan.

Wilder, L. I. 1953. *Little house in the Big Woods*. New York: Harper.

Williams, T. 1989. *The cocaine kids*. Reading, MA: Addison-Wesley.

Willimack, D. K., H. Schuman, B. Pennell, & J. M. Lepkowski. 1995. Effects of a prepaid monetary incentive on response rates and response quality in a face-to-face survey. *Public Opinion Quarterly* 59: 78–92.

Wolcott, H. F. 1982. Differing styles for on-site research, or, "If it isn't ethnography, what is it?" *Review Journal of Philosophy and Social Science* 7: 154–169.

World almanac and book of facts. 1995. New York: Newspaper Enterprise Association.

World Bank. 1995a. *World tables various editions*. Baltimore: The Johns Hopkins University Press.

World Bank. 1995b. *World debt tables*. Washington, D.C.: Oxford University Press.

World Bank. 1995c. *World development report various years*. Washington, D.C. Oxford University Press.

World Bank. 1995d. *Social indicators of development*. Baltimore: The Johns Hopkins University Press.

World Health Organization. 1992. *World health statistics annual*. Geneva: World Health Organization.

Wysong, E., & D. W. Wright. 1995. A decade of DARE: Efficacy, politics and drug education. *Sociological Focus* 28: 283–311.

Yin, R. K. 1984. *Case study research design and methods*. Beverly Hills: Sage.

Yolen, J. 1987. *Owl moon*. New York: Philomel.

Index